# Cutting-Edge Agri-Technologies for Next-Generation Farming


*Editors*

**Mahendra Rai**

Department of Biotechnology
Sant Gadge Baba Amravati University
Amravati, Maharashtra, India

**Graciela Avila Quezada**

Facultad de Ciencias Agrotecnologicas
Universidad Autonoma de Chihuahua
Chihuahua, Mexico



CRC Press
Taylor & Francis Group
Boca Raton London New York

CRC Press is an imprint of the
Taylor & Francis Group, an **informa** business
A SCIENCE PUBLISHERS BOOK

Cover illustrations reproduced by kind courtesy of Elanit Rubalcava Avila and Yossef Rubalcava Avila, Chihuahua City, Mexico.

First edition published 2027
by CRC Press
2385 NW Executive Center Drive, Suite 320, Boca Raton FL 33431

and by CRC Press
4 Park Square, Milton Park, Abingdon, Oxon, OX14 4RN



*CRC Press is an imprint of Taylor & Francis Group, LLC*



*Library of Congress Cataloging-in-Publication Data (applied for)*

ISBN: 978-1-041-08145-6 (hbk)
ISBN: 978-1-041-11421-5 (pbk)
ISBN: 978-1-003-65990-7 (ebk)

DOI: 10.1201/9781003659907

Typeset in Times New Roman
by Prime Publishing Services

# Preface

Agriculture, the cornerstone of human civilization, is now at the cusp of a revolutionary transformation. In the face of escalating challenges, climate variability, soil degradation, increasing global population, and the urgent need for food security, traditional farming methods alone are no longer sufficient. There is a critical need to reimagine agriculture through the lens of innovation, sustainability, and technology.

*Cutting-Edge Agri-Technologies for Next-Generation Farming* brings together pioneering research and forward-thinking approaches that illustrate how advanced technologies are reshaping the agricultural landscape. This volume explores the synergistic potential of nanotechnology, CRISPR-Cas9 gene editing, and artificial intelligence, offering an integrated perspective on how these tools can enhance productivity, reduce environmental impact, and promote resilient farming systems.

Divided into four thematic sections, the book starts with a general overview of technological convergence in agriculture. The subsequent sections delve into specific domains, novel nanotechnologies, genome editing, and AI applications, each addressing critical aspects of plant health, crop improvement, precision farming, and smart decision-making. The contributing authors, an esteemed consortium of scientists from around the globe, present both conceptual frameworks and practical applications, supported by case studies and the latest research findings.

What distinguishes this volume is its interdisciplinary approach, bridging biology, engineering, data science, and ethical discourse to present a holistic view of agri-innovation. Whether its using nanosensors to monitor plant health, editing crop genomes for improved yield, or deploying deep learning for disease forecasting, each chapter emphasizes sustainable solutions for future-ready agriculture.

This book is intended for a wide readership, including researchers, students, agri-technologists, and policymakers. It aims not only to inform but also to inspire innovation that empowers farmers, benefits consumers, and safeguards the planet.

We hope this compendium will serve as a valuable reference and catalyst for further research, collaboration, and practical implementation in the field of smart and sustainable agriculture.

Editors

Mahendra Rai
Graciela Avila Quezada

# Contents

## Section IV: Artificial Intelligence Technology

# Section I
# General

# Chapter 1

# Nanotechnology and AI in Next-Generation Agriculture

*Graciela Avila Quezada*[1,*] and *Mahendra Rai*[2,3]

## 1. Introduction

Agriculture in the 21st century is undergoing an accelerated transformation, driven by emerging technologies (Dayıoğlu and Turker 2021). Among these, nanotechnology and artificial intelligence (AI) stand out for their ability to address crucial challenges such as climate change, natural resource scarcity, and the need for sustainable, efficient and resilient agricultural production (Zhang et al. 2021).

Nanotechnology, based on the use of materials at the nanometer scale (between 1 and 100 nanometers) (Avila-Quezada and Rai 2023), is opening up new possibilities for the development of innovative, precise agricultural solutions that have less environmental impact. This discipline makes it possible to intervene at the molecular level in key processes, which translates into a significant improvement in production efficiency.

In agriculture applications, nanotechnology is used to create materials, devices or systems developed at the molecular scale to improve key agricultural processes to address critical needs such as the efficient use of water and fertilizers, and pests/pathogens control without contaminating the soil (Avila-Quezada et al. 2022a, b, Wahab et al. 2024). Notable examples include nanofertilizers and nanoformulated pesticides, and nanofungicides that allow for targeted slow-release delivery, minimizing waste release and improving plant absorption (Periakaruppan et al. 2023, Shen et al. 2023), while supporting crop health and environmental sustainability.

Artificial intelligence (AI) defined as the capacity of machines to perform tasks that typically require human intelligence, such as pattern recognition, prediction, and decision-making (Korteling et al. 2021), has advanced rapidly in recent decades. Since the 2000s, rapid advances in big data, machine learning, and deep neural networks have made AI a transformative force across sectors, including agriculture (Misra et al. 2020).

In agriculture, AI enables real-time processing of data from multiple sources, such as drones, satellites, sensors, and weather models, to support more informed, automated, and adaptive farming

[1] Facultad de Ciencias Agrotecnologicas, Universidad Autonoma de Chihuahua 31000, Mexico.
[2] Department of Chemistry, Natural Science Center, Federal University of Piaui - UFPI, 64049-550, Teresina, Piauí, Brazil.
[3] Department of Biotechnology, Sant Gadge Baba Amravati University, Amravati 444602, India.
Email: mahendrarai@sgbau.ac.in
* Corresponding authors: gdavila@uach.mx

DOI: 10.1201/9781003659907-2

practices. This helps in early detection of stress conditions, dynamic input adjustment, and overall optimization of agricultural productivity (Kumar et al. 2020).

AI plays a critical role in precision agriculture, a data-driven approach that allows farmers to use water, fertilizer, and agrochemicals only where and when needed, reducing waste and maximizing yield. By using GPS systems, remote sensing, and drones, AI can create field maps for soil variability, monitor plant development, and track climate dynamics, enabling farmers to act precisely and efficiently (Han et al. 2024).

Tools such as GPS, remote sensing, and drone imagery allow for the creation of field variability maps, monitoring of crop development, and adaptation to climate fluctuations (Inoue 2020). Sensor networks, satellite imagery, and Internet of Things (IoT) devices feed real-time data into AI systems, which process this information to guide timely and site-specific interventions (Ali et al. 2025).

AI is also revolutionizing the early detection of plant diseases and pests. By using high-resolution imaging, computer vision, and pattern recognition algorithms, AI systems can detect symptoms of biotic and abiotic stress—often before they are visible to the human eye. This enables early intervention, reducing crop loss and minimizing pesticide use. These detection capabilities are further enhanced when integrated with biosensors and nanosensors, to provide more precise diagnostics (Thabit and Moursy 2023).

In addition, AI is used to predict agricultural yields by integrating variables such as climate, soil conditions, crop type, and genetic data (Xu et al. 2022). These predictive models assist farmers and stakeholders in planning harvests, managing logistics, and making strategic decisions in line with environmental and market dynamics. Crop modeling simulations also provide insights into plant behavior under varying environmental scenarios, which is essential for climate-resilient agriculture.

Smart irrigation systems powered by AI have also emerged as a crucial tool for water conservation (Obaideen et al. 2022). By analyzing data from soil moisture sensors, weather forecasts, and crop requirements, AI can determine optimal irrigation schedules and amounts (Kumar et al. 2023). These systems reduce water waste and improve plant health by preventing over- or under-watering. When coupled with IoT infrastructure, smart irrigation can operate autonomously and adapt in real time to changing field conditions (García et al. 2020).

While nanotechnology operates at the molecular scale, its transformative potential is significantly amplified when integrated with AI, robotics, and remote sensing technologies such as drones and satellites (Ashique et al. 2025). Nanosensors embedded in soil or plants can continuously monitor parameters like moisture, nutrient levels, and disease markers. The data they generate can be processed by AI to support timely, data-driven decision-making (Shitharth et al. 2021). For example, AI systems can detect early signs of plant stress and trigger the release of nanofertilizers or nanopesticides only in the affected areas, reducing resource use and environmental impact (Thabit and Moursy 2023).

Advanced platforms like Microsoft FarmVibes demonstrate this integration by combining AI models, satellite imagery, and field sensor data to optimize planting, fertilization, and pesticide application (Jeyalakshmi et al. 2025). Besides, innovative tools for cloud removal and image fusion—such as those combining SCATSAT-1 and MODIS data or using cycle-consistent generative adversarial networks (GANs) to merge optical and radar imagery (Ebel et al. 2020)—are enhancing the usability and accuracy of remote sensing for precision agriculture.

Ultimately, the convergence of nanotechnology and AI enables hyper-local, efficient, and sustainable agricultural practices, contributing to food security while minimizing environmental impact. These synergistic technologies represent a paradigm shift in how we grow, monitor, and manage crops in an era of increasing ecological and socioeconomic pressures.

The objective of this chapter is to explore the current and emerging applications of nanotechnology and artificial intelligence in agriculture, examining their roles, specific uses in precision farming, and disease detection, and analyzing how their integration is shaping the future of sustainable and intelligent agriculture.

### *1.1 Smart Monitoring and Precision Detection of Pathogens*

Recent advances in nanosensor technology have enabled real-time monitoring of key agricultural parameters, contributing significantly to precision crop management. These nanosensors can detect plant physiological changes, humidity levels, the presence of pathogens and pests, and variations in soil conditions (Lee et al. 2023, Lu et al. 2020, Awaludin et al. 2020).

For instance, Lu et al. (2020) developed a flexible $ZnIn_2S_4$ nanosheet-based sensor capable of monitoring plant health by measuring humidity, light intensity, and transpiration. This sensor demonstrates a fast response time of approximately four minutes and maintains operational stability for over f days.

Love et al. (2021) reviewed diverse sensor platforms based on metal oxides, conductive polymers, and carbon allotropes. These systems offer a wide dynamic range and the capability to detect multiple gases including water vapor, even in complex gas environments, thanks to their cross-sensitivity features. Moreover, Awaludin et al. (2020) developed a fluorescence-based nanosensor immunoassay for early detection of bacterial blight in rice caused by *Xanthomonas oryzae* pv. *oryzae*. This system achieved a detection limit of 22 CFU $mL^{-1}$ and showed high specificity under field-like conditions using actual rice leaf samples.

### *1.2 Nanopriming an Effective Technique for Seed Germination, Protection and Improvement of Abiotic or Biotic Stress*

Nanopriming is a technique to enhance seed germination and seedling vigor in crops. For instance, nanopriming with mesoporous silica ($SiO_2$) and zinc oxide (ZnO) nanoparticles has shown significant improvement in germination percentage, seedling growth, and vigor index compared to unprimed seeds (Chourasiya et al. 2021, Rai-Kalal et al. 2021). These nanoparticles enhance water uptake, α-amylase activity, and photosynthetic performance (Rai-Kalal et al. 2021). Additionally, nanopriming with $SiO_2$ has demonstrated improved drought tolerance in wheat by maintaining photosynthetic efficiency and balancing antioxidant enzyme activity (Rai-Kalal et al. 2021, Rai Kalal et al. 2023). When compared to traditional priming techniques like hydropriming and magneto-priming, nanopriming exhibited superior results in enhancing germination efficiency and drought stress tolerance (Rai Kalal et al. 2023). These findings suggest that nanopriming, particularly with $SiO_2$ and ZnO nanoparticles, is a promising technique for improving wheat seed performance and stress tolerance.

Studies have demonstrated that nanopriming, particularly with zinc oxide nanoparticles, enhances seed germination, seedling growth, and overall plant performance under saline conditions (Mu et al. 2024, Khan et al. 2023). The mechanisms behind nanopriming's effectiveness include improved antioxidant enzyme activities, better maintenance of ion homeostasis, and enhanced physiological processes (Mu et al. 2024, Khan et al. 2025). Nanopriming offers advantages over conventional methods, such as using fewer materials and minimizing environmental concerns (Khan et al. 2025). However, further research is needed to fully understand the molecular and physiological mechanisms involved, explore diverse nanoparticle types, and assess long-term environmental implications (Khalid et al. 2023). Despite these knowledge gaps, nanopriming shows great potential as an eco-friendly and effective approach to boost crop resilience against salt stress and other abiotic stressors (Khalid et al. 2023, Khan et al. 2025) (Figure 1).

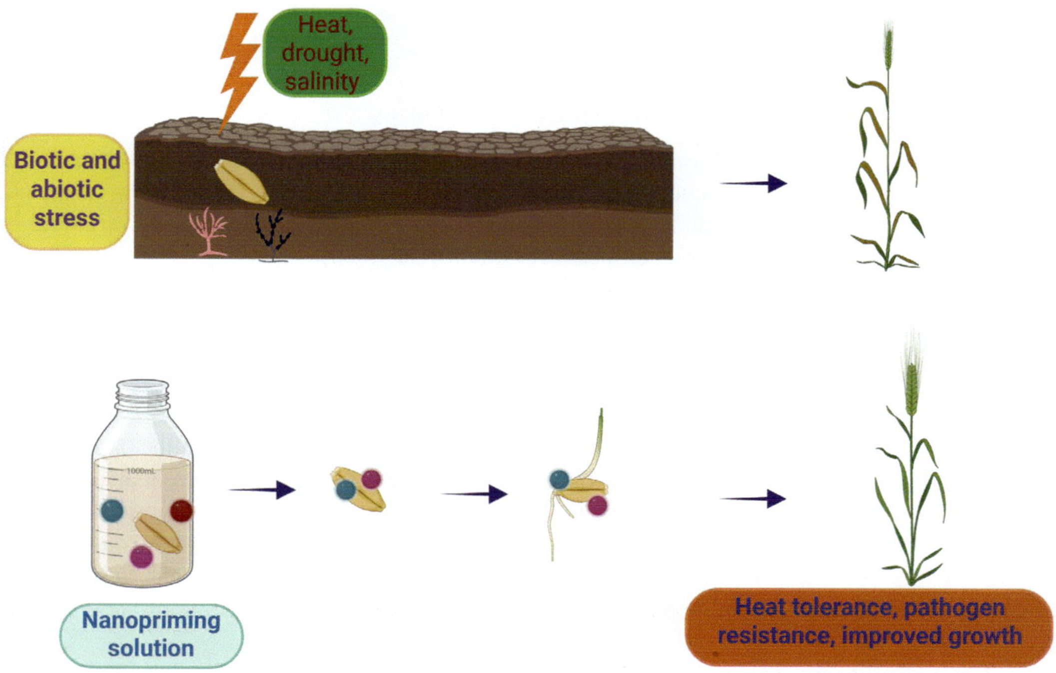


**Figure 1.** Nanopriming: enhancing seed performance under stress.

### *1.3 Nanoformulations for Crop Protection*

Nanoformulations play an important role in achieving sustainable farming and effective management of plant pathogens. Biodegradable materials such as starch, cellulose, and chitosan are commonly used in these formulations (Kumari et al. 2022). Among them, chitosan stands out as the second most important biopolymer due to its biodegradability, biocompatibility, non-toxic nature, and multifaceted antimicrobial activity. These properties make it a promising candidate for both plant disease management and growth promotion (Bhatt et al. 2024, Nandhini et al. 2025).

Various types of nanoformulations such as metal-based nanopesticides, nanoemulsions, nanofungicides, and nanofertilizers have been developed for agricultural applications (Chaud et al. 2021, Bhaskar et al. 2023). These nanoformulations enhance crop yield and productivity by improving the bioavailability of nutrients and fertilizers while reducing soil contamination caused by excessive use of synthetic agrochemicals. Furthermore, they facilitate targeted and controlled release of active ingredients, minimizing chemical wastage and supporting sustainable agricultural practices.

Mathur et al. (2025) reported the development of several types of nanoformulations, including polymeric nanoparticles, liposomes, solid lipid nanoparticles, nanogels, and magnetic nanoparticles. In addition, RNA-based nanotechnologies employing intracellular delivery of messenger RNA (mRNA) are being explored for the effective management of pests, pathogens, and abiotic stress.

More recently, Ghosh et al. (2025) synthesized silver nanoparticles (AgNPs) using an extract of *Ocimum sanctum*. This eco-friendly nanobiopesticide was effectively used for managing the hairy caterpillar (*Spilosoma obliqua*, Lepidoptera: Erebidae), a major jute pest. The study concluded that AgNPs, in combination with bioactive compounds from *O. sanctum*, exhibited synergistic pesticidal activity.

Besides, smart nanocapsules are structures that transport bioactive compounds—such as hormones, vitamins or phytohormones—directly to plant cells (Yadav et al. 2024), optimizing their absorption and effectiveness. On the other hand, nanometric coatings applied on seeds or plants protect against extreme conditions, diseases or environmental stress. For example, silver nanoparticles (AgNPs) coating on wheat seeds protect against fungal diseases (Karimi et al. 2012). At 50 ppm, AgNPs controlled pathogens like *Sclerotium rolfsii* and promoted growth (Desai et al. 2021). Chitosan-coated biogenic AgNPs also combated fungi like *Fusarium oxysporum* in wheat seedlings (Mondéjar-López et al. 2023).

### *1.4 Advantages of Nanoformulated Pesticides and Fungicides*

Nanotechnology-based delivery systems offer several advantages for modern agriculture by improving the precision, efficiency, and safety of agrochemical applications. One benefit is targeted delivery, which enables site-specific release of active compounds, thereby minimizing off-target effects and reducing environmental exposure (Beckers et al. 2021).

In addition, these systems allow for a reduced dosage while enhancing bioefficacy, meaning lower quantities of active agents are needed to achieve the desired effect (Kapeleka and Mwema 2024). This optimizes resource use and also lowers input costs for farmers.

Another significant advantage is controlled release, which enables the sustained delivery of agrochemicals over time, reducing the frequency of applications and ensuring more consistent protection or nutrient availability (Shen et al. 2023). From an environmental perspective, nanodelivery systems promote environmental safety by minimizing issues such as leaching into the soil, volatilization into the atmosphere, and toxicity to non-target organisms (Chaud et al. 2021). These innovative formulations contribute to resistance management by helping delay the emergence of resistant pest or pathogen populations, ensuring longer-term effectiveness of agricultural treatments (Malandrakis et al. 2022).

### *1.5 Nanofertilizer Formulation and Enhanced Nutrient Uptake*

Nanofertilizers are advanced formulations designed to improve nutrient delivery and uptake in plants. Their development involves integrating nanoparticles with substances such as surfactants and polymers to enhance efficacy (Tarafder et al. 2020). These nanoparticles are often coated with polymers, surfactants, or organic compounds to increase their stability, longevity, and enable targeted delivery to plant roots. They can be formulated as suspensions, emulsions, or gels, making them easier to handle and apply in agricultural settings (Bogusz et al. 2021).

One of the key advantages of nanofertilizers lies in their ability to enhance nutrient uptake and use efficiency. By improving the bioavailability and transport of nutrients within plant systems, they support better plant growth and higher yields while minimizing environmental nutrient losses. Nanomaterials play a central role in these improvements, contributing to more sustainable and efficient nutrient management (Singh et al. 2024, Zhang et al. 2024).

Although these technologies are showing interesting results in controlled environments, their large-scale adoption faces several challenges such as high production costs, the need for scalability and the lack of regulations on the use of nanoparticles in food, as well as their final destination in soil and groundwater bodies.

## 2. Artificial Intelligence in Agriculture

### *2.1 Disease and Pest Detection*

Artificial intelligence, understood as the ability of machines to perform tasks that require human intelligence (Korteling et al. 2021), has evolved greatly since its conception in the 20th century. Its roots date back to the work of British mathematician Alan Turing, who in 1950 published the article "Computing Machinery and Intelligence", in which he proposed the Turing Test (Turing 2009). This idea laid the foundation for automation and computational intelligence. The term artificial intelligence was coined in 1956 by John McCarthy during a conference at Dartmouth, in

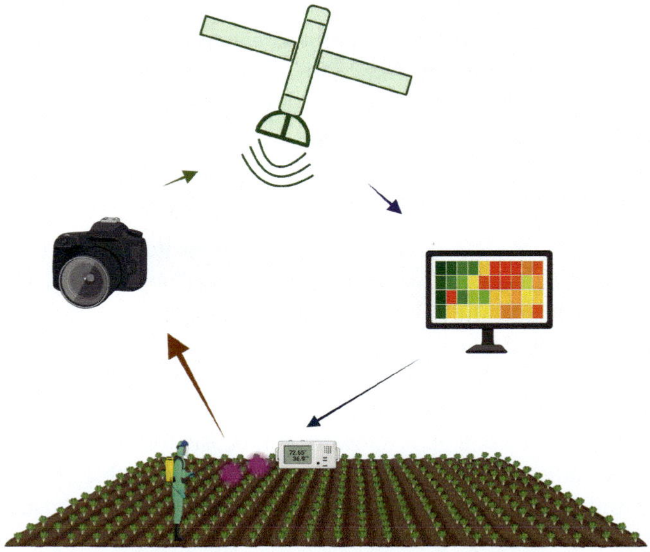

**Figure 2.** Drone imaging uses image recognition and deep learning models to monitor plant health.

collaboration with Marvin Minsky and Claude Shannon (Rajaraman 2014). Since the 2000s, the advancement of big data and the development of deep neural networks (deep learning) have driven the application of AI in multiple fields, including agriculture (Misra et al. 2020). In the agricultural sector, AI can transform data into decisions, being applied in the early detection of plant diseases and nutritional deficiencies through high-resolution image analysis. AI can also be used to optimize inputs such as pesticides, which it does through algorithms, as well as remote crop monitoring using sensors, drones and satellites to record key variables in real time (Kumar et al. 2020) (Figure 2).

Although nanotechnology acts at the molecular level, its potential grows when combined with other technologies such as artificial intelligence, agricultural robotics, and remote monitoring systems, including drones and satellites. Data generated by nanosensors can be integrated into intelligent systems (Shitharth et al. 2021) to transform the information into useful knowledge for agronomic decision-making. For example, AI systems can analyze in real-time, the data collected by nanosensors installed in the soil or on plants, making it possible to detect nutritional deficiencies, water stress or the appearance of diseases before symptoms are even visible (Thabit and Moursy 2023).

## *2.2 From Prediction to Precision*

AI is also used to predict agricultural yields from models that combine climatic, soil and genetic data (Xu et al. 2022). The convergence of nanotechnology and AI in sustainable agriculture holds immense promise by enhancing efficiency and environmental stewardship. AI can analyze large datasets, including satellite images, sensor data, and climate models, to predict yield and then optimize the use of nanomaterials in agriculture, such as nanofertilizers and nanopesticides that must be applied to increase yield. Another example is the control of weeds with systems that apply herbicides (Dasgupta et al. 2020) only where necessary.

For instance, AI can predict the nutrient needs of crops, allowing nanomaterials to deliver precise quantities of nutrients or protect against pests with minimal environmental impact (Zhang et al. 2021). Nanotechnology, in turn, provides AI with powerful tools for real-time monitoring and control, such as smart sensors embedded in soil or plants to track moisture levels, nutrient uptake, or disease indicators.

Together, these technologies can reduce resource waste, enhance crop yield, and promote environmentally sustainable farming practices by minimizing the use of chemicals and improving water and soil management (Zhang et al. 2021). In addition, predictive algorithms and platforms such as Microsoft FarmVibes integrate satellite images, field sensors and climate models to optimize decisions such as planting, fertilization and agrochemical application (Jeyalakshmi et al. 2025). This integration makes it possible to act in a precise and localized way, applying nanoparticles only in the areas that require it and avoiding excessive use of chemicals.

### *2.3 Smart Irrigation Systems*

Smart irrigation systems represent a transformative approach to sustainable agriculture, leveraging cutting-edge technologies to optimize water use and enhance crop productivity. These systems utilize IoT-based platforms, real-time environmental sensors, artificial intelligence (AI), and embedded systems to monitor key parameters such as soil moisture, temperature, and weather conditions, enabling precise, automated irrigation control (Morchid et al. 2024, Shilpa et al. 2024).

Studies demonstrate that smart irrigation can significantly outperform traditional methods. For example, crop productivity has been shown to increase by up to 34.9%, while water consumption may be reduced by 30–47.8% (Pratama and Mandela 2024, Diallo et al. 2024). The Smart Irrigation Yield Optimization (SIYO) system, in particular, enhances water use efficiency by 0.30% compared to furrow irrigation and improves nutrient uptake by 0.25% relative to overhead systems (Shilpa et al. 2024).

Autonomous agricultural robotics are also playing an important role in this domain, carrying out planting, irrigation, and harvesting tasks with high efficiency and minimal human intervention (Gupta and Gupta 2024). These robotic systems, when integrated with AI and smart irrigation frameworks, further enhance precision and reduce labor demands.

A vital component of smart irrigation lies in the integration of multisource data, including satellite imagery, ground sensors, and aerial platforms. Techniques such as cloud removal from satellite images and data fusion significantly improve the usability and accuracy of remote sensing in agriculture. For instance, merging data from sensors like SCATSAT-1 and MODIS, using methods such as auto-associative neural networks or stationary wavelet transforms, has improved image clarity (Kaur et al. 2023, Sahoo and Patnaik 2008). Additionally, radio frequency-based reconstruction techniques (Zhao et al. 2023) and advanced models like cycle-consistent GANs have enabled the combination of optical and radar images without the need for paired training data (Ebel et al. 2020), enhancing remote monitoring capabilities.

These advancements are highly compatible with the incorporation of nanotechnology, which brings an additional layer of precision to smart irrigation. For example, nanosensors embedded in crops can detect nutrient deficiencies or early signs of stress, triggering the site-specific release of nanofertilizers, fungicides, or biocontrol agents only where and when they are needed (Yadav et al. 2023).

Despite their benefits, challenges remain. High initial investment costs, the need for reliable infrastructure, and limited accessibility in rural or resource-constrained areas continue to hinder widespread adoption—particularly among smallholder farmers (Diallo et al. 2024). Overcoming these barriers will require targeted support through public policy, education, and infrastructure development.

Smart irrigation systems offer a highly promising path toward sustainable and resource-efficient agriculture. When integrated with robotics, remote sensing, AI, and nanotechnology, they have the potential to revolutionize how water and inputs are managed in farming, making precision agriculture both smarter and more accessible.

## 3. Convergence of Nanotechnology and Artificial Intelligence as a Transformative Science

Although nanotechnology acts at the molecular level, its potential grows when combined with other technologies such as artificial intelligence, agricultural robotics, and remote monitoring systems, including drones and satellites. Data generated by nanosensors can be integrated into intelligent systems (Shitharth et al. 2021) to transform information into useful knowledge for agronomic decision-making. For example, AI systems can analyze in real-time the data collected by nanosensors installed in the soil or on plants, making it possible to detect nutritional deficiencies, water stress or the appearance of diseases before symptoms are even visible (Thabit and Moursy 2023).

The convergence of nanotechnology and AI in sustainable agriculture holds immense promise by enhancing both efficiency and environmental stewardship. AI can analyze large datasets, including satellite images, sensor data, and climate models, to optimize the use of nanomaterials in agriculture, such as nanofertilizers, nanosensors, and nanopesticides (Devarajan 2025). For instance, AI can predict the nutrient needs of crops, allowing nanomaterials to deliver precise quantities of nutrients or protect against pests with minimal environmental impact. Nanotechnology, in turn, provides AI with powerful tools for real-time monitoring and control, such as smart sensors embedded in soil or plants to track moisture levels, nutrient uptake, or disease indicators (Paul et al. 2022). Together, these technologies can reduce resource waste, enhance crop yield, and promote environmentally sustainable farming practices by minimizing the use of chemicals and improving water and soil management (Zhang et al. 2021).

In addition, predictive algorithms and platforms such as Microsoft FarmVibes integrate satellite images, field sensors and climate models to optimize decisions such as planting, fertilization and agrochemical application (Jeyalakshmi et al. 2025). This integration makes it possible to act in a precise and localized way, applying nanoparticles only in the areas that require it and avoiding excessive use of chemicals.

People have adopted advanced technologies that combine AI, sensors, and satellite data (Javaid et al. 2023). Although his primary focus is artificial intelligence, the model of agriculture can be developed and is highly compatible with the incorporation of nanotechnology-based solutions. Tools that remove clouds from satellite images, and fuse data from ground and aerial sensors, could be integrated with information generated by nanosensors installed on crops (Mutunga et al. 2024).

Some examples of approaches include the merging data from different sensors like SCATSAT-1 and MODIS (Kaur et al. 2023), which employ auto-associative neural networks with stationary wavelet transforms (Sahoo and Patnaik 2008), and leveraging radio frequency signals to reconstruct occluded regions in multispectral images (Zhao et al. 2023). These methods aim to enhance the quality and usability of satellite data for applications such as agriculture. Besides, models like cycle-consistent GANs have been used to combine optical and radar imagery, reducing the need for strictly paired cloudy and cloud-free images in training (Ebel et al. 2020). These cloud removal techniques have shown promising results, overcoming traditional methods in terms of accuracy and visual quality, and offer the potential for improving remote sensing applications. Thus, by detecting conditions such as deficiencies, it would be possible to trigger the release of nanofertilizers, fungicides, and other agrochemicals or biocontrol agents, at the time and on the plants that need them.

## Conclusions

The integration of nanotechnology and artificial intelligence (AI) will revolutionize modern agriculture, offering innovative tools to enhance productivity, sustainability, and resilience. From smart monitoring and precision detection of plant pathogens to nanopriming techniques that boost seed germination and fortify crops against environmental stresses, nanotechnology has demonstrated remarkable potential in strengthening agricultural foundations. Nanoformulations, especially nano-enabled pesticides, fungicides, and fertilizers, not only improve the efficacy of crop protection and nutrient delivery but also reduce environmental impact, paving the way for more sustainable farming practices.

Meanwhile, AI contributes to agriculture by interpreting vast and complex datasets, enabling accurate disease and pest detection, predictive modelling, and real-time decision-making. Smart irrigation systems, powered by AI algorithms, further optimize water usage and enhance crop performance under varying climatic conditions.

The convergence of nanotechnology and AI represents a transformative leap in agricultural science, merging material-level precision with data-driven intelligence. This synergy fosters a new era of precision agriculture, where proactive, resource-efficient, and environmentally conscious farming becomes achievable. As these technologies continue to evolve and integrate, they will be instrumental in addressing global food security challenges, adapting to climate change, and supporting the development of sustainable agricultural systems for future generations.

## References

Ali, G., Mijwil, M. M., Adamopoulos, I. and Ayad, J. (2025). Leveraging the internet of things, remote sensing, and artificial intelligence for sustainable forest management. Babylonian Journal of Internet of Things 2025: 1–65. https://doi.org/10.58496/BJIoT/2025/001.

Ashique, S., Raikar, A., Jamil, S., Lakshminarayana, L., Gajbhiye, S. A., De, S. et al. (2025). Artificial intelligence integration with nanotechnology: A new frontier for sustainable and precision agriculture. Current Nanoscience 21(2): 242–273. https://doi.org/10.2174/0115734137275111231206072049.

Avila-Quezada, G. D., Golinska, P. and Rai, M. (2022a). Engineered nanomaterials in plant diseases: can we combat phytopathogens? Applied Microbiology and Biotechnology 106(1): 117–129. https://doi.org/10.1007/s00253-021-11725-w.

Avila-Quezada, G. D., Ingle, A. P., Golińska, P. and Rai, M. (2022b). Strategic applications of nano-fertilizers for sustainable agriculture: Benefits and bottlenecks. Nanotechnology Reviews 11(1): 2123–2140. https://doi.org/10.1515/ntrev-2022-0126.

Avila-Quezada, G. D. and Rai, M. (2023). Nanophytopathology. CRC Press. https://doi.org/10.1201/9781003344513.

Awaludin, N., Abdullah, J., Salam, F., Ramachandran, K., Yusof, N. A. and Wasoh, H. (2020). Fluorescence-based immunoassay for the detection of *Xanthomonas oryzae* pv. *oryzae* in rice leaf. Analytical Biochemistry 610: 113876. https://doi.org/10.1016/j.ab.2020.113876.

Beckers, S. J. J., Staal, A. H., Rosenauer, C., Srinivas, M., Landfester, K. and Wurm, F. R. (2021). Targeted drug delivery for sustainable crop protection: Transport and stability of polymeric nanocarriers in plants. Advanced Science 8(11): 2100067. https://doi.org/10.1002/advs.202100067.

Bhaskar, M., Kumar, A. and Rani, R. (2023). Application of nano formulations in agriculture. Biocatalysis and Agricultural Biotechnology 54: 102934. https://doi.org/10.1016/j.bcab.2023.102934.

Bhatt, S., Pathak, R., Punetha, V. D. and Punetha, M. (2024). Chitosan nanocomposites as a nano-bio tool in phytopathogen control. Carbohydrate Polymers 331: 121858. https://doi.org/10.1016/j.carbpol.2024.121858.

Bogusz, P., Rusek, P. and Brodowska, M. S. (2021). Suspension fertilizers: How to reconcile sustainable fertilization and environmental protection. Agriculture 11(10): 1008. https://doi.org/10.3390/agriculture11101008.

Chaud, M., Souto, E. B., Zielinska, A., Severino, P., Batain, F., Oliveira-Junior, J. et al. (2021). Nanopesticides in agriculture: Benefits and challenge in agricultural productivity, toxicological risks to human health and environment. Toxics 9(6): 131. https://doi.org/10.3390/toxics9060131.

Chourasiya, V. K., Nehra, A., Shukla, P. S., Singh, K. P. and Singh, P. (2021). Impact of mesoporous nano-silica ($SiO_2$) on seed germination and seedling growth of wheat, pea and mustard seed. Journal of Nanoscience and Nanotechnology 21(6): 3566–3572. https://doi.org/10.1166/jnn.2021.19013.

Dasgupta, I., Saha, J., Venkatasubbu, P. and Ramasubramanian, P. (2020). AI crop predictor and weed detector using wireless technologies: A smart application for farmers. Arabian Journal for Science and Engineering 45: 11115–11127. https://doi.org/10.1007/s13369-020-04928-2.

Dayıoğlu, M. A. and Turker, U. (2021). Digital transformation for sustainable future-agriculture 4.0: A review. Journal of Agricultural Sciences 27(4): 373–399. https://doi.org/10.15832/ankutbd.986431.

Desai, P., Jha, A., Markande, A. and Patel, J. (2021). Silver nanoparticles as a fungicide against soil-borne *Sclerotium rolfsii*. *In*: Sarma, H., Joshi, S. J., Prasad, R. and Jampilek, J. (eds.). Biobased Nanotechnology for Green Applications. Nanotechnology in the Life Sciences. Springer, Cham. https://doi.org/10.1007/978-3-030-61985-5_18.

Devarajan, Y. D. C. S. (2025). Investigation of emerging technologies in agriculture: An in-depth look at smart farming, nano-agriculture, AI, and Big data. Journal of Biosystems Engineering 50: 170–192. https://doi.org/10.1007/s42853-025-00258-z.

Diallo, M. C. A., Santos, R. C., Gomes, E. P., Machado, C. A. C., Días, C. R. A., Santos, E. C. dos et al. (2024). Challenges of smart irrigation implementation in water optimization and agricultural sustainability. Contribuciones a las Ciencias Sociales 17(13): e13723. https://doi.org/10.55905/revconv.17n.13270.

Ebel, P., Schmitt, M. and Zhu, X. X. (2020). Cloud removal in unpaired Sentinel-2 imagery using cycle-consistent GAN and SAR-optical data fusion. pp. 2065–2068. In IGARSS 2020–2020 IEEE International Geoscience and Remote Sensing Symposium, Waikoloa, HI, USA, 2020, https://doi.org/10.1109/IGARSS39084.2020.9324060.

García, L., Parra, L., Jimenez, J. M., Lloret, J. and Lorenz, P. (2020). IoT-based smart irrigation systems: An overview on the recent trends on sensors and IoT systems for irrigation in precision agriculture. Sensors 20(4): 1042. https://doi.org/10.3390/s20041042.

Ghosh, A., Majumdar, D., Biswas, H., Chowdhury, A. and Podder, S. (2025). Nano-biopesticide formulation comprising of silver nanoparticles anchored to *Ocimum sanctum*: A sustainable approach to pest control in jute farming. Scientific Reports 15(1): 1–18. https://doi.org/10.1038/s41598-025-87727-9.

Gupta, N. and Gupta, P. K. (2024). Robotics and artificial intelligence (AI) in agriculture with major emphasis on food crops. pp. 577–605. *In*: Priyadarshan, P. M., Jain, S. M., Penna, S. and Al-Khayri, J. M. (eds.). Digital Agriculture. Springer, Cham. https://doi.org/10.1007/978-3-031-43548-5_19.

Han, H., Liu, Z., Li, J. and Zeng, Z. (2024). Challenges in remote sensing based climate and crop monitoring: navigating the complexities using AI. Journal of Cloud Computing 13(1): 1–14. https://doi.org/10.1186/s13677-023-00583-8.

Inoue, Y. (2020). Satellite-and drone-based remote sensing of crops and soils for smart farming—a review. Soil Science and Plant Nutrition 66(6): 798–810. https://doi.org/10.1080/00380768.2020.1738899.

Javaid, M., Haleem, A., Khan, I. H. and Suman, R. (2023). Understanding the potential applications of Artificial intelligence in agriculture sector. Advanced Agrochem 2(1): 15–30. https://doi.org/10.1016/j.aac.2022.10.001.

Jeyalakshmi, J., Vijay, K., Berna, I. E. and Sowmia, K. R. (2025). AI in optimisation of agricultural output in precision agriculture. pp. 221–237. *In*: Pethuru Raj, Gayathri, N. and Jaspher Willsie Kathrine, G. (eds.). Artificial Intelligence for Precision Agriculture. Auerbach Publications. Taylor & Francis.

Kapeleka, J. A. and Mwema, M. F. (2024). State of nano pesticides application in smallholder agriculture production systems: Human and environmental exposure risk perspectives. Heliyon 10(20): e39225. https://doi.org/10.1016/j.heliyon.2024.e39225.

Karimi, N., Minaei, S., Almassi, M. and Shahverdi, A. R. (2012). Application of silver nano-particles for protection of seeds in different soils. African Journal of Agricultural Research 7(12): 1863–1869. https://doi.org/10.5897/AJAR11.1150.

Kaur, R., Maini, R. and Tiwari, R. K. (2023). Cloud removal with multisource data fusion of remote sensing images for agricultural applications. pp. 120–124. In 2023 IEEE 9th International Women in Engineering (WIE) Conference on Electrical and Computer Engineering (WIECON-ECE). IEEE. https://doi.org/10.1109/WIECON-ECE60392.2023.10456404.

Khalid, F., Asif, K., Rasheed, Y., Ashraf, H., Maqsood, M. F., Rana, S. et al. (2023). Nano priming for boosting growth and resilience in crops under abiotic stresses. Biocatalysis and Agricultural Biotechnology 53: 102892. https://doi.org/10.1016/j.bcab.2023.102892.

Khan, M. N., Fu, C., Li, J., Tao, Y., Li, Y., Hu, J. et al. (2023). Seed nanopriming: How do nanomaterials improve seed tolerance to salinity and drought? Chemosphere 310: 136911. https://doi.org/10.1016/j.chemosphere.2022.136911.

Khan, M. N., Li, Y., Mu, Y., Sultan, H., Baloch, A., Din, I. et al. (2025). Recent advances in nano-enabled plant salt tolerance: Methods of application, risk assessment, opportunities and future prospects. Journal of Integrative Agriculture 24(5): 1611–1630. https://doi.org/10.1016/j.jia.2024.05.028.

Korteling, J. H., van de Boer-Visschedijk, G. C., Blankendaal, R. A., Boonekamp, R. C. and Eikelboom, A. R. (2021). Human-versus artificial intelligence. Frontiers in Artificial Intelligence 4: 622364. https://doi.org/10.3389/frai.2021.622364.

Kumar, P., Udayakumar, A., Anbarasa Kumar, A., Senthamarai Kannan, K. and Krishnan, N. (2023). Multiparameter optimization system with DCNN in precision agriculture for advanced irrigation planning and scheduling based on soil moisture estimation. Environmental Monitoring and Assessment 195(1): 13. https://doi.org/10.1007/s10661-022-10529-3.

Kumar, R., Yadav, S., Kumar, M., Kumar, J. and Kumar, M. (2020). Artificial intelligence: new technology to improve Indian agriculture. International Journal of Chemical Studies 8(2): 2999–3005. https://doi.org/10.22271/chemi.2020.v8.i2at.9208.

Kumari, S. V. G., Pakshirajan, K. and Pugazhenthi, G. (2022). Recent advances and future prospects of cellulose, starch, chitosan, polylactic acid and polyhydroxyalkanoates for sustainable food packaging applications. International Journal of Biological Macromolecules 221: 163–182. https://doi.org/10.1016/j.ijbiomac.2022.08.203.

Lee, G., Hossain, O., Jamalzadegan, S., Liu, Y., Wang, H., Saville, A. C. et al. (2023). Abaxial leaf surface-mounted multimodal wearable sensor for continuous plant physiology monitoring. Science Advances 9(15): eade2232. https://doi.org/10.1126/sciadv.ade2232.

Love, C., Nazemi, H., El-Masri, E., Ambrose, K., Freund, M. S. and Emadi, A. (2021). A review on advanced sensing materials for agricultural gas sensors. Sensors 21(10): 3423. https://doi.org/10.3390/s21103423.

Lu, Y., Xu, K., Zhang, L., Deguchi, M., Shishido, H., Arie, T. et al. (2020). Multimodal plant healthcare flexible sensor system. ACS Nano 14(9): 10966–10975. https://doi.org/10.1021/acsnano.0c03757.

Malandrakis, A. A., Kavroulakis, N. and Chrysikopoulos, C. V. (2022). Metal nanoparticles against fungicide resistance: alternatives or partners? Pest Management Science 78(10): 3953–3956. https://doi.org/10.1002/ps.7014.

Mathur, S., Chaturvedi, A. and Ranjan, R. (2025). Advances in RNAi-based nanoformulations: Revolutionizing crop protection and stress tolerance in agriculture. Nanoscale Advances 7(7): 1768. https://doi.org/10.1039/D5NA00044K.

Misra, N. N., Dixit, Y., Al-Mallahi, A., Bhullar, M. S., Upadhyay, R. and Martynenko, A. (2020). IoT, big data, and artificial intelligence in agriculture and food industry. IEEE Internet of Things Journal 9(9): 6305–6324. https://doi.org/10.1109/JIOT.2020.2998584.

Mondéjar-López, M., López-Jimenez, A. J., Ahrazem, O., Gómez-Gómez, L. and Niza, E. (2023). Chitosan coated-biogenic silver nanoparticles from wheat residues as green antifungal and nanoprimig in wheat seeds. International Journal of Biological Macromolecules 225: 964–973. https://doi.org/10.1016/j.ijbiomac.2022.11.159.

Morchid, A., Jebabra, R., Alami, R. E., Charqi, M. and Boukili, B. (2024). Smart agriculture for sustainability: the implementation of smart irrigation using real-time embedded system technology. pp. 1–6. In 2024 4th International Conference on Innovative Research in Applied Science, Engineering and Technology (IRASET). IEEE. https://doi.org/10.1109/IRASET60544.2024.10548972.

Mu, Y., Li, Y., Zhang, Y., Guo, X., Song, S., Huang, Z. et al. (2024). A comparative study on the role of conventional, chemical, and nanopriming for better salt tolerance during seed germination of direct seeding rice. Journal of Integrative Agriculture 23(12): 3998–4017. https://doi.org/10.1016/j.jia.2023.12.013.

Mutunga, T., Sinanovic, S. and Harrison, C. S. (2024). Integrating wireless remote sensing and sensors for monitoring pesticide pollution in surface and groundwater. Sensors 24(10): 3191. https://doi.org/10.3390/s24103191.

Nandhini, R., Rajeswari, E., Harish, S., Sivakumar, V., Selvi, G. and Jayasundrasharmila, D. (2025). Role of chitosan nanoparticles in sustainable plant disease management. Journal of Nanoparticle Research 27(13). https://doi.org/10.1007/s11051-024-06203-z.

Obaideen, K., Yousef, B. A., AlMallahi, M. N., Tan, Y. C., Mahmoud, M., Jaber, H. et al. (2022). An overview of smart irrigation systems using IoT. Energy Nexus 7: 100124. https://doi.org/10.1016/j.nexus.2022.100124.

Paul, K., Chatterjee, S. S., Pai, P., Varshney, A., Juikar, S., Prasad, V. et al. (2022). Viable smart sensors and their application in data driven agriculture. Computers and Electronics in Agriculture 198: 107096. https://doi.org/10.1016/j.compag.2022.107096.

Periakaruppan, R., Romanovski, V., Thirumalaisamy, S. K., Palanimuthu, V., Sampath, M. P., Anilkumar, A. et al. (2023). Innovations in modern nanotechnology for the sustainable production of agriculture. ChemEngineering 7(4): 61. https://doi.org/10.3390/chemengineering7040061.

Pratama Julianto, A. and Mandela, R. (2024). Evaluating the effectiveness of smart irrigation systems in improving agricultural productivity. Agriculture Power Journal 1(4): 1–9. https://doi.org/10.70076/apj.v1i4.43.

Rai-Kalal, P., Tomar, R. S. and Jajoo, A. (2021). Seed nanopriming by silicon oxide improves drought stress alleviation potential in wheat plants. Functional Plant Biology 48(9): 905–915. https://doi.org/10.1071/FP21079.

Rai-Kalal, P. R., Mishra, S. and Jajoo, A. (2023). Exploring the potential of nano-priming and magneto-priming in enhancing drought tolerance potential of wheat seeds. International Journal Plant Biol Res 11(2): 1137. https://doi.org/10.47739/2333-6668.plantbiology.1137.

Rajaraman, V. (2014). JohnMcCarthy-Father of artificial intelligence. Resonance 19: 198–207. https://doi.org/10.1007/s12045-014-0027-9.

Sahoo, T. and Patnaik, S. (2008). Cloud removal from satellite images using auto associative neural network and stationary wevlet transform. pp. 100–105. In 2008 First International Conference on Emerging Trends in Engineering and Technology. IEEE. https://doi.org/10.1109/ICETET.2008.99.

Shen, M., Liu, S., Jiang, C., Zhang, T. and Chen, W. (2023). Recent advances in stimuli-response mechanisms of nano-enabled controlled-release fertilizers and pesticides. Eco-Environment & Health 2(3): 161–175. https://doi.org/10.1016/j.eehl.2023.07.005.

Shilpa, M., Ravi, P., Sharmila, N., Mallikarjunaswamy, S., Deepak, B. L., Pavithra, G. S. et al. (2024). Enhancing crop yield and growth prediction using IoT-Based smart irrigation systems and machine learning algorithms. pp. 1–5. In 2024 Second International Conference on Networks, Multimedia and Information Technology (NMITCON). IEEE. https://doi.org/10.1109/NMITCON62075.2024.10698825.

Shitharth, S., Meshram, P., Kshirsagar, P. R., Manoharan, H., Tirth, V. and Sundramurthy, V. P. (2021). Impact of big data analysis on nanosensors for applied sciences using neural networks. Journal of Nanomaterials 2021(1): 4927607. https://doi.org/10.1155/2021/4927607.

Singh, A., Sharma, A., Singh, O., Rajput, V. D., Movsesyan, H., Minkina, T. et al. (2024). In-depth exploration of nanoparticles for enhanced nutrient use efficiency and abiotic stresses management: Present insights and future horizons. Plant Stress 14: 100576. https://doi.org/10.1016/j.stress.2024.100576.

Tarafder, C., Daizy, M., Alam, M. M., Ali, M. R., Islam, M. J., Islam, R. et al. (2020). Formulation of a hybrid nanofertilizer for slow and sustainable release of micronutrients. ACS Omega 5(37): 23960–23966. https://doi.org/10.1021/acsomega.0c03233.

Thabit, F. N. and Moursy, A. R. (2023). Sensors efficiency in smart management of the environmental resources. pp. 1–40. *In*: Ali, G. A. M., Chong, K. F. and Makhlouf, A. S. H. (eds.). Handbook of Nanosensors. Springer, Cham. Nature Switzerland. https://doi.org/10.1007/978-3-031-16338-8_51-1.

Turing, A. M. (2009). Computing machinery and intelligence. pp. 23–65. *In*: Copeland, B. J. (ed.). The Essential Turing. Springer Netherlands. https://doi.org/10.1007/978-1-4020-8860-5_5.

Wahab, A., Muhammad, M., Ullah, S., Abdi, G., Shah, G. M., Zaman, W. et al. (2024). Agriculture and environmental management through nanotechnology: Eco-friendly nanomaterial synthesis for soil-plant systems, food safety, and sustainability. Science of the Total Environment 926: 171862. https://doi.org/10.1016/j.scitotenv.2024.171862.

Xu, Y., Zhang, X., Li, H., Zheng, H., Zhang, J., Olsen, M. S. et al. (2022). Smart breeding driven by big data, artificial intelligence, and integrated genomic-enviromic prediction. Molecular Plant 15(11): 1664–1695. https://doi.org/10.1016/j.molp.2022.09.001.

Yadav, A., Alhan, S. and Dhania, G. (2024). Nanobiostimulants for enhancing plant growth and yield. pp. 111–134. *In*: Singh, V., Bhat, R. A. and Dar, G. H. (eds.). Nanobiostimulants: Emerging Strategies for Agricultural Sustainability. Springer, Cham. https://doi.org/10.1007/978-3-031-68138-7_6.

Yadav, A., Yadav, K., Ahmad, R. and Abd-Elsalam, K. A. (2023). Emerging frontiers in nanotechnology for precision agriculture: Advancements, hurdles and prospects. Agrochemicals 2(2): 220–256. https://doi.org/10.3390/agrochemicals2020016.

Zhang, H., Zheng, T., Wang, Y., Li, T. and Chi, Q. (2024). Multifaceted impacts of nanoparticles on plant nutrient absorption and soil microbial communities. Frontiers in Plant Science 15: 1497006. https://doi.org/10.3389/fpls.2024.1497006.

Zhang, P., Guo, Z., Ullah, S., Melagraki, G., Afantitis, A. and Lynch, I. (2021). Nanotechnology and artificial intelligence to enable sustainable and precision agriculture. Nature Plants 7(7): 864–876. https://doi.org/10.1038/s41477-021-00946-6.

Zhao, M., Olsen, P. and Chandra, R. (2023). Seeing through clouds in satellite images. IEEE Transactions on Geoscience and Remote Sensing 61: 1–16. https://doi.org/10.1109/TGRS.2023.3239592.

# Section II

# Novel Nanotechnologies for Shaping Farming

# Chapter 2

# Nanomaterials in Agriculture

## Enhancing Soil Health and Plant Growth

*Amanpreet K. Sidhu,*[1,*] *Sanskruthi B. Agrawal,*[2] *Naveen Verma*[3] and *Priya Kaushal*[4]

## 1. Introduction

The Food and Agriculture Organization (FAO) has reported that, at the midpoint of the 2030 Agenda, progress on food and agriculture-related Sustainable Development Goals (SDGs) remains significantly off track due to the persistent effects of COVID-19, accelerating climate change, and geopolitical conflicts. In 2022, global hunger affected between 691 million and 783 million people, while food insecurity worsened, reversing decades of progress. As the global population is projected to reach nearly 9.1 billion by 2050, the challenge of ensuring food security becomes increasingly critical. However, achieving this goal is complicated by shrinking arable land, depleting natural resources, and the escalating impact of climate change on agricultural productivity. FAO estimates that global food production must increase by approximately 50% to meet future demands, a target that conventional agricultural practices alone cannot fulfill (Shukla et al. 2019, FAO 2025). These challenges necessitate transformative agricultural strategies that enhance productivity while maintaining ecological sustainability.

One of the most promising advancements in modern agriculture is the integration of nanotechnology, particularly nanomaterials, into agricultural systems. The application of nanotechnology offers novel solutions to enhance crop productivity, improve soil health, and mitigate environmental stressors. Over the past decades, various nanomaterials including metal and metal oxide nanoparticles (da Silva et al. 2022, Sidhu and Kaushal 2023), carbon-based nanomaterials (Nepal et al. 2023), polymeric nanoparticles (Dat et al. 2024b), and natural nanoparticles (Raut et al. 2024) have been extensively characterized for their potential applications in agriculture. These nanomaterials have demonstrated significant roles in enhancing abiotic stress tolerance (Verma et al. 2023), such as improving plant resilience against salt stress (Rossi et al. 2017), heat stress (Thakur et al. 2021), and drought stress (Ayyaz et al. 2022). Additionally, they have been explored for their ability to enhance biotic stress tolerance, improving plant defense against microbial pathogens and pests (Chun and Chandrasekaran 2019, Tortella et al. 2023, Verma et al. 2024). Furthermore,

[1] Department of Sciences, Khalsa College for Women, Amritsar, Punjab-143002.
[2] Department of Biotechnology, K.T.H.M. College, Nashik, Maharashtra, 422001 India.
[3] Department of Biotechnology, Khalsa College, G.T. Road, Amritsar, Punjab, 143002, India.
[4] Department of Biotechnology, Guru Nanak Dev University, Amritsar, Punjab, 143001 India.
* Corresponding author: aman.preet1807@gmail.com

DOI: 10.1201/9781003659907-4

nanomaterials have shown promise in boosting crop yields and productivity through improved nutrient uptake efficiency, stress mitigation, and targeted delivery of agrochemicals (Ditta et al. 2015, Guleria et al. 2023). Beyond agriculture, their applications extend to therapeutic medicine, demonstrating their versatility across multiple scientific disciplines (Sidhu et al. 2022, Bernela et al. 2023).

In addition to their direct impact on plant growth and resilience, nanoparticles have emerged as a crucial tool for soil enhancement (Kah et al. 2018), offering diverse applications such as improving fertilizer efficiency (Naderi and Abedi 2012), controlling pests and diseases (Rai et al. 2021), facilitating soil remediation (Thirumalaisamy et al. 2022), and enhancing soil structure and water retention (Saberi Riseh et al. 2024). These advancements have the potential to revolutionize soil management strategies, ensuring long-term agricultural sustainability. However, despite their numerous benefits, the increasing utilization of nanomaterials raises concerns regarding their environmental and human health implications (Handy and Shaw 2007). Issues such as nanoparticle accumulation in soil and water systems, potential toxicity to beneficial soil microbiota, and risks associated with bioaccumulation in food chains require careful evaluation to ensure their safe and sustainable use (Zhang et al. 2020).

This chapter presents a comprehensive review of the application of nanoparticles in soil health management, with a focus on their mechanisms of nanomaterial interaction with soil and plants, their diverse roles in agricultural sustainability, and their concern about potential risks to environmental and human health. Through an extensive literature survey, this work aims to provide a holistic understanding of the opportunities and challenges associated with nanotechnology in agriculture, paving the way for future research and practical implementation.

## 2. Types of Nanomaterials Used for Soil Enhancement

### *2.1 Metal and Metal Oxide Nanoparticles*

Metallic nanoparticles (Me-NPs) can be applied in agriculture for different purposes, mainly to increase crop production (nano-fertilizers, nano-pesticides, growth stimulators, plant protection, seed priming, biosensing), animal husbandry, and remediate soil and water affected by agricultural practices (da Silva et al. 2022). Me-NPs used in fertilizers can be composed mainly of calcium, potassium, zinc, magnesium, silver, and iron. Metal oxide nanoparticles are increasingly used to treat iron shortage in plants due to their distinct physicochemical properties. The following metal and metal oxide particles can be used in soil enhancement.

#### *2.1.1 Iron and Iron Oxide Nanoparticles*

Different crops suffer from iron deficiency; major Fe content remains in soil in the form of insoluble $Fe^{3+}$. Plants usually absorb iron in the form $Fe^{2+}$ from soil, thus plants become iron deficient (Rui et al. 2016). Usually, iron oxide nanoparticles enter plants through the root and then move on to the shoot. They also help in soil remediation as well as interaction with soil microorganisms, enhancing their activity (Perea Velez et al. 2021). $Fe_3O_4$ nanoparticles not only degrade some organic pollutants, but also show good removal ability for heavy metal ions.

Trace elements like lead contaminate the soil and harm the ecosystem. Oxide nanoparticles (IONPs) can interact with pollutants in soils such as heavy metals, pesticides, and other contaminants and stabilize trace elements (Mounier et al. 2023). Ahmad et al. (2024) showed that Fe nanoparticles reduce the absorption of cadmium from the roots of tomato plants and improve the growth and yield of the plant.

#### *2.1.2 Zinc and Zinc Oxide Nanoparticles*

Zinc oxide nanoparticles (ZnO NPs) have been utilized as nano-fertilizers to help mitigate different stresses encountered by plants, such as heavy metals, salinity, drought, extreme warmth and cold, and primarily nutrient deficiency, while positively impacting plant yield (Hanif et al. 2024b). Soil application of Zn is typically viewed as the most effective method for delivering macronutrients to plants. A study by Mi et al. (2023a) showed ZnO NPs are effective in increasing rice grain yield in Zn-deficient soils. The application of these particles to soil can substantially increase plant growth, yield, physiological and metabolic activity, and crop quality. They also help in mitigating saline stress and enhancing crop productivity in saline soils (Seleiman et al. 2023a). They increase soil moisture, water-holding capacity and also improve soil porosity, thus reducing bulk density, which further enhances plant root growth (Strekalovskaya et al. 2024).

#### *2.1.3 Copper and Copper Oxide Nanoparticles*

CuO NPs have enhanced biomass production and help in increasing N and K levels in the soil, thus increasing soil fertility (Phung et al. 2023). By enhancing photosynthesis, nutrient uptake, and root growth CuO NPs can enhance plant growth (Feigl et al. 2023). Shafiq et al. (2024a) evaluated the potential of foliar application of CuO-NPs on decreasing the negative effect of salinity on crops. They can substantially decrease soil redox potential and increase electrical conductivity (Shi et al. 2018). The small size and gradual release of copper ions from copper-based nanoparticles (NPs) facilitate their uptake by plants and extend their effectiveness in the soil. This enhances their bioavailability and reduces the necessary application rate, helping to minimize the harmful effects of overuse and runoff on ecosystems, especially on soils (Su et al. 2024).

#### *2.1.4 Silver Nanoparticles*

Silver nanoparticles (AgNPs) possess unique properties and a wide array of applications, especially in agriculture and plant biotechnology. Research has demonstrated that they can improve seed germination, promote plant growth, and boost photosynthetic efficiency (Khan et al. 2023). They can enhance the availability and absorption of essential nutrients from the soil, making them effective and safe as both nano fertilizers and pesticides. Additionally, AgNPs could stimulate a range of physiological and biochemical processes that support plants in managing stress, including increased tolerance to abiotic stresses such as drought, salinity, and heavy metal contamination (Przemieniecki et al. 2024a). Bacterial diseases are a major cause of substantial crop yield loss globally. AgNPs are effective against plant pathogenic bacteria. These particles have demonstrated antifungal activity against pathogenic fungi like *Fusarium oxysporum*, *Alternaria alternata* and *Aspergillus flavus* (Kale et al. 2021).

### *2.2 Carbon-Based Nanomaterials*

Carbon-based nanomaterials commonly applied to agricultural systems include fullerenes (bucky balls), single-walled and multi-walledcarbon nanotubes (CNTs), nanographenes, quantum and carbon dots, nano diamonds, and nano horns (Nepal et al. 2023). Carbon NPs are made up of pure carbon and have high stability and low toxicity, thus they are environmentally friendly. They have been used to boost crop production, seed germination, cell division, water uptake, and photosynthesis. In one of the studies, treating tomato plants with carbon nanotubes (CNTs) doubled the number of fruits and flowers. Graphene seed priming in tomatoes increased chlorophyll, vitamin C, β-carotene, phenols, flavonoids, and $H_2O_2$ content (Zhao et al. 2021). CNTs are regarded as an innovative fertilizer due to their ability to act as both a slow-release fertilizer and a plant growth enhancer. In agriculture, CNTs can serve as intelligent delivery systems for nano-encapsulated agrochemicals, providing controlled, self-regulating, and targeted release for substances like fertilizers and pesticides (Safdar et al. 2022a). CNT's surface hydroxyl ions react with excess $H^+$ ions in acidic soil and help in neutralizing the pH of soil (Alam et al. 2022a). They improved the aggregation of a sandy loam soil and can increase the water holding capacity of soil by creating more surface area, which can help plants cope with water loss and drought stress (Sharma and Kumar 2024). The graphene nanoparticles, when added to soil can increase soil clay content, improve soil texture and the soil's capacity to absorb and store nutrients, thus reducing nutrient loss by run off (Wang et al. 2023).

## 2.3 Polymeric Nanoparticles

Polymeric nanoparticles (PNPs) have gained attention in agriculture. They help in improving plant nutrition by controlling the release of agrochemicals and managing abiotic/biotic stress (Vodyashkin et al. 2022, Dat et al. 2024a). Different types of biopolymers like polysaccharides (cellulose, chitosan, pectin, starch, alginate, hyaluronic acid), and proteins (collagen, gelatin, albumin) are commonly used in the synthesis of PNPs (Pu et al. 2018). Xin et al. (2020) used PNPs to improve seed germination and plant growth under copper stress. This polymeric compound can bond with other natural polymers and in combination with fertilizers and nutritional elements can help to improve soil texture (Riseh et al. 2023a). Its cationic nature builds strong interactions with negatively charged soil particles, improving nutrient retention and soil health (Atanda et al. 2025). Water-absorbing and soil-binding properties of cellulose nanofibers (CNF) help soil to enhance soil water retention and aggregate stability (Ngo et al. 2024). Lignocellulosic nanoparticles contain many hydroxyl groups, enabling them to form strong hydrogen bonds with water and making them ideal candidates for surface modification. These properties allow them to create hydrogels, which are beneficial for soil conditioning (Vinzat et al. 2023). It is otherwise known as 'man-made polymers', which are synthesized by polymerization of chemical molecules. It has been used to improve the properties of soil such as strength, stiffness, permeability, erosion resistance, water stability, volume changes, etc. (Almajed et al. 2022). Poly (lactic-*co*-glycolic acid) PLGA NPs are synthetic NPs that affect the nitrification and essential nitrifying bacteria in soil environments (Nishu et al. 2020).

## 2.4 Natural Nanoparticles

Naturally occurring nanoparticles (NONPs), also known as environmental nanoparticles, are particles that exist in the environment without human intervention or artificial creation. These nanoparticles are formed through natural processes such as volcanic eruptions, forest fires, ocean water evaporation, and dust storms, leading to their widespread distribution throughout the environment. They exhibit distinctive characteristics such as a high surface area, strong adsorption capabilities, electron-donating properties, and robust electrostatic interactions. These features, along with their unique optical, magnetic, biological, and electronic properties, contribute to their widespread applications across various fields (Raut et al. 2024). Clay minerals, including phyllosilicates and oxides, are fundamental components of natural soils. Soil nanoparticles (NPs), the smallest fraction of soil solids, are characterized by high organic carbon content and large specific surface areas. These nanoparticles exhibit strong colloidal mobility, enabling their presence in surface runoff and downstream water bodies, which influences the distribution and transport of nutrients and contaminants within ecosystems. Under natural conditions, colloidal particles and NPs tend to aggregate, a process that significantly affects their size, morphology, dispersibility, stability, and the formation and stability of soil aggregates (Li et al. 2025). Biochars are carbon-rich materials produced through the pyrolysis of biomass—heating organic matter in the absence of oxygen. They possess a porous structure, abundant surface functional groups, and a high negative charge density. When applied to soil, biochars enhance soil physical and chemical properties, aid in immobilizing potentially toxic elements, and sequester carbon, thereby mitigating greenhouse gas emissions. Upon soil application, biochars decompose into colloidal particles and nanoparticles, influencing soil aggregation and stability (Qambrani et al. 2017). Humic substances (HSs) act as multifunctional natural catalysts in sustainable agriculture and help to enhance soil health, plant productivity, and environmental resilience (Maffia et al. 2025).

Humic acids enhance soil physical and biochemical properties by improving structure, texture, water holding capacity, and supporting microbial populations. They increase the availability of soil nutrients, particularly micronutrients, by chelating and facilitating their transport to plants.

Additionally, humic acids reduce the mobility of toxic heavy metals by precipitating them, thereby decreasing their uptake by plants (Ampong et al. 2022).

## 3. Mechanisms of Nanomaterial Interaction with Soil and Plants

### *3.1 Nanomaterial-Soil Interactions*

The distribution and fate of nanomaterials in soil are governed by their physicochemical properties, including morphology, surface coating, surface functionality, size, and composition. Various soil characteristics, such as microbial community composition, pH, porosity, moisture content, hydraulic conductivity, ionic strength, organic matter content, and redox potential, regulate the bioavailability, transformation, and interaction of nanomaterials within soil matrices. Once introduced into the soil, nanomaterials undergo physical, chemical, and biological transformations that modulate their environmental behavior and functional activity (Zhang et al. 2020, Chanu et al. 2021).

Physical transformation is primarily driven by homoaggregation and heteroaggregation of nanomaterials with soil particles, leading to alterations in their dispersion and mobility. Additionally, nanomaterials interact with humic substances in the soil, modifying their surface properties and influencing their reactivity and stability (Hotze et al. 2010, Lundqvist et al. 2011). Chemical and biological transformations further affect the fate of nanomaterials in soil systems. Soil microorganisms secrete organic acids that facilitate oxidation, reduction, chelation, and phosphorylation of nanomaterials, thereby altering their chemical state and bioavailability. Furthermore, the decomposition of biotic components, including plants, animals, and microbial biomass, leads to the release of enzymatic mediators such as reductases, oxidases, and hydrolases, which contribute to the biotransformation of nanomaterials (Benckiser and Schnell 2006). In addition to these transformation pathways, nanomaterials undergo interactions within the soil via adsorption and sequestration mechanisms, which influence their mobility, persistence, and potential ecological impact.

### *3.2 Nanomaterial-Plant Interactions*

Nanomaterial-plant interactions are influenced by various factors, including plant species, physicochemical properties of nanomaterials (such as composition, size, stability, concentration, and surface characteristics), and the mode of entry into plant tissues. The uptake of nanomaterials varies depending on the route of application: foliar-applied nanomaterials primarily penetrate through stomata, stigmas, hydathodes, trichomes, and the leaf cuticle, whereas soil-applied nanomaterials enter via root uptake, and seed-applied nanomaterials permeate through cotyledon diffusion or the parenchymatic tissues of the seed coat (Tripathi et al. 2017). Among these, foliar and root applications remain the most widely adopted strategies.

Once internalized, nanomaterials are translocated within plant tissues through two primary pathways: (a) the apoplastic pathway and (b) the symplastic pathway. The apoplastic pathway involves the movement of nanomaterials through extracellular spaces, the cell walls of adjacent cells, and xylem vessels. Notably, root surface morphology plays a crucial role in nanomaterial interactions, as root hairs secrete mucilage and organic acids that attract positively charged nanomaterials (de la Rosa et al. 2021). However, the transport of nanomaterials through the apoplastic route is restricted by the casparian strip in the endodermis, necessitating subsequent translocation via the xylem (Kumar et al. 2024). Conversely, the symplastic pathway facilitates nanomaterial transport through endocytosis, plasmodesmata, ion channels, carrier proteins, and nanopore formation, enabling direct cytoplasmic movement (Pérez-de-Luque 2017). Collectively, the efficiency of nanomaterial uptake and distribution in plants is dictated by the interplay between plant physiology and nanomaterial properties, which in turn governs their potential applications in agriculture and environmental remediation.

## 4. Applications of Nanomaterials in Soil Enhancement

Nanomaterials have developed as novel techniques for improving soil quality, increasing nutrient availability, and supporting environmentally friendly agriculture practices. Their unique physicochemical qualities, such as large surface area, reactivity, and controlled release mechanisms, allow them to improve soil fertility, reduce environmental stress, and promote plant growth (Kah et al. 2018). Metal and metal oxide nanoparticles, carbon-based nanomaterials, polymeric nanoparticles, and natural nanoparticles have all been widely used to improve soil characteristics and nutrient delivery (Seleiman et al. 2023b) (Figure 1). These nanomaterials help improve fertilizer efficiency, control pests and diseases, remediate contaminated soils, and improve soil water retention and structure (Raliya et al. 2018). Their capacity to interact with soil particles, microbial communities, and plant roots makes them extremely successful at encouraging sustainable soil management techniques. While nanomaterials have numerous advantages, it is critical to evaluate their long-term influence on soil ecosystems to assure environmental safety and sustainability (Naderi and Abedi 2012).

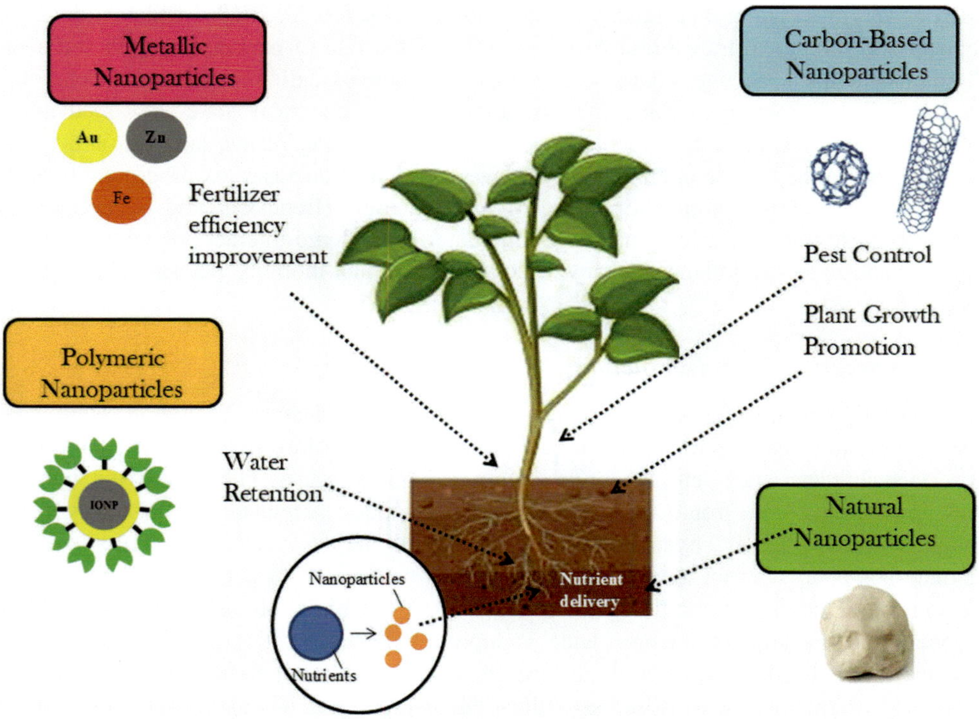


**Figure 1.** Various nanoparticles and their application in soil enhancement.

### *4.1 Fertilizer Efficiency Improvement*

Nanotechnology has changed fertilizer efficiency by enabling regulated and targeted nutrient delivery systems that reduce nitrogen loss while increasing plant uptake. Nanofertilizers made from nanoparticles such as iron, zinc, copper, and polymeric materials have demonstrated amazing promise for increasing nutrient use efficiency, lowering environmental pollution, and improving soil fertility (Table 1) (Kah et al. 2018). Nanoparticles can be used in two ways: as fertilizers by giving vital nutrients directly to plants, or as carriers to distribute fertilizers in a regulated and sustained manner (Naderi and Abedi 2012). For instance, zinc oxide nanoparticles (ZnO NPs) have

**Table 1.** Characteristics and functions of nanoparticles in fertilizer efficiency enhancement for various crops.

| Nanoparticle | Characteristics | Plant/Target Crop | Activity/Function | References |
|---|---|---|---|---|
| Zinc Oxide (ZnO NPs) | High solubility, antimicrobial properties, and slow release | Rice, Wheat, Maize | Mitigates zinc deficiency, increases yield, enhances stress tolerance | Mi et al. 2023a, Hanif et al. 2024a |
| Iron Oxide ($Fe_3O_4$ NPs) | Magnetic properties, oxidative degradation, enhance microbial activity | Tomato, Maize | Provides iron, improves soil fertility, and enhances microbial activity | Perea et al. 2021, Safdar et al. 2022a |
| Copper Oxide (CuO NPs) | Antimicrobial, increases soil conductivity, and controls ion release | Wheat, Tomato | Enhances nutrient uptake, increases soil fertility, and boosts plant growth | Feigl et al. 2023, Shafiq et al. 2024b |
| Silver Nanoparticles (AgNPs) | Strong antimicrobial properties, high surface area | Tomato, Rice | Enhances nutrient uptake, improves growth, and reduces biotic stress | Khan et al. 2023, Przemieniecki et al. 2024b |
| Carbon Nano-tubes (CNTs) | High surface area, adsorptive capacity, controlled release | Tomato, Wheat | Enhances nutrient absorption, mitigates stress, and improves soil porosity | Alam et al. 2022a, Safdar et al. 2022b |
| Polymeric Nanoparticles (PNPs) | Biodegradable, slow-release system, nutrient retention | Rice, Wheat | Improves nutrient release, enhances water retention, and supports microbial growth | Nishu et al. 2020, Riseh et al. 2023a |
| Poly(lactic-co-glycolic acid) (PLGA) NPs | Biodegradable, pH-sensitive release, affects nitrification | Soil Microbial Communities | Controls nitrogen release, reduces nitrification, and minimizes nitrogen loss | Nishu et al. 2020, Riseh et al. 2023b |
| Graphene Oxide (GO) | High surface area, adsorptive capacity, improves nutrient retention | Tomato, Maize | Increases soil moisture, enhances nutrient absorption, and improves soil quality | Safdar et al. 2022a, Alam et al. 2022b |

been utilized to treat zinc shortage in plants, resulting in improved growth and yields, particularly in zinc-deficient soils (Mi et al. 2023b). Similarly, iron oxide nanoparticles ($Fe_3O_4$ NPs) not only give iron to plants but also boost soil microbial activity, resulting in better soil structure and nutrient availability (Perea et al. 2021).

As carrier nanoparticles, such as carbon-based nanomaterials, including carbon nanotubes (CNTs) and graphene oxide, have emerged as viable fertilizer carriers due to their large surface area and ability to absorb and release nutrients under controlled conditions. These nanomaterials minimize nutrient loss through leaching and volatilization by enhancing soil nutrient retention and allowing for progressive nutrient release (Safdar et al. 2022b). CNTs have been proven in studies to improve nitrogen, phosphorus, and potassium uptake, resulting in improved plant growth and soil fertility (Alam et al. 2022b). Furthermore, polymeric nanoparticles (PNPs) made from biodegradable polymers such as chitosan and alginate have been utilized to encapsulate fertilizers and agrochemicals, assuring long-term nutrient release and lowering application frequency (Riseh et al. 2023b). Poly (lactic-co-glycolic acid) (PLGA) nanoparticles have been shown to alter nitrification processes in soil by limiting the release of nitrogenous chemicals, reducing nitrogen losses, and increasing soil fertility (Nishu et al. 2020).

The incorporation of nanotechnology into fertilizers improves not just nutrient uptake and utilization but also soil water retention and microbial activity, creating an optimal environment for plant growth. However, despite their great potential, the long-term effects of nanofertilizers on soil health, microbial diversity, and environmental safety are still being investigated. To ensure the safe and sustainable use of nanofertilizers, further study is required to analyze their environmental impact and optimize their application in a variety of agricultural contexts (Kah et al. 2018).

## *4.2 Pest and Disease Control*

Nanomaterials have shown enormous promise in pest and disease control, providing targeted, efficient, and environmentally benign alternatives to traditional pesticides. Nanoparticles can work as direct antimicrobial agents or as carriers for regulated pesticide administration, increasing efficacy while lowering environmental impact (Rai et al. 2021). Metal and metal oxide nanoparticles, such as silver (AgNPs), copper oxide (CuO NPs), and zinc oxide (ZnO NPs), show excellent antibacterial, antifungal, and antiviral properties against a variety of plant diseases. Silver nanoparticles, recognized for their broad-spectrum antibacterial characteristics, have been utilized to suppress bacterial and fungal infections in crops by breaking cell membranes and producing reactive oxygen species (ROS), which cause oxidative damage to microbial cells (Khan et al. 2023). AgNPs have been demonstrated to be effective against *Fusarium oxysporum, Alternaria alternata,* and *Aspergillus flavus,* resulting in considerable reductions in fungal infections in a variety of crops.

Similarly, copper oxide nanoparticles have strong antifungal and antibacterial capabilities, which improve plant defense mechanisms and reduce the spread of plant diseases. CuO NPs have been shown in studies to limit the growth of fungal infections, like *Phytophthora infestans* and *Botrytis cinerea,* preventing common plant diseases, such as late blight and gray mold (Shafiq et al. 2024b). Zinc oxide nanoparticles not only supply important micronutrients to plants, but they also have excellent antibacterial properties, breaking microbial cell walls and decreasing enzyme activity, hence preventing bacterial and fungal diseases in crops (Hanif et al. 2024a).

In addition to their direct antimicrobial characteristics, nanoparticles can be used as carriers to distribute insecticides, herbicides, and fungicides in a controlled and targeted manner. Polymeric nanoparticles, such as chitosan-based formulations, have been utilized to encapsulate biopesticides and chemical pesticides, resulting in a delayed and sustained release that improves efficacy while reducing environmental toxicity (Riseh et al. 2023a). Chitosan nanoparticles have natural antibacterial characteristics and can promote systemic resistance in plants, which improves their defensive mechanisms against pathogens (Dat et al. 2024a). Furthermore, carbon-based nanomaterials such as carbon nanotubes (CNTs) and graphene oxide can act as smart delivery systems for nano-encapsulated agrochemicals, allowing for controlled, targeted, and self-regulated pesticide release, resulting in reduced pesticide use and environmental contamination (Safdar et al. 2022a).

The use of nanotechnology in pest and disease management not only enhances crop health and productivity, but it also minimizes the use of chemical pesticides, reducing their negative influence on the environment. However, despite their promising potential, more research is required to determine the long-term ecological impact of nanoparticle-based pest and disease management systems in order to maintain sustainable agricultural practices (Rai et al. 2021).

## *4.3 Soil Remediation*

Rapid industrialization and globalization have significantly contributed to the release of electronic waste and various toxic compounds, including heavy metal ions, phenolics, hydrocarbons, pesticides, and dyes, into the environment, which in turn adversely impact soil quality and health (Thirumalaisamy et al. 2022). In response, nanoremediation has emerged as an advanced and promising technology for soil decontamination. The unique physicochemical properties of nanomaterials, such as high surface area, enhanced surface-to-volume ratio, abundant active sites, selectivity, specificity, and superior adsorption capacities, make them viable alternatives for effective soil remediation (Mukhopadhyaya et al. 2021, Rajput et al. 2023).

Several studies have demonstrated the efficiency of nanomaterials in the remediation of organic and inorganic contaminants from soil. For instance, research conducted by Soto-Vázquez et al. (2019) highlighted the application of titanium dioxide nanowires for the photodegradation of benzophenone-4, achieving a 90% removal efficiency within three hours. Similarly, Al-Qassar Bani Al-Marjeh et al. (2019) reported the effectiveness of polylactic acid-coated nanofibers in the removal of methyl orange contamination, achieving a 70% removal efficiency. Furthermore, nanomaterials have shown significant potential for the remediation of heavy metal-contaminated soils. Studies by Gaur et al. (2014) and Wang et al. (2014) reported the adsorption and immobilization of heavy metals, such as cadmium (Cd), arsenic (As), and chromium (Cr), using ferric oxide nanoparticles and nanohydroxyapatite. Additionally, Su et al. (2016) demonstrated the high Cr immobilization efficiency of biochar-coated nanoscale zero-valent iron (nZVI) in soil. Usman et al. (2020) further highlighted the efficacy of sodium carboxymethyl cellulose-stabilized nZVI in remediating Cr-contaminated soil, achieving an 80% removal efficiency.

In conclusion, nanoremediation presents a highly effective and innovative approach for soil decontamination, offering superior removal efficiencies for both organic and inorganic pollutants. The versatility and efficiency of nanomaterials in adsorption, immobilization, and degradation of contaminants underscore their potential in sustainable environmental remediation.

### *4.4 Water Retention and Soil Structure Improvement*

Traditional agricultural practices, climate change, diminishing arable land, excessive application of chemical fertilizers, and high water consumption, have led to the deterioration of soil properties, ultimately affecting agricultural productivity (Saberi Riseh et al. 2024). Among the critical soil parameters, water availability and aggregate stability play a vital role in crop growth and yield by improving water-holding capacity, nutrient cycling, soil aeration, and overall soil health (Kang et al. 2022).

Nanomaterials have emerged as a promising approach to enhance soil structure and aggregation due to their unique physicochemical properties. Recent studies have highlighted the beneficial impact of organic nanoparticles on soil quality. Soliman and Mansour (2024) investigated the effects of polyvinyl alcohol (PVA) polymerized with chitosan (CS) and pectin (PC) nanoparticles on soil organic carbon content and water retention. Their findings indicate that the application of CS/PVA and PC/PVA nanomaterials, combined with NPK fertilizers, increased soil carbon cycling activity within a range of 9.71 to 11.68 g $kg^{-1}$ and enhanced soil water-holding capacity by 18.23% compared to the control (NPK alone). Similarly, Ngo et al. (2024) demonstrated that a 1% dosage of cellulose nanofibers (CNF) significantly improved soil water retention capacity (156–186%) and extended water retention duration to 16–22 days in sandy and silt loam soils. Additionally, CNF treatment also facilitated macro-aggregate formation, increasing aggregation by 48% and 59% in masa and paddy soils, respectively, thereby supporting plant growth in drought-prone regions.

Furthermore, Jaramillo-Quiceno et al. (2024) evaluated the role of hydrogels (HFT) synthesized from carboxymethylcellulose, polyvinyl alcohol, and silk sericin in enhancing soil water retention. Their study revealed that a 1% HFT application in sandy loam soil (under conditions of 40°C temperature and 50% relative humidity) improved soil moisture content by 7% after thirty-two hours. Additionally, Naik et al. (2025) reported that biobased nano-additives enhanced soil aggregate stability from 45% to 68% within six weeks of application, while water retention capacity increased from 15% to 22%.

These findings collectively emphasize the potential of nanomaterials as a sustainable strategy for mitigating soil degradation, enhancing water conservation, and promoting crop resilience under changing environmental conditions.

## 5. Environmental and Health Implications

Nanomaterials have gained significant attention due to their extensive applications across various industrial (electronics, optics, textiles, cosmetics, and food), medical (biosensors, drug delivery systems), and agricultural (nanofertilizers, nanopesticides) sectors. The global production of nanomaterials has witnessed a remarkable 30-fold increase between 2011 and 2015 (Madannejad et al. 2019) to meet growing industrial demands. However, concerns regarding their potential environmental and human health toxicity remain a critical challenge (Handy and Shaw 2007). The toxicity of nanomaterials is largely governed by their physicochemical properties, including structural composition, surface chemistry, stability, charge, dosage, and their tendency to undergo agglomeration (Odaudu et al. 2022). The persistent nature of nanomaterials and their bioaccumulation potential in ecosystems can lead to significant alterations in soil structure, nutrient bioavailability, and microbiome composition. Furthermore, the infiltration of nanomaterials into the food chain poses serious risks to higher trophic levels, including plants, aquatic organisms, and human health (Zhang et al. 2020). In plants, exposure to high concentrations of nanomaterials induces oxidative stress, nuclear and cellular damage, and protein denaturation, ultimately inhibiting growth and reducing productivity (Zaman et al. 2025).

Nanomaterials may enter the human body through multiple pathways, including inhalation, ingestion, and dermal absorption. Due to their nanoscale size, they can easily cross biological barriers and enter the bloodstream, leading to cytotoxic effects and adverse impacts on vital organs such as the brain, heart, lungs, gastrointestinal tract, and reproductive system (Ayanda et al. 2024). Given these potential risks, it is imperative to develop biodegradable and eco-friendly nanomaterials while conducting comprehensive time-dependent risk assessments. Standardized *in vitro* and *in vivo* assays should be routinely employed to evaluate cytotoxicity, genotoxicity, and oxidative stress responses, which will provide crucial insights into the biodistribution, pharmacokinetics, and metabolic fate of nanomaterials (Gupta et al. 2024).

Despite the growing concerns surrounding the environmental and health implications of nanomaterials, significant knowledge gaps persist in the implementation of regulatory frameworks governing their safe use. Future research must focus on designing biodegradable and biocompatible nanomaterials while integrating advanced toxicological assessments to evaluate their long-term ecological and physiological impacts. The establishment of stringent guidelines, standardized toxicity assessment protocols, and regulatory policies will be essential to mitigate the risks associated with nanomaterials and ensure their sustainable application across industries.

## 6. Future Prospects and Challenges

The use of nanomaterials in agriculture, particularly for soil enhancement and plant growth, has enormous potential for increasing crop productivity, soil fertility, and resource efficiency. Nanofertilizers, nano-based soil conditioners, and precision nutrient delivery systems have the potential to transform traditional agricultural practices by increasing nutrient uptake efficiency, reducing nutrient losses, and lowering environmental pollution (Singh et al. 2023). Furthermore, the use of smart nanocarriers capable of controlled and targeted agrochemical release can improve plant growth while reducing soil degradation (Mishra 2024). Despite these benefits, several obstacles prevent the widespread adoption of nanotechnology in agriculture. Nanoparticles' potential toxicity and environmental fate raise concerns about the long-term impact on soil microbial diversity, plant health, and human safety. Furthermore, the high cost of nanomaterial synthesis, regulatory constraints, and a lack of standardized testing protocols all present significant barriers to widespread application. Future research should concentrate on environmentally friendly, biodegradable, and low-cost nanomaterials that promote sustainable agricultural development while also addressing safety concerns. Additionally, comprehensive risk assessment studies and regulatory frameworks must be developed to assess the environmental and health consequences of nanoparticle use in agriculture, ensuring their safe and responsible deployment (Ahmad et al. 2024). Collaboration among researchers, policymakers, and agricultural industries is required to realize the full potential of nanotechnology in sustainable farming.

## 7. Conclusion

The use of nanomaterials in agriculture has emerged as a promising approach for improving soil health, plant growth, and crop productivity. Researchers have shown that the unique physicochemical properties of nanoparticles can improve fertilizer efficiency, pest and disease control, soil remediation, and water retention. Nanoparticles can serve as fertilizers themselves or as carriers for controlled nutrient release, ensuring better nutrient availability and uptake by plants. Furthermore, nanomaterials, including metal and metal oxide nanoparticles, carbon-based nanomaterials, polymeric nanoparticles, and naturally occurring nanoparticles have shown the capability to improve soil structure, increase microbial activity, and reduce environmental stressors.

Despite its numerous benefits, the widespread use of nanotechnology in agriculture raises concerns about environmental safety, potential toxicity to soil microbes and plants, regulatory barriers, and cost-effectiveness. More research is needed to determine the long-term impact of nanomaterials on agroecosystems and to create eco-friendly, biodegradable nanomaterials that have fewer negative environmental consequences. Furthermore, advancements in nano-formulation strategies and precision delivery systems will be critical for maximizing agricultural benefits.

In conclusion, while nanotechnology has enormous potential to revolutionize agricultural practices, a multidisciplinary approach involving material science, agronomy, environmental science, and regulatory frameworks is required to fully realize its potential in a sustainable and responsible manner. Future research should concentrate on improving nanoparticle formulations, determining their long-term environmental impact, and ensuring their safe and efficient integration into modern agricultural systems.

## References

Ahmad, A., Javad, S., Iqbal, S., Shahid, T., Naz, S., Shah, A. A. et al. (2024). Efficacy of soil drench and foliar application of iron nanoparticles on the growth and physiology of Solanum lycopersicum L. exposed to cadmium stress. Scientific Reports 14(1): 27920.

Alam, K., Mukherjee, S., Adak, S. and Roy, D. (2022a). Carbon nanotubes: their role in the modification of soil environment to facilitate sustainable crop production. pp 62–67. In book: Futuristic Trends in Agriculture Engineering & Food Sciences. Iterative International Publishers 2(9).

Alam, M. S., Khan, S., Molla, A. H. and Rahman, M. S. (2022b). Application of carbon nanotubes to improve nutrient availability and plant growth: A review. Journal of Nanobiotechnology 20(1): 45. https://doi.org/10.1186/s12951-022-01287-3.

Almajed, A., Lemboye, K. and Moghal, A. A. B. (2022). A critical review on the feasibility of synthetic polymers inclusion in enhancing the geotechnical behavior of soils. Polymers 14(22): 5004.

Al-Qassar Bani Al-Marjeh, R., Atassi, Y., Mohammad, N. and Badour, Y. (2019). Adsorption of methyl orange onto electrospun nanofiber membranes of PLLA coated with pTSA-PANI. Environmental Science and Pollution Research 26: 37282–37295.

Ampong, K., Thilakaranthna, M. S. and Gorim, L. Y. (2022). Understanding the role of humic acids on crop performance and soil health. Frontiers in Agronomy 4: 848621.

Atanda, S. A., Shaibu, R. O. and Agunbiade, F. O. (2025). Nanoparticles in agriculture: balancing food security and environmental sustainability. Discover Agriculture 3(1): 26.

Ayanda, O. S., Mmuoegbulam, A. O., Okezie, O., Durumin Iya, N. I., Mohammed, S. A. E., James, P. H. et al. (2024). Recent progress in carbon-based nanomaterials: critical review. Journal of Nanoparticle Research 26(5): 106.

Ayyaz, A., Fang, R., Ma, J., Hannan, F., Huang, Q., Sun, Y. et al. (2022). Calcium nanoparticles (Ca-NPs) improve drought stress tolerance in *Brassica napus* by modulating the photosystem II, nutrient acquisition and antioxidant performance. NanoImpact 28: 100423.

Benckiser, G. and Schnell, S. (eds.). (2006). Biodiversity in Agricultural Production Systems. CRC Press, pp 429.

Bernela, M., Kaushal, P., Verma, N., Thakur, R., Ahuja, M. and Kaur, P. (2023). Anti-inflammatory therapeutics: conventional concepts and future with nanotechnology. Recent Advances in Inflammation & Allergy Drug Discovery 17(1): 7–19.

Chanu, N. B., Alice, A. K., Thokchom, A., Singh, M. C., Chanu, N. T. and Singh, Y. D. (2021). Engineered nanomaterial and their interactions with plant–soil system: a developmental journey and opposing facts. Nanotechnology for Environmental Engineering 6: 1–15.

Chun, S. C. and Chandrasekaran, M. (2019). Chitosan and chitosan nanoparticles induced expression of pathogenesis-related proteins genes enhances biotic stress tolerance in tomato. International Journal of Biological Macromolecules 125: 948–954.

da Silva Júnior, A. H., Mulinari, J., de Oliveira, P. V., de Oliveira, C. R. S. and Júnior, F. W. R. (2022). Impacts of metallic nanoparticles application on the agricultural soils microbiota. Journal of Hazardous Materials Advances 7: 100103.

Dat, M. T., Nguyen, D. T. and Tran, Q. K. (2024a). Chitosan-based nanocarriers for the delivery of biopesticides in sustainable agriculture. Journal of Agricultural and Food Chemistry 72(3): 589–602. https://doi.org/10.1021/jafc.3c02456.

Dat, T. Q. M., Nhut, D. T. and Long, V. B. (2024b). Application of polymeric nanoparticles on ameliorating crop productivity. pp. 83–103. In Metal Nanoparticles in Plant Cell, Tissue and Organ Culture. Singapore: Springer Nature Singapore.

de la Rosa, G., Vázquez-Núñez, E., Molina-Guerrero, C., Serafín-Muñoz, A. H. and Vera-Reyes, I. (2021). Interactions of nanomaterials and plants at the cellular level: current knowledge and relevant gaps. Nanotechnology for Environmental Engineering 6: 1–19.

Ditta, A., Arshad, M. and Ibrahim, M. (2015). Nanoparticles in sustainable agricultural crop production: applications and perspectives. Nanotechnology and Plant Sciences: Nanoparticles and their Impact on Plants 55–75.

Feigl, G. (2023). The impact of copper oxide nanoparticles on plant growth: a comprehensive review. Journal of Plant Interactions 18(1): 2243098.

Gaur, N., Flora, G., Yadav, M. and Tiwari, A. (2014). A review with recent advancements on bioremediation-based abolition of heavy metals. Environmental Science: Processes & Impacts 16(2): 180–193.

Guleria, G., Thakur, S., Shandilya, M., Sharma, S., Thakur, S. and Kalia, S. (2023). Nanotechnology for sustainable agro-food systems: The need and role of nanoparticles in protecting plants and improving crop productivity. Plant Physiology and Biochemistry 194: 533–549.

Gupta, R. K., Guha, P. and Srivastav, P. P. (2024). Investigating the toxicological effects of nanomaterials in food packaging associated with human health and the environment. Journal of Hazardous Materials Letters 5: 100125.

Handy, R. D. and Shaw, B. J. (2007). Toxic effects of nanoparticles and nanomaterials: implications for public health, risk assessment and the public perception of nanotechnology. Health, Risk & Society 9(2): 125–144.

Hanif, M., Ahmad, R., Nisar, M. F. and Ali, M. (2024a). Role of zinc oxide nanoparticles in plant stress mitigation and nutrient management. Environmental Science and Pollution Research 31(5): 4512–4525. https://doi.org/10.1007/s11356-023-29087-6.

Hanif, S., Javed, R., Cheema, M., Kiani, M. Z., Farooq, S. and Zia, M. (2024b). Harnessing the potential of zinc oxide nanoparticles and their derivatives as nanofertilizers: Trends and perspectives. Plant Nano Biology 100110.

Hotze, E. M., Phenrat, T. and Lowry, G. V. (2010). Nanoparticle aggregation: challenges to understanding transport and reactivity in the environment. Journal of Environmental Quality 39(6): 1909–1924.

https://openknowledge.fao.org/server/api/core/bitstreams/3a8baf8c-960d-4105-9538-e4bd9b1d4503/content/index.html.

Jaramillo-Quiceno, N., Carmona, A. S. D., Serna, J. S., Carmona, D. M., Torres-Taborda, M., Hincapié-Llanos, G. A. et al. (2024). Enhancing soil water retention and plant growth with thermal crosslinked silk sericin-based hydrogel. Journal of Environmental Chemical Engineering 12(2): 112260.

Kah, M., Kookana, R. S., Gogos, A. and Bucheli, T. D. (2018). A critical evaluation of nanopesticides and nanofertilizers against their conventional analogues. Nature Nanotechnology 13(8): 677–684. https://doi.org/10.1038/s41565-018-0131-1.

Kale, S. K., Parishwad, G. V. and Patil, A. S. H. A. S. (2021). Emerging agriculture applications of silver nanoparticles. ES Food & Agroforestry 3: 17–22.

Kang, M. W., Yibeltal, M., Kim, Y. H., Oh, S. J., Lee, J. C., Kwon, E. E. et al. (2022). Enhancement of soil physical properties and soil water retention with biochar-based soil amendments. Science of the Total Environment 836: 155746.

Khan, S., Zahoor, M., Khan, R. S., Ikram, M. and Islam, N. U. (2023). The impact of silver nanoparticles on the growth of plants: The agriculture applications. Heliyon 9(6).

Kumar, D., Singh, R., Upadhyay, S. K., Verma, K. K., Tripathi, R. M., Liu, H. et al. (2024). Review on interactions between nanomaterials and phytohormones: Novel perspectives and opportunities for mitigating environmental challenges. Plant Science 340: 111964.

Li, Q. R., Xu, C. Y., Geng, Z. C., Hu, F. N. and Wang, C. L. (2025). Biochar addition changes the aggregation of clay mineral and natural soil nanoparticles via different mechanisms. European Journal of Soil Science 76(2): e70056.

Lundqvist, M., Stigler, J., Cedervall, T., Berggard, T., Flanagan, M. B., Lynch, I. et al. (2011). The evolution of the protein corona around nanoparticles: a test study. ACS Nano 5(9): 7503–7509.

Madannejad, R., Shoaie, N., Jahanpeyma, F., Darvishi, M. H., Azimzadeh, M. and Javadi, H. (2019). Toxicity of carbon-based nanomaterials: Reviewing recent reports in medical and biological systems. Chemico-biological Interactions 307: 206–222.

Maffia, A., Oliva, M., Marra, F., Mallamaci, C., Nardi, S. and Muscolo, A. (2025). Humic substances: bridging ecology and agriculture for a greener future. Agronomy 15(2): 410.

Mi, G., Chen, F., Zhang, F. and Zhao, Y. (2023a). Enhancing zinc uptake and grain yield in rice using zinc oxide nanoparticles. Frontiers in Plant Science 14: 1–12. https://doi.org/10.3389/fpls.2023.1065289.

Mi, K., Yuan, X., Wang, Q., Dun, C., Wang, R., Yang, S. et al. (2023b). Zinc oxide nanoparticles enhanced rice yield, quality, and zinc content of edible grain fraction synergistically. Frontiers in Plant Science 14: 1196201.

Mishra, H. (2024). Nanobiostimulants and precision agriculture: a data-driven approach to farming and market dynamics. pp. 365–398. In Nanobiostimulants: Emerging Strategies for Agricultural Sustainability. Cham: Springer Nature Switzerland.

Mounier, L., Pédrot, M., Bouhnik-Le-Coz, M. and Cabello-Hurtado, F. (2023). Impact of iron oxide nanoparticles on a lead-polluted water–soil–plant system under alternating periods of water stress. Environmental Science: Advances 2(5): 767–779.

Mukhopadhyay, R., Sarkar, B., Khan, E., Alessi, D. S., Biswas, J. K., Manjaiah, K. M. et al. (2021). Nanomaterials for sustainable remediation of chemical contaminants in water and soil. Critical Reviews in Environmental Science and Technology 52(15): 2611–2660. https://doi.org/10.1080/10643389.2021.1886891.

Naderi, M. R. and Abedi, A. (2012). Application of nanotechnology in agriculture and refinement of environmental pollutants. Journal of Nanotechnology 2012: 1–5. https://doi.org/10.1155/2012/269687.

Naik, N., Noonia, A., Maddodi, B. S., Hiremath, P. and Guo, Z. (2025). Multifaceted improvements in soil characteristics and molecular dynamics through inorganic and bio-based nano additives. ES Food & Agroforestry 19: 1378.

Nepal, J., Xin, X., Maltais-Landry, G., Ahmad, W., Pereira, J., Santra, S. et al. (2023). Carbon nanomaterials are a superior soil amendment for sandy soils than biochar based on impacts on lettuce growth, physiology and soil biochemical quality. NanoImpact 31: 100480.

Ngo, A. T., Mori, Y. and Bui, L. T. (2024). Effects of cellulose nanofibers on soil water retention and aggregate stability. Environmental Technology & Innovation 35: 103650.

Nishu, A., Roy, S. and Kumar, S. (2020). Impact of poly(lactic-co-glycolic acid) nanoparticles on nitrification and nitrifying bacteria in soil environments. Environmental Science and Pollution Research 27(21): 26892–26903. https://doi.org/10.1007/s11356-020-09122-7.

Nishu, S. D., Park, S., Ji, Y., Han, I., Key, J. and Lee, T. K. (2020). The effect of engineered PLGA nanoparticles on nitrifying bacteria in the soil environment. Journal of Industrial and Engineering Chemistry 84: 297–304.

Odaudu, O. R. and Akinsiku, A. A. (2022, September). Toxicity and cytotoxicity effects of selected nanoparticles: a review. In IOP Conference Series: Earth and Environmental Science (Vol. 1054, No. 1, p. 012007). IOP Publishing.

Perea Velez, Y. S., Carrillo-González, R. and Gonzalez-Chavez, M. D. C. A. (2021). Interaction of metal nanoparticles–plants–microorganisms in agriculture and soil remediation. Journal of Nanoparticle Research 23(9): 206.

Perea, L. M., Rojas, L. A., González, A. M. and Fernández, A. (2021). Iron oxide nanoparticles enhance soil microbial activity and plant growth. Applied Soil Ecology 168: 103932. https://doi.org/10.1016/j.apsoil.2021.103932.

Pérez-de-Luque, A. (2017). Interaction of nanomaterials with plants: what do we need for real applications in agriculture?. Frontiers in Environmental Science 5: 12.

Phung, L. D., Afriani, S. D., Pertiwi, P. A. P., Ito, H., Kumar, A. and Watanabe, T. (2023). Effects of CuO nanoparticles in composted sewage sludge on rice-soil systems and their potential human health risks. Chemosphere 338: 139555.

Przemieniecki, S. W., Dworzański, J., Piotrowska, M. and Węglarz-Tomczak, E. (2024a). Antifungal activity of silver nanoparticles against fungal pathogens in agriculture. Journal of Plant Pathology 106(2): 245–259. https://doi.org/10.1007/s42161-024-01458-.

Przemieniecki, S. W., Ruraż, K., Kosewska, O., Oćwieja, M. and Gorczyca, A. (2024b). The impact of various forms of silver nanoparticles on the rhizosphere of wheat (*Triticum aestivum* L.)—Shifts in microbiome structure and predicted microbial metabolic functions. Science of The Total Environment 914: 169824.

Pu, W., Shen, C., Wei, B., Yang, Y. and Li, Y. (2018). A comprehensive review of polysaccharide biopolymers for enhanced oil recovery (EOR) from flask to field. J. IndEngChem. 61: 1–11.

Qambrani, N. A., Rahman, M. M., Won, S., Shim, S. and Ra, C. (2017). Biochar properties and eco-friendly applications for climate change mitigation, waste management, and wastewater treatment: A review. Renewable and Sustainable Energy Reviews 79: 255–273.

Rai, M., Ingle, A. P., Pandit, R., Paralikar, P. and Bonde, S. (2021). Role of nanotechnology in crop protection: Innovations and future perspectives. Applied Microbiology and Biotechnology 105(12): 4921–4934. https://doi.org/10.1007/s00253-021-11442-5.

Rajput, V. D., Kumari, A., Minkina, T., Barakhov, A., Singh, S., Mandzhieva, S. S. et al. (2023). A practical evaluation on integrated role of biochar and nanomaterials in soil remediation processes. Environmental Geochemistry and Health 45(12): 9435–9449.

Raliya, R., Saharan, V., Dimkpa, C. and Biswas, P. (2018). Nanofertilizer for precision and sustainable agriculture: Current state and future perspectives. Journal of Agricultural and Food Chemistry 66(26): 6487–6503. https://doi.org/10.1021/acs.jafc.7b02178.

Raut, S. S., Singh, R. and Lekhak, U. M. (2024). Naturally occurring nanoparticles (NONPs): A review. Next Sustainability 3: 100037.

Riseh, R. S., Ali, M., Sabir, M. and Khan, A. R. (2023a). Role of polymeric nanoparticles in sustainable agriculture: Applications and future perspectives. International Journal of Agricultural Sciences 19(4): 34–47. https://doi.org/10.1016/j.ijas.2023.104556.

Riseh, R. S., Vazvani, M. G. and Kennedy, J. F. (2023b). The application of chitosan as a carrier for fertilizer: A review. International Journal of Biological Macromolecules 252: 126483.

Rossi, L., Zhang, W. and Ma, X. (2017). Cerium oxide nanoparticles alter the salt stress tolerance of *Brassica napus* L. by modifying the formation of root apoplastic barriers. Environmental Pollution 229: 132–138.

Rui, M., Ma, C., Hao, Y., Guo, J., Rui, Y., Tang, X. et al. (2016). Iron oxide nanoparticles as a potential iron fertilizer for peanut (*Arachishypogaea*). Frontiers in Plant Science 7: 815.

Saberi Riseh, R., Hassanisaadi, M., Vatankhah, M., Varma, R. S. and Thakur, V. K. (2024). Nano/micro-structural supramolecular biopolymers: innovative networks with the boundless potential in sustainable agriculture. Nano-Micro Letters 16(1): 147.

Safdar, H., Hussain, S., Raza, M. A., Iqbal, N., Saleem, M. F. and Ali, M. A. (2022a). Carbon-based nanomaterials for sustainable agriculture: Mechanisms and potential applications. Journal of Nanomaterials 2022: 1–12. https://doi.org/10.1155/2022/9893067.

Safdar, M., Kim, W., Park, S., Gwon, Y., Kim, Y. O. and Kim, J. (2022b). Engineering plants with carbon nanotubes: a sustainable agriculture approach. Journal of Nanobiotechnology 20(1): 275.

Seleiman, M. F., Ahmad, A., Battaglia, M. L., Bilal, H. M., Alhammad, B. A. and Khan, N. (2023a). Zinc oxide nanoparticles: A unique saline stress mitigator with the potential to increase future crop production. South African Journal of Botany 159: 208–218.

Seleiman, M. F., Al-Suhaibani, N., Ali, N., Akmal, M., Alotaibi, M., Refay, Y. et al. (2023b). Nanofertilizers and their roles in sustainable agriculture: Advancement and applications. Plants 12(4): 789. https://doi.org/10.3390/plants12040789.

Shafiq, H., Shani, M. Y., Ashraf, M. Y., De Mastro, F., Cocozza, C., Abbas, S. et al. (2024a). Copper oxide nanoparticles induced growth and physio-biochemical changes in maize (*Zea mays* L.) in saline soil. Plants 13(8): 1080.

Shafiq, M., Raza, M. A. and Nazir, I. (2024b). Copper oxide nanoparticles as potential antifungal agents in plant disease management. Plant Pathology Journal 40(1): 23–35. https://doi.org/10.1007/s11032-024-01545-2.

Sharma, R. and Kumar, V. (2024). Nano enabled agriculture for sustainable soil. Waste Management Bulletin 2(1): 152–161.

Shi, J., Ye, J., Fang, H., Zhang, S. and Xu, C. (2018). Effects of copper oxide nanoparticles on paddy soil properties and components. Nanomaterials 8(10): 839.

Shukla, P., Chaurasia, P., Younis, K., Qadri, O. S., Faridi, S. A. and Srivastava, G. (2019). Nanotechnology in sustainable agriculture: studies from seed priming to post-harvest management. Nanotechnology for Environmental Engineering 4: 1–15.

Sidhu, A. K. and Kaushal, P. (2023). Biosynthesis of iron oxide nanoparticles using fungi. pp. 81–98. In Mycosynthesis of Nanomaterials. CRC Press.

Sidhu, A. K., Verma, N. and Kaushal, P. (2022). Role of biogenic capping agents in the synthesis of metallic nanoparticles and evaluation of their therapeutic potential. Frontiers in Nanotechnology 3: 801620.

Singh, A., Rawat, S., Rajput, V. D., Ghazaryan, K., Minkina, T., Al Tawaha, A. R. M. et al. (2023). Impact of nanofertilizers for the mitigation of multiple environmental stresses. pp. 431–454. In Nanofertilizers for Sustainable Agroecosystems: Recent Advances and Future Trends. Cham: Springer Nature Switzerland.

Soliman, E. and Mansour, M. M. (2024). Enhancing soil organic carbon content and water retention using polyvinyl alcohol cross-linked with chitosan and pectin. Journal of Soil Science and Plant Nutrition 24(1): 791–803.

Soto-Vázquez, L., Rolón-Delgado, F., Rivera, K., Cotto, M. C., Ducongé, J., Morant, C. et al. (2019). Catalytic use of $TiO_2$ nanowires in the photodegradation of Benzophenone-4 as an active ingredient in sunscreens. Journal of Environmental Management 247: 822–828.

Strekalovskaya, E. I., Perfileva, A. I. and Krutovsky, K. V. (2024). Zinc oxide nanoparticles in the "Soil–Bacterial Community–Plant" system: Impact on the stability of soil ecosystems. Agronomy 14(7): 1588.

Su, C., Chen, A., Liang, W., Xie, W., Xu, X., Zhan, X. et al. (2024). Copper-based nanomaterials: Opportunities for sustainable agriculture. Science of The Total Environment 171948.

Su, H., Fang, Z., Tsang, P. E., Fang, J. and Zhao, D. (2016). Stabilisation of nanoscale zero-valent iron with biochar for enhanced transport and in-situ remediation of hexavalent chromium in soil. Environ. Pollut. 214: 94–100.

Thakur, S., Asthir, B., Kaur, G., Kalia, A. and Sharma, A. (2021). Zinc oxide and titanium dioxide nanoparticles influence heat stress tolerance mediated by antioxidant defense system in wheat. Cereal Research Communications 1–12.

Thirumalaisamy, R., Suriyaprabha, R., Prabhu, M. and Thesai, A. S. (2022). Role of nanomaterials in environmental remediation: recent advances—a review. *In*: Aravind, J., Kamaraj, M. and Karthikeyan, S. (eds.). Strategies and Tools for Pollutant Mitigation. Springer, Cham. https://doi.org/10.1007/978-3-030-98241-6_3.

Tortella, G., Rubilar, O., Pieretti, J. C., Fincheira, P., de Melo Santana, B., Fernández-Baldo, M. A. et al. (2023). Nanoparticles as a promising strategy to mitigate biotic stress in agriculture. Antibiotics 12(2): 338.

Tripathi, D. K., Singh, S., Singh, V. P., Prasad, S. M., Dubey, N. K. and Chauhan, D. K. (2017). Silicon nanoparticles more effectively alleviated UV-B stress than silicon in wheat (*Triticum aestivum*) seedlings. Plant Physiology and Biochemistry 110: 70–81.

Usman, M., Farooq, M., Wakeel, A., Nawaz, A., Cheema, S. A., Rehman, H. u. et al. (2020). Nanotechnology in agriculture: current status, challenges and future opportunities, The Science of the Total Environment 721: 137778, https://doi.org/10.1016/j.scitotenv.2020.1377.

Verma, N., Kaushal, P. and Sidhu, A. K. (2024). Harnessing biological synthesis: Zinc oxide nanoparticles for plant biotic stress management. Frontiers in Chemistry 12: 1432469.

Verma, N., Kaushal, P., Gahalot, D., Sidhu, A. K. and Kaur, K. (2023). Mechanistic aspect of zinc oxide nanoparticles in alleviating abiotic stress in plants—a sustainable agriculture approach. BioNanoScience 13(4): 1645–1661.

Vinzant, K., Rashid, M. and Khodakovskaya, M. V. (2023). Advanced applications of sustainable and biological nano-polymers in agricultural production. Frontiers in Plant Science 13: 1081165.

Vodyashkin, A. A., Kezimana, P., Vetcher, A. A. and Stanishevskiy, Y. M. (2022). Biopolymeric nanoparticles–multifunctional materials of the future. Polymers 14(11): 2287.

Wang, Y., Fang, Z., Liang, B. and Tsang, E. P. (2014). Remediation of hexavalent chromium contaminated soil by stabilized nanoscale zero-valent iron prepared from steel pickling waste liquor. Chemical Engineering Journal 247: 283–290, https://doi.org/10.1016/j.cej.2014.03.011, 2-s2.0-84897454878.

Wang, S., Liu, Y., Wang, X., Xiang, H., Kong, D., Wei, N. et al. (2023). Effects of concentration-dependent graphene on maize seedling development and soil nutrients. Scientific Reports 13(1): 2650.

Xin, X., Zhao, F., Rho, J. Y., Goodrich, S. L., Sumerlin, B. S. and He, Z. (2020). Use of polymeric nanoparticles to improve seed germination and plant growth under copper stress. Science of the Total Environment 745: 141055.

Zaman, W., Ayaz, A. and Park, S. (2025). Nanomaterials in agriculture: a pathway to enhanced plant growth and abiotic stress resistance. Plants 14(5): 716.

Zhang, P., Guo, Z., Zhang, Z., Fu, H., White, J. C. and Lynch, I. (2020). Nanomaterial transformation in the soil–plant system: implications for food safety and application in agriculture. Small 16(21): 2000705.

Zhao, F., Xin, X., Cao, Y., Su, D., Ji, P., Zhu, Z. et al. (2021). Use of carbon nanoparticles to improve soil fertility, crop growth and nutrient uptake by corn (*Zea mays* L.). Nanomaterials 11(10): 2717.

# Chapter 3

# Seed Nanopriming for Sustainable Crop Production

*Joanna Trzcińska-Wencel,*[1] *Julia Nadrowska,*[1] *Aniket Gade,*[1,2] *Sudhir S. Shende,*[1] *Mahendra Rai*[3,4] and *Patrycja Golińska*[1,*]

## 1. Introduction

World food production must increase by 50% by 2050 to meet the needs of an expected nine billion people. The growing food demand and rapidly changing climatic conditions worldwide are the reasons for looking for technological solutions to provide food security for future generations (FAO 2009). Seeds are a primary requirement for crop production and determine the ultimate productivity (FAO 2025).

Different chemical-based fertilizers and pesticides are used extensively in agriculture to increase seed vigor and crop production. However, due to the environmental risks associated with their use, including leaching, degradation, hydrolysis, and pollution, there is an urgent need to develop a sustainable technology that can contribute to the green revolution to address these growing concerns and to restore the damage caused to the ecosystem (Tudi et al. 2021, Shelar et al. 2021).

Different seed treatments are given to seeds before planting to enhance their quality and potential yield. The process of seed priming involves soaking seeds in water or limiting the imbibition time so that germination of the seed becomes possible, but does not allow the radicle to protrude through the seed coat (Imtiaz et al. 2023). Using this technique, seeds are advanced to an equal germination stage to facilitate emergence from the seed quickly and uniformly. Thus, seed priming is critical to ensure productivity (Pagano et al. 2023). Seed priming methods include hydro-priming, halo-priming, osmo-priming, hormonal priming, solid matrix priming, bio-priming, and nanopriming to stimulate seeds, encourage germination, improve seedling growth and crop resilience, and reduce environmental stress (Nile et al. 2022, Pagano et al. 2023).

The latest innovative technique with potential application in seed priming is seed nanopriming. This technique uses nanoparticles (NPs) or nanocarriers (e.g., silica based ones) to accelerate

[1] Department of Microbiology, Faculty of Biological and Veterinary Sciences, Nicolaus Copernicus University in Toruń, Toruń, Poland.
[2] Department of Biological Sciences and Biotechnology, Institute of Chemical Technology, Matunga, Mumbai, India.
[3] Nanobiotechnology Laboratory, Department of Biotechnology, Sant Gadge Baba Amravati University, Amravati, India.
[4] Department of Chemistry, Federal University of Piaui (UFPI), Teresina, Piauí, Brazil.
* Corresponding author: golinska@umk.pl

DOI: 10.1201/9781003659907-5

germination and improve seedling vigor by enhancing water absorption, nutrient uptake such as nitrogen, phosphorus, and potassium, and stress resistance to various abiotic and biotic stresses, such as drought, salinity, heat, and diseases by increasing antioxidant capacity and activating defense mechanisms that lead to better plant growth and development (Sharma et al. 2021, Zhao et al. 2024, Cao et al. 2025, Trzcińska-Wencel et al. 2025a). Therefore, nanopriming can reduce the need for pesticides and fertilizers while increasing yields, ultimately contributing to sustainable and environmentally friendly agriculture. This technology can potentially contribute to more sustainable and productive food systems (Zhao et al. 2024, Cao et al. 2025, Trzcińska-Wencel et al. 2025a).

The stimulating effect on seed germination has been demonstrated mainly by NPs, such as iron oxide nanoparticles (FeONPs), calcium oxide (CaONPs), magnesium oxide (MgONPs), selenium (SeNPs), zinc oxide (ZnONPs), and copper oxide (CuNPs) which are based on micro- and macro-elements required for plant growth and development (Shinde et al. 2020, Afzal et al. 2021, Mazhar et al. 2022, Trzcińska-Wencel et al. 2023, García-Locascio et al. 2024, Gomes et al. 2024, Ashwini et al. 2024). However, the potential of silver nanoparticles (AgNPs) and carbon-based materials such as fullerenes, nanotubes, graphene, and biochar NPs for positive stimulation of seed germination has also been reported (Seddighinia et al. 2020, Sonawane et al. 2021, Fonseca et al. 2024, Trzcińska-Wencel et al. 2025a).

Moreover, some reports indicate the potential of combined use of nanomaterials (NMs), such as molybdenum nanoparticles (MoNPs), ZnONPs, iron NPs with various microorganisms (e.g., *Bacillus subtilis*, *Lysinibacillus macrolides*, *Mesorhizobium ciceri*, *Azospirillum lipoferum*, *Azospirillum brasilense*, and many more) for pre-sowing seed treatment. Their effectiveness as growth stimulators is demonstrated by modified rhizosphere microbiome and exudates composition, increased chlorophyll content and tolerance to abiotic stress, and carbon sequestration or microelements uptake thus significantly improving soil quality and plant productivity (Shcherbakova et al. 2017, Akhtar et al. 2021, 2022, Yang et al. 2023, AbdElgawad et al. 2023, Joel et al. 2023).

The effect of seed nanopriming depends on the physicochemical properties of the nanomaterials used, including size, shape, surface charge, stability, or capping agents, as these parameters affect their translocation, transport, and undergo accumulation or biotransformation upon interaction with plants (Shelar et al. 2023).

Although much attention has recently shifted to the development and optimization of NMs for application in agriculture, including improved growth, plant protection, and overall performance, agricultural NMs are still in their infancy in terms of their application to sustainable agriculture (Ioannou et al. 2020, Trzcińska-Wencel et al. 2025b).

## 2. Seed Nanopriming Technology

The aging process and biotic and abiotic stresses reduce the quality of seeds, thereby limiting both the growth and yield of crop plants (El-Saadony et al. 2022a). Seed priming technologies such as hydro-priming, osmo-priming, and bio-priming are designed to enable quick and uniform germination of the seed (Pagano et al. 2023). The priming process primarily aims to hydrate the seed and facilitate germination, albeit until the root breaks through the seed coat (Imtiaz et al. 2023). Consequently, seed priming is used to improve the vigor of the seed and enhance the germination process, which is essential for seedling establishment and subsequent development (Pagano et al. 2023). More recently, nanopriming procedures using nanoscale materials have become an innovative approach to increase germination or improve seed storage by providing nutrients, biostimulants, and other active substances (Nile et al. 2022). NMs applied in seed conditioning include nanostructures and nanocomposites as systems for the sustained and efficient release of active ingredients (Imtiaz et al. 2023).

## *2.1 NMs for Seed Priming*

The unique properties of NMs, in comparison to their bulk counterparts, make them indispensable and superior in agricultural applications (Barman 2024). Using NMs for seed treatment aims to stimulate germination and plant growth and/or prevent the development of seed-associated pathogens (Nile et al. 2022). A stimulating effect on seed germination is primarily demonstrated by NPs based on micro- and macronutrients represented by FeONPs (Afzal et al. 2021), CaONPs (Mazhar et al. 2022), MgONPs (Shinde et al. 2020), SeNPs (El-Badri et al. 2021, García-Locascio et al. 2024), ZnONPs (Trzcińska-Wencel et al. 2023, Ashwini et al. 2024) or CuNPs (Gomes et al. 2024). Moreover, some reports indicate the potential of AgNPs or carbon-based materials such as fullerenes, nanotubes, graphene, and biochar NPs as they positively affected seed germination (Seddighinia et al. 2020, Sonawane et al. 2021, Fonseca et al. 2024, Trzcińska-Wencel et al. 2025a).

The effect of seed treatment with NMs depends on the physicochemical properties of the NMs used, including size, shape, surface charge, stability, or capping agents (Shelar et al. 2023). Typically, the shape, size, and composition of metal and metal-oxide NPs are controlled during the synthesis process and its conditions (e.g., precursor and reducing agent concentration, temperature). In contrast, polymeric ones require templates or post-synthesis manipulation (Kinnear et al. 2017). NMs are tiny and have many morphologies, including spherical, triangular, conical, tubular, hollow-core, and many others. Furthermore, NMs demonstrate a variety of crystalline structures (single- or multi-crystalline solids with a well-defined arrangement of atoms or amorphous bodies). The bioactivity of NMs is influenced by their higher surface-to-volume ratio and surface properties, such as charge or the presence of capping molecules, as these determine stability or the tendency to form aggregates (Mamidi et al. 2025). Several studies have shown the significance of the physicochemical properties of NMs regarding their translocation, transport, and undergoing accumulation or biotransformation upon interaction with plants (Shelar et al. 2023). The size is a crucial determinant of NPs activity, as described by Ranaware et al. (2025), who reported optimized watermelon seeds treatment with AgNPs of size 51.57 nm for improved germination in comparison with other sizes of 142.8, 123.4, and 22.04 nm. Besides, smaller citric-coated AgNPs with sizes between 10, 20, and 40 nm reduced coleoptile development in wheat, while no adverse effect was observed when AgNPs of 60 nm and 100 nm were applied (Lahuta et al. 2022). In contrast, Lau et al. (2020) found the surface charge of magnetic FeONPs to be a significant factor affecting the NP uptake, transport, and effect on tomato (*Solanum lycopersicum*) development after seed treatment. At the same time, the application of differently surface-modified ZnONPs (bare ZnONPs, ZnO core, $Zn_3(PO_4)_2$ shell, dextran (DEX)-coated, and dextran sulfate ($DEX(SO_4)$)-coated ZnONPs) increased Zn concentration in wheat. Among them, positively charged ZnONPs, namely bare ZnONPs (Zeta potential of 29.1 mV) and DEX-ZnO NPs (Zeta potential 19.5 mV), displayed the highest efficiency in increasing Zn uptake in shoots and roots, respectively (Elhaj Baddar and Unrine 2018).

## *2.2 Nanocarriers*

Seed coating technologies involve the use of thin layers of adhesive materials, which are loaded with various additives such as macro- and micronutrients, agrochemicals, plant growth regulators, plant growth-promoting bacteria, or biostimulators (Zaim et al. 2023). Many nanomaterials are used as nanocarriers due to their high adsorption and adhesion capacity, non-toxicity, and biodegradability (Zhao et al. 2021, Javed et al. 2022). The primary purpose of these nanocomposites in seed treatment is to ensure the controlled release of active ingredients (Xu et al. 2020). Among nanocarriers are porous materials such as mesoporous silica NPs, porous carbon materials, nanoclays, zeolite, apatite, or calcium carbonate (Sharma et al. 2021). Given their widespread availability and cost-effective production, as well as their unique properties that include large surface area, modifiable surface chemistry, and the capacity to adsorb a wide range of substances, these porous NMs show great potential for use as carriers of bioactive agents in targeted and specific delivery (Giannandrea et al. 2025). Their performance highly depends on the design of the active agents, surface area, and release/diffusion kinetics (Sharma et al. 2021). Researchers recently used a nanocomposite made of nanoclay/starch hydrogel to coat rice seeds for the prolonged release of zinc and improved water retention (Kumari et al. 2024). In a study reported by Bravo Cadena et al. (2018), mesoporous silica NPs with a size of ~ 100 nm and pores of around 2–2.8 nm in composite with alginate as carriers for cinnamon essential oil were used to protect seeds against pathogens.

Moreover, natural-origin polymers in nanoform such as chitosan, lignin, alginate, or zein, as well as biocompatible synthetic ones, for instance, polylactic acid, show potential for application as carriers of agrochemicals (Pereira et al. 2019, Falsini et al. 2019, Lau et al. 2020, Bonser et al. 2023, Gomes et al. 2024). Some reports indicate the potential of polymeric nanocarriers for efficient delivery of pesticides with controlled release (Saharan et al. 2015, Sathiyabama and Muthukumar 2020). For instance, highly porous Cu-chitosan NPs with an average hydrodynamic diameter of 374.3 nm and Zeta potential of +22.6 mV showed slow release of Cu and antifungal effect against *Fusarium* in tomato seeds (Saharan et al. 2015). In another study, the guar-chitosan nanoparticles (CGNPs) with sizes ranging from 300 to 800 nm and antimicrobial activity against *Pyricularia grisea* and *Xanthomonas oryza* improved the nutrient uptake and accelerated the growth rate of rice (Sathiyabama and Muthukumar 2020). In addition, some reports highlight the potential of different nanoformulations for controlled nutrient delivery (Lau et al. 2020, Steven et al. 2024).

## 3. Mechanisms of Seed Nanopriming

Seed germination is a three-stage process and includes imbibition (phase I), activation or lag phase (phase II), and germination (phase III). The influence of nanomaterials on germination is associated with their impact on all three phases, particularly water uptake, reserve mobilization, antioxidant mechanisms, cell cycle activation, and repair systems activity (Fonseca et al. 2024). Figure 1 illustrates the effect of NMs with different physicochemical properties on bioprocesses of plants at different growth stages.

NMs affect multiple cellular and metabolic processes, display significant redox capabilities, act as nanocatalysts to enhance enzyme activity, modulate gene expression, or serve as a source of elements constituting enzyme cofactors and building blocks of biomolecules (Shelar et al. 2021, Maity et al. 2022). Nanopriming increases the activity of enzymes, including lipases, amylases, and proteases, in the embryo, resulting in reduced germination time and accelerated seedling establishment (Hemalatha et al. 2024, Yang et al. 2024). A study by Khan et al. (2021) found that a higher water uptake rate and up-regulation of genes responsible for α-amylase activity were the nanopriming mechanisms for improved rapeseed germination after nanoceria treatment. Table 1 summarizes a wide range of NMs with specific physicochemical properties, seed accumulation, and effects on crop plant germination. Studies on the seed treatment with NPs are mainly based on the pre-sowing treatment without drying seeds before sowing in comparison to conventional priming technologies, in which seeds are re-dried to their original moisture (Pagano et al. 2023, Imtiaz et al. 2023). During the nanopriming procedure, NMs directly interact with the seed coat and, depending on their properties, adhere to the surface of seeds or are incorporated within the internal tissues. The priming of seeds with the SeNPs, $CeO_2$NPs, and AgNPs resulted in their uptake and accumulation in various tissues of the tomato, cotton, and bean seeds, respectively, suggesting that the internalization of NMs, improves water uptake during the imbibition process (Mahakham et al. 2017, Afzal et al. 2021). While $CeO_2$ NPs used for seed preparation accumulated in the seed coat or inner integument; cotyledon, and radicle, suppressed ROS accumulation and initiated signaling mechanisms, including channels controlling the pathway of the stress response via Mg and Ca ions (An et al. 2020). However, the cell wall and plasma membrane limit the uptake of NMs into the plant cells. This indicates the formation of a coating on the surface of the seed and activity based on the release and penetration of ions (Afzal et al. 2021, Gomes et al. 2024). In a study by Afzal et al. (2021), priming rice seeds with FeONPs caused a higher iron ion content in the seed coat,

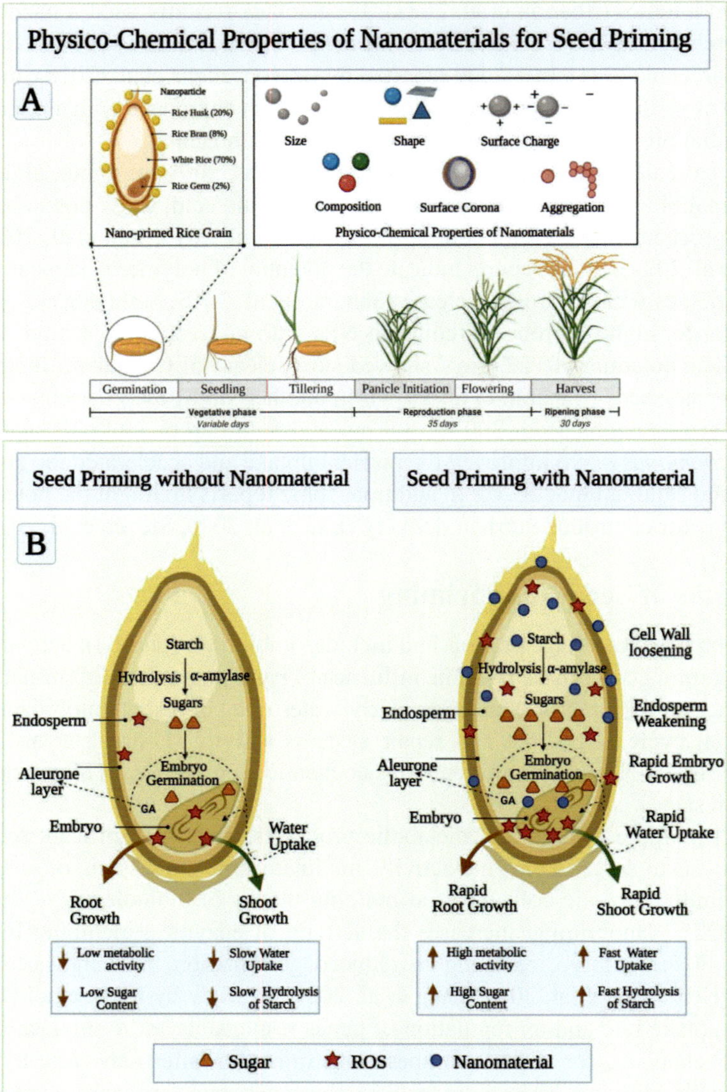


**Figure 1.** Influence of the physicochemical properties of NMs on the effects seed nanopriming on plant growth. (A) the diversity of physicochemical properties (size, shape, and surface chemistry) used for seed treatment. (B) the catalytic activity of NMs and their role in promoting enzymatic and biochemical pathways to promote root/shoot growth (reprinted from Shelar et al. 2021, under the terms and conditions of the Creative Commons Attribution (CC BY) license (https://creativecommons.org/licenses/by/4.0/)).

the endosperm, and embryonic tissues. This treatment resulted in increased germination associated with enhanced water diffusion, higher levels of $H_2O_2$, and increased starch metabolism in seeds. Similarly, the application of AgNPs biosynthesized from kaffir lime leaf extract resulted in the internalization of $Ag^+$ in rice-aged seeds. Furthermore, AgNP-priming increased α-amylase activity, upregulated aquaporin genes, and altered the levels of reactive oxygen species and activity of antioxidant systems, including the generation of hydroxyl radicals, which are involved in loosening the cell wall (Mahakham et al. 2017).

**Table 1.** Nanomaterials for seeds treatments: uptake and accumulation in plants, their effect on germination and plant growth.

| Type of nanomaterial | Properties (size, Zeta potential) | Concentration applied | Plant | Detection of nanomaterials and other alterations in seeds | Effect on plant development | Reference |
|---|---|---|---|---|---|---|
| AgNPs | 15 nm | 1, 10, 100 mg $L^{-1}$ | bean (*Phaseolus vulgaris*) | AgNPs accumulated in seed coat, undergo oxidative dissolution and form thiolate complexes in the internal layer of seeds | Unchanged germination and hypocotyl length | Savassa et al. (2021) |
| $Ag_2SNPs$ | 42 nm | 1, 10, 100 mg $L^{-1}$ | bean (*Phaseolus vulgaris*) | Deposited clusters of $Ag_2SNPs$ observed around hilum, micropyle, and lens | Unchanged germination, increased hypocotyl length | Savassa et al. (2021) |
| AgNPs | 6–36 nm | 10 and 20 mg $L^{-1}$ | jasmine rice (*Oryza sativa* L. cv. KDML105) | AgNPs penetrated the seed coat into endosperm and embryonic tissues | Improved imbibition and germination, higher degrees of starch decomposition | Mahakham et al. (2017) |
| $CeO_2NPs$ | 1.8 nm (TEM), 2.1 nm (DLS), –51.7 mV | 500 mg $L^{-1}$ | cotton | $CeO_2NPs$ accumulated in the seed coat inner tegmen, cotyledon and radicle | Improved growth under salinity stress | An et al. (2020) |
| CuNPs | 83.2 nm (TEM); 322.5 nm (DLS) –20.2 mV | 58 mg Cu $L^{-1}$ | maize (*Zea mays* cv. Balu 787) | Two and three-fold higher copper content in seedling developing in normal and water restriction conditions | Improved seedling dry and fresh weight | Gomes et al. (2024) |
| FeONPs | 20–50 nm (SEM) –22.5 mV | 20 mg $L^{-1}$, 40 mg $L^{-1}$ | rice (*Oryza sativa* L.) | Penetration of Fe ions through seed coat into endosperm and embryonic tissues | Enhanced seed germination and starch metabolism, increased diffusion of water and $H_2O_2$ level | Afzal et al. (2021) |
| citrate-coated $CoFe_2O_4NPs$ | average size 13 ± 5 nm, hydrodynamic diameter of 216 ± 10 nm, +10.5 ± 6.6 mV, | 10, 20, and 40 mg NPs $L^{-1}$ | bean (*Phaseolus vulgaris* L.) | NPs localized in the hilum and seed coat | Increased K and Zn content in progeny seeds | Perea-Vélez et al. (2024) |
| SeNPs | 143.9 nm (DLS) | 1, 10 and 50 ppm | tomato (*Solanum lycopersicum*) | SeNPs detected around the plastids of the seed cells, increased Se content in seedlings | Improved germination rate (10 ppm) and germination trays (1 ppm), increased Mg, K, Mn, Zn, Fe, Cu and Mo uptake into plants | García-Locascio et al. (2024) |

*Table 1 contd. ...*

*...Table 1 contd.*

| Type of nanomaterial | Properties (size, Zeta potential) | Concentration applied | Plant | Detection of nanomaterials and other alterations in seeds | Effect on plant development | Reference |
|---|---|---|---|---|---|---|
| SeNPs, | 10–55 nm (TEM), 120–260 nm (DLS), –32.4 mV | 150 µmol $L^{-1}$ | rapeseed (*Brassica napus*) | Alterations in seed microstructure, increased gaps between oil bodies | Enhanced germination rate and final germination percentage, increased SOD, POD and APX activities | El-Badri et al. (2021) |
| poly acrylic acid coated selenium doped carbon dots (PAA@Se-CDs) | 3.8 ± 0.2 nm (TEM) –30 mV | 50 mg $L^{-1}$ | rapeseed (*Brassica napus*) | Accumulation in seed compartments (seed coat, cotyledon, and radicle) | Increased germination and fresh weight of seedlings under saline conditions | Nauman Khan et al. (2024) |
| ZnONPs | 25nm (XRD) 20 nm (TEM) | 100 mg $L^{-1}$ | rapeseed (*Brassica napus*) | Alterations in seed microstructure, increased gaps between oil bodies | Enhanced germination rate and final germination percentage, increased SOD, POD and APX activities | El-Badri et al. (2021) |
| ZnONPs | 45.90 –140.70 nm, with an average size of 63.60 nm (DLS) –20.4 mV | 1000 ppm | groundnut | ZnONPs uniformly distributed across the surface and internal tissues | Improved germination rate, increased root length, shoot length, and seedling vigor index | Ashwini et al. (2024) |
| ZnONPs | 50 nm, Zeta potential between –20 and –23 mV | 25 µg $mL^{-1}$, 50 µg $mL^{-1}$ | lettuce | ZnONPs adsorbed into seed, without morphological changes in seed tissue structures | Enhanced germination and biomass production | Rawashdeh et al. (2020) |

### *3.1 Effect of Nanopriming on Seed Germination, Plant Growth, and Physiology*

As mentioned above, the results of seed treatment with NMs on plants depend on the properties of the NMs. However, they are considerably diverse depending on treatment conditions and the plant species, and are revealed in various ways at different stages of plant development (Nile et al. 2022, Trzcińska-Wencel et al. 2024). Seed nanopriming can affect seed germination, plant growth, and multiple physiological parameters of plants, as summarized in Table 2.

Nanopriming aims to affect plant growth at various developmental stages of the life cycle (germination, vegetative growth, fructification) (Shelar et al. 2023). Studies on the effects of treating seeds with NMs most often examine germination rates and performance and the long-term effects at various stages of plant development up to full maturity and yield. However, the non-standardized conditions of the treatments indicate the challenges of clearly reaching conclusions about the impact of NMs and mechanisms of action in the treatment of the seeds (Nile et al. 2022, Shelar et al. 2023, Cao et al. 2025). The following sections discuss more detailed data. They include the effects on germination, growth, and the morpho-physiological attributes of seedlings. These seedlings develop from seeds subjected to various NMs.

### *3.2 Seed Germination Efficiency and Growth Parameters*

Germination is a multi-stage process that begins with the activation of the seed through water uptake and is completed by the development of the embryonic axis. This crucial stage for plants is influenced by several internal and external factors (e.g., water availability, temperature, or oxygen) (Rifna et al. 2019). Furthermore, several studies indicate that NMs used for seed nanopriming may positively affect seed germination efficiency and rate and, in subsequent growth stages, enhance biomass production, plant vigor, and yield. Thus, the germination of mung bean (*Vigna radiata*) seeds increased from 76.67% in the control to 88.33% after treatment with CuONPs at a concentration of 48 ppm (Sarkar et al. 2021). While treatment of tomato (*Solanum lycopersicum*) with SeNPs (1 ppm) improved germination rate and potential by 32.5%, germination index by 54.16%, and vigor index by 57.23% (García-Locascio et al. 2024). Multiwall carbon nanotubes functionalized with carboxylic acid (MWCNT-COOH) increased seed germination percentage and seedling vigor index of *Rhododendron groenlandicum* L. at 20 $\mu g\ mL^{-1}$ (Ali et al. 2020). Afzal et al. (2021) found that priming *Oryza sativa* L. seeds with FeONPs at 20 mg $L^{-1}$ enhanced the rate of early germination and final germination percentage and resulted in the increase of radicle and plumule length by more than 50 and 22%, respectively, and improved fresh and dry weight of the plant. Application of MWCNTs and $SiO_2$ NPs, depending on the dosage (50 and 100 $\mu g\ mL^{-1}$, respectively) resulted in increased siliquae length and number of seeds per siliqua of Indian mustard (*Brassica juncea*) in the field experiment (Dhingra et al. 2022). Interestingly, Trzcińska-Wencel et al. (2023) found that seed treatment with ZnONPs had no significant effect on germination performance. However, applying ZnONPs to *Zea mays* grains at concentration of 32–256 $\mu g\ mL^{-1}$ improved seedling length by 13–34%, fresh weight by 24–40%, and dry weight by 16–26% were achieved. According to the literature, nanomaterials demonstrate potential for agricultural applications, although most studies highlight the requirement to optimize pre-sowing seed treatment, particularly with the applied concentration.

#### *3.2.1 Effect of Nanopriming on Plant Physiology*

The role of the antioxidant system activity in scavenging reactive oxygen species (ROS) is essential in each stage of plant development due to the ROS-signaling functions in physiological processes of plant development, but also in the response to stress factors (Considine and Foyer 2021). Therefore, several studies have shown increased activity of the antioxidant system, including enzymes activity (catalase (CAT), peroxidase (POX), ascorbate peroxidase (APX) dehydroascorbate reductase (DHR), glutathione reductase (RG), and superoxide dismutase (SOD), non-enzymatic antioxidants (ascorbate, glutathione, tocopherols)) during germination and early seedling growth after treatment

**Table 2.** The effects of seed nanopriming with different nanomaterials on seed germination, growth, and physiology of plants.

| Nanomaterial | Plant | Treatment conditions | Growth condition | Effects on germination and growth parameters | References |
|---|---|---|---|---|---|
| CuONPs; triethoxysilane (APTES)-coated CuO (CuO@ APTES) | moong bean (*Vigna radiata*) | 16, 48 and 80 ppm, soaking 12 h | overnight germination | Improved imbibition and seed germination percentage | Sarkar et al. (2021) |
| Multiwall carbon nanotubes functionalized with carboxylic acid | bog birch (*Betula pumila* L.) labrador tea (*Rhododendron groenlandicum* L.) | 20 and 40 μg/mL, 24 h, RT, cold stratified at 2–4°C for 15 days | 3 weeks, 25°C | Improved seed germination percentage and seedling vigor index, mediated seed dormancy by remodeling the seed membrane lipids | Ali et al. (2020) |
| SeNPs | tomato (*Solanum lycopersicum*) | 1, 10 and 50 ppm, 250 rpm for 180 min, RT | 7 DAS (germination) 21 DAS (growth) | Increased germination rate (1 and 10 ppm), germination potential, and vigor index | García-Locascio et al. (2024) |
| ZnONPs AgNPs | maize (*Zea mays*) | 1, 8, 32, 64, 128, and 256 μg/mL, 30 min | 3 days (germination) 10 DAS (growth), 22°C ± 2°C | Unaffected germination, dose-dependent effect on growth parameter, namely stimulation of growth by ZnONPs (> 8 μg/mL) and improved seedling length and DW by AgNPs (32–128 μg/mL) and reduced FW by AgNPs treatment at concentration of 256 μg/mL | Trzcińska-Wencel et al. (2023) |
| AgNPs | maize (*Zea mays*) | 32, 128 and 512 μg/mL, 30 min, RT | 14 DAS | Unaffected germination, enhanced FW, DW and length of shoots (512 μg/mL), and DW of leaves (32 and 128 μg/mL), affected redox homeostasis and chlorophyll content | Trzcińska-Wencel et al. (2025) |
| AgNPs AuNPs | onion (*Allium cepa* L.) | 31.3 ppm (AgNPs), 5.4 ppm (AuNPs), 12 h at ambient temperature | 6 and 21 DAS | Improved seedling emergence in ten-year old onion seeds | Acharya et al. (2019) |
| Thymol nanoemulsion | soybean (*Glycine max*) | 0.03–0.06%, v/v for 4h | 35 DAS, 25°C | Lowered severity of the disease caused by *Xanthomonas axonopodis* | Kumari et al. (2018) |
| Chitosan NPs | sea daffodil (*Pancratium maritimum* L.) | 0.1, 0.5, 1 mg/mL for 4, 8, 12 h | Germination: 22–27°C/14–18°C (day/night) Growth: 20±2°C, 14 h/10 h light/dark photoperiod 99 DAS | Higher germination efficiency and decreased mean germination time, increased seedling vigor (4 and 8 h priming) and decreased after 12 h-priming, decreased alkaloids and increased pancratistatin, lycorine and antioxidant content | Allam et al. (2024) |

*Table 2 contd. ...*

...*Table 2 contd.*

| Nanomaterial | Plant | Treatment conditions | Growth condition | Effects on germination and growth parameters | References |
|---|---|---|---|---|---|
| $MnO_2$NPs stabilized with cocamidopropyl betaine | pea (*Pisum sativum* L. ) | 0.1, 1, 10 mg/mL | 7 days, 20°C | Higher germination energy (91.25%), germinability (95.60%) (0.1 mg/L); adverse effect at higher concentrations (1–10 mg/L), increased content of soluble sugars, chlorophyll, and carotenoids | Nagdalian et al. (2024) |
| ZnONPs | okra (*Abelmoschus esculentus*) | 0, 20 and 40 ppm, 18 and 24 h, at room temperature (28–30°C) | 30 days | Improved germination, growth parameters and increased chlorophyll and carotenoids, flavonoids, (20 pmm, 18 h), increased proline content (40 ppm, 18 h), higher antioxidant enzymes activity in both treatments | Ramzan et al. (2024) |
| ZnO@LNP (ZnO@lignin nanoparticles) | maize (*Zea mays*) | 3.2, 12.8, 51.2, 204.8 and 819.2 mg of ZnO/mL; 24 h under slow stirring | 24 ± 2°C (germination) 14 days, (growth): 12/12 h photopheriod | Improved germination and growth parameters, higher content of chlorophyll, carotenoids, anthocyanins, total phenols, and a better antioxidant activity | Del Buono et al. (2022) |
| FeONPs | rice (*Oryza sativa* L.) | 20 and 40 mg/L for 24 h with constant aeration at RT (28–30°C). | 6 days (germination) | Increased radicle and plumule length, and fresh and dry weight | Afzal et al. (2021) |

DAS: days after sowing; DW: dry weight; FW: fresh weight; RT: room temperature

with NMs (Allam et al. 2024, Chandrasekaran et al. 2020, Hemalatha et al. 2024, Hussain et al. 2024, Trzcińska-Wencel et al. 2025a). Recently, Ashwini et al. (2024) reported that ZnONPs treatment of groundnut seeds improved germination and biomass production and enhanced CAT, SOD, APX, and GPX activities in the leaves of the seedlings. Sonawane et al. (2021) described the concentration-dependent effects of biogenic $TiO_2$NPs and AgNPs on the germination and emergence of tomato hypocotyl. Their study showed a stimulatory potential of $TiO_2$NPs in the concentration range from 25 to 100 $\mu g\ mL^{-1}$ and adverse effects of AgNPs at concentrations of 100–400 $\mu g\ mL^{-1}$ and 25–400 $\mu g\ mL^{-1}$ on germination and seedling vigor, respectively. The application of each type of NPs increased SOD, POX, and CAT activity, as well as protein, carotenoid, and proline levels (Sonawane et al. 2021). The stimulatory effect of grain treatment with AgNPs (32 $\mu g\ mL^{-1}$) on maize growth, along with increased total ascorbate content (by 48.1%) in leaves, was reported by Trzcińska-Wencel et al. (2025a). Moreover, several studies indicate that treating seeds with NMs contributes to increasing the performance of photosynthesis, which is a key process for plant growth and, thereby, the efficiency of agricultural crop production (Nile et al. 2022). Seed nanopriming has shown potential in improving the content of photosynthetic pigments, as reported for seed treatments with ZnONPs, MnONPs, TiONPs, and SeNPs (Rai-Kalal and Jajoo 2021, Kasote et al. 2021, García-Locascio et al. 2024, Anbalagan et al. 2025). For instance, ZnONPs at a concentration of 20 mg $L^{-1}$ increased the content of chlorophyll a (by 42.4%), chlorophyll b (by 79.3%), and carotenoids (by 35.2%) in *Abelmoschus esculentus* (Ramzan et al. 2024). Nagdalian et al. (2024) reported that $MnO_2$ nanoparticles at 1 mg $L^{-1}$ increased the content of chlorophyll a and b, as well as carotenoids in *Pisum sativum* L. A similar effect was reported after the use of lignin/zinc oxide hybrid nanoparticles (ZnO@LNP) on *Zea mays* seeds, which resulted in higher content of chlorophyll a, chlorophyll b, and carotenoids (Del Buono et al. 2022).

## 3.3 Nanopriming-Induced Plant Responses Against Abiotic and Biotic Stresses

Nowadays, several adverse factors and environmental conditions reduce worldwide crop production. Therefore, seed nanopriming provides a potential solution to overcome the adverse effects of biotic and abiotic stresses in an environmentally friendly, efficient, and cost-effective approach (El-Saadony et al. 2022a). Studies on nanopriming-mediated alleviation of symptoms of various environmental stresses in plants by influencing a wide range of physiological, morphological, biochemical, and molecular mechanisms is demonstrated in Table 3.

Noteworthy is the fact that the effect of nanopriming on the overall physiology of plants depends on several factors and undergoes modification at the molecular level, including modification of gene expression (Yang et al. 2024). Due to the complexity of plant responses to abiotic and biotic stresses, no general mechanism is responsible for improving plant tolerance by seed nanopriming. However, several studies indicate alterations in various stress-related signaling pathways, phytohormone synthesis, ROS scavenging processes, and many others (Shelar et al. 2021, Singh et al. 2024).

### *3.3.1 Salinity Stress*

Seed nanopriming has been found to alleviate plant salinity stress by increasing water absorption, promoting the production of osmoregulating compounds, or balancing ion uptake (Chattha et al. 2022, Farooq et al. 2022). Multiple biochemical mechanisms of response to salinity stress after ZnNPs treatment have been found in wheat (*Triticum aestivum*), including promoted accumulation of osmoregulatory compounds (proline, soluble sugars, soluble proteins, free amino acids) and increased uptake of Ca, K, Mg, N and P as unlike to uptake of Na and Cl ions. ZnNPs-priming increased water content, photosynthetic rate, antioxidant activity, level of gibberellic acid (GA) and indole acetic acid (IAA) and reduced electrolyte leakage (Al-Salama 2022, Chattha et al. 2022). Moreover, some studies showed NPs effect on antioxidative mechanisms by upregulation of specific gene expression and enhancement in the activity of antioxidants in plants exposed to salinity stress (Khan et al. 2021, El-Badri et al. 2021, Akhtar et al. 2022, Cembrowska-Lech and Rybak 2023). Khan et al. (2021) reported higher water absorption, balanced maintenance of $Na^+/K^+$ ratio, and higher antioxidant enzymes (SOD and POX) activity along with lower ROS accumulation in *Brassica napus* L. developed from seeds treated with $CeO_2$ NPs coated with polyacrylic acid under salinity stress. While, alleviation of adverse effects of salinity stress in barley after seeds treatment with AgNPs was related to enhanced activity of SOD, CAT, GR, and GPX and higher glutathione synthesis, as also observed through upregulated expression of genes related to antioxidant enzymes (Cembrowska-Lech and Rybak 2023), in another study priming of rapeseed seed with SeNPs and ZnONPs resulted in higher metabolic activity and modulation of the expression of stress-responsive genes (antioxidant defense-responsive genes, ABA-related genes), which improved the seedling development under stress (El-Badri et al. 2021). Moreover, the application of Se-chitosan NPs for bitter melon seeds reduced the adverse effect of salt stress and induced alterations in expression patterns of stress-related genes (*WRKY1*, *SOAR1*, *Mc5PTase7*) and genes responsible for regulation of ions homeostasis (*SOS1*, *PM $H^+$-ATPase*, and *SKOR*) (Sheikhalipour et al. 2023). Furthermore, poly(acrylic acid)-coated cerium oxide nanoparticles (PNC) improved salt-tolerance in cotton (*Gossypium hirsutum* L.) by modulation of ROS level and conserved $Ca^{2+}$ plant signaling pathways, leading to improved root development (An et al. 2020).

**Table 3.** The effects of seed nanopriming with different nanomaterials on plants grown under environmental stresses.

| Stress | Nanomaterial | Plant | Effects | Reference |
|---|---|---|---|---|
| Salinity | Cerium oxide NPs coated with polyacrylic acid | *Brassica napus* L. | Higher water absorption, lower ROS accumulation, higher antioxidant enzyme (SOD, POD) activity, better maintenance of $Na^+/K^+$ ratio | Khan et al. (2021) |
| Salinity | ZnONPs | *Brassica napus* L. | Improved germination parameters, increased shoot and root length, improved biomass, increased content of proline, soluble sugar, and soluble protein, reduction of accumulated ROS | El-Badri et al. (2021) |
| Salinity | Chitosan and green tea nanocomposite | *Triticum aestivum* L. | Increased protein content, increased photosynthetic pigments, upregulated antioxidant enzymes (CAT, POD), improved germination and vigor, reduced germination time, reduced MDA content | Farooq et al. (2022) |
| Salinity | ZnNPs | *Triticum aestivum* L. | Improved yield, reduced content of abscisic acid, MDA, and $H_2O_2$, reduced electrolyte leakage, increased water content and photosynthetic rate, improved antioxidant activity, accumulation of osmoregulating compounds, increased level of gibberellic acid and indole acetic acid, increased uptake of Ca, K, Mg, N, and P | Chattha et al. (2022) |
| Salinity | Polyvinylpyrrolidone-coated AgNPs | *Hordeum vulgare* | Improved germination, increased antioxidants expression | Cembrowska-Lech and Rybak (2023) |
| Drought | SiNPs | *Triticum aestivum* L. | Increased grain weight, grain yield, and plant height, improved chlorophyll concentration, diminished electrolyte leakage, increased activity of SOD and POD | Raza et al. (2023) |
| Drought | ZnONPs | *Oryza sativa* L. | Increased plant height, fresh and dry weight, and chlorophyll content, reduced MDA content, enhanced level of proline, higher activity of SOD, CAT, and POD | Mazhar et al. (2022) |
| Drought | ZnONPs | *Triticum aestivum* L. | Increased activity of POD, APX, and CAT, increased content of phenolic compounds and flavonoids, higher lipid peroxidation inhibition, higher content of free amino acids and soluble carbohydrates | El-Shazoly et al. (2025) |
| Drought | $Fe_3O_4$NPs | *Medicago sativa* L. | Enhanced seed vigor, germination rate, and α-amylase activity, increased root surface | Wang et al. (2025) |
| Drought | n-$Fe_2O_3$NPs | *Zea mays* L. | Improved content of water, chlorophyll, flavonoids, and ascorbic acid, accelerated functioning of SOD and POD, decreased levels of MDA and $H_2O_2$, improved cob length and kernel weight | Mazhar et al. (2024) |
| Moisture | Cysteine-capped ZnONPs | *Vigna radiata* | Improved germination and seedling vigor, increased hydrolytic and antioxidant enzyme activity | Kalimuthu et al. (2023) |
| Low temperature | Zn-salicylic acid-chitosan bioconjugate | *Triticum aestivum* L. | Reduced emergence time, increased germination rate, higher activity of reserve-mobilizing enzymes, increased levels of antioxidant enzymes (SOD and CAT), reduced levels of MDA and $H_2O_2$ | Mohan et al. (2024) |

*Table 3 contd. ...*

*...Table 3 contd.*

| Stress | Nanomaterial | Plant | Effects | Reference |
|---|---|---|---|---|
| Co | ZnONPs | *Zea mays* L. | Improved plant growth, biomass, and photosynthetic machinery, reduced ROS and MDA content, decreased Co uptake, higher antioxidant enzymes content | Salam et al. (2022) |
| Cd | ZnONPs | *Oryza sativa* L. | Improved seedling growth and seedling fresh weight, increased SOD and POD activity, increased metallothionein content, improved α-amylase activity | Li et al. (2021) |
| Cd | SiNPs | *Triticum aestivum* L. | Decreased Cd and ROS content, increased antioxidant enzyme activity (SOD, POD, CAT), enhanced chlorophyll content, growth, biomass, and yield, improved gas exchange parameters, decreased MDA and $H_2O_2$ content, decreased electrolyte leakage | Hussain et al. (2019a) |
| Cd | Iron oxide NPs (IONPs) and SiNPs | *Phaseolus vulgaris* L. | Enhanced growth, net photosynthetic rate, and gas exchange, increased $K^+$ content, decreased MDA content and electrolyte leakage, higher spermidine and putrescine levels (alleviate Cd stress), increased activity of antioxidant enzymes (SOD and CAT) | Koleva et al. (2022) |
| Pb | $Fe_3O_4$NPs | *Basella alba* L. | Increased germination and root length, decreased Pb content, increased antioxidant enzymes (SOD, CAT, POD) and proline content, decreased $H_2O_2$ and MDA content | Gupta et al. (2022) |
| Bacteria | $SiO_2$NPs | *Beta vulgaris* L. | Reduced soft rot and root rot symptoms in plants infected with *Pectobacterium betavasculorum* | Khan and Siddiqui (2020) |
| Fungi | SeNPs | *Triticum aestivum* L. | Lower disease incidence in plants infected with *Fusarium culmorum* and *F. graminearum* | El-Saadony et al. (2021) |
| Fungi | Chitosan NPs | *Pennisetum glaucum* | Induced systemic resistance against downy mildew caused by *Sclerospora graminicola* | Siddaiah et al. (2018) |
| Fungi | $SiO_2$NPs | *Beta vulgaris* L. | Reduced soft rot and root rot symptoms in plants infected with *Rhizoctonia solani* | Khan and Siddiqui (2020) |
| Nematodes | $SiO_2$NPs | *Beta vulgaris* L. | Reduction in galling and nematode multiplication in plants inoculated with *Meloidogyne incognita* | Khan and Siddiqui (2020) |

### *3.3.2 Drought Stress*

Drought primarily affects plant growth through reduced water-use efficiency and impaired enzymatic activity, resulting in a loss of turgor pressure or a reduced rate of cell division, which adversely affects the proliferation and elongation of leaves, stems, and roots (Imtiaz et al. 2023). Drought conditions can be highly damaging, especially during germination and seedling growth. However, nanopriming of seeds demonstrates the potential to mitigate their adverse effects on plants by improving water uptake, facilitating root development, or reducing oxidative stress (Imtiaz et al. 2023, Elkelish et al. 2025). The improved plant's tolerance to drought stress after AgNP, ZnONPs, and n-$Fe_2O_3$NPs priming of seeds was reported in wheat, rice and maize, as manifested by improved germination, seedling growth, roots development photosynthesis efficiency and adjusted antioxidants machinery to maintain ROS homeostasis (Mazhar et al. 2022, Mazhar et al. 2024, Elkelish et al. 2025, El-Shazoly et al. 2025). Hence, the role of AgNPs in improving drought stress response by maintaining drought-responsive genes, regulating ABA signaling pathways, and scavenging ROS has been reported in wheat (Elkelish et al. 2025). Moreover, $SiO_2$NPs increased photosynthesis efficiency under drought stress by preventing chlorophyll degradation and impairment in quantum yields and enhancing electron transport rate (Kalal et al. 2022). Mazhar et al. (2024) reported an improved content of water, chlorophyll, flavonoids, and ascorbic acid, as well as decreased levels of MDA and $H_2O_2$ along with higher antioxidant enzymes activity (SOD and POD) in maize after the treatment of grains with n-$Fe_2O_3$NPs.

### *3.3.3 Heavy Metal Stress*

Seed nanopriming can mitigate the effects of heavy metal (HMs) stress and improve physiological and growth parameters of plants by the activation of the antioxidant system, modifications of nutrient and water uptake in plants, e.g., counteracting heavy metal or by improved detoxification of HMs in cells (Saboor et al. 2019). Therefore, in a study by Salam et al. (2022) ZnONPs priming treatment improved the growth and biomass of maize (*Zea mays* L.) under cobalt stress, increased zinc uptake by 52 and 73% in the roots and shoots respectively, decreased Co uptake, reduced ROS and MDA content, and increased the levels of antioxidant enzymes (APX, CAT, SOD) (Salam et al. 2022). Hussain et al. (2019) used SiNPs for nanopriming of seeds of wheat (*Triticum aestivum* L.) grown under cadmium stress. They reported decreased Cd content (especially in grains), improved yield, gas exchange parameters, and chlorophyll content, decreased electrolyte leakage, MDA and $H_2O_2$ content, as well as increased antioxidant enzyme activity (SOD, POD, CAT). While $SiO_2$ NPs increased *Brassica napus* tolerance to ultrastructural damage under chromium stress by activating the antioxidant defense system, which increased transcription of antioxidants (SOD, CAT, APX, GR) or defense-related genes (phenylalanine ammonia lyase*; CAD:* cinamyl alcohol dehydrogenase*, PPO:* polyphenol oxidase*; PAO:* polyamine oxidase*; MT-1:* metallothionein-1). In addition, $SiO_2$NP priming of seeds altered subcellular Cr distribution, increased its accumulation in walls, and promoted the synthesis of GSH involved in vacuolar Cr sequestration (Ulhassan et al. 2023).

### *3.3.4 Extreme Temperature Stress*

The improvement of plant tolerance to heat and cold stresses by NMs involves the enhanced synthesis of biomolecules such as antioxidants, protection-related proteins, or accumulation of sugars that act as osmolytes, cryoprotectants, and reserve substances (Gusain et al. 2023). Hence, the improvement of morphophysiological parameters of wheat seedlings under heat stress conditions after grain treatment with AgNPs, ZnONPs, and $TiO_2$NPs (Iqbal et al. 2019, Thakur et al. 2022). Furthermore, chitosan based nanocomposites improved heat tolerance in wheat (Al Masruri et al. 2023) and cotton (Savani et al. 2023). Priming seeds with Zn-salicylic acid-chitosan bionanoconjugates (ZnSA-BNCs) has shown positive effects in mitigating wheat's low-temperature conditions, as reported by Mohan et al. (2024). Seeds primed with ZnSA-BNC demonstrated improved germination rate and shorter emergence time under chilling stress. At the same time, lower MDA and $H_2O_2$ content, increased activity of antioxidant enzymes (SOD and CAT), and reserve mobilizing enzymes (α-amylase and protease) were found in established seedlings (Mohan et al. 2024).

### *3.3.5 Biotic Stress*

After exposure to NMs during seed priming, several studies have shown improved plant defense mechanisms against biotic stresses (micro- and macro-organisms, including bacteria, fungi, and nematodes). Hence, the $SiO_2$NPs priming of *Beta vulgaris* L. seeds reduced soft rot and root rot symptoms caused by *Pectobacterium betavasculorum* and reduced galling and nematode multiplication (*Meloidogyne incognita*) (Khan and Siddiqui 2020). Hussain et al. (2024) reported that priming of *Brassica rapa* seeds with melatonin and CuONPs resulted in reduction of clubroot disease symptoms and higher activity of SOD (by 25/29%), CAT (by 83/55%), and APX (by 83/46%), as well as increased the content of salicylic acid (by 98/82%), and jasmonic acid (by 145/82%) in the roots/shoots, respectively. SeNPs at 100 μg $mL^{-1}$ significantly reduced disease incidence in *Triticum aestivum* L. infected with *Fusarium culmorum* and *F. graminearum* (El-Saadony et al. 2021). In another study, Siddaiah et al. (2018) found that chitosan NPs induced systemic resistance in pearl millet (*Pennisetum glaucum*) through earlier and higher expression of PR1 and PR5 pathogenesis-related proteins, facilitating improved defense against mildew caused by *Sclerospora graminicola*. Improved wheat tolerance to fungal infection (*Aspergillus niger*) after the application of silver nanochitosan, as a priming agent for grains, was associated with stimulation of defense enzymes (CAT, POD, SOD and NOX) (Chouhan et al. 2024).

## 4. Effect of Combined Treatment of Seeds with Nanomaterials and Microorganisms

Plant growth-promoting bacteria, rhizobia, and arbuscular mycorrhizal fungi are an alternative to the excessive use of chemical fertilizers and pesticides, which are important for crop plant growth, but pose a risk of soil degradation and adverse environmental consequences (El-Saadony et al. 2022b). The most relevant microorganisms that increase plant productivity include species of *Azospirillum*, *Achromobacter*, *Bacillus*, *Bradyrhizobium, Streptomyces*, and fungi such as *Trichoderma* or *Glomeromycota*. Microorganisms beneficial for plants enhance the availability of specific nutrients, increase water uptake, synthesize the phytohormones, regulate the ionic equilibrium, and are antagonists to the number of phytopathogens (Acharya et al. 2024). In addition, seed-associated microbial communities are important to maintain the quality of stored seeds and to ensure the healthy development of the next generation of plants (Taheri et al. 2025). Therefore, some reports indicate the potential of combining NMs with various microorganisms for pre-sowing seed treatment to enhance plant growth (Akhtar et al. 2021, AbdElgawad et al. 2023, Joel et al. 2023). For instance, combined treatment of microbial inoculants of *Mesorhizobium ciceri* ST282 and *Bacillus subtilis* Ch13 and molybdenum nanoparticles (MoNPs) altered the composition of root exudates and rhizosphere microbiome structure of chickpea (*Cicer arietinum* L.) plants (Shcherbakova et al. 2017). Applying ZnONPs with *Bacillus cereus* and *Lysinibacillus macrolides* enhanced the growth and increased chlorophyll contents in rice grown in heavy metal stress conditions (Akhtar et al. 2022). In another study, Akhtar et al. (2021) improved drought stress tolerance in wheat by co-application of $SiO_2$NPs with plant growth-promoting rhizobacteria, i.e., *Bacillus* sp., *Azospirillum lipoferum,* and *Azospirillum brasilense*. In the study by Yang et al. (2023), priming of maize grains with nanoscale zero-valent iron (nZVI) improved rhizosphere colonization by arbuscular mycorrhizal fungi of *Funneliformis mosseae*. This combined system increased carbon sequestration and Fe uptake in the soil, significantly improving semiarid soil quality. In addition, there is a potential to provide beneficial microorganisms in seed coatings with nanomaterials as delivery systems (Hussain et al. 2019, Sohail et al. 2022, Bhutto et al. 2024). The application of a gelatin film reinforced with cellulose nanocrystals (CNC) and *Trichoderma harzianum* spores as seed coating material improved the germination and growth of maize (Dogaru et al. 2021). Rhizobacteria *Pantoea agglomerans* ISIB55, *Burkholderia caribensis* ISIB40, and *Bradyrizhobium* strains immobilized on nanofibers were used for soybean (*Glycine max* L.) seeds coating. Nanoimmobilization of plant growth-stimulating bacteria is an effective system for the precise delivery of microorganisms that increases their viability and enables colonization, thereby increasing plant productivity (Damasceno et al. 2013, Gregorio et al. 2017).

## 5. Conclusion and Future Prospects

Seed nanopriming is emerging as a transformative technique in sustainable agriculture, offering a powerful means to enhance seed performance by accelerating germination, reducing chemical inputs, and improving crop resilience to both biotic and abiotic stresses. By harnessing the unique physicochemical properties of nanomaterials, nanoprimed seeds can deliver nutrients and protection more efficiently, and in significantly smaller quantities, thereby reducing reliance on conventional fertilizers and pesticides.

However, to fully realize the potential of seed nanopriming, comprehensive research is essential to evaluate the long-term effects of nanomaterials on plant physiology, soil health, and microbial communities, especially those in the rhizosphere. Moreover, the development of standardized regulatory frameworks and rigorous safety assessments will be crucial to ensure responsible application and public acceptance.

With its interdisciplinary nature, seed nanopriming stands at the intersection of nanotechnology, plant science, and environmental sustainability. Through collaborative research, supportive policy frameworks, and in-depth toxicological evaluations, this technique can be integrated into eco-friendly farming systems, helping to meet the increasing global food demand while safeguarding ecological balance.

## Acknowledgement

JTW acknowledges the National Science Centre, Poland for financial support under grant no. 2022/45/N/NZ9/01483. AKG acknowledges the National Science Centre and the European Union Framework Programme (Project No. UMO-2022/45/P/NZ9/01571), co-funded by for Research and Innovation Horizon 2020 under the Marie Skłodowska-Curie grant agreement No. 945339. SS and PG acknowledge the YUFE4Postdocs (Y4P) programme, financially supported by European Union funds as a part of the Marie Sklodowska-Curie Actions Cofund programme (Agreement No. 101081327), Ministry of Science and Higher Education under the "International Co-financed Projects" programme (Agreement No. 5678/HE/2023/2024/2). MR thankfully acknowledges Federal University of Piaui (UFPI), Teresina, for the Visiting Professorship.

## References

AbdElgawad, H., Magdy Korany, S., Reyad, A. M., Zahid, I., Akhter, N., Alsherif, E. et al. (2023). Synergistic impacts of plant-growth-promoting bacteria and selenium nanoparticles on improving the nutritional value and biological activities of three cultivars of brassica sprouts. ACS Omega 8: 26414–26424. https://doi.org/10.1021/acsomega.3c02957.

Acharya, B. R., Gill, S. P., Kaundal, A. and Sandhu, D. (2024). Strategies for combating plant salinity stress: the potential of plant growth-promoting microorganisms. Frontiers in Plant Sciences 15: 1406913. https://doi.org/10.3389/fpls.2024.1406913.

Afzal, S., Sharma, D. and Singh, N. K. (2021). Eco-friendly synthesis of phytochemical-capped iron oxide nanoparticles as nano-priming agent for boosting seed germination in rice (*Oryza sativa* L.). Environmental Science and Pollution Research 28: 40275–40287. https://doi.org/10.1007/s11356-020-12056-5.

Akhtar, N., Ilyas, N., Mashwani, Z. R., Hayat, R., Yasmin, H., Noureldeen, A. et al. (2021). Synergistic effects of plant growth promoting rhizobacteria and silicon dioxide nano-particles for amelioration of drought stress in wheat. Plant Physiology and Biochemistry 166: 160–176. https://doi.org/10.1016/j.plaphy.2021.05.039.

Akhtar, N., Khan, S., Rehman, S. U., Rha, E. S. and Jamil, M. (2022). Combined effect of zinc oxide nanoparticles and bacteria on osmolytes and antioxidative parameters of rice (*Oryza sativa* L.) plant grown in heavy metal-contaminated water. Adsorption Science & Technology 2022: 4148765. https://doi.org/10.1155/2022/4148765.

Al Masruri, M. H. K., Ullah, A. and Farooq, M. (2023). Application of nano chitosan-glycinebetaine for improving bread wheat performance under combined drought and heat stresses. Journal of Soil Science and Plant Nutrition 23: 3482–3499. https://doi.org/10.1007/s42729-023-01265-9.

Ali, M. H., Sobze, J.-M., Pham, T. H., Nadeem, M., Liu, C., Galagedara, L. et al. (2020). Carbon nanotubes improved the germination and vigor of plant species from peatland ecosystem via remodeling the membrane lipidome. Nanomaterials 10: 1852. https://doi.org/10.3390/nano10091852.

Allam, E., El-Darier, S., Ghattass, Z., Fakhry, A. and Elghobashy, R. M. (2024). Application of chitosan nanopriming on plant growth and secondary metabolites of *Pancratium maritimum* L. BMC Plant Biology 24: 466. https://doi.org/10.1186/s12870-024-05148-8.

Al-Salama, Y. (2022). Effect of seed priming with zno nanoparticles and saline irrigation water in yield and nutrients uptake by wheat plants. Environmental Sciences Proceedings 16: 37. https://doi.org/10.3390/environsciproc2022016037.

An, J., Hu, P., Li, F., Wu, H., Shen, Y. C., White, J. et al. (2020). Emerging investigator series: molecular mechanisms of plant salinity stress tolerance improvement by seed priming with cerium oxide nanoparticles. Environmental Science: Nano 7: 2214–2228. https://doi.org/10.1039/D0EN00387E.

Anbalagan, A., Yadav, S., Choudhary, R., Sushma, M. K., Yadav, A., Singh, D. et al. (2025). Nanopriming with $TiO_2$ enhances germination by affecting chlorophyll content, osmolytes and antioxidant defense system in lentil (*Lens culinaris* Medik.). Russian Journal of Plant Physiology 72: 12. https://doi.org/10.1134/S1021443724608310.

Ashwini, M. N., Gajera, H. P., Hirpara, D. G., Savaliya, D.D. and Kandoliya, U. K. (2024). Comparative impact of seed priming with zinc oxide nanoparticles and zinc sulphate on biocompatibility, zinc uptake, germination, seedling vitality, and antioxidant modulation in groundnut. Journal of Nanoparticle Research 26: 235. https://doi.org/10.1007/s11051-024-06141-w.

Barman, M. K. (2024). Modern agricultural breakthroughs made using metal-based nanoparticles: an environmentally friendly strategy. New Journal of Chemistry 48: 19337–19354. https://doi.org/10.1039/D4NJ04053H.

Bhutto, M. A., Bhutto, M. A., Mangrio, G. S., Charan, T. R. and Tunio, A. A. (2024). Study on the viability and sustainable release of rice rhizobacteria (*Paenibacillus* IBGE-MAB1) immobilized in nanofibers for enhanced rice seed coating and germination. BioNanoScience 14: 3274–3285. https://doi.org/10.1007/s12668-024-01460-7.

Bonser, C. A. R., Borgatta, J., White, J. C., Astete, C. E., Sabliov, C. M. and Davis, J. A. (2023). Impact of zein and lignin-PLGA biopolymer nanoparticles used as pesticide nanocarriers on soybean growth and yield under field conditions. Agrosystems, Geosciences & Environment 6: e20350. https://doi.org/10.1002/agg2.20350.

Bravo Cadena, M., Preston, G. M., Van der Hoorn, R. A. L., Flanagan, N. A., Townley, H. E. and Thompson, I. P. (2018). Enhancing cinnamon essential oil activity by nanoparticle encapsulation to control seed pathogens. Industrial Crops and Products 124: 755–764. https://doi.org/10.1016/j.indcrop.2018.08.043.

Cao, Y., Turk, K., Bibi, N., Ghafoor, A., Ahmed, N., Azmat, M. et al. (2025). Nanoparticles as catalysts of agricultural revolution: enhancing crop tolerance to abiotic stress: a review. Frontiers in Plant Sciences 15: 1510482. https://doi.org/10.3389/fpls.2024.1510482.

Cembrowska-Lech, D. and Rybak, K. (2023). Nanopriming of barley seeds—a shotgun approach to improve germination under salt stress conditions by regulating of reactive oxygen species. Plants 12: 405. https://doi.org/10.3390/plants12020405.

Chandrasekaran, U., Luo, X., Wang, Q. and Shu, K. (2020). Are there unidentified factors involved in the germination of nanoprimed seeds? Frontiers in Plant Sciences 11: 832. https://doi.org/10.3389/fpls.2020.00832.

Chattha, M. U., Amjad, T., Khan, I., Nawaz, M., Ali, M., Chattha, M. B. et al. (2022). Mulberry based zinc nano-particles mitigate salinity induced toxic effects and improve the grain yield and zinc bio-fortification of wheat by improving antioxidant activities, photosynthetic performance, and accumulation of osmolytes and hormones. Frontiers in Plant Sciences 13: 920570. https://doi.org/10.3389/fpls.2022.920570.

Chouhan, D., Mandal, P., Choudhuri, C. and Mathur, P. (2024). Impact of silver nanochitosan in protecting wheat seeds from fungal infection and increasing growth parameters. Plant Nano Biology 10: 100116. https://doi.org/10.1016/j.plana.2024.100116.

Considine, M. J. and Foyer, C. H. (2021). Oxygen and reactive oxygen species-dependent regulation of plant growth and development. Plant Physiology 186: 79–92. https://doi.org/10.1093/plphys/kiaa077.

Damasceno, R., Roggia, I., Pereira, C. and de Sá, E. (2013). Rhizobia survival in seeds coated with polyvinyl alcohol (PVA) electrospun nanofibres. Canadian Journal of Microbiology 59: 716–719. https://doi.org/10.1139/cjm-2013-0508.

Del Buono, D., Luzi, F., Tolisano, C., Puglia, D. and Di Michele, A. (2022). Synthesis of a lignin/zinc oxide hybrid nanoparticles system and its application by nano-priming in maize. Nanomaterials 12: 568. https://doi.org/10.3390/nano12030568.

Dhingra, P., Sharma, S., Singh, K. H., Kushwaha, H. S., Barupal, J. K., Haq, S. et al. (2022). Seed priming with carbon nanotubes and silicon dioxide nanoparticles influence agronomic traits of Indian mustard (*Brassica juncea*) in field experiments. Journal of King Saud University - Science 34: 102067. https://doi.org/10.1016/j.jksus.2022.102067.

Dogaru, B.-I., Stoleru, V., Mihalache, G., Yonsel, S. and Popescu, M.-C. (2021). Gelatin reinforced with CNCs as nanocomposite matrix for *Trichoderma harzianum* KUEN 1585 spores in seed coatings. Molecules 26: 5755. https://doi.org/10.3390/molecules26195755.

El-Badri, A. M., Batool, M., Wang, C., Hashem, A. M., Tabl, K. M., Nishawy, E. et al. (2021). Selenium and zinc oxide nanoparticles modulate the molecular and morpho-physiological processes during seed germination of *Brassica napus* under salt stress. Ecotoxicology and Environmental Safety 225: 112695. https://doi.org/10.1016/j.ecoenv.2021.112695.

Elhaj Baddar, Z. and Unrine, J. M. (2018). Functionalized-ZnO-nanoparticle seed treatments to enhance growth and Zn content of wheat (*Triticum aestivum*) seedlings. Journal of Agricultural and Food Chemistry 66: 12166–12178. https://doi.org/10.1021/acs.jafc.8b03277.

Elkelish, A., Alqudah, A. M., Alammari, B. S., Alsubeie, M. S., Hamed, S. M. and Thabet, S. G. (2025). Exploring genetic determinants of silver oxide nanoparticle-induced seed priming for drought tolerance in wheat. Genetic Resources and Crop Evolution 72: 3203–3218. https://doi.org/10.1007/s10722-024-02138-5.

El-Saadony, M. T., Saad, A. M., Najjar, A. A., Alzahrani, S. O., Alkhatib, F. M., Shafi, M. E. et al. (2021). The use of biological selenium nanoparticles to suppress *Triticum aestivum* L. crown and root rot diseases induced by *Fusarium* species and improve yield under drought and heat stress. Saudi Journal of Biological Sciences 28: 4461–4471. https://doi.org/10.1016/j.sjbs.2021.04.043.

El-Saadony, M. T., Saad, A. M., Soliman, S. M., Salem, H. M., Ahmed, A. I., Mahmood, M. et al. (2022a). Role of nanoparticles in enhancing crop tolerance to abiotic stress: A comprehensive review. Frontiers in Plant Sciences 13: 946717. https://doi.org/10.3389/fpls.2022.946717.

El-Saadony, M. T., Saad, A. M., Soliman, S. M., Salem, H. M., Ahmed, A. I., Mahmood, M. et al. (2022b). Plant growth-promoting microorganisms as biocontrol agents of plant diseases: Mechanisms, challenges and future perspectives. Frontiers in Plant Sciences 13: 923880. https://doi.org/10.3389/fpls.2022.923880.

El-Shazoly, R. M., Othman, A. A., Zaheer, M. S., Al-Hossainy, A. F. and Abdel-Wahab, D. A. (2025). Zinc oxide seed priming enhances drought tolerance in wheat seedlings by improving antioxidant activity and osmoprotection. Scientific Reports 15: 3863. https://doi.org/10.1038/s41598-025-86824-z.

Falsini, S., Clemente, I., Papini, A., Tani, C., Schiff, S., Salvatici, M. C. et al. (2019). When sustainable nanochemistry meets agriculture: Lignin nanocapsules for bioactive compound delivery to plantlets. ACS Sustainable Chemistry & Engineering 7: 19935–19942. https://doi.org/10.1021/acssuschemeng.9b05462.

Farooq, T., Akram, M. N., Hameed, A., Ahmed, T. and Hameed, A. (2022). Nanopriming-mediated memory imprints reduce salt toxicity in wheat seedlings by modulating physiobiochemical attributes. BMC Plant Biology 22: 540. https://doi.org/10.1186/s12870-022-03912-2.

Fonseca, J. D. da S., Wojciechowska, E., Kulesza, J. and Barros, B. S. (2024). Carbon nanomaterials in seed priming: Current possibilities. ACS Omega 9: 44891–44906. https://doi.org/10.1021/acsomega.4c07230.

Food and Agriculture Organization of the United Nations (FAO). (2009). FAO's Director-General on How to Feed the World in 2050. Population and Development Review 35: 837–839. https://doi.org/10.1111/j.1728-4457.2009.00312.x.

Food and Agriculture Organization of the United Nations (FAO) Seeds. https://www.fao.org/seeds/en. Accessed 17 Jun 2025.

García-Locascio, E., Valenzuela, E. I. and Cervantes-Avilés, P. (2024). Impact of seed priming with selenium nanoparticles on germination and seedlings growth of tomato. Scientific Reports 14: 6726. https://doi.org/10.1038/s41598-024-57049-3.

Giannandrea, T., Benjamin, J., Nicolini, A. T., Alvarez, V. A., Mansilla, A. Y. and Primiano, R. P. O. (2025). Pilot-scale production of functionalized nanoclays for sustainable agricultural applications: process scale-up and optimization. Discover Chemical Engineering 5: 6. https://doi.org/10.1007/s43938-025-00078-9.

Gomes, D. G., Pieretti, J. C., Lange, C. N., Santana, B. de M., Batista, B. L., Parreira, P. S. et al. (2024). Seed nanopriming with green-synthesized Cu-based nanoparticles for maize plants with improved tolerance to hydric stress. ACS Applied Nano Materials 7: 20058–20070. https://doi.org/10.1021/acsanm.4c02607.

Gregorio, P. R. D., Michavila, G., Muller, L. R., Borges, C. de S., Pomares, M. F., Sá, E. L. S. et al. (2017). Beneficial rhizobacteria immobilized in nanofibers for potential application as soybean seed bioinoculants. PLOS ONE 12: e0176930. https://doi.org/10.1371/journal.pone.0176930.

Gupta, N., Singh, P. M., Sagar, V., Pandya, A., Chinnappa, M., Kumar, R. et al. (2022). Seed priming with ZnO and $Fe_3O_4$ nanoparticles alleviate the lead toxicity in *Basella alba* L. through reduced lead uptake and regulation of ROS. Plants 11: 2227. https://doi.org/10.3390/plants11172227.

Gusain, S., Joshi, S. and Joshi, R. (2023). Sensing, signalling, and regulatory mechanism of cold-stress tolerance in plants. Plant Physiology and Biochemistry 197: 107646. https://doi.org/10.1016/j.plaphy.2023.107646.

Hemalatha, M., Hilli, J. S., Chandrashekhar, S. S., Vijayakumar, A. G., Reddy, U. G. and Tippannavar, P. S. (2024). Application of green synthesized Ag and Cu nanoparticles for the control of bruchids and their impact on seed quality and yield in greengram. Heliyon 10. https://doi.org/10.1016/j.heliyon.2024.e31551.

Hussain, A., Rizwan, M., Ali, Q. and Ali, S. (2019). Seed priming with silicon nanoparticles improved the biomass and yield while reduced the oxidative stress and cadmium concentration in wheat grains. Environmental Science and Pollution Reserach 26: 7579–7588. https://doi.org/10.1007/s11356-019-04210-5.

Hussain, I., Zhao, T., Wang, Y., Lei, N., Liu, K., Yu, H. et al. (2024). Melatonin and copper oxide nanoparticles synergistically mitigate clubroot disease and enhance growth dynamics in *Brassica rapa*. Plant Physiology and Biochemistry 215: 109020. https://doi.org/10.1016/j.plaphy.2024.109020.

Hussain, Z., Khan, M. A., Iqbal, F., Raffi, M. and Hafeez, F. Y. (2019). Electrospun microbial-encapsulated composite-based plasticized seed coat for rhizosphere stabilization and sustainable production of canola (*Brassica napus* L.). Journal of Agricultural and Food Chemistry 67: 5085–5095. https://doi.org/10.1021/acs.jafc.8b06505.

Imtiaz, H., Shiraz, M., Mir, A. R., Siddiqui, H. and Hayat, S. (2023). Nano-priming techniques for plant physio-biochemistry and stress tolerance. Journal of Plant Growth Regulation 42: 6870–6890. https://doi.org/10.1007/s00344-023-10981-6.

Ioannou, A., Gohari, G., Papaphilippou, P., Panahirad, S., Akbari, A., Dadpour, M. R. et al. (2020). Advanced nanomaterials in agriculture under a changing climate: The way to the future? Environmental and Experimental Botany 176: 104048. https://doi.org/10.1016/j.envexpbot.2020.104048.

Iqbal, M., Raja, N. I., Mashwani, Z.-U.-R., Hussain, M., Ejaz, M. and Yasmeen, F. (2019). Effect of silver nanoparticles on growth of wheat under heat stress. Iranian Journal of Science and Technology, Transactions A: Science 43: 387–395. https://doi.org/10.1007/s40995-017-0417-4.

Javed, T., Afzal, I., Shabbir, R., Ikram, K., Saqlain Zaheer, M., Faheem, M. et al. (2022). Seed coating technology: An innovative and sustainable approach for improving seed quality and crop performance. Journal of Saudi Society of Agricultural Sciences 21: 536–545. https://doi.org/10.1016/j.jssas.2022.03.003.

Joel, J. M., Johnson, R. and Puthur, J. T. (2023). Co-application of arbuscular mycorrhizal fungi and engineered nanomaterials: A promising strategy for crop resilience against abiotic stresses. South African Journal of Botany 162: 314–323. https://doi.org/10.1016/j.sajb.2023.09.022.

Kalal, P. R., Tomar, R. S. and Jajoo, A. (2022). $SiO_2$ nanopriming protects PS I and PSII complexes in wheat under drought stress. Plant Nano Biology 2: 100019. https://doi.org/10.1016/j.plana.2022.100019.

Kalimuthu, R., Meenachi Sellan, K., Antony, D., Rajaprakasam, S., Chokkalingam, V., Chidambaram, P. et al. (2023). Nanopriming action of microwave-assisted biofunctionalized ZnO nanoparticles to enhance the growth under moisture stress in *Vigna radiata*. ACS Omega 8: 28143–28155. https://doi.org/10.1021/acsomega.3c01329.

Kasote, D. M., Lee, J. H. J., Jayaprakasha, G. K. and Patil, B. S. (2021). Manganese oxide nanoparticles as safer seed priming agent to improve chlorophyll and antioxidant profiles in watermelon seedlings. Nanomaterials 11: 1016. https://doi.org/10.3390/nano11041016.

Khan, M. N., Li, Y., Khan, Z., Chen, L., Liu, J., Hu, J. et al. (2021). Nanoceria seed priming enhanced salt tolerance in rapeseed through modulating ROS homeostasis and α-amylase activities. Journal of Nanobiotechnology 19: 276. https://doi.org/10.1186/s12951-021-01026-9.

Khan, M. R. and Siddiqui, Z. A. (2020). Use of silicon dioxide nanoparticles for the management of *Meloidogyne incognita*, *Pectobacterium betavasculorum* and *Rhizoctonia solani* disease complex of beetroot (*Beta vulgaris* L.). Scientia Horticulturae 265: 109211. https://doi.org/10.1016/j.scienta.2020.109211.

Kinnear, C., Moore, T. L., Rodriguez-Lorenzo, L., Rothen-Rutishauser, B. and Petri-Fink, A. (2017). Form follows function: Nanoparticle shape and its implications for nanomedicine. Chemical Reviews 117: 11476–11521. https://doi.org/10.1021/acs.chemrev.7b00194.

Koleva, L., Umar, A., Yasin, N. A., Shah, A. A., Siddiqui, M. H., Alamri, S. et al. (2022). Iron oxide and silicon nanoparticles modulate mineral nutrient homeostasis and metabolism in cadmium-stressed *Phaseolus vulgaris*. Frontiers in Plant Sciences 13: 806781. https://doi.org/10.3389/fpls.2022.806781.

Kumari, P., Kumari, N., Mohan, C., Chinglenthoiba, C. and Amesho, K. T. T. (2024). Environmentally benign approach to formulate nanoclay/starch hydrogel for controlled release of zinc and its application in seed coating of *Oryza sativa* plant. International Journal of Biological Macromolecules 257: 128278. https://doi.org/10.1016/j.ijbiomac.2023.128278.

Kumari, S., Kumaraswamy, R. V., Choudhary, R. C., Sharma, S. S., Pal, A., Raliya, R. et al. (2018). Thymol nanoemulsion exhibits potential antibacterial activity against bacterial pustule disease and growth promotory effect on soybean. Scientific Reports 8: 6650. https://doi.org/10.1038/s41598-018-24871-5.

Lahuta, L. B., Szablińska-Piernik, J., Stałanowska, K., Głowacka, K. and Horbowicz, M. (2022). The size-dependent effects of silver nanoparticles on germination, early seedling development and polar metabolite profile of wheat (*Triticum aestivum* L.). International Journal of Molecular Sciences 23: 13255. https://doi.org/10.3390/ijms232113255.

Lau, E., Carvalho, L. B., Pereira, A. E., Montanha, G. S., Corrêa, C. G., Carvalho, H. W. et al. (2020). Localization of coated iron oxide ($Fe_3O_4$) nanoparticles on tomato seeds and their effects on growth. ACS Applied Bio Materials 3: 4109–4117. https://doi.org/10.1021/acsabm.0c00216.

Li, Y., Liang, L., Li, W., Ashraf, U., Ma, L., Tang, X. et al. (2021). ZnO nanoparticle-based seed priming modulates early growth and enhances physio-biochemical and metabolic profiles of fragrant rice against cadmium toxicity. Journal of Nanobiotechnology 19: 75. https://doi.org/10.1186/s12951-021-00820-9.

Mahakham, W., Sarmah, A. K., Maensiri, S. and Theerakulpisut, P. (2017). Nanopriming technology for enhancing germination and starch metabolism of aged rice seeds using phytosynthesized silver nanoparticles. Scientific Reports 7: 8263. https://doi.org/10.1038/s41598-017-08669-5.

Maity, D., Gupta, U. and Saha, S. (2022). Biosynthesized metal oxide nanoparticles for sustainable agriculture: Next-generation nanotechnology for crop production, protection and management. Nanoscale 14: 13950–13989. https://doi.org/10.1039/D2NR03944C.

Mamidi, N., Delgadillo, R. M. V., Sustaita, A. O., Lozano, K. and Yallapu, M. M. (2025). Current nanocomposite advances for biomedical and environmental application diversity. Medical Research Review 45: 576–628. https://doi.org/10.1002/med.22082.

Mazhar, M. W., Ishtiaq, M., Maqbool, M. and Akram, R. (2022). Seed priming with calcium oxide nanoparticles improves germination, biomass, antioxidant defence and yield traits of canola plants under drought stress. South African Journal of Botany 151: 889–899. https://doi.org/10.1016/j.sajb.2022.11.017.

Mazhar, M. W., Ishtiaq, M., Maqbool, M., Muzammil, K., Mohieldin, A., Dawria, A. et al. (2024). Optimizing water relations, gas exchange parameters, biochemical attributes and yield of water-stressed maize plants through seed priming with iron oxide nanoparticles. BMC Plant Biology 24: 624. https://doi.org/10.1186/s12870-024-05324-w.

Mohan, N., Pal, A., Saharan, V., Kumar, A., Vashishth, R. and Prince, S. E. (2024). Development, characterization, and evaluation of Zn-SA-chitosan bionanoconjugates on wheat seed, experiencing chilling stress during germination. Heliyon 10: e31708. https://doi.org/10.1016/j.heliyon.2024.e31708.

Nagdalian, A., Blinov, A., Gvozdenko, A., Golik, A., Rekhman, Z., Rzhepakovsky, I. et al. (2024). Effect of $MnO_2$ nanoparticles stabilized with cocamidopropyl betaine on germination and development of pea (*Pisum sativum* L.) seedlings. Nanomaterials 14: 959. https://doi.org/10.3390/nano14110959.

Nauman Khan, M., Fu, C., Liu, X., Li, Y., Yan, J., Yue, L. et al. (2024). Nanopriming with selenium doped carbon dots improved rapeseed germination and seedling salt tolerance. The Crop Journal 12: 1333–1343. https://doi.org/10.1016/j.cj.2024.03.007.

Nile, S. H., Thiruvengadam, M., Wang, Y., Samynathan, R., Shariati, M. A., Rebezov, M. et al. (2022). Nano-priming as emerging seed priming technology for sustainable agriculture—recent developments and future perspectives. Journal of Nanobiotechnology 20: 254. https://doi.org/10.1186/s12951-022-01423-8.

Pagano, A., Macovei, A. and Balestrazzi, A. (2023). Molecular dynamics of seed priming at the crossroads between basic and applied research. Plant Cell Reports 42: 657–688. https://doi.org/10.1007/s00299-023-02988-w.

Perea-Vélez, Y. S., Carrillo-González, R., González-Chávez Ma del, C. A., Vangronsveld, J., Tapia Maruri, D. and López-Luna, J. (2024). Nano-priming of *Phaseolus vulgaris* OTI cultivar with cobalt ferrite nanoparticles enhances the mineral composition of progeny seeds. Journal of Nanoparticle Research 26: 205. https://doi.org/10.1007/s11051-024-06101-4.

Pereira, A. do E. S., Oliveira, H. C. and Fraceto, L. F. (2019). Polymeric nanoparticles as an alternative for application of gibberellic acid in sustainable agriculture: a field study. Scientific Reports 9: 7135. https://doi.org/10.1038/s41598-019-43494-y.

Rai-Kalal, P. and Jajoo, A. (2021). Priming with zinc oxide nanoparticles improve germination and photosynthetic performance in wheat. Plant Physiology and Biochemistry 160: 341–351. https://doi.org/10.1016/j.plaphy.2021.01.032.

Ramzan, M., Parveen, M., Naz, G., Sharif, H. M. A., Nazim, M., Aslam, S. et al. (2024). Enhancing physio-biochemical characteristics in okra genotypes through seed priming with biogenic zinc oxide nanoparticles synthesized from halophytic plant extracts. Scientific Reports 14: 23753. https://doi.org/10.1038/s41598-024-74129-6.

Ranaware, A. S., Kushwaha, S. B., Kunchge, N., Prakash, G. and Lele, S. S. (2025). Effect of different size silver nanoparticles synthesized at varying pH of plant extract for germination improvement of tetraploid watermelon (*Citrullus lanatus*). Journal of Cluster Science 36: 101. https://doi.org/10.1007/s10876-025-02818-x.

Rawashdeh, R. Y., Harb, A. M. and AlHasan, A. M. (2020). Biological interaction levels of zinc oxide nanoparticles; lettuce seeds as case study. Heliyon 6. https://doi.org/10.1016/j.heliyon.2020.e03983.

Raza, M. A. S., Zulfiqar, B., Iqbal, R., Muzamil, M. N., Aslam, M. U., Muhammad, F. et al. (2023). Morpho-physiological and biochemical response of wheat to various treatments of silicon nano-particles under drought stress conditions. Scientific Reports 13: 2700. https://doi.org/10.1038/s41598-023-29784-6.

Rifna, E. J., Ratish Ramanan, K. and Mahendran, R. (2019). Emerging technology applications for improving seed germination. Trends in Food Science & Technology 86: 95–108. https://doi.org/10.1016/j.tifs.2019.02.029.

Saboor, A., Mustafa, G., Arshad, M., Ahmad, M., Hussain, S., Ahmed, N. et al. (2019). Seed priming and metal/metalloid stress tolerance in plants. pp. 287–311. *In*: Hasanuzzaman, M. and Fotopoulos, V. (eds.). Priming and Pretreatment of Seeds and Seedlings: Implication in Plant Stress Tolerance and Enhancing Productivity in Crop Plants. Springer, Singapore.

Saharan, V., Sharma, G., Yadav, M., Choudhary, M. K., Sharma, S. S., Pal, A. et al. (2015). Synthesis and *in vitro* antifungal efficacy of Cu–chitosan nanoparticles against pathogenic fungi of tomato. International Journal of Biological Macromolecules 75: 346–353. https://doi.org/10.1016/j.ijbiomac.2015.01.027.

Salam, A., Khan, A. R., Liu, L., Yang, S., Azhar, W., Ulhassan, Z. et al. (2022). Seed priming with zinc oxide nanoparticles downplayed ultrastructural damage and improved photosynthetic apparatus in maize under cobalt stress. Journal of Hazardous Materials 423: 127021. https://doi.org/10.1016/j.jhazmat.2021.127021.

Sarkar, N., Sharma, R. S. and Kaushik, M. (2021). Innovative application of facile single pot green synthesized CuO and CuO@APTES nanoparticles in nanopriming of *Vigna radiata* seeds. Environmental Science and Pollution Research 28: 13221–13228. https://doi.org/10.1007/s11356-020-11493-6.

Sathiyabama, M. and Muthukumar, S. (2020). Chitosan guar nanoparticle preparation and its in vitro antimicrobial activity towards phytopathogens of rice. International Journal of Biological Macromolecules 153: 297–304. https://doi.org/10.1016/j.ijbiomac.2020.03.001.

Savani, K. R., Gajera, H. P., Hirpara, D. G., Savaliya, D. D. and Kandoliya, U. K. (2023). Salicylic acid-functionalised chitosan nanoparticles restore impaired sucrose metabolism in the developing anther of cotton (*Gossypium hirsutum*) under heat stress. Functional Plant Biology 50: 736–751. https://doi.org/10.1071/FP22309.

Savassa, S. M., Castillo-Michel, H., del Real, A. E. P., Reyes-Herrera, J., Marques, J. P. R. and de Carvalho, H. W. P. (2021). Ag nanoparticles enhancing *Phaseolus vulgaris* seedling development: understanding nanoparticle migration and chemical transformation across the seed coat. Environmental Science: Nano 8: 493–501. https://doi.org/10.1039/D0EN00959H.

Seddighinia, F. S., Iranbakhsh, A., Oraghi Ardebili, Z., Nejad Satari, T. and Soleimanpour, S. (2020). Seed priming with cold plasma and multi-walled carbon nanotubes modified growth, tissue differentiation, anatomy, and yield in bitter melon (*Momordica charantia*). Journal of Plant Growth Regulation 39: 87–98. https://doi.org/10.1007/s00344-019-09965-2.

Sharma, S., Sahu, B. K., Cao, L., Bindra, P., Kaur, K., Chandel, M. et al. (2021). Porous nanomaterials: Main vein of agricultural nanotechnology. Progress in Materials Science 121: 100812. https://doi.org/10.1016/j.pmatsci.2021.100812.

Shcherbakova, E. N., Shcherbakov, A. V., Andronov, E. E., Gonchar, L. N., Kalenskaya, S. M. and Chebotar, V. K. (2017). Combined pre-seed treatment with microbial inoculants and Mo nanoparticles changes composition of root exudates and rhizosphere microbiome structure of chickpea (*Cicer arietinum* L.) plants. Symbiosis 73: 57–69. https://doi.org/10.1007/s13199-016-0472-1.

Sheikhalipour, M., Mohammadi, S. A., Esmaielpour, B., Spanos, A., Mahmoudi, R., Mahdavinia, G. R. et al. (2023). Seedling nanopriming with selenium-chitosan nanoparticles mitigates the adverse effects of salt stress by inducing multiple defence pathways in bitter melon plants. International Journal of Biological Macromolecules 242: 124923. https://doi.org/10.1016/j.ijbiomac.2023.124923.

Shelar, A., Nile, S. H., Singh, A. V., Rothenstein, D., Bill, J., Xiao, J. et al. (2023). Recent advances in nano-enabled seed treatment strategies for sustainable agriculture: challenges, risk assessment, and future perspectives. Nano-Micro Letters 15: 54. https://doi.org/10.1007/s40820-023-01025-5.

Shelar, A., Singh, A. V., Maharjan, R. S., Laux, P., Luch, A., Gemmati, D. et al. (2021). Sustainable agriculture through multidisciplinary seed nanopriming: prospects of opportunities and challenges. Cells 10: 2428. https://doi.org/10.3390/cells10092428.

Shinde, S., Paralikar, P., Ingle, A. P. and Rai, M. (2020). Promotion of seed germination and seedling growth of Zea mays by magnesium hydroxide nanoparticles synthesized by the filtrate from *Aspergillus niger*. Arabian Journal of Chemistry 13: 3172–3182. https://doi.org/10.1016/j.arabjc.2018.10.001.

Siddaiah, C. N., Prasanth, K. V. H., Satyanarayana, N. R., Mudili, V., Gupta, V. K. et al. (2018). Chitosan nanoparticles having higher degree of acetylation induce resistance against pearl millet downy mildew through nitric oxide generation. Scientific Reports 8: 2485. https://doi.org/10.1038/s41598-017-19016-z.

Singh, K. M., Baksi, S., Rani, S., Jha, A. B., Dubey, R. S. and Sharma, P. (2024). NanoBoost: Maximizing crop resilience and yield via nanopriming under salt stress. Environmental and Experimental Botany 226: 105937. https://doi.org/10.1016/j.envexpbot.2024.105937.

Sohail, M., Pirzada, T. H., Opperman, C. A. and Khan, S. (2022). Recent advances in seed coating technologies: transitioning toward sustainable agriculture. Green Chemistry 24: 6052–6085. https://doi.org/10.1039/D2GC02389J.

Sonawane, H., Arya, S., Math, S. and Shelke, D. (2021). Myco-synthesized silver and titanium oxide nanoparticles as seed priming agents to promote seed germination and seedling growth of *Solanum lycopersicum*: a comparative study. International Nano Letters 11: 371–379. https://doi.org/10.1007/s40089-021-00346-w.

Steven, S., Islam, M. S., Ghimire, A., Methela, N. J., Kwon, E.-H., Yun, B.-W. et al. (2024). Chitosan-GSNO nanoparticles and silicon priming enhance the germination and seedling growth of soybean (*Glycine max* L.). Plants 13: 1290. https://doi.org/10.3390/plants13101290.

Taheri, P., Puopolo, G. and Santoyo, G. (2025). Plant growth-promoting microorganisms: New insights and the way forward. Microbiological Research 297: 128168. https://doi.org/10.1016/j.micres.2025.128168.

Thakur, S., Asthir, B., Kaur, G., Kalia, A. and Sharma, A. (2022). Zinc oxide and titanium dioxide nanoparticles influence heat stress tolerance mediated by antioxidant defense system in wheat. Cereal Research Communications 50: 385–396. https://doi.org/10.1007/s42976-021-00190-w.

Trzcińska-Wencel, J., Golińska, P., Gade, A., Ingle, P. U., Shende, S. S. and Rai, M. (2025b). Microbial biosynthesis of biostimulant nanomaterials and nanofertilizers. pp. 301–356. *In*: Juárez-Maldonado, A., Benavides-Mendoza, A., Ojeda-Barrios, D. L., Tortella Fuentes, G. and Seabra, A. B. (eds.). Plant Biostimulation with Nanomaterials. Springer Nature, Singapore.

Trzcińska-Wencel, J., Golińska, P. and Rai, M. (2024). Toxicity of nanomaterials to plants. pp. 369–389. *In*: Rai, M. and Avila-Quezada, G. (eds.). Nanotechnology in Plant Health. CRC Press.

Trzcińska-Wencel, J., Mucha, N., Rai, M., Tyburski, J. and Golińska, P. (2025a). Pre-sowing grain treatment with bio-AgNPs stimulates plant growth and affects redox homeostasis in maize. Frontiers in Plant Sciences 16. https://doi.org/10.3389/fpls.2025.1494741.

Trzcińska-Wencel, J., Wypij, M., Terzyk, A. P., Rai, M. and Golińska, P. (2023). Biofabrication of novel silver and zinc oxide nanoparticles from *Fusarium solani* IOR 825 and their potential application in agriculture as biocontrol agents of phytopathogens, and seed germination and seedling growth promoters. Front Chem 11. https://doi.org/10.3389/fchem.2023.1235437.

Tudi, M., Daniel Ruan, H., Wang, L., Lyu, J., Sadler, R., Connell, D. et al. (2021). Agriculture development, pesticide application and its impact on the environment. International Journal of Environmental Research and Public Health 18: 1112. https://doi.org/10.3390/ijerph18031112.

Ulhassan, Z., Yang, S., He, D., Khan, A. R., Salam, A., Azhar, W. et al. (2023). Seed priming with nano-silica effectively ameliorates chromium toxicity in *Brassica napus*. Journal of Hazardous Materials 458: 131906. https://doi.org/10.1016/j.jhazmat.2023.131906.

Wang, X., Ge, M. and He, X. (2025). Effect of green synthesized $Fe_3O_4$NP priming on alfalfa seed germination under drought stress. Plants 14: 1236. https://doi.org/10.3390/plants14081236.

Xu, T., Ma, C., Aytac, Z., Hu, X., Ng, K. W., White, J.C. et al. (2020). enhancing agrichemical delivery and seedling development with biodegradable, tunable, biopolymer-based nanofiber seed coatings. ACS Sustainable Chemistry & Engineering 8: 9537–9548. https://doi.org/10.1021/acssuschemeng.0c02696.

Yang, L., Zhang, L., Zhang, Q., Wei, J., Zhao, X., Zheng, Z. et al. (2024). Nanopriming boost seed vigor: Deeper insights into the effect mechanism. Plant Physiology and Biochemistry 214: 108895. https://doi.org/10.1016/j.plaphy.2024.108895.

Yang, Y.-M., Naseer, M., Zhu, Y., Wang, B.-Z., Wang, S., Ma, Y. et al. (2023). Priming effects of nZVI on carbon sequestration and iron uptake are positively mediated by AM fungus in semiarid agricultural soils. Science of The Total Environment 882: 163632. https://doi.org/10.1016/j.scitotenv.2023.163632.

Zaim, N. S., Tan, H. L., Rahman, S. M., Abu Bakar, N. F., Osman, M. S., Thakur, V. K. et al. (2023). Recent advances in seed coating treatment using nanoparticles and nanofibers for enhanced seed germination and protection. Journal of Plant Growth Regulation 42: 7374–7402. https://doi.org/10.1007/s00344-023-11038-4.

Zhao, L., Zhou, X., Kang, Z., Peralta-Videa, J. R. and Zhu, Y.-G. (2024). Nano-enabled seed treatment: A new and sustainable approach to engineering climate-resilient crops. Science of The Total Environment 910: 168640. https://doi.org/10.1016/j.scitotenv.2023.168640.

Zhao, X., Chen, Y., Li, H. and Lu, J. (2021). Influence of seed coating with copper, iron and zinc nanoparticles on growth and yield of tomato. IET Nanobiotechnology 15: 674–679. https://doi.org/10.1049/nbt2.12064.

# Chapter 4

# Emerging Nanopesticides

## Opportunities and Challenges

*Kamarshi Yashkamal*[1] and *Muthugounder Subramaniam Shivakumar*[2,*]

## 1. Introduction

Agricultural productivity is affected due to abiotic and biotic factors, the chief among them being damage caused by pests to standing crops and to stored grains. Insects are one of the most important pests which due to their numbers and diversity have become difficult to control. It is estimated that there is about 10 to 28% crop loss (Savary et al. 2019) due to insects. Further, their secondary loss due to the development of fungal and bacterial growth also renders food unfit for human and animal consumption (Deligeorgakis et al. 2023). The crop pests of vegetables include *Plutella xylostella* which damages the cruciferous crops, *Spodoptera litura*, *S. exiuga, Helicoverpa armigera*, aphids and others which cause significant damage to a variety of agricultural crops. Further, between 25 to 30% of the stored grains are damaged due to insect pests like *Rhyzopertha dominica*, *Sitophilus granaries*, and *Sitophilus oryzae* (FAO 2019). Various strategies have been employed for the control of insect pests. Chief among them is the use of chemical insecticides belonging to neonicotinoids, synthetic pyrethroids, organophosphates and organochlorines besides use of the insecticides of newer chemistries (Muthusamy and Shivakumar 2015). Nearly four million tonnes of pesticides are used annually in insect pest control of which only a fraction reaches target pest (Zhang 2018). More than 95% of the pesticide used remains in the soil or is washed off to aquatic bodies resulting in environmental contamination.

Since the start of nanotechnology, there has been an explosion in the number of applications and the types of nanomaterials (NMs) synthesized. These NMs find biological applications in medicine, drug delivery, bioinstrumentation, environmental remediation, clean energy and in use of insect pest control. Today several classes of nanomaterials are being used as nanopesticides. These can be broadly classified into organic NMs, Inorganic NMs and Hybrid NMs (Figure 1). These classes in turn contain several types of NMs.

The developments in nanopesticides have arisen from the fact that chemical insecticides have a large environmental footprint and are also toxic to non-target biota. While botanicals and microbial

[1] Department of Biotechnology, K.S. Rangasamy College of Arts and Science (Autonomous), Tiruchengode – 637 215, Tamil Nadu, India.

[2] Department of Biotechnology, Periyar University, Salem – 636011, Tamil Nadu, India.

Email: k.yashkamal@ksrcas.edu

* Corresponding author: skentomology@gmail.com; sk24@periyaruniversity.ac.in

DOI: 10.1201/9781003659907-6

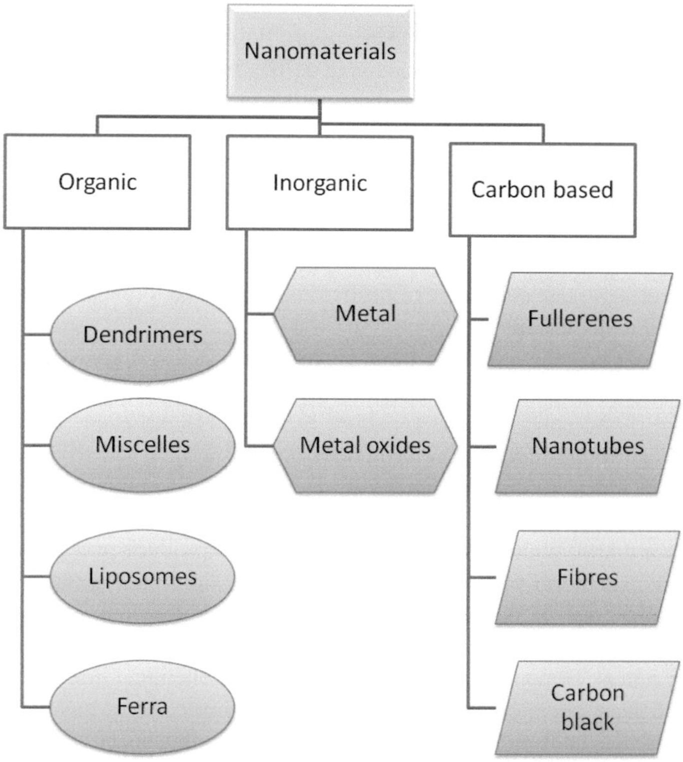


**Figure 1.** Types of nanomaterials used in nanopesticides.

insecticides are either unstable or have lower efficacy and hence require periodic inundation/spraying which in turn increases cost, nanopesticides solve both these problems by having a lower environmental footprint, making botanicals more stable and solving the problem of periodic release by smart/sustained/delayed release methods. This has made NMs a vital component in the development of nanopesticides. In this chapter, we explore the recent development in the field of nanoparticles. The term nanopesticide is given to materials synthesized using nanotechnology for controlling insect pests of crops and diseases. We also compare the conventional insecticides with nanopesticides. Several studies have also pointed that NPs are also toxic to the organisms and have environmental footprints. We also discuss the studies dealing with toxicity of NMs from the perspective of regulatory frame work for nanopesticides and at the end we talk about the public perception and acceptance of nanopesticides.

## 2. Current Control Measures

Since the advent of crop protection, the use of pesticides has shown an increase over time. As per reports by Food and Agriculture Organization (FAO) published in the year 2022, use of pesticides has doubled since 1990s. A comparison in the pesticide application in the last decade with 1990s shows an increase of 121% in the usage of herbicides, 54% increase of fungicides and bactericides and 48% in application of insecticides. Regionally the United States of America recorded the highest use of pesticides with an increase of 10% in total pesticide usage between 2020 and 2022. Asia showed a 76% increase in pesticide usage since 1990s and a 1% decrease compared with 2021. In Europe the application of pesticides was slightly more than the world average on per hectare basis for the entire study period since 1990. However, the overall usage of pesticides showed a steady decline of approximately 5% with maximum reduction seen in the last decade (7%). The chief reason could be stringent regulation by European Common Agriculture Policy which monitors and controls the use of pesticides (Alexoaei et al. 2022). Africa showed a 26% increase in the use of pesticides over the last decade although it applied lowest levels of pesticides per area of cropland. There was a marginal decrease in total pesticide consumption between 2021 and 2022 (0.1%). Pesticides usage in Oceania remained stable from 2021 to 2022 even though it had nearly tripled since 1990. As per the global trend, the use of herbicides increased by 4%, while it declined for insecticides by 3% with a slight decrease of 1% for fungicides and bactericides. In 2022, Brazil recorded the highest consumption of pesticides for agricultural use at 801 kilo tonnes, which was nearly 70% more than the United States of America, the second highest consumer of pesticides at 468 kilo tonnes. The next three users were Indonesia, Argentina and China.

Interestingly, there is a change in the trend since around 2010. A growing awareness regarding the ill-effects of indiscriminate pesticide use has led to use of chemicals and a gradual shift towards environment friendly and safer options (Kishore et al. 2024). In 2024, for the first time, FAO report added data regarding the use of biopesticides on global scale due to its increased popularity. Among the different categories of pesticides used, there was a reduction in use of insecticides (4%), fungicides and bactericides (3%) while there was a slight increase in use of herbicides (10%), hinting towards the popularity of organically grown produce and growing costs of labour for field sanitation. Synthetic pyrethroids are one of the dominant groups of insecticides like beta-cyfluthrin, gama- and lamda-cyhalothrin; organphosphates like chlorpyrifos, dimethoate, malathion, methyl parathion and carbamates like carbaryl, along with spinosad, are some of the highly used pesticides for wheat. Depending on the crop, the type of chemical insecticides differs (US EPA 2022). For most of these chemical insecticides, the target insect pests have developed resistance. Resistance to insecticides has emerged as a major constraint for the successful control of insect pests (Mota-Sanchez and Wise 2020).

## *2.1 Insecticide Resistance*

Insecticide resistance poses significant threat to the successful control of insect pests in agriculture and in stored products (Wang and Wang 2024). The continuous use of chemical insecticides has resulted in the development of insecticide resistance among insect pests. As per pesticide resistance database *Cydia pomonella, Earias vitella, Helicoverpa armigera, H. zea, Heliothis virescens, Pectinophora gossypiella, Plutella xylostella, Spodoptera exigua, S. litura, S. frugipersa, Leptinotarsa decemlineata, Rhyzopertha dominica, Tribolium castaneus, Aphis gossypii, Bemesia tabaci, Myzus persicae, Nilparvata lugens, Phenacoccus solenopsis* and *Trialeurodes varorariorum* are some of the insect pests which have been reported in several studies to be resistant to all the major classes of insecticides used in the control of insect pests (Mota-Sanchez and Wise 2020). From a single incidence in 1914, the cases of infectious disease resistance have increased to 16920 incidences in 2020 (Mota-Sanchez and Wise 2020). This exponential increase in insecticide resistance has been attributed to the introduction of chemical insecticides in agriculture. The management of this has become of pivotal importance for sustainable agriculture (Sparks et al. 2021).

Key resistance mechanisms by which insects develop resistance include metabolic resistance, target-site mutations, reduced penetration, and behavioural changes (Muthusamy and Shivakumar 2015). Recent research has identified new mechanisms, such as sequestration resistance, where overexpressed olfactory proteins bind and sequester insecticides (Pu and Chung 2024). Transcriptional and post-transcriptional regulatory pathways, including AhR/ARNT, Nuclear receptors, and CncC/Keap1, play crucial roles in regulating detoxification genes (Bavithra et al. 2023). Non-coding RNAs and epitranscriptomics also contribute to resistance development (Bavithra et al. 2023). Effective management strategies emphasize Integrated Pest Management (IPM), combining chemical, biological, and cultural control methods (Shudeer et al. 2024). Continuous monitoring, research, and education are essential for developing sustainable pest management solutions and overcoming the challenges posed by insecticide resistance (Shudeer et al. 2024). Due to insecticide resistance, the use of chemical insecticides has increased in terms of the quantity and frequency of application. This has also resulted in secondary problems of non-target toxicity, environmental contamination, and biomagnification of pesticides in the food chain. All this has necessitated the importance of sustainable insect control measures, which are eco-friendly.

## *2.2 Botanicals and Biopesticides*

Botanical insecticides are gaining importance as alternatives to synthetic pesticides due to their efficacy, biodegradability and low toxicity to non-target organisms (Lengai et al. 2020). These include plant-derived secondary metabolites and essential oils. Neem (*Azadirachta indica*) is a prominent botanical insecticide, containing active compounds like azadirachtin that affect insect feeding, oviposition, and development (Nouri-Ganbalani et al. 2016). Plants such as garlic, chilli pepper, cassava, and tobacco have also shown insecticidal properties (Dougoud et al. 2019, Manjula et al. 2020). Several non-edible plants such as *Elaeagnus indica, Prosopis juliflora, Acanthospermum hispidum,* and *Glycosmis pentaphylla* (Srinivasan et al. 2019, Dhivya et al. 2018, Vivekanandhan et al. 2018, Ramkumar et al. 2016) have also been shown to have insecticidal properties. In total, more than 2000 plant species have been shown to have active principles which can be used in insect control (Shaalan et al. 2005). However, the efficacy of homemade botanical insecticides can be variable due to differences in active ingredient content and preparation methods (Dougoud et al. 2019). While botanical insecticides show promise for sustainable pest management, especially in developing countries, their commercialization has been slow in regions with stricter regulatory requirements (Isman 2020). Despite challenges in formulation and adoption, botanical pesticides can contribute to integrated pest management systems and sustainable agricultural production (Lengai et al. 2020).

Biopesticides are emerging as sustainable alternatives to chemical pesticides for insect pest control in agriculture (Parajuli et al. 2022). They are derived from natural sources such as plants, microorganisms, and genetically modified organisms (Prabha et al. 2016). Biopesticides offer several advantages, including environmental safety, minimal impact on non-target organisms, and reduced production costs (Parajuli et al. 2022). Various botanical and microbial biopesticides have shown efficacy against a wide range of insect pests (Parajuli et al. 2022). Examples include neem, chili pepper, and garlic as botanical sources, and *Bacillus thuringiensis, Beauveria bassiana*, and baculoviruses as microbial agents (Parajuli et al. 2022). The global use of biopesticides is increasing by approximately 10% annually (Prabha et al. 2017). However, further research is needed in areas such as bioformulation and commercialization to make biopesticides more accessible and affordable for farmers (Mishra et al. 2020).

## *2.3 Current Control Efficacy and Cost*

Chemical insecticides have been widely used in agriculture for pest control, demonstrating higher efficacy against various pests like painted bugs, aphids, and other insects (Patel et al. 2018). Studies have shown that insecticides such as imidacloprid, spinosad, and acetamiprid are particularly effective in controlling pests and improving crop yields (Singh et al. 2016). However, the use of chemical pesticides has raised concerns about environmental and health impacts (Pathak et al. 2022). As a result, there is growing interest in biopesticides as a more sustainable alternative. Biopesticides offer advantages such as cost-effectiveness and reduced environmental impact (Chaturvedi and Khandelwal 2019). Despite these concerns, chemical insecticides remain an important tool in pest management due to their high efficacy and ease of use by farmers (Pathak et al. 2022). The choice between chemical and biological pesticides involves balancing effectiveness, cost, and environmental considerations.

### *2.4 Need for a New Approach in Pest Management*

Recent research highlights the need for new approaches in insect pest management due to the limitations of traditional methods. Next-generation strategies focus on eco-friendly and targeted approaches, including biopesticides, semiochemicals, and biotechnology-based methods like RNA interference (RNAi) and CRISPR/Cas9 gene editing (Belagalla et al. 2024, Jaba and Sharma 2017). Nanotechnology offers promising applications in developing efficient pest management techniques, though more research is needed in this area (De et al. 2014). Modern innovations incorporate drones and artificial intelligence for precise pest monitoring and targeted pesticide application, enhancing the effectiveness of integrated pest management (IPM) strategies (Kariyanna and Sowjanya 2024). These advanced approaches aim to reduce environmental impact, combat insecticide resistance, and promote sustainable food production. However, challenges remain in regulatory approval, public acceptance, and further research and development to optimize these new pest management techniques (Belagalla et al. 2024).

## 3. Nanomaterials used for Insecticide Synthesis

Nanomaterials are emerging as promising alternatives for insecticide synthesis and pest management in agriculture. Their unique physical and chemical properties make them excellent candidates for developing new insecticides (Kapinder et al. 2021). Various nanoformulations, including nanoemulsions and nanosuspensions, have been developed to combat storage insect pests, offering better efficacy than conventional pesticides due to their small size and reduced harmful effects on non-target species (Jasrotia et al. 2022). Cu-doped ZnO nanoparticles have shown significant insecticidal properties against *Spodoptera littoralis*, with a 95% population reduction at 5% concentration (El-Latef et al. 2023). Micronutrient-based nanopesticides, such as Ag, ZnO, $TiO_2$, Cu, and $SiO_2$, have demonstrated enhanced pest control and improved crop productivity (Gutiérrez-Ramírez et al. 2021). While concerns exist regarding nanomaterial synthesis and usage, their potential in sustainable agriculture and pest management warrants further research and responsible implementation (Gutiérrez-Ramírez et al. 2021).

### *3.1 History and Development of Nanomaterials as Pesticides*

The use of nanomaterials has been in place for about 4000 years, but the exact science behind it was not known. The modern use of nanotechnology started with the idea of physicist Richard Feynman in 1959, that it would be possible to manipulate the materials on the nanoscale. It took almost twenty-six years before this became a reality. Since the breakthrough, nanotechnology has found applications in virtually every domain of human innovation. Engineered nanomaterials/nanoparticles have potential applications in many biological applications like environmental remediation, pharmaceuticals, medical devices, and biosensors. Recently, the use of these nanomaterials in pesticide development has received a lot of attention. Nanopesticide is a term that denotes the use of nanotechnology for the production of plant protection products (Kah et al. 2013, Gogos et al. 2012, Khot et al. 2012). Nanopesticides have a very small amount of active ingredient as compared to conventional insecticides (Bergeson 2010). The advantage of nanopesticides is their efficacy, reduced cost of active ingredients and lower environmental footprint. There are several types of nanopesticides which are currently being developed (Figure 2). These include nanoemulsions, nanocapsules having polymers and inorganic engineered nanomaterials such as metals, metal oxides, and nanoclays (Kah et al. 2013). Some of the nanometals and metal oxides themselves have antimicrobial, which can help in dal benefit of targeting plant pathogens which are usually bacteria or fungi (Khot et al. 2012).

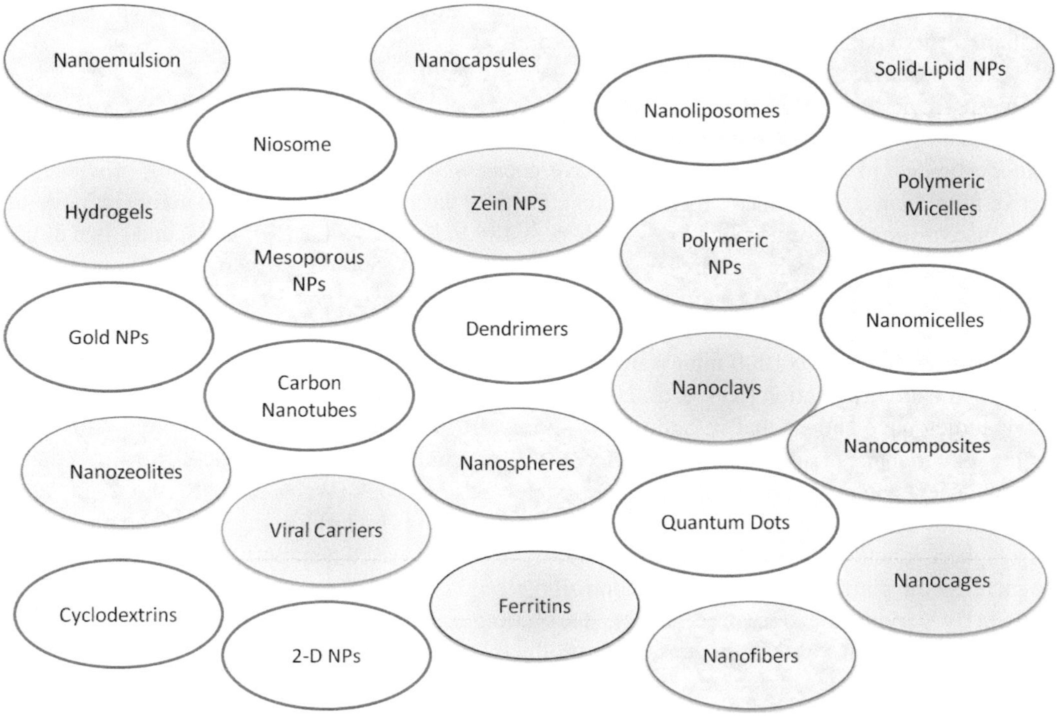


**Figure 2.** Types of nanomaterials used for nanopesticide synthesis.

## *3.2 Nanopesticide Classification and Types*

Nanopesticides can be broadly classified into two groups. Group one includes metal and metal oxides, while group two includes polymers and non-metals (Wang et al. 2022). Group one includes metals like zinc, copper, iron, palladium and their salts. These nanopesticides have a high dissolution rate and can form ions in aquatic medium (Zanoni et al. 2022). They have a high antimicrobial property, are cytotoxic, and generate ROS (Albalghiti et al. 2020). The nanopesticides of group one find applications in antimicrobial creams, in gels in pharmaceutical products, and are also used for the control of microbial pathogens in agriculture (Ma et al. 2020). Group one nanopesticides are also used for the control of insect pests (Kadir et al. 2025). Group one nanopesticides have a very short half-life, and usually within a few months, they completely dissociate. Metals and metal oxides are most popular for green synthesis using plant secondary metabolites, plant essential oils, and bacterial metabolites (Vivekanandhan et al. 2023, Manimegalai et al. 2023, Adhavan et al. 2024). This is since the metal ions can interact with the anions generated in the secondary metabolites of plants and microbes (Marslin et al. 2018).

Group two nanopesticides can be further subdivided into natural polymers and clay-based nanopesticides. These pesticides are considered important as carriers of active ingredients. These nanopesticides especially polymers are cheaper, biocompatible and are an excellent choice for delayed release and stimuli-based release of active ingredients (Wang et al. 2022). Polymer-based nanopesticides use cellulose, poly lactic acid, and chitosan as the base materials. These are used widely in the preparation of hydrogels, nanocapsules, nanomicelles, and nanospheres for carrying active ingredients. The clay-based nanopesticides include mesoporous silica nanoparticles and montmorillonites, which are also used for nanoencapsulation of the active ingredients. Currently there are about 1094 types of group two nanopesticides as per the available literature (Wang et al.

2022). Group two nanopesticides are widely used as encapsulating chemical insecticides such as deltamethrin, botanicals, herbicides like glyphosate, etc.

Nanopesticides have a larger surface-to-volume ratio, smaller size and surface area, which increases their interaction with the cellular systems and are easily taken up by the cells. This makes nanopesticides more efficient as compared to their conventional counterparts. Group two nanopesticides, in addition to the above, have encapsulation, loading and releasing efficiency of active ingredients, which makes them a better choice for the sustainable release of active ingredients. Their size is more uniform with a low polydispersivity index below 0.2 (Mohanraj and Chen 2006). Polymer-based nanopesticides are also relatively stable in colloidal suspensions for months as compared to the group one nanopesticides (Pasquoto-Stigliani 2017). Usually, nanomaterials are defined in the size range of 1 to 100 nm, group two nanopesticides are usually larger with a size ranging from 500 nm to 1000 nm (Wang et al. 2022) and comes within the range as suggested by European Union regulation (Kah et al. 2018, Hardy et al. 2018).

Studies have shown that in addition to the control of insect pests and microbial pathogens, group one nanopesticides can also increase the vitamins, proteins, organic acid, antioxidant, micronutrients, and photosynthetic pigment content of the crops (Wang et al. 2020, Gomez et al. 2021). They also improve the bioavailability of micronutrients by influencing the cell signalling and rhizosphere function (Sun et al. 2020). In field studies, metal-based nanopesticides have been shown to increase the plant immunity besides controlling the soil-based plant pathogens (Elmer and White 2016). By varying the size, morphology, dissolution and composition of nanopesticides, besides suppressing the soil-borne pathogens, plant productivity can be increased.

Group two nanopesticides also increase the plant antioxidant defense system against diseases. Chitosan-coated mesoporous nanosilica has been shown to increase the fruit yield in watermelons (Buchman et al. 2019). Further, group two based nanopesticides have shown to have a more controlled release, making them more effective in lower doses for the control of insect pests (Chaud et al. 2021). Besides the above, group two nanopesticides are advantageous for the release of active ingredients based on the environmental parameters, for example, nanofungicide hymexazol when combined with temperature-sensitive polydompamine–grapheme based nanocarrier, releases more hymexazol at higher temperatures, which corresponds with the plant growth (Tong et al. 2018). Mesoporous nanosilica combined with 2,4-D nanoherbicide releases more when pH becomes alkaline (Cao et al. 2018). While comparing the two groups, each of the groups has its own advantages, like group one is more effective in the control of pathogens and insect pests, while group two is effective for the controlled release of active ingredients and is effective in a wide range of temperatures and pH. Nano-formulated pesticides are increasingly desired to control insect pests and plant disease with superior efficacy for guaranteeing high yield and quality in crop production. The shape of the nanocarriers determines the bioavailability, wash resistance and resistance to photo-degradation. One-dimensional worm-like nanopesticides exhibit an extremely high retention of 80% on leaves and a good resistance to photodegradation, while spherical nanocarriers and spherical micelles have a lower resistance (Yang et al. 2025).

### *3.3 Nanopesticide Formulations used in the Field*

The versatility of nanotechnology allows for the development of diverse nanopesticide formulations, including nanoemulsions, nanocapsules, nanospheres, and nanocomposites. Nanoemulsions, for example, are stable dispersions of oil and water phases stabilised by surfactants, offering improved solubility and bioavailability of lipophilic active ingredients. Nanocapsules consist of a core containing the active ingredient surrounded by a polymeric shell, providing controlled release and protection from degradation. Nanospheres are solid matrix systems where the active ingredient is dispersed or dissolved within the nanoparticle matrix, also offering controlled release properties. Nanocomposites involve the incorporation of active ingredients into nanoscale matrices made of materials like clays or carbon nanotubes, providing enhanced stability and targeted delivery potential (Figure 2).

There are several types of nanomaterials, which include nanoemulsions, nanopowders, essential oils, metal-organic frameworks (MOFs), nanomicelles, and nanospheres. These find applications in various fields. Essential oils are of biological origin, primarily from plants and find use in pharmaceuticals, beverages, and perfumes. Most of the nanopesticides in the market are in the form of nanoemulsions, consisting of an aqueous-aqueous mixture. Nanoemulsions are also used for drug delivery and in food preservation. Nanopowders, due to their larger surface area, can interact with biomolecules. So, these find applications in drug delivery, biomedical diagnostics and tissue engineering. Metal organic frameworks (MOFs) find wide application in the conjugation of chemical insecticides with metal ions for efficient delivery. These MOFs are increasingly being researched for preparing nanopesticides. Nanomicelles and nanospheres are other types of nanomaterials which find applications in drug delivery and targeted therapy (Figure 1).

Nanopesticides in either of the forms discussed above, due to their versatility, are used in a variety of ways for the control of insect pests. Arcot et al. (2024) have extensively reviewed the effect of foliar nanopesticides on plant cuticle and movement to insect cuticle. Soil drenching of nanopesticides is another way to enhance the systemic uptake by the plants (Ahmad et al. 2024). Seed coating protects the seeds from pest attack and also increases their germinating ability (do Espirito Santo Pereira et al. 2021), and systemic application/injection into trees has been done to protect them from fungal and nematode pathogens (Shangguan et al. 2024).

### *3.4 Chemical Insecticides used in Nanopesticides*

Nanoparticles and nanopolymers are also used for conjugating/loading chemical insecticides for use as nanopesticides. Chitosan NPs and silica NPs, loaded with deltamethrin, a synthetic pyrethroid, have been used for the control of mosquitoes (Khalifa et al. 2022). Cellulose network carrier-based nanopesticide loaded with selenium nanoparticles enables the precise release of pesticides under natural conditions, and further improves pesticide permeability and deposition in pine trees, helping in increasing the effective control of the nematode (Lin et al. 2024). Combining copper oxide nanoparticles with deltamethrin significantly enhanced the insecticide's efficacy against *B. oleae* adults (Malandrakis et al. 2024). The application of a chlorfenapyr-loaded iron-based metal organic framework to maize mixed with surfactant aerosol exhibits excellent wetting, deposition and adhesion with maize leaves, significantly improving insecticidal activity on *Spodoptera litura* (Chen et al. 2024). Polydopamine-capped cyclodextrin-modified abamectin-loaded hollow mesoporous silica nanoparticles displayed a significantly foliar affinity and improved rainfastness on lotus leaves. The results of the field trial demonstrated a significantly higher control efficacy against *Spodoptera litura* compared to abamectin (Li et al. 2024). Indendicarb encapsulated in star polymer nanocarrier increases the mortality at lower doses against *Plutella xylostella* and *Pieris rapae* (Chen et al. 2024). Graphene oxide (GO) and silver nanoparticle concentrations could bring about moderate changes in the digestive activity of *Acheta domesticus* (Alian et al. 2024). Combined use of graphene-based spinosad can reduce insecticide resistance in *Tribolium castaneum, Oryzaephilus surinamensis* and *Rhyzopertha dominica* (Lampiri et al. 2025)

Nanosuspensions having Chlorantraniliprole have higher bioavailability and are efficient in the control of *Cnaphalocrocis medinalis*, and at the same time have lower toxicity on non-target organisms (Ding et al. 2024). Graphene-loaded nanocarrier containing synthetic pyrethroids like lambda-cyhalothrin and cyfluthrin shows higher control efficiency as compared to conventional insecticides (Chen et al. 2022). Nanoformulations can improve efficacy, reduce effective doses, and extend shelf life, while controlled-release products can improve the delivery of pesticides (Mubeen et al. 2023). Microencapsulated lambda-cyhalothrin, have been shown to be very effective against sucking pests of crops and at the same time are relative non-toxic on soil invertebrate *Enchytraeus crypticus* (Gomes et al. 2023). Nanoencapsulated dimethoate has been shown to have reduced effects on zebrafish larvae (Vallim et al. 2022). The effectiveness and functionality of nanopesticides can be increased with precise design. Nanocarriers loaded with imidacloprid having rod-like shapes improve the stability, dispersibility, foliar wettability, deposition, retention and at the same time have higher toxicity on *Aphis gossypii* (Lin et al. 2025).

### *3.5 Botanicals used in Nanopesticides*

Biologically synthesized nanoparticles using microbial and plant secondary metabolites have been extensively studied for the control of insect pests and microbial pathogens of crops. The secondary metabolites of biological origin have been shown to be ideal for synthesizing nanopesticides, as their phytochemical groups act as reducing and stabilizing agents for reducing the metal oxides and metal salts (Baliyarsingh and Pradhan 2023). Silver nanoparticles synthesised using bacterial metabolites from bacteria *Xenorhabdus nematophila* have insecticidal activity against *Spodoptera litura* (Lalitha et al. 2022). Similarly, copper nanoparticles synthesised using bacterial metabolite from *Xenorhabdus* sp. show higher toxicity on *Spodoptera litura* larvae (Guru-Bharathi et al. 2022). Metallic silver and copper oxide nanoparticles coated with gibberellic acid inhibit the detoxification enzymes in brown plant hopper and also the hormone promotes paddy plant growth (Abou et al. 2022). Cry1Ac protein from *Bacillus thuringiensis* Kurstaki encapsulated with zeolite imidazolate framework-8 (ZIF-8) has improved stability to heat exposure, UV irradiation, internal absorption in maize leaves and long-term storage without affecting the toxicity on *Ostrinia furnacalis, Helicoverpa armigera* and *Spodoptera litura* (Lu et al. 2024). Novel green synthesized $NiFe_2O_4$ nanoparticles produce oxidative damage resulting in male reproductive system dysfunction in *Blaps polychresta* beetle (Arafat et al. 2025). Zinc oxide nanoparticles synthesised using *Eranthemum roseum* have been shown to have insecticidal activity on mosquito larvae (Adhavan et al. 2024). Fungal metabolites are also ideal for the synthesis of nanopesticides as their free functional groups have reducing properties. Fungal-synthesised titanium and zinc oxide bimetallic nanoparticles produce significant toxicity on *Spodoptera frugiperda* (Kumaravel et al. 2021). Selenium nanoparticles synthesised using *Trichoderma* sp. metabolites are toxic to *Spodoptera litura* larvae (Arunthirumeni et al. 2021). Bimetalic nanoparticles synthesised using cell-free extracts of *Penicillium citrinum* have higher insecticidal activity (Balumahendhiran et al. 2024).

In addition, biologically synthesized nanoparticles are effective in controlling stored grain insect pests like *Callosobruchus chinensis* and *Callosobruchus maculatus* (Helmy et al. 2023, San et al. 2023). Graphene formulations can provide long-lasting control of stored grain pests like *Sitophilus oryzae* and *Rhyzopertha dominica*, serving as an alternative to chemical insecticides (Lampiri et al. 2025). Copper/Graphene oxide core-shell nanoparticles have been effective in the management of ticks such as *Rhipicephalus rutilus* and *Rhipicephalus turanicus* (Senbill et al. 2025). Graphene powder-based contact nanopesticides provide effective control of stored grain insect pests like *S. oryzae, R. dominica* and *Prostephanus truncatus* (Lampiri et al. 2024). Metal organic framework loaded with emamectin benzoate has high nematocidal ability and can be used for the control of pine wilt disease caused by *Bursaphelenchus xylophilus* (Teng et al. 2025).

Metal nanoparticles also improve plant growth by improving their tolerance towards stress (Mahawar et al. 2023). Copper nanoparticles are effective antimicrobial agents for the control of diseases in crops (Liu et al. 2024). Hollow mesoporous organosilica nanoparticles loaded with abamectin have improved affinity to the leaf surface, as well as enhanced insecticidal activity against *Meloidogyne incognita* (Ma et al. 2023). Nanosuspensions of essential oils of *Eucalyptus globulus* and *Myristica fragrans* have been effectively used for the control of the termite *O. formosanus* (Nasser et al. 2024). Nanoemulsions using plant essential oils have higher mosquito control efficacy as compared to only essential oils (Vivekanandhan et al. 2023).

## 3.6 Target Pests of Nanopesticides

Nanopesticides are used for a plethora of pests and diseases due to their unique properties. There are chiefly four ways in which nanopesticides are used (i) nanocompounds directly being used as pesticides, (ii) active ingredients directly milled into a nanosuspension, (iii) pesticides delivered through nanocarriers and (iv) RNAi-based approach.

Silica and graphene are toxic to insects only in the form of nanoparticles. For instance, Silica nanoparticles, if sprayed on the plants, form a thin protective layer and prevent insect and larval development, while in adult insects they block the trachea, affecting respiration and causing mortality (Shoaib et al. 2018, Saw et al. 2023). Chlorantraniliprole nanosuspension shows high efficacy in the control of rice insect pests, while exhibiting safety towards Zebra fish (Ding et al. 2024). Similar studies conducted by Cui et al. (2019) showed high efficacy of abamectin nanosuspension against *Plutella xylostella*. Pesticides encapsulated in nano-formulations such as dendritic mesoporous silica nanoparticles enable sustained release and are biocompatible with pesticides. These have been used for the control of pine wood nematodes, grapevine fungal diseases, and aquatic weeds like water hyacinth (Wang et al. 2023, Chen et al. 2024, Fischer et al. 2019, Hou et al. 2021). Pesticides encapsulated in nano-structured matrix are designed using sensitive polymers that exhibit controlled release of pesticides, which are sensitive to pH, light, temperature, humidity or ions. Such formulations have been developed to control a wide array of pests like nematodes, lepidopterans, mites, grubs, fungal diseases like blight, wilt and weeds like (Shangguan et al. 2024, Chaud et al. 2021, Wang et al. 2022).

RNAi-based nanopesticides are under development and might become the next revolutionary technology in the crop protection industry. RNAi conjugated with nanoparticles can inhibit the expression of target genes through silencing. dsRNA, when applied topically on plant surfaces, has shown highly specific activity against target fungal pathogens and pests. However, dsRNA has a short period of environmental stability and variable uptake rates in pathogens and pests. Efforts are being made to use nanoparticles or nano-encapsulation techniques to stabilize the dsRNA upon internalization into insect pests. Nanocarriers can facilitate the release of RNA from the nanoparticles sensitive to changes in pH induced by pathogen invasion, and other such specific conditions enabling direct and targeted delivery of dsRNA to the pathogen. By the time this article is published, there is a possibility that this technology might get commercialized.

## 3.7 Mode of Translocation of Nanopesticides

Interaction of nanopesticides with plants, soil microbiota, aquatic non-target organisms and their environmental behavior are important aspects to be considered for their use in sustainable agriculture (Figure 3). When nanopesticides are applied by foliar spray or via soil, they enter the plants through the cuticle and epidermis (14–32 nm) and reach the mesophyll and phloem tissues (Yin et al. 2023). Plant hydathodes allow entry of large-sized nanoparticles (256–345 nm). Once inside, nanopesticides are transported by apoplastic and symplastic pathways. Within the cells smaller sized nanoparticles (5–20 nm) readily pass through the pores. While the large-sized nanopesticides are taken up by endocytotic pathways. Damaged cell junctions and tissues also allow the entry of nanopesticides into the tissues. Once inside the cell, the nanopesticides pass through and interact with mitochondria, chloroplasts, proteins and DNA and affect the physiological and Biochemical functioning of the cell (Li et al. 2020, Fincheira et al. 2020). Trophic transfer of $TiO_2$ nanopesticides in the algae–Daphnia food chain has been reported to increase with decreasing particle size (from 100 to 25 nm, and to 5 nm) (Chen et al. 2019).

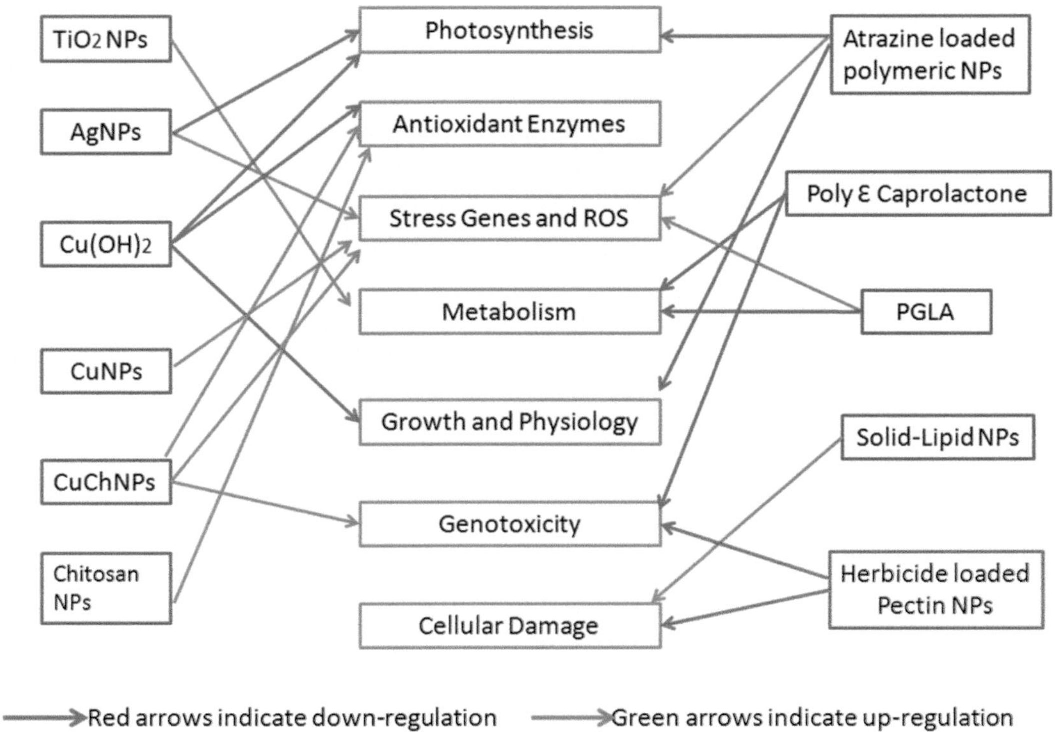


**Figure 3.** Mode of action of nanopesticides in plants.

### *3.8 Mechanism of Action of Nanopesticides on Insect Pests*

Nanopesticides enter the insect body in several ways, including through disturbance in the pigmentation of the cuticle, by the disruption of the laccase enzyme, inducing immune responses, producing metabolic alterations in carbohydrate, protein and lipid metabolism pathways, inducing cell toxicity and genotoxicity. They also impair development and reproduction in insects (Figure 4). Studies have shown that metal and metal oxide nanoparticles can themselves act as insecticides (Nie et al. 2023) or they can act as carriers of bioactive molecules (Shabatina et al. 2022). Silver and graphene nanoparticles affect the antioxidant and detoxification enzyme system (Benelli 2018). Silver nanoparticles affect acetylcholine esterase enzyme, which is an important enzyme for nerve conduction. Polystyrene NPs reduced the activity of CYP450 enzymes. Metal nanoparticles also embed themselves in S and P in Proteins and Nucleic acids, reducing the membrane permeability and promoting apoptosis (Shukla et al. 2021). Silver NPs are also known to reduce protein synthesis and lower gonadotropin-releasing hormone, thus interfering in development and reproduction (Chimkhan et al. 2022). At a cellular level, AgNPs can bioaccumulate within the cells and lead to apoptosis. In addition, they can damage the stratum corneum (Sultana et al. 2018). Silica and aluminium nanoparticles bind with the cuticle and cause damage. $TiO_2$ NPs cause oxidative stress (Demir et al. 2020) and result in DNA damage (Chen et al. 2022). Cadmium and Silver NPs in *Chironomus riparius* disrupt the expression of the GST enzyme system and reproductive-related genes (Nair and Choi 2011). SPc nanocomposites reduce the expression of cuticle-related genes (Aksakal and Arslan 2020) and increase endocytotic-related genes (Figure 4) (Zhang et al. 2022).

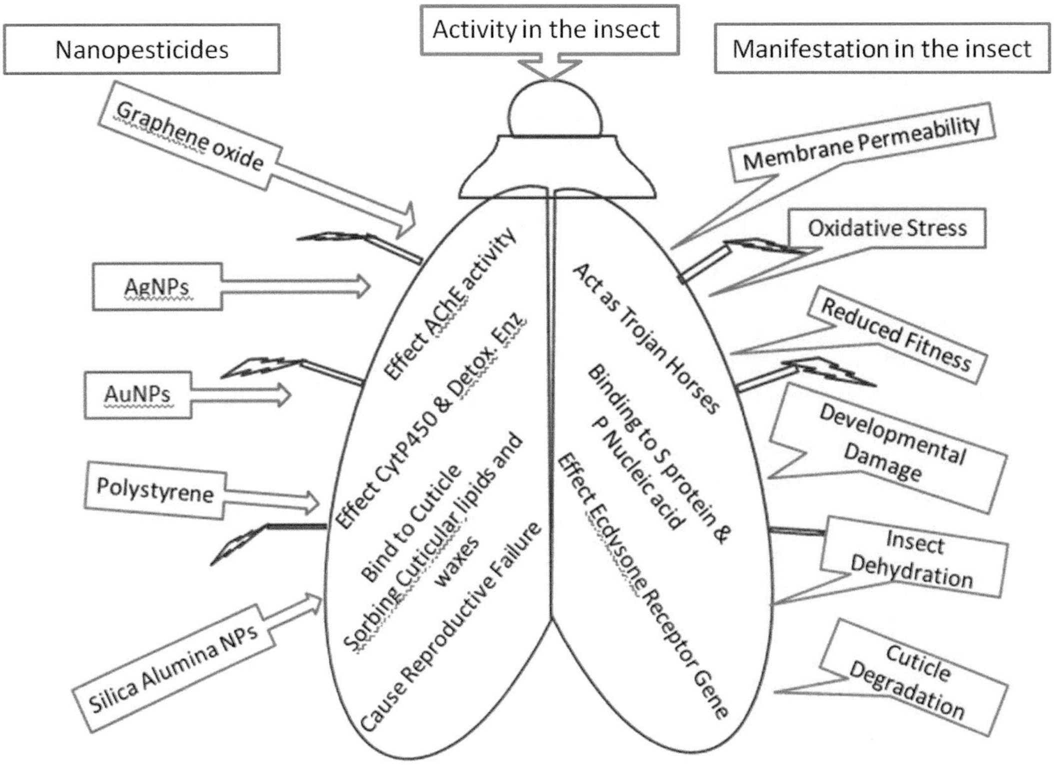


**Figure 4.** Mode of action nanopesticides in insects.

## 4. Environmental Toxicity and Concerns

The increasing use of these nanopesticides has raised environmental concerns as there are very few regulatory frameworks which address for truly assess the toxicity of these novel materials. The mode of action and environmental behavior of nanomaterials may differ significantly from those of conventional insecticides, highlighting the need for specific regulatory measures to address these concerns. The REACH framework for chemical assessment is considering amendments to its annexes to account for the unique properties of engineered nanomaterials (ENPs). Regarding pesticides, the Scientific Advisory Panel under the United States Federal Insecticide, Fungicide, and Rodenticide Act (FIFRA-SAP 2009) has also evaluated these issues to guide regulatory approaches. Similarly, the European union regulation 1107/2009 for marketing authorization of plant protection products in Europe also requires the novel characteristics of a nanopesticide be considered in the assessment process as it calls for due consideration of the interactions between the ai and safener (chemical substances used in combination with pesticides to make them safer and more targeted), synergists and co-formulations during the evaluation of plant protection products.

Based on the behaviour of the nanopesticides, Kookana et al. (2014) have categorized the nanopesticides into three categories, namely (i) In cases where the nanoformulation is used to enhance the apparent solubility or dispersion of the pesticide, or to protect it from degradation within the formulation, the nanocarrier may no longer remain associated with the active ingredient (AI) after application; (ii) where the ENP is utilized to alter either the toxicokinetics or toxicodynamics of the pesticide, and as such is associated with the AI at least until it reaches the site of action; and (iii) where either the ENP itself or the nanopesticide complex is itself the active ingredient. It is the last category that presents the most complex scenario from a regulatory perspective. In all cases, there exists also the potential for mixtures of "free" and nanoparticle-associated AI to coexist in the environment, the ratio of which may not be constant but change over time or as a function of the formulation and environmental conditions, thereby confounding measurements of efficacy and toxicity. Therefore, a key criterion for risk assessment purposes is the "durability" of the nano-ai complex following application of the nanopesticides

Adverse effects of nanopesticides are likely to occur on terrestrial and aquatic ecosystems. Studies have shown that copper-based nanopesticides on the marine benthic community leads to localised mechanical injury leads to physiological changes (Ho et al. 2018). Metal and metal oxide-based nanopesticides, besides being effective in the control of plant pathogens and insect pests, also have some detrimental effects on the soil biota. The deleterious effects of metal-based nanopesticides are more pronounced in the aquatic systems (Carley et al. 2020). Polymer-based nanopesticides have a lower non-target toxicity as compared to their metal counterparts, primarily due to their larger size (Guilger et al. 2017). Lambda-cyhalothrin nanocapsules synthesized using electrostatic self-assembly with iron mineralization increase the expression of detoxification, immune enzymes in honeybees and disrupt the gut community structure of honey bees. So, the potential impact of nanopesticides on beneficial organisms is essential (Guo et al. 2024). Metal oxide framework nanoparticles like Zeolitic Imidazolate Framework 8, 90, Cu-MOF, Fe-MOF, and Zr-MOF have differential effects on the histopathology and physiology of zebra fish. These metal oxide framework nanoparticles significantly altered the gut microbiome of zebra fish, suggesting that environmental effects of these nanoparticles should be considered (Yang et al. 2024). Ecotoxicity of nanoencapsulated imidacloprid on non-target freshwater organisms, embryonic *Danio rerio, Daphnia magna*, and *Chironomus kiinensis* shows that nanoencapsulated imidacloprid produced five times higher toxicity than imidacloprid. Ecotoxicological considerations are important (Wu et al. 2024). The toxicity of iron-based metal organic frameworks loaded with chlorfenapyr on earthworms using filter paper and soil environment bioassay shows higher contact toxicity than residual contact in soil. The gut microbiome was also significantly altered. Soil microbial communities, especially those participating in the nitrogen cycle, are severely affected due to MOF-loaded chlorfenapyr (Chen et al. 2024).

## *4.1 Comparison of Nanopesticides with Conventional Pesticides*

Nanopesticides are a safe and economically viable solution as compared to conventional insecticides (Dangi and Verma 2021). Nanopesticide formulation has improved stability, efficiency, dispersion, and regulated release. As a result, fewer pesticides may be applied to crops worldwide. In addition, an adequate dispersion of the active ingredient may lead to interactions between nanopesticides and non-target organisms, implying persistence (Yan et al. 2021). Though there are several advantages that exist with nanopesticides over conventional pesticides, such as improved stability, regulated active ingredient release, higher efficacy, reduced dose requirements, proper dispersion, and reduced residues; the disadvantages of conventional pesticides include high organic content, dust drift, poor dispersibility, capacity to survive for a long time in the soil (Zhao et al. 2017). Though nanopesticides are effective, there are still several scattered studies that have shown that nanopesticides also exhibit toxicity, and their impact on non-target organisms and the environment has not been fully appreciated. There is a lack of any comprehensive study on any of the nanopesticides. There are a few studies which shows that chronic bioaccmulation of nanopesticides in non target organisms can lead to genotoxicity, activation of gene expression of immune, developmental and reproduction related genes. On a cellular level, we find that nanopesticides, due to their charge and shape, can affect cellular transport and impair cell metabolic processes (Behzadi et al. 2017). In plants, it has been shown to affect the photosynthetic process and can bio-accumulate and has a chance of being biomagnified in the food chain. Comprehensive studies are needed to understand the fate of nanopesticides within biological systems and the environment.

Alternatively, active ingredients (AI) in the form of botanicals and biopesticides can be explored in place of chemical AIs. Plant and microbial diversity can be explored for identifying newer compounds having insecticidal properties. Further, other conventional approaches like the use of RNAi technology need to be developed and integrated with nanotechnology. There is a need to develop smart nanomaterials with reduced environmental footprints. These nanopesticides should have characteristics like being low-cost, easily removable from the environment, and integrating biotechnology (RNAi and CRISPR technologies) with nanotechnology. Further, there is a need to develop newer protocols for testing of nanoformulation before registration. Extensive toxicity testing is needed, which should include extra components or tests that are exclusive to nanopesticides. Strong regulatory guidelines should be prepared for the registration and release of nanomaterials in the environment.

## *4.2 Regulatory Issues and Public Perceptions*

Risk assessment for chemical pesticides is required prior to registration. In the case of nanopesticides, the risk assessment protocols of conventional pesticides are used. The agencies fail to take into consideration the unique properties of their nanoscale. FIFRA, a U.S.-based pesticide regulatory agency, states that a pesticide in its nanoscale also has potential health hazards and environmental risk, and conventional risk assessment protocols may not be sufficient (FIFRA-SAP 2010). In Europe, any chemical or nanomaterial manufactured or imported has to be certified by the EU's chemical legislation, REACH (REACH Regulation (EC) No. 1907/2006). This agency does risk assessment for workers and the environment.

Further, the registration of an AI does not extend to its nanoform and the individual components, like surfactants, stabilizers, are not tested individually or together (Kookana et al. 2014). In India, the government published guidelines for nanotechnology products in agriculture (India, Government 2019), but the framework for the same needs to be developed. Similarly, in Australia, the government published guidelines for nanoinsecticides, addressing risk assessment. As this topic is new, these guidelines are still in their discussion stage. These government discussions are also hindered due to a lack of data on the effect of nanopesticides on organisms and environment (Santiago et al. 2020). The physico-chemical properties of nanopesticides, like size, pH, and solubility, are different. This makes it difficult to identify their biokinetic behaviour and their movement in environment (Vryzas 2016). Hence, as per '*EFSA Opinion',* it is important that the risk assessment of nanomaterials should have been laid out (EFSA 2018).

As against conventional insecticides, risk assessment of nanopesticides requires additional data on physico-chemical characteristics of active ingredients and formulation, estimated concentration in environment, and ecotoxicology in different species. The current risk assessment strategy for conventional insecticides evaluates a single representative formulation of AI. This may not sufficiently address the effects of different dispersion or formulations used on the properties, behaviors, and toxicity effects of nanopesticides (EFSA 2018). Analytical methods for testing of nanoparticles in the bound, aggregated or agglomerated stages require sophisticated analytical instruments which may not be accessible for all laboratories. These are required for drawing meaningful information on fate, behaviour, mobility, persistence and ecotoxicology of conventional insecticides in its nanoform. KoW (octanol water partition coefficient) is a gold standard for partitioning nanoparticles into aqueous and organic phases. This may not be possible or nanoparticles. This can be a hindrance in risk assessment. In the absence of access to high-end instruments and a lack of adequate methods to address the risk posed by nanopesticides to health and environment, computational approaches can be used (Villaverde et al. 2017). *In silico* methods like Nano-QSAR and Nano-QSPR have been proposed (Abramenko et al. 2020). These models have chemical and nanodescriptions and ecotoxicological endpoints. The chemical descriptors include melting point, boiling point, vapor pressure, density, surface tension, partition coefficient, LD50, LC50, and qualitative descriptors like flammability, toxicity, explosive, mutagens, etc. These computation tools, though in their infancy, have to be rigorously tested and validated for application in risk assessment. Given the tremendous number of nanopesticide formulations in the market, this tool can help in faster assessment of risk and can be substantiated with wet lab protocols for additional confirmation.

## 5. Conclusion

Nanotechnology is at the forefront of developing nanopesticides and will continue to do so in the near future. Concerns regarding the environmental fate of nanopesticides, their mode of action in target insect species, and in non-target animals should be studied in depth. Such studies should be done in wide-ranging agroecosystems and environment for understanding the potential health and environmental impacts. A new regulatory framework should be enacted for the registration of nanopesticides due to their unique behaviour as compared to conventional insecticides. Bioinformatics and OMICS tools should be incorporated for understanding the toxicity of nanopesticides to provide quick insights regarding their environmental fate. Smart agriculture using AI technologies with smart nanopesticides is the future of sustainable crop protection. Safer and smarter nanopesticides, which are released in response to micro-environmental changes resulting from pest attack, will minimize the input of chemical insecticides. Newer technologies should be researched for integrating bioactive principles from plant and microbial sources for developing commercially viable insecticides. Nanocarriers from natural polymers such as chitosan, cellulose and zein, along with essential oils, will dominate the insecticide market in the near future and will have wider public acceptance as safer alternatives for integrated pest management.

The developments in nanotechnology have led to several new nanopesticides in the market, and this is growing at a fast pace. The size and charge of nanopesticides make them very active in integrating with biological molecules and in their movement in soil and water. The existing knowledge about the regulatory ecotoxicology of nanopesticides is few. There are several lacunae that need to be addressed quickly to make nanopesticides environmentally sustainable. Environmental fate and ecotoxicological effects of nanopesticides need to be studied thoroughly, which should include formulation, and other materials added to the nanopesticides should also be tested individually and in combination. Ageing of nanopesticides and their fate in soil and water should be studied in detail. New protocols for assessing risk and ecotoxicology should be formulated. Smart nanopesticides should be developed for addressing pest control and increasing agricultural productivity in a sustainable manner. Nanomaterials, which can be reused, for example, nanomaterials like magnetic nanoparticles, can be developed so that environmental bioaccumulation can be reduced. Nanoparticles used in genome editing technologies such as CRISPR and RNAi require further research to minimize their environmental footprint and unintended effects on non-target organisms.

## References

Abou El-Ela, A. S., Ntiri, E. S., Munawar, A., Shi, X. X., Zhang, C., Pilianto, J. et al. (2022). Silver and copper-oxide nanoparticles prepared with $GA_3$ induced defense in rice plants and caused mortalities to the brown planthopper, *Nilaparvata lugens* (Stål). NanoImpact. 28: 100428.

Abramenko, N., Kustov, L., Metelytsia, L., Kovalishyn, V., Tetko, I. and Peijnenburg, W. (2020). A review of recent advances towards the development of QSAR models for toxicity assessment of ionic liquids. Journal of Hazardous Materials 384: 121429

Adhavan, R., Selvam, K., Prakash, P., Manimegalai, P., Kirubakaran, D. and Shivakumar, M. S. (2024). Bioefficacy of Zinc oxide nanoparticle synthesis and their Biological, Environmental applications from *Eranthemum roseum*. Toxicology Reports 13: 101758.

Ahmad, A., Javad, S., Iqbal, S., Shahid, T., Naz, S., Shah, A. A. et al. (2024). Efficacy of soil drench and foliar application of iron nanoparticles on the growth and physiology of *Solanum lycopersicum* L. exposed to cadmium stress. Scientific Reports 14: 27920.

Aksakal, F. I. and Arslan, H. (2020). Detoxification and reproductive system-related gene expression following exposure to Cu(OH)(2) nanopesticide in water flea (*Daphnia magna* Straus 1820). Environmental Science and Pollution Research 27: 6103–6111.

Albalghiti, E., Stabryla, L. M., Gilbertson, L. and Zimmerman, J. (2020). Towards resolution of antibacterial mechanisms in metal and metal oxide nanomaterials: a meta-analysis of the influence of study design on mechanistic conclusions. Environmental Science Nano 8: 37–66.

Alexoaei, A. P., Robu, R. G., Cojanu, V., Miron, D. and Holobiuc, A. M. (2022). Good practices in reforming the common agricultural policy to support the european green deal—a perspective on the consumption of pesticides and fertilizers. Amfiteatru. Economy 24: 525–545.

Alian, S. R., Flasz, B., Kędziorski, A., Majchrzycki, Ł. and Augustyniak, M. (2024). Concentration- and time-dependent dietary exposure to graphene oxide and silver nanoparticles: effects on food consumption and assimilation, digestive enzyme activities, and body mass in Acheta domesticus. Insects 2024 Jan 29; 15(2): 89. doi: 10.3390/insects15020089.

An, C., Sun, C., Li, N., Huang, B., Jiang, J., Shen, Y. et al. (2022). Nanomaterials and nanotechnology for the delivery of agrochemicals: strategies towards sustainable agriculture. Journal of Nanobiotechnology. doi: 10.1021/acssuschemeng.9b04800.

Arafat, E. A., Eltaweil, A. S., Abd El-Monaem, E. M., Elhenawy, H. I., Hussein, H. K., El-Samad, L. M. et al. (2025). Fabrication of biosynthesized nickel ferrites nanoparticles and evaluation of their insecticidal efficacy on beetles (Blaps polychresta) testicular integrity. Scientific Reports 15: 7214.

Arcot, Y., Monica Iepure, Li Hao, Younjin Min, Spencer T. Behmer and Mustafa Akbulut. (2024). Interactions of foliar nanopesticides with insect cuticle facilitated through plant cuticle: Effects of surface chemistry and roughness-topography-texture. Plant Nano Biology 7: 100062. https://doi.org/10.1016/j.plana.2024.100062.

Arunthirumeni, M., Veerammal, V. and Shivakumar, M. S. (2021). Biocontrol efficacy of mycosynthesized selenium nanoparticle using Trichoderma sp. on insect pest *Spodoptera litura*. Journal of Cluster Science 33: 1645–1653.

Baliyarsingh, B. and Pradhan, C. K. (2023). Prospects of plant-derived metallic nanopesticides against storage pests—A review. Journal of Agriculture and Food Research 14: 2666–1543.

Balumahendhiran, K., Sharmila, M., Sethuraman, V., Arunthirumeni, M. and Shivakumar, M. S. (2024). Potential efficacy of bimetallic nanoparticles synthesized from extracellular cell-free extract of *Penicillium citrinum* against Culex quinquefasciatus mosquito larvae, and their photocatalytic activity. Inorganic and Nano-Metal Chemistry 55: 548–558.

Bavithra, M. L. C., Murugan, M., Pavithran, S. and Naveena, K. (2023). Enthralling genetic regulatory mechanisms meddling insecticide resistance development in insects: role of transcriptional and post-transcriptional events. Frontiers in Molecular Bioscience 10: 1257859.

Behzadi, S., Serpooshan, V., Tao, W., Hamaly, M. A., Alkawareek, M. Y., Dreaden, E. C. et al. (2017). Cellular uptake of nanoparticles: journey inside the cell. Chemical Society Reviews 46: 4218–4244.

Belagalla, N., Kaur, R. and Abhishek, G. (2024). Eco-friendly and targeted through next-generation approaches to insect pest management. Uttar Pradesh Journal of Zoology 45: 73–99.

Benelli, G. A. (2018). Mode of action of nanoparticles against insects. Environmental Science and Pollution Research 25: 12329–12341.

Bergeson, L. L. (2010). Nanosilver: US EPA's pesticide office considers how best to proceed? Environmental Quality Management 19: 79–85.

Buchman, J. T., Elmer, W., Ma, C., Landy, K., White, J. C. and Haynes, C. L. (2019). Chitosan-coated mesoporous silica nanoparticle treatment of *Citrullus lanatus* (watermelon): enhanced fungal disease suppression and modulated expression of stress-related genes. ACS Sustainable Chemical Engineering 7: 19649–19659.

Cao, L., Zhou, Z., Niu, S., Cao, C., Li, X., Shan, Y. et al. (2018). Positive-charge functionalized mesoporous silica nanoparticles as nanocarriers for controlled 2,4-dichlorophenoxy acetic acid sodium salt release. Journal of Agricultural and Food Chemistry 66: 6594–6603.

Carley, L. N., Panchagavi, R., Song, X., Davenport, S., Bergemann, C. M., McCumber, A. W. et al. (2020). Long-term effects of copper nanopesticides on soil and sediment community diversity in two outdoor mesocosm experiments. Environmental Science Technology 54: 8878–8889.

Chaud, M., Souto, E. B., Zielinska, A., Severino, P., Batain, F., Oliveira-Junior, J. et al. (2021). Nanopesticides in agriculture: benefits and challenge in agricultural productivity, toxicological risks to human health and environment. Toxics 9: 131.

Chen, H., Yang, L., Zhao, S., Xu, H. and Zhang, Z. (2024). Long-term toxic effects of iron-based metal-organic framework nanopesticides on earthworm-soil microorganism interactions in the soil environment. Science of Total Environment 917: 170146.

Chen, H., Yang, L., Wu, P., Liu, P., Xu, H. and Zhang, Z. (2024). Combined application of surfactants and iron-based metal–organic framework nanoparticles for targeted delivery of insecticides. Chemical Engineering Journal 488: 151193. doi.org/10.1016/j.cej.2024.151193.

Chen, L., Hu, J., Pang, H., Yin, S., Liu, H., Tian, Y. et al. (2024). Sustainable DMSNs nano-biopesticide platform built by a "one-pot" method focusing on injury-free drug demonstration of pine wood nematodes. Environmental Science Nano 11: 363–372.

Chen, M., Zhang, J., Wang, H., Li, L., Yin, M., Shen, J. et al. (2024). Preparation of nanoscale indoxacarb by using star polymer for efficiency pest management. Agriculture 14: 1006.

Chen, X., Zhu, Y., Yang, K., Zhu, L. and Lin, D. (2019). Nanoparticle $TiO_2$ size and rutile content impact bioconcentration and biomagnification from algae to daphnia. Environmental Pollution 247: 421–430.

Chen, Z., Shi, J., Zhang, Y., Han, S., Zhang, J. and Jia, G. (2022). DNA oxidative damage as a sensitive genetic endpoint to detect the genotoxicity induced by titanium dioxide nanoparticles. Nanomaterials 12: 2616.

Chen, Z., Zhao, J., Liu, Z., Bai, X., Li, W., Guan, Z. et al. (2022). Graphene-delivered insecticides against cotton bollworm. Nanomaterials (Basel) 12: 2731.

Chimkhan, N., Thammasittirong, S. N. R., Roytrakul, S., Krobthong, S. and Thammasittirong, A. (2022). Proteomic response of *Aedes aegypti* larvae to silver/silver chloride nanoparticles synthesized using *Bacillus thuringiensis* subsp. *Israelensis* Metabolites. Insects 13: 641.

Cui, B., Lv, Y., Gao, F., Wang, C., Zeng, Z., Wang, Y. et al. (2019). Improving abamectin bioavailability via nanosuspension constructed by wet milling technique. Pest Management Science 75: 2756–276.

Dangi, K. and Verma, A. K. (2021). Efficient & eco-friendly smart nano-pesticides: Emerging prospects for agriculture. Materials Today: Proceedings 45: 3819–3824.

De, R. B., Kumar, A. and Mozumdar, S. (2014). Management of insect pests using nanotechnology: as modern approaches. pp. 29–32. *In*: De, A., Bose, R., Kumar, A. and Mozumdar, S. (eds.). Targeted Delivery of Pesticides Using Biodegradable Polymeric Nanoparticles. Springer: New Delhi.

Deligeorgakis, C., Magro, C., Skendi, A., Gebrehiwot, H. H., Valdramidis, V. and Papageorgiou, M. (2023). Fungal and toxin contaminants in cereal grains and flours: systematic review and meta-analysis. Foods 12: 4328.

Demir, E. (2020). Drosophila as a model for assessing nanopesticide toxicity. Nanotoxicology 14: 1271–1279.

Dhivya, K., Vengateswari, G., Arunthirumeni, M., Karthi, S., Senthil-Nathan, S. and Shivakumar, M. S. (2018). Bioprospecting of *Prosopis juliflora* (Sw.) DC seed pod extract effect on antioxidant and immune system of *Spodoptera litura* (Lepidoptera: Noctuidae). Physiological Molecular Plant Pathology 101: 45–53.

Ding, X., Guo, L., Du, Q., Wang, T., Zeng, Z., Wang, Y. et al. (2024). Preparation and comprehensive evaluation of the efficacy and safety of chlorantraniliprole nanosuspension. Toxics 12: 78. doi.org/10.3390/toxics12010078.

do Espirito, S. P. A., Oliveira, C. H., Fraceto, F. L. and Santaella, C. (2021). Nanotechnology potential in seed priming for sustainable agriculture. Nanomaterials (Basel) 11: 267.

Dougoud, J. S., Toepfer, M. and Jenner, B. W. (2019). Efficacy of homemade botanical insecticides based on traditional knowledge. A review. Agronomy Sustainable Development 39: 37.

EC. (2009). Regulation (EC) No. 1107/2009 of the European Parliament and the Council of 21 October 2009 concerning the placing of plant protection products on the market and repealing Council Directives 79/117/EEC and 91/414/EEC. Off. J. Eur. Union 309: 1–50.

EFSA (European Food Safety Authority). (2018). Guidance on risk assessment of the application of nanoscience and nanotechnologies in the food and feed chain: Part 1, human and animal health. EFSA Journal 16: 5327.

El-Latef, A., Adel, E., Wahba, M., Mousa, S., El-Bassyouni, G. and El-Shamy, A. (2023). Cu-doped ZnO-nanoparticles as a novel eco-friendly insecticide for controlling *Spodoptera littoralis*. Biocatalysis and Agricultural Biotechnology 52: 102823. doi:10.1016/j.bcab.2023.102823.

Elmer, W. H. and White, J. C. (2016). The use of metallic oxide nanoparticles to enhance growth of tomatoes and eggplants in disease infested soil or soilless medium. Environmental Science Nano 3: 1072–1079.

European Food Safety Authority. (2013). Panel on Plant Protection Products and their Residues. DRAFT Guidance Document on tiered risk assessment for plant protection products for aquatic organisms in edge-of-field surface waters. EFSA Journal 11: 3290.

FAO. (2019). The State of Food and Agriculture 2019. Moving Forward on Food Loss and Waste Reduction. Rome.

FAO. (2024). Pesticides use and trade – 1990–2022. FAOSTAT Analytical Briefs, No. 89. Rome. https://doi.org/10.4060/cd1486en.

FIFRA-SAP. (2009). Evaluation of the hazard and exposure associated with nanosilver and other nanometal pesticide products. U.S. Federal Insecticide, Fungicide, and Rodenticide Act (FIFRA) Scientific Advisory Panel (SAP); http://www.epa.gov/scipoly/sap/meetings/2009/November/110309ameetingminutes.pdf.

FIFRA-SAP. (2010). Evaluation of the hazard and exposure associated with nanosilver and other nanometal pesticide products. U.S. Federal Insecticide, Fungicide, and Rodenticide Act (FIFRA) Scientific Advisory Panel (SAP); SAP Minutes 2010–01. http://www.epa.gov/scipoly/sap/meetings/2009/november/110309ameetingminutes.pdf.

Fincheira, P., Tortella, G., Duran, N., Seabra, A. B. and Rubilar, O. (2020). Current applications of nanotechnology to develop plant growth inducer agents as an innovation strategy. Critical Reviews Biotechnology 40: 15–30.

Fischer, J., Beckers, S. J., Yiamsawas, D., Thines, E., Landfester, K. and Wurm, F. R. (2019). Targeted drug delivery in plants: enzyme-responsive lignin nanocarriers for the curative treatment of the worldwide grapevine trunk disease esca. Advanced Science 6: 1802315.

Geraldin, M., Lengai, W., Muthomi and Mbega, E. (2020). Phytochemical activity and role of botanical pesticides in pest management for sustainable agricultural crop production. Scientific African 7: e00239. doi.org/10.1016/j.sciaf.2019.e00239.

Gogos, A., Knauer, K. and Bucheli, T. (2012). Nanomaterials in plant protection and fertilization: current state, foreseen applications, and research priorities. Journal of Agricultural and Food Chemistry 60: 9781–9792.

Gomes, S. I. L., Chidiamassamba, S. B., Scott-Fordsmand, J. J. and Amorim, M. J. B. (2023). Environmental hazards of nanopesticides to non-target soil species - commercial nanoformulation versus its active substance (Karate Zeon® and lambda-cyhalothrin). Science of Total Environment 891: 164664.

Gomez, A., Narayan, M., Zhao, L., Jia, X., Bernal, R. A., Lopez-Moreno, M. L. et al. (2021). Effects of nano-enabled agricultural strategies on food quality: Current knowledge and future research needs. Journal of Hazardous Material 401: 123385.

Guilger, M., Pasquoto-Stigliani, T., Bilesky-Jose, N., Grillo, R., Abhilash, P. C., Fraceto, L. F. et al. (2017). Biogenic silver nanoparticles based on trichoderma harzianum: synthesis, characterization, toxicity evaluation and biological activity. Scientific Reports 7: 44421.

Guo, D., Li, Z., Zhang, Y., Zhang, W., Wang, C., Zhang, D. X. et al. (2024). The effect of lambda-cyhalothrin nanocapsules on the gut microbial communities and immune response of the bee elucidates the potential environmental impact of emerging nanopesticides. Journal of Hazardous Material 479: 135650.

Guru-Bharathi, B., Lalitha, K. and Shivakumar, M. S. (2022). Biosynthesis of copper nanoparticles using symbiotic bacterium Xenorhabdus sp, isolated from entomopathogenic nematode and its antimicrobial and insecticidal activity against *Spodoptera litura.* Inorganic Nano-Metal Chemistry 54: 858–870.

Gutiérrez-Ramírez, J. A., Betancourt-Galindo, R., Aguirre-Uribe, L. A., Cerna-Chávez, E., Sandoval-Rangel, A., Ángel, E. C. et al. (2021). Insecticidal effect of zinc oxide and titanium dioxide nanoparticles against Bactericera cockerelli Sulc. (Hemiptera: Triozidae) on Tomato Solanum lycopersicum. Agronomy 11: 1460.

Hardy, A., Benford, D., Halldorsson, T., Jeger, M. J., Knutsen, H. K., More, S. et al. (2018). Guidance on risk assessment of the application of nanoscience and nanotechnologies in the food and feed chain: Part 1, human and animal health. EFSA Journal 16: e05327.

Helmy, E. A. M., San, P. P., Zhang, Y. Z., Adarkwah, C. and Tuda, M. (2023). Entomotoxic efficacy of fungus-synthesized nanoparticles against immature stages of stored bean pests. Scientific Reports 13: 8508. doi: 10.1038/s41598-023-35697-1.

Ho, K. T., Portis, L., Chariton, A. A., Pelletier, M., Cantwell, M., Katz, D. et al. (2018). Effects of micronized and nano-copper azole on marine benthic communities. Environmental Toxicology and Chemistry 37: 362–375.

Hou, R., Wu, J., Yang, L., Zhao, K., Huang, S., Kaziem, A. E. et al. (2021). Preparation of alginate-chitosan floating granules loaded with 2-methyl-4-chlorophenoxy acetic acid (MCPA) and their bioactivity on water hyacinth. Pest Management Science 77: 3942–3951.

India, Government. (2019). Guidelines for Evaluation of Nano-Agri input Products and Nano- Nano-Agriproducts, 1. Department of Biotechnology, Ministry of Science & Technology, Government of India, India.

Isman, M. B. (2020). Botanical insecticides in the twenty-first century- fulfilling their promise? Annual Review of Entomology 65: 233–249.

Jaba, J. and Sharma, P. (2017). Advance towards host mediated RNA interference insect pest management. Advances in Biotechnology and Microbiology 5: 1–6.

Jasrotia, P., Nagpal, M., Mishra, C., Sharma, A., Kumar, S., Umesh, R. et al. (2022). Nanomaterials for postharvest management of insect pests: current state and future perspectives. Frontiers in Nanotechnology. https://doi.org/10.3389/fnano.2021.811056.

Kadir, M. L., Dageri, A. and Aslan, T. N. (2025). Nanopesticides for managing primary and secondary stored product pests: Current status and future directions, Heliyon 11: e42341.

Kah, M., Beulke, S., Tiede, K. and Hofmann, T. (2013). Nano-pesticides: state of knowledge, environmental fate and exposure modelling. Critical Reviews in Environmental Science and Technology 43: 1823–1867.

Kah, M., Kookana, R. S., Gogos, A. and Bucheli, T. D. (2018). A critical evaluation of nanopesticides and nanofertilizers against their conventional analogues. Nature Nanotechnology 13: 677–684.

Kapinder, K., Dangi, A. and Verma, K. (2021). Efficient and eco-friendly smart nano-pesticides: Emerging prospects for agriculture. Materials Today: Proc. 45: 3819–3824.

Kariyanna, B. and Sowjanya, M. (2024). Unravelling the use of artificial intelligence in management of insect pests. Smart Agricultural Technology 8: 100517.

Khalifa, A. G., Moselhy, W. A., Mohammed, H. M., Khalil, F., Shaban, M., El-Nahass, E. S. et al. (2022). Deltamethrin and its nanoformulations induce behavioral alteration and toxicity in rat brain through oxidative stress and JAK2/STAT3 signaling pathway. Toxics 10: 303.

Khot, L. R., Sankaran, S., Maja, J. M., Ehsani, R. and Schuster, E. (2012). Application of nanomaterials in agricultural production and crop protection: a review. Crop Protection 35: 64–70.

Kishore, S. M., Sujith, K. M., Shalini, K. S. and Bharthisha, S. M. (2024). Navigating the threat: invasive insect pests and their impact on indian agriculture. Annual Research Review in Biology 39: 50–59.

Kookana, R. S., Boxall, A. B., Reeves, P. T., Ashauer, R., Beulke, S., Chaudhry, Q. et al. (2014). Nanopesticides: guiding principles for regulatory evaluation of environmental risks. Journal of Agricultural and Food Chemistry 62(19): 4227–4240. https://doi.org/10.1021/jf500232f.

Kumaravel, J., Lalitha, K., Arunthirumeni, M. and Shivakumar, M. S. (2021). Mycosynthesis of bimetallic zinc oxide and titanium dioxide nanoparticles for control of Spodoptera frugiperda. Pesticide Biochemistry and Physiology 178: 104910.

Lalitha, K., Nithya, K., Venkatesan, S. and Shivakumar, M. S. (2022). Biosynthesis and characterization of silver nanoparticles from symbiotic bacteria Xenorhabdus nematophila and testing its insecticidal efficacy on Spodoptera litura larvae. Biometals 35: 795–812.

Lampiri, E., Losic, D. and Athanassiou, C. G. (2025). Long-term effectiveness of two graphene formulations against two major stored product insects. Journal of Stored Product Research 112: 102631.

Lampiri, E., Losic, D. and Athanassiou, C. G. (2025). Synergetic enhancement of insecticidal efficacy of graphene with spinosad against major stored grain insects, Journal of Stored Product Research 112: 102623.

Lampiri, E., Yap, P. L., Berillis, P., Athanassiou, C. G. and Losic, D. (2024). Graphene powders as new contact nanopesticides: Revealing key parameters on their insecticidal activity for stored product insects. Chemosphere 364: 143200.

Lengai, G. M. W., Muthomi, W. and Mbega, E. R. (2020). Phytochemical activity and role of botanical pesticides in pest management for sustainable agricultural crop production Scientific African 7: e00239, https://doi.org/10.1016/j.sciaf.2019.e00239.

Li, J., Li, D., Zhang, Z., Yu, C., Sun, D., Mo, Z. et al. (2024). Smart and sustainable crop protection: design and evaluation of a novel α-amylase-responsive nanopesticide for effective pest control. Journal of Agriculture and Food Chemistry 72: 12146–12155.

Li, L., Luo, Y., Li, R., Zhou, Q., Peijnenburg, W. J. G. M., Yin, N. et al. (2020). Effective uptake of submicrometre plastics by crop plants via a crack-entry mode. Nature Sustainability 3: 929–937.

Li, X., Chen, Y., Xu, J., Lynch, I., Guo, Z., Xie, C. et al. (2023). Advanced nanopesticides: Advantage and action mechanisms. Plant Physiology and Biochemistry 203: 108051.

Lin, H., Ning, L., Dai, T., Wei, W., Ma, N., Hong, M. et al. (2024). Application of bio-based nanopesticides for pine surface retention and penetration, enabling effective control of pine wood nematodes. Chemical Engineering Journal 502: 158009,

Lin, H., Ning, L., Wei, W., Ji, X., Wang, F. and You, C. (2025). Chitosan- and sodium alginate-coated dendritic mesoporous organosilica nanoparticles improve pesticide adhesion on leaves and enable dual-stimulus-responsive release. International Journal of Biological Macromolecules 299: 140211.

Liu, S., Tian, W., Liu, Z., Wei, X., Yuan, K., Du, W. et al. (2024). Biosynthesis of cupric oxide nanoparticles: its antiviral activities against TMV by directly destroying virion and inducing plant resistance. Phytopathology Research 6: 30.

Lu, Y., Sui, L., Dai, C., Zheng, W., Zhao, Y., Li, Q. et al. (2024). Immobilization of *Bacillus thuringiensis* Cry1Ac in metal-organic frameworks through biomimetic mineralization for sustainable pest management. International Journal of Biological Macromolecules 274: 133388.

Ma, C., Borgatta, J., Hudson, B. G., Tamijani, A. A., De La Torre-Roche, R., Zuverza-Mena, N. et al. (2020). Advanced material modulation of nutritional and phytohormone status alleviates damage from soybean sudden death syndrome. Nature Nanotechnology 15: 1033–1042.

Ma, E., Fu, Z., Sun, L., Chen, K., Liu, Z., Wei, Z. et al. (2023). Organosilica-based deformable nanopesticides with enhanced insecticidal activity prepared by flash nanoprecipitation. Reaction Chemical Engineering 8: 1457–1463.

Mahawar, L., Ramasamy, K. P., Suhel, M., Prasad, S. M., Živčák, M., Brestic, M. et al. (2023). Silicon nanoparticles: Comprehensive review on biogenic synthesis and applications in agriculture. Environmental Research 232: 116292.

Malandrakis, A. A., Varikou, K., Kavroulakis, N., Nikolakakis, A., Dervisi, I., Reppa, C. I. et al. (2024). Copper nanoparticles interfere with insecticide sensitivity, fecundity and endosymbiont abundance in olive fruit fly Bactrocera oleae (Diptera: Tephritidae). Pest Management Science 80: 3640–3649.

Manimegalai, P., Selvam, K., Loganathan, S., Kirubakaran, D., Shivakumar, M. S., Govindsamy, M. et al. (2023). Green synthesis of zinc oxide (ZnO) nanoparticles using aqueous leaf extract of Hardwickia binata: their characterizations and biological applications. Biomass Conversion and Biorefinery 14: 12559–12574.

Manjula, P., Lalitha, K. and Shivakumar, M. S. (2020). Effect of Manihot esculenta (Crantz) leaf extracts on Antioxidant and Immune system of Spodoptera litura (Lepidoptera: Noctuidae). Biocatalysis and Agricultural Biotechnology. doi.org/10.1016/j.bcab.2019.101476.

Marslin, G., Siram, K., Maqbool, Q., Selvakesavan, R. K., Kruszka, D., Kachlicki, P. et al. (2018). Secondary metabolites in the green synthesis of metallic nanoparticles. Materials (Basel) 11: 940.

Mishra, J., Dutta and Arora, N. K. (2020). Biopesticides in India: technology and sustainability linkages. 3 Biotech 10: 10.

Mohanraj, V. J. and Chen, Y. (2006). Nanoparticles—a review. Trop. J. Pharm. Res. 5: 561–573.

Mota-Sanchez, D. and Wise, J. C. (2020). Arthropod Pesticide Resistance Database (ARPD), Michigan State University. http://www.pesticideresistance.org/.

Mubeen, I., Mfarrej, F. B. M., Razaq, Z., Iqbal, S., Naqvi, S. A. H., Hakim, F. et al. (2023). Nanopesticides in comparison with agrochemicals: Outlook and future prospects for sustainable agriculture. Plant Physiology and Biochemistry 198: 107670.

Muthusamy, R. and Shivakumar, M. S. (2015). Involvement of metabolic resistance and F1534C *kdr* mutation in the pyrethroid resistance mechanisms of *Aedes aegypti* in India. Acta Tropica 148: 137–141.

Muthusamy, R. and Shivakumar, M. S. (2015). Resistance selection and molecular mechanisms of cypermethrin resistance in red hairy caterpillar (*Amsacta albistriga* walker). Pesticide Biochemistry and Physiology 117: 54–61.

Nair, P. M. G. and Choi, J. (2011). Identification, characterization and expression profiles of Chironomus riparius glutathione S-transferase (GST) genes in response to cadmium and silver nanoparticles exposure. Aquatic Toxicology 101: 550–560.

Nasser, R., Ibrahim, E., Fouad, H., Ahmad, F., Li, W., Zhou, Q. et al. (2024). Termiticidal, biochemical, and morpho-histological effects of botanical based nanoemulsion against a subterranean termite, *Odontotermes Formosanus* Shiraki. Frontiers in Plant Science 14: 1292272.

Nie, D., Li, J., Xie, Q., Ai, L., Zhu, C., Wu, Y. et al. (2023). Nanoparticles: A potential and effective method to control insect-borne diseases. Bioinorganic Chemistry Application 2023: 5898160.

Nouri-Ganbalani, G., Borzoui, E., Abdolmaleki, A., Abedi, Z. and Kamita, S. G. (2016). Individual and combined effects of Bacillus Thuringiensis and Azadirachtin on Plodia Interpunctella Hübner (Lepidoptera: Pyralidae), Journal of Insect Science 16: 95.

Parajuli, S., Shrestha, J., Subedi, S. and Pandey, M. (2022). Biopesticides: a sustainable approach for pest management. SAARC Journal of Agriculture. doi.org/10.3329/sja.v20i1.60526.

Pasquoto-Stigliani, T., Campos, E. V. R., Oliveira, J. L., Silva, C. M. G., Bilesky-José, N., Guilger, M. et al. (2017). Nanocapsules containing Neem (Azadirachta Indica) oil: Development, characterization, and toxicity evaluation. Scientific Reports 7: 5929. doi: 10.1038/s41598-017-06092-4.

Patel, A., Chandra, R. and Upadhyay, M. (2018). Efficacy of different fungicides against *Alternaria brassicae* caused Alternaria leaf spot of cauliflower. The Pharma Innov. J. 7: 231–234.

Pathak, V. M., Verma, V. K., Rawat, B. S., Kaur, B., Babu, N., Sharma, A. et al. (2022). Current status of pesticide effects on environment, human health and it's eco-friendly management as bioremediation: A comprehensive review. Frontiers in Microbiology 13: 962619.

Pu, J. and Chung, H. (2024). New and emerging mechanisms of insecticide resistance. Current Opinion in Insect Science 63: 101184.

Ramkumar, G., Karthi, S., Muthusamy, R., Suganya, P., Natarajan, D., Kweka, E. J. et al. (2016). Mosquitocidal effect of Glycosmis pentaphylla leaf extracts against three Mosquito Species (Diptera: Culicidae). PLoS ONE 11(7): e0158088.

San, P. P., Yao Zhuo Zhang, Eman Ahmed Mohamed Helmy, Charles Adarkwah and Midori Tuda. (2023). Sex-dependent effects of biosynthesized nanoparticles on stored bean pests and their non-target parasitoid. Crop Protection 173: 106352. https://doi.org/10.1016/j.cropro.2023.106352.

Santiago, E. F., Pontes, M. S., Arruda, G. J., Caires, A. R. L., Colbeck, I., MaldonadoRodriguez, I. et al. (2020). Understanding the interaction of nanopesticides with plants. pp. 69–109. *In*: Fraceto, L. F., de Castro, V. L. S. S., Grillo, R., Avila, D., Oliveira, H. C. and Lima, R. (eds.). Nanopesticides: From Research and Development to Mechanisms of Action and Sustainable Use in Agriculture. Springer Nature, Switzerland.

Sarika Chaturvedi and Shikha Khandelwal. (2019). A Review on comparative analytical approach towards Chemical pesticides to Bio pesticides. Int. Res. J. Pharm. 10(9): 1–12. http://dx.doi.org/10.7897/2230-8407.1009254.

Savary, S., Willocquet, L., Pethybridge, S. J., Esker, P., McRoberts, N. and Nelson, A. (2019). The global burden of pathogens and pests on major food crops. Nature Ecology and Evoltion 3: 430–439.

Saw, G., Nagdev, P., Jeer, M. and Murali-Baskaran, R. K. (2023). Silica nanoparticles mediated insect pest management. Pesticide Biochemistry and Physiology 194: 105524. https://doi.org/10.1016/j.pestbp.2023.105524.

Senbill, H., Gangan, A., Saeed, A. M., Gad, M. E., Zeb, J. and Fahmy, A. (2025). Effects of copper/graphene oxide core-shell nanoparticles on Rhipicephalus ticks and their detoxification enzymes. Scientific Reports 15: 3334. doi: 10.1038/s41598-025-86560-4.

Shaalan, E. A., Canyon, D., Younes, M. W. F., Abdel-Wahab, H. and Mansour, A. H. (2005). A review of botanical phytochemicals with mosquitocidal potential. Environment International 31: 1149–1166.

Shabatina, T., Vernaya, O., Shumilkin, A., Semenov, A. and Melnikov, M. (2022). Nanoparticles of bioactive metals/metal oxides and their nanocomposites with antibacterial drugs for biomedical applications. Materials 15: 3602.

Shangguan, W., Chen, H., Zhao, P., Cao, C., Yu, M., Huang, Q. et al. (2024). Scenario-oriented nanopesticides: Shaping nanopesticides for future agriculture. Advanced Agrochemicals 3: 265–278.

Shashi Prabha, Ashwani Yadav, Ashwani Kumar, Anuj Yadav, Hemant Kumar Yadav, Sujit Kumar, R. S. et al. (2016). Biopesticides—an alternative and eco-friendly source for the control of pests in agricultural crops. Plant Archieves 16(2): 902–906.

Shoaib, A., Elabasy, A., Waqas, M., Lin, L., Cheng, X., Zhang, Q. et al. (2018). Entomotoxic effect of silicon dioxide nanoparticles on Plutella xylostella (L.) (Lepidoptera: Plutellidae) under laboratory conditions. Toxicology and Environmental Chemistry 100: 80–91.

Shudeer, Balaji, B. N. and Pandit. 2024. Insecticide resistance management: a detailed review. Uttar Pradesh Journal of Zoology 45(17): 513–521. https://doi.org/10.56557/upjoz/2024/v45i174395.

Shukla, R. K., Badiye, A., Vajpayee, K. and Kapoor, N. (2021). Genotoxic potential of nanoparticles: structural and functional modifications in DNA. Frontiers in Genetics 12: 728250.

Singh, P., Kim, Y.-J., Zhang, D. and Yang, D.-C. (2016). Biological synthesis of nanoparticles from plants and microorganisms. Trends Biotechnol. 34(7): 588599.

Sparks, T. C., Storer, N., Porter, A., Slater, R. and Nauen, R. (2021). Insecticide resistance management and industry: the origins and evolution of the Insecticide Resistance Action Committee (IRAC) and the mode of action classification scheme. Pest Management Science 77: 2609–2619.

Srinivasan, R., Aruna, A., Manigandan, K., Pugazhendhi, A., Kima, M., Shivakumar, M. S. et al. (2019). Phytochemical, antioxidant, antimicrobial and antiproliferative potential of Elaeagnus indica. Biocatalysis and Agricultural Biotechnology 20: 101265.

Sultana, N., Raul, P. K., Goswami, D., Das, B., Gogoi, H. K. and Srinivas Raju, P. (2018). Nanoweapon: control of mosquito breeding using carbon-dot-silver nanohybrid as a biolarvicide. Environmental Chemistry Letters 16: 1017–1023.

Sun, L., Wang, Y., Wang, R., Zhang, P., Ju, Q. and Xu, J. (2020). Physiological, transcriptomic and metabolomic analyses reveal zinc oxide nanoparticles modulate plant growth in tomato. Environmental Science Nano 7: 3587–3604.

Teng, F., Wang, Z., Xue, M., Li, H., Wen, Q., Xu, P. et al. (2025). Development of metal-organic framework-based Emamectin Benzoate nanopesticides for effective control of Bursaphelenchus xylophilus. Langmuir 41: 7455–7466.

Tong, Y., Shao, L., Li, X., Lu, J., Sun, H., Xiang, S. et al. (2018). Adhesive and stimulus-responsive polydopamine-coated graphene oxide system for pesticide-loss control. Journal of Agriculture and Food Chemistry 66: 2616–2622.

US EPA. (2022). Balancing Wildlife Protection and Responsible pesticide use: How EPA's program will meet its endandered act obligations (www.epa.gov/endangered-species). https://www.epa.gov/ingredients-used-pesticide-products/what-are-biopesticides.

Vallim, J. H., Clemente, Z., Castanha, R. F., do Espírito Santo Pereira, A., Campos, E. V. R., Assalin, M. R. et al. (2022). Chitosan nanoparticles containing the insecticide dimethoate: A new approach in the reduction of harmful ecotoxicological effects. NanoImpact. 27: 100408. doi: 10.1016/j.impact.2022.100408.

Villaverde, J. J., Sevilla-Moran, B., Lopez-Goti, C., Alonso-Prados, J. L. and Sandin-Espana, P. (2017). Computational methodologies for the risk assessment of pesticides in the European Union. Journal of Agricultural and Food Chemistry 65: 2017–2018.

Vivekanandhan, P., Panikar, S., Sethuraman, V., Usha-Raja-Nanthini, A. and Shivakumar, M. S. (2023). Toxic and synergetic effect of plant essential oils along with nano-emulsion for control of three mosquito species. Journal of Natural Pesticide Research 5: 100045.

Vivekanandhan, P., Senthil-Nathan, S. and Shivakumar, M. S. (2018). Larvicidal, pupicidal and adult smoke toxic effects of Acanthospermum hispidum (DC) leaf crude extracts against mosquito vectors. Physiological Molecular Plant Pathology 101: 156–162.

Vryzas, Z. (2016). The plant as metaorganism and research on next-generation systemic pesticides—prospects and challenges. Frontiers in Microbiology 7, https://doi.org/10.3389/fmicb.2016.01968.

Wang, D., Saleh, N. B., Byro, A., Zepp, R., Sahle-Demessie, E., Luxton, T. P. et al. (2022). Nano-enabled pesticides for sustainable agriculture and global food security. Nature Nanotechnology 17: 347–360.

Wang, G., Xu, X., Cheng, Q., Hu, J., Xu, X., Zhang, Y. et al. (2023). Preparation of sustainable release mesoporous silica nano-pesticide for control of Monochamus alternates. Sustainable Material Technology 35: e00538.

Wang, G., Xu, X., Cheng, Hu, J., Xu, X., Zhang, Y. et al. (2023). Preparation of sustainable release mesoporous silica nano-pesticide for control of Monochamus alternatus, Sustainable Materials Technology 35: e00538.

Wang, X. and Wang, R. (2024). Insecticide resistance: Monitoring, mechanism and management. J. Appl. Entomol. 148: 735–737. https://doi.org/10.1111/jen.13322.

Wang, Z., Yue, L., Dhankher, O. P. and Xing, B. (2020). Nano-enabled improvements of growth and nutritional quality in food plants driven by rhizosphere processes. Environment International 142: 105831.

Wu, F., Zhang, S., Li, H., Liu, P., Su, H., Zhang, Y. et al. (2024). Toxicokinetics explain differential freshwater ecotoxicity of nanoencapsulated imidacloprid compared to its conventional active ingredient. Environmental Science and Technology 58: 9548–9558.

Yan, S., Cheng, W. Y., Han, Z. H., Wang, D., Yin, M. Z. and Du, X. G. (2021). Nanometerization of thiamethoxam by a cationic star polymer nanocarrier efficiently enhances the contact and plant-uptake dependent stomach toxicity against green peach aphids. Pest Manag. Sci. 77: 1954–1962.

Yang, G., Ren, H., Li, Q., Bi, F., Chen, G., Yu, D. et al. (2025). Enhancing environment resistance and bioactivity of pesticide enabled by structure-controllable polymer nanocarriers: emphasizing the role of morphology. Small 21: e2409537. doi: 10.1002/smll.202409537.

Yang, L., Chen, H., Kaziem, A. E., Miao, X., Huang, S., Cheng, D. et al. (2024). Effects of exposure to different types of metal-organic framework nanoparticles on the gut microbiota and liver metabolism of adult zebrafish. ACS Nano 18: 25425–25445.

Yin, J., Su, X., Yan, S. and Shen, J. (2023). Multifunctional nanoparticles and nanopesticides in agricultural application. Nanomaterials (Basel) 13: 1255. doi: 10.3390/nano13071255.

Zanoni, I., Keller, J. G., Sauer, U. G., Müller, P., Ma-Hock, L., Jensen, K. A. et al. (2022). Dissolution rate of nanomaterials determined by ions and particle size under lysosomal conditions: contributions to standardization of simulant fluids and analytical methods. Chemical Research in Toxicology 35: 963–980.

Zhang, L., Yan, S. and Li, M. (2022). Nanodelivery system alters an insect growth regulator's action mode: from oral feeding to topical application. ACS Applied Materials and Interfaces 14: 35105–35113.

Zhang, W. (2018). Global pesticide use: profile, trend, cost/benefit and more. Proceedings of International Academy of Ecology and Environment Sciences 8: 1–27.

Zhao, X., Cui, H., Wang, Y., Sun, C., Cui, B. and Zeng, Z. (2017). Development strategies and prospects of nano-based smart pesticide formulation. Journal of Agricultural and Food Chemistry 66: 6504–6512.

# Chapter 5

# Nanofertilizers for Sustainable Agriculture

## Limitations and Future Trends

*Arun Ghuge,*[1] *Harris Panakkal,*[2] *Rahul Bhagat,*[2] *Mahendra Rai*[3] and *Indarchand Gupta*[2,*]

## 1. Introduction

The agriculture sector is the lifeline for human and animal sustainability on earth. Globally it is major sector supplying food and feed to humans and animals respectively. As far as the compelling rise in global human population is concerned, it is estimated to be 9.6 billion by 2050. However, the continuous decrease in the land under agriculture (Zhang et al. 2015) is one of the hurdles in the way of maintaining the food supply chain. Climate change is another problem which has been caused mainly due to human activities. According to Intergovernmental Panel on Climate Change (IPCC), since the year 1880 the global temperature has increased by 1.2°C (Saraiva et al. 2023). The warming beyond 1.5°C can have irreversible and permanent consequences for the sustainability of humanity (Saraiva et al. 2023). It is increasingly showing a huge impact on the ecosystem, resulting in depletion of freshwater, land and other natural resources. It has also caused biotic and abiotic stress to crop plants, resulting in reduced or complete failure of agricultural production (Verma et al. 2022).

In order to cater to the huge population, it is necessary to have advancements in the agricultural sector with the aim of enhanced agricultural productivity and providing global food security. To increase the agricultural productivity, chemical fertilizers are being preferred. They have been reported to increase the agriculture produce to a greater extent. Chemical fertilizers may be expensive for farmers (Verma et al. 2022). However, prolonged intensive applications of chemical conventional fertilizers led to the emergence of potential environmental issues viz. degradation of soil

[1] Department of Forensic Biology, Government Institute of Forensic Science, Civil Lines, Nagpur, (M.S.)- India.
[2] Department of Biotechnology, Government Institute of Science, Nipatniranjan Nagar, Caves Road, Chhatrapati Sambhajinagar- 431 004 (M.S.) – India.
[3] Department of Chemistry and Pharmacy, The Federal University of Piaui (UFPI), Teresina, Brazil.
* Corresponding author: indarchandgupta@gmail.com

DOI: 10.1201/9781003659907-7

quality, air and ground water pollution (Congreves and Van Eerd 2015). Moreover, efficient nutrient utilization by plants appears to be important factor for their enhanced productivity. Therefore, there is always a pressing need to have novel alternatives which could provide a better solution to enhance farm production and environmental sustainability. This can be achieved by nanotechnology-based interventions in the form of nanofertilizers.

Nanotechnology is the interdisciplinary branch of science having applications in various industries including electronics, medical, agricultural, etc., to solve the problems. It involves the manipulation and control of the material at the nano level to develop the new material and applications in the agricultural industries (Kumar et al. 2025). It has significant potential to enhance agricultural production and quality. There are various applications of nanotechnology in agriculture such as the use of nanoparticles (NP) for nanofertilizers, nanopesticides, nano-based soil remediation, nanosensor for precision farming, nanoformulation for seed germination, Food packaging, etc. (Prachi and Mirza 2025, Suganthi et al. 2025).

Nanofertilizers are the revolutionary forms of composts used to improve crop yield, nutrient delivery and reduce environmental impact using nanotechnology (Munir et al. 2024). Nanofertilizers such as nanoparticle-based fertilizers, nanoscale coated fertilizers, nanoscale encapsulated fertilizers generally release nutrients in a controlled manner, enhancing its efficacy by avoiding leaching and volatilization (Sales et al. 2024). The major impact of using nanofertilizers is that it minimizes energy use by increasing nutrient uptake efficiency along with reducing losses and leads to productivity enhancement (Channab et al. 2024). It is the solution to improved agricultural sustainability, by enhancing nutrient efficiency and reducing environmental pollution. Alongside the development of nanotechnology-based interventions, it is also equally important to follow certain guidelines provided by regulatory authorities integrating the scientific and public view points for eco-friendly commercial applications of nanofertilizers. It will be required to assess the benefits versus environmental risk associated with the use of nanofertilizers. The present chapter hence aim to discuss the advantages and limitations of nanofertilizers for sustainable agriculture. Moreover, the chapter also highlights the future trends, development and upgradation of existing regulatory guidelines for making this technology more promising for the stakeholders.

## 2. Advantages of Nanofertilizers

The rising global demand for food increased chemical fertilizer use. Chemical fertilizers meet the demands by increasing plant growth and output to some extent, but they can harm the soil, the environment, and health (Kumar et al. 2025). There is an urgent need for alternatives that can solve these issues. Nanofertilizers are the alternative solution as they provide higher nutritional management, owing to their ability to increase the nutrient uptake efficiency by targeted delivery mechanisms, controlled release of nutrients through nanoscale active substances, nontoxic, and being less harmful to living organisms (Nisha Raj et al. 2021). Nanofertilizers play a vital role in maintaining soil fertility. All the advantages of nanofertilizers are illustrated in Figure 1.

The main advantages of nanofertilizers are controlled and slow release of nutrients, lower dose requirements, improved soil health, enhanced plant growth and productivity, better nutrient absorption, resistance to loss, and cost effectiveness (Kumar et al. 2025). They are promising solutions for enhancing nutrient uptake/enhanced nutrient efficiency because of their larger surface area and targeted delivery property (Singh et al. 2024). They can enhance nutrient retention by reducing leaching, avoiding volatilization and controlling release of nutrients to minimize losses (Yadav et al. 2023). Nanofertilizers are playing a key role in controlling environmental pollution. Conventional fertilizers quickly dissolve, wash away with irrigation and run off into the water reservoirs and pollute the environment (Chhipa 2017). Instead, nanofertilizers release nutrients in a controlled manner, and reduce runoffs. Nano fertilizers improve soil fertility (Congreves and Van Eerd 2015) and avoid chemical buildup (Rautela et al. 2021).

**Figure 1.** Advantages of nanofertilizers.

## 3. Limitation of Nanofertilizers

Although nanofertilizers play a significant role in enhancing nutrient delivery for improving plant growth, it is also necessary to investigate their environmental implications. The study of the fate and transport of nanofertilizers, along with their harmful effects, needs to be considered, which could help in enhancing the benefits of nanofertilizers to enhance crop yield.

### *3.1 Concerns about their Fate and Transport*

The fate and transport of nanofertilizers in agricultural systems are required to know their environmental impact. While applying nanoparticles, their transformation causes reduction in efficiency for their use as nanofertilizers (Ahmed et al. 2021). While making suspension in media such as water, they intend to get dissolute, thus releasing ions (Mbanga et al. 2022). The interaction of these ions with soil components and plants leads to the harmful effects on the plant and plant ecosystem. Nanoparticles and the ion released thereof interact with various molecules especially the enzymes (Ahmed et al. 2021). The nanoparticle/ion-enzyme interaction disturbs the normal metabolism of a cell (Habas et al. 2021). It also induces the accumulation of reactive oxygen species in the cell. Further, it may lead to the growth inhibition and apoptosis (Gupta et al. 2012).

The nanoparticle-soil interaction significantly influences their stability and transport. It finally affects the nanoparticle safety in the plant and environment. The major concern regarding nanofertilizers arises due to their excessive accumulation in the soil and water. This may result in soil contamination causing the potential disruption of the soil ecosystem (Zulfiqar et al. 2019). Furthermore, nanoparticles may accumulate in human and animals via transfer through food chain (Zulfiqar et al. 2019), posing a health risk.

While using nanofertilizers for their desired application, the nanoparticles first get deposited on the plant surfaces such as on leaf, stem or root. Later on, they pass through the cuticle and epidermis, followed by their transport within the plant tissues. The plant leaves have various components such as stomata, mesophyll and vascular tissue (Wang et al. 2023). They have larger pores, which can allow the NP penetration. Depending upon the environmental conditions and plant physiology, the stomatal opening facilitates the entry of NP inside plant (Avellan et al. 2021). As mentioned earlier, NPs uptake through plant roots are another way of their entry. Among plants, monocots possess fibrous root system whereas eudicots have a long-lived primary root. The high surface of roots in monocots make them prone to NPs accumulation (Pan et al. 2023).

### *3.2 Ecotoxicological Impact*

For nanomaterials the soil is a sink into which they can enter via direct release and accidently pose a threat to the ecosystem and human health (Capek et al. 2021). The NPs can disrupt the soil microbial content (Gupta et al. 2015), hampering the decomposition of organic matter and lowering soil nutrient content (Shah et al. 2022). This mechanism affects the essential function of soil microflora and therefore affects soil fertility. Moreover, NPs get transformed in soil to form aggregates and alternatively they may get dissolute to release ions. These processes alter the overall nature, composition and property of the soil.

From soil, the nanofertilizers can be released into the water bodies. In water bodies the nanomaterial component of nanofertilizers can lead to acute and chronic toxicity to aquatic flora and fauna. In this regard, nanofertilizers were reported to exhibit moderate toxicity on zebrafish embryos (Fadzil et al. 2022). Susana et al. (2023) has assessed ecotoxic effects of copper nanoparticles (CuNPs) and zinc nanoparticles (ZnNPs) on soil invertebrate *Enchytraeus crypticus*. The study revealed that both in combination exhibited higher toxicity as compared to the individual nanoparticles. They were found to exhibit low toxicity when exposed for short term duration. Whereas, they showed significant toxicity when exposed for a long term. This indicates that nanofertilizers can show significant impact on soil environment through prolonged exposure. According to Shah et al. (2022) nanofertilizers significantly reduce the colony-forming units of fungi and bacteria which are vital for nutrient cycling and soil health. The study reported that zinc oxide nanoparticles (ZnO NPs) and $Fe_2O_3$NPs reduce the beneficial bacteria and fungi of plants by 32% and 28% respectively. The nano component of nanofertilizers get accumulated in distinct parts of the plant to generate reactive oxygen species (ROS), inhibit their growth and ultimately result in plant cell death (Qureshi et al. 2018).

### *3.3 Uneven Distribution of Nanoparticles*

Nanomaterials used in nanofertilizers may form agglomerate (Adhikari 2025). The agglomeration occurring inside the plant cells can result in uneven distribution. In such a scenario nanoparticles may remain clustered in certain areas of the soil and plants instead of being distributed uniformly (Sharma et al. 2023). Lopez-Moreno and group (2010) investigated the uptake and accumulation of 8 nm sized ZnO NPs in soybean (*Glycine max*). The study reported that ZnNPs uptake by soybean seedlings was significantly higher at the concentration of 500 mg $L^{-1}$ as compared to 1000 mg $L^{-1}$. This might be due to the phenomenon that nanoparticles form lesser aggregation at lower concentrations as compared to higher concentrations. Hence, the passage of oversized agglomerates of ZnONPs through the plant cell wall pores become problematic (Lopez-Moreno et al. 2010, Kumar et al. 2021). Moreover, the slow release of nutrients from agglomerated nanoparticles could be less efficient as compared to the well-dispersed nanofertilizers (Alfalahi et al. 2021). Additionally, the uneven distribution of nanofertilizers could lead to nutrient wastage, increased environmental pollution of nanomaterials in soil ecosystem. Hence, it may show adverse effects on flora and fauna, thereby raising the potential impact on ecosystem and human health (Mahesha et al. 2023).

## 4. Future Trends for Using Nanofertilizers for Sustainable Agriculture

Although the use of nanofertilizers is a better option to increase agricultural yield but their use must be controlled. Moreover, the technology used for generating nanofertilizers must be cost-effective and eco-friendly. In this regard, many studies have suggested making use of the green synthesis approach. Moreover, there is also a need to have the control of various regulatory authorities which can keep an eye to make sure of the better use of nanofertilizers.

### *4.1 Improved Methods of NP Synthesis by Green Synthesis Approaches*

Nanomaterials can be produced using a variety of methods, but are generally created using two approaches classified as top-down synthesis and bottom-up synthesis. The top-down method consists of breaking down bigger materials into nanoparticles. However, the latter approach involves the building of nanoparticles from smaller components such as molecules or atoms. The commonly employed synthesis methods include laser ablation, hydrothermal synthesis, sol-gel, chemical vapor deposition and biological synthesis, etc. These methods have some limitations such as high production cost and being labor-intensive and non-eco-friendly. The green synthesis methods offer greater advantages including low cost, being suitable for large-scale synthesis, environmentally friendly, and also not necessitating the use of higher temperatures, hazardous chemicals or excessive energy (Saxena et al. 2025). Green synthesis is a sustainable and environmentally friendly route which involves the use of several components including plants, bacteria, fungi and algae, etc. The synthesis of nanoparticles from biological sources offers greater advantages compared to chemical methods. The NP produced via green route often results in less toxic and enhanced biocompatibility over those created using conventional methods. Additionally, these nanoparticles possess unique shapes and size, and also composition, thus rendering them suitable for several applications in various fields including agriculture, medicine, bioenergy, catalysis and environmental remediation. The various parts of the plants such as leaves, fruits, flowers, bark, shoots and roots are exploited for the synthesis of different types of nanomaterials. The extract of plants parts contains divers bioactive compounds that can act as reducing and stabilizing agents in the synthesis process (Deka et al. 2025, Jabeen et al. 2025). Plant-based green synthesis of nanomaterials is most efficient because of higher yield and more stability. Some of the examples of phytosynthesis of NPs are *Piper nigram* utilized for synthesis of silver nanoparticles (Slman et al. 2025), ZnO NPs produced using plant extract of *Fumaria officinalis* and *Peganum harmala* (Hayat et al. 2025) and $TiO_2$ NP from *Ocimum sanctum* (Ahmad et al. 2022).

Various microbial sources have been tested for the production of nanoparticles as they are more biocompatible, safe and allow stable production. Biosynthesis of nanoparticles using microbes offers reliable and sustainable approaches compared to current synthetic processes. The microbes serve as reducing and capping agents due to the presence of peptides, poly (amino acids) enzymes, polysaccharides, polyhydroxyalkanoate (Jadoun et al. 2022). Moreover, microbe-based synthesis is an eco-friendly and economical method since microorganisms like bacteria, yeast, fungi, and viruses act as nano-factories for synthesis of materials at a nano scale. In the presence of reductase enzymes, microbial entity can reduce metallic salts to nanomaterials (Alsaiari et al. 2023). Jadoun et al. (2022) reviewed the synthesis of various nanoparticles including iron oxide, zinc oxide, titanium dioxide selenium, palladium, copper, silver, gold and platinum from several microscopic organisms. Recently, El-Khawaga et al. (2025) studied the green synthesis of ZnO nanoparticles from yeast and also investigated their antibacterial activity and photocatalytic degradation. The study by Khan et al. (2022) reveals the production of silver and gold nanoparticles from *algae*.

Emphasis is growing on developing nanofertilizers that are safe-by-design, meaning they achieve the desired agronomic effects while using materials that are environmentally benign. This includes using biodegradable or biogenic nanomaterials (e.g., silica from rice husk, chitosan nanoparticles, or biochar-based carriers) instead of persistent inorganic nanoparticles. As mentioned above, it also involves "green" synthesis methods – for example, using plant extracts or microbes to synthesize nanoparticles – which reduce toxic chemical use during production (Ude 2024). Adopting such eco-friendly synthesis can lower the overall environmental impact from the manufacturing stage itself and ensure that the end product is more likely to decompose harmlessly after fulfilling its purpose (Saurabh et al. 2024). Research in the past couple of years has reported progress in this area, such as nanofertilizer formulations using zinc and iron nanoparticles derived via biological routes that are more easily taken up by plants and leave less residue (Ude 2024).

## *4.2 Waste Utilization Methods*

Nowadays much emphasis is given on the synthesis of nanomaterials through sustainable and eco-friendly green synthesis route which utilizes agricultural and industrial waste. This approach also helps to control and manage waste, use renewable resources and non-toxic reagents. Waste management is one of the key issues due to non-economic methods, requiring more time, improper practices, associated health hazards, and its impact on environment, etc. Whereas, waste from several sectors such as industry, agriculture, and municipal waste can be utilized for the production of nanomaterials. The synthesis of nanoparticles using waste biomass involves the conversion of numerous forms of waste into valuable precursors of nanomaterials. Moreover, these methods lessen the environmental burden, have low costs and provide sustainable substitutes to traditional synthetic routes. Zamani et al. (2019) demonstrated the use of lignin, hemicellulose, and cellulose-rich agricultural residues and other waste material including wheat straw, bamboo leaves, rice husks, tea waste, sugarcane bagasse, walnut shells, and eggshells for economical nanoparticle synthesis. The ZnO nanoparticles synthesized using banana peels and date seed extract shows significant antimicrobial potency (Hussien 2023). Recently, Sharma et al. (2025) investigated the use of green coconut shells, which remains as waste after coconut water consumption for ZnO NPs synthesis. Waste banana peel extracts serve as excellent precursor for the Ag/Cu bimetallic nanoparticles via eco-friendly and sustainable green synthesis method. Another study by Idris et al. (2025) reported the synthesis of bimetallic nanoparticles consisting of silver oxide-nickel oxide ($Ag_2O$-NiO) using discarded orange peels which acts as a stabilizing and reducing agent. Aboelghait et al. (2024) applied the AgNP produced via environment friendly approach using waste of *Murcott mandarin* peel for efficient removal of heavy metals from industrial waste water. Industrial and agricultural waste is being produced in a large scale and the utilization of such waste material for value added products is one of the key areas for proper waste management and sustainability. In recent years, precursors based on waste materials are being used for nanoparticle synthesis. An abundant amount of waste is generated from different sectors like industries, and municipalities and agriculture. Such sources act as cost-effective and easily available large-scale templates for nanoparticle synthesis (Aswathi et al. 2023). Industrial and municipal waste is already being used for production of nanomaterials. Samaddar et al. (2018) studied the synthesis of nanoparticles from industrial waste such as tires, batteries, waste water and sludge. Municipal waste containing vegetable peels, fruits and food waste can be utilized for biogenic synthesis for NPs offering economical and environment benign nanomaterial synthesis (Aswathi et al. 2023, Verma et al. 2024). Thus, biogenic nanoparticles possess remarkable properties and can be used in diverse applications.

## *4.3 Regulatory Frameworks*

Global regulatory frameworks for nanotechnology in agriculture are still evolving, with various countries and international bodies working to ensure nanofertilizers are used safely and effectively. Some key elements of these frameworks are mentioned below.

### *4.3.1 Regulatory Oversight*

Most nations regulate nano-agricultural inputs under existing chemical or agricultural laws. For example, in the United States nano-pesticides fall under the EPA and USDA, while in the European Union nanomaterials in fertilizers or pesticides are overseen via existing chemical and food safety agencies (e.g., REACH through ECHA and food/feed via EFSA) (Kumari et al. 2023). This means nanofertilizers must comply with the same safety and approval processes as conventional products, even if explicit "nano-specific" fertilizer laws are still lacking in many jurisdictions (Kumari et al. 2023).

### *4.3.2 Risk Assessment Requirements*

Regulators universally demand thorough risk assessments before nanofertilizers can be approved for the market. Pre-market evaluations typically examine the nanomaterial's toxicity, potential environmental release pathways, and human health impacts (Kumari et al. 2023). Standard toxicological tests are being adapted for nanoscale materials to account for unique properties (e.g., high reactivity or novel exposure routes). In practice, this means a nano-formulated fertilizer undergoes additional scrutiny to ensure its nanoscale features do not introduce unforeseen hazards.

### *4.3.3 Labeling and Transparency*

Many regulatory agencies mandate labeling of agricultural products that contain engineered nanomaterials (Kumari et al. 2023). Such labeling guidelines aim to inform end-users (farmers and consumers) about the presence of nano-ingredients, promoting transparency and enabling informed decision-making. For instance, the EU has considered nano-labeling in food/feed, and countries like the US require pesticide labels to reflect novel ingredients, including nanoscale materials (Kumari et al. 2023).

### *4.3.4 International Standards and Testing Protocols*

International standards organizations and cooperative frameworks have begun issuing guidelines specific to nanotech in agriculture. The International Organization for Standardization (ISO) has published terminology and definition standards (e.g., ISO/TS 80004-1) for nanomaterials to ensure a common language for regulation (Kumari et al. 2023). Standardized testing protocols for efficacy and safety are also being developed. Notably, the OECD in 2022 adopted Test Guideline No. 125 to harmonize measurement of nanoparticle size distribution, a critical parameter for characterizing nanofertilizer materials (Stoll 2022). Such test guidelines extend the existing safety testing framework (originally developed for conventional chemicals) to better capture nano-specific behaviors, ensuring consistent and reliable assessment of nanofertilizer properties. Similarly, efforts are underway to adapt toxicity, environmental fate, and ecotoxicity test methods for nanomaterials so that regulators globally can rely on a common set of scientific protocols.

### *4.3.5 Nanotechnology Safety Guidelines*

Regulators and occupational safety agencies have issued guidelines for the safe handling and use of nanomaterials in agriculture. These often cover workplace safety (to protect workers manufacturing or applying nanofertilizers) and environmental precautions. For example, the UK's Health and Safety Executive released a Nanotechnology Safety Guide for industry (though earlier in 2011), and many of these principles (like minimizing inhalation or dermal exposure, using protective equipment, etc.) are being integrated into current agricultural safety standards (Kumari et al. 2023). At the international level, the Food and Agriculture Organization (FAO) and World Health Organization (WHO) have also reviewed risk assessment approaches for nano-based food and agri-products, concluding that existing risk assessment paradigms are generally applicable but may require additional data for nanoscale materials (Department of Biotechnology (DBT) Govt. of India 2020). These reviews and technical papers form the basis of guidance for national authorities developing their own safety protocols.

International cooperation is a hallmark of regulating nanotechnology in agriculture. Organizations such as the OECD facilitate collaboration among countries to research and manage nanomaterial risks. The OECD's Working Party on Manufactured Nanomaterials has created a platform for governments to share data and develop consensus guidelines so that safety evaluation of nanofertilizers can be globally aligned (German Environment Agency (UBA) 2022). This includes developing testing methods, databases on nano-material effects, and even a "safer innovation approach" encouraging safety-by-design in nano-agri products (German Environment Agency (UBA) 2022). Likewise, Codex Alimentarius (joint FAO/WHO) has begun examining nano-related food standards to guide international food trade involving nano-enabled products (Department of Biotechnology (DBT) Govt. of India 2020).

At the country-specific level, policies are rapidly emerging to address nanofertilizers. For instance, India has taken a proactive stance by issuing *Guidelines for Evaluation of Nano-Based Agri-Input and Food Products* in 2020, which supplement existing regulations for fertilizers, pesticides, and food safety (Department of Biotechnology (DBT) Govt. of India 2020). These Indian guidelines explicitly integrate nano-specific requirements into the approval process (e.g., requiring characterization of nano ingredients, additional toxicity tests, etc.) and have been harmonized with international frameworks such as OECD, US EPA, EU REACH, and FAO/WHO principles (Department of Biotechnology (DBT) Govt. of India 2020). This ensures that nanofertilizer products like India's recently approved nano-urea are evaluated rigorously for quality, safety, and efficacy before widespread use. Similarly, the United States treats nanofertilizers under laws like the Toxic Substances Control Act (for novel substances) and FIFRA (for pesticidal uses), with the EPA reviewing data on environmental and health effects as part of registration (Kumari et al. 2023). The European Union, while not having a separate nanofertilizer law, requires compliance with REACH chemical regulations and has guidance from the European Food Safety Authority for nanomaterials in the food chain (Kumari et al. 2023). Overall, many governments are in the process of updating their regulatory definitions to include nanomaterials (e.g., defining what counts as a "nano" fertilizer) and establishing clear approval pathways. A common theme is the call for more scientific data: regulators often encourage further research and even post-market monitoring to continually assess nanofertilizer safety in real-world conditions (Kumari et al. 2023).

Major international efforts are shaping the regulatory landscape. The FAO and WHO have convened expert meetings to examine nanotech applications in food and agriculture, providing recommendations that help countries formulate policies (Department of Biotechnology (DBT) Govt. of India 2020). The OECD's cooperative research not only produces test guidelines but also compiles dossiers on various nanomaterials (including those used in fertilizers) to support risk assessors globally. ISO and national standard bodies (like ASTM) are developing standard reference materials and metrology techniques to ensure that when a nanofertilizer is tested for, say, particle size or solubility, the results are reliable and comparable across labs. While explicit "nanofertilizer laws" may still be nascent, the existing global regulatory frameworks—backed by safety guidelines, standardized testing protocols, and international coordination—are progressively being adapted to accommodate nanofertilizers. This evolving mosaic of regulations and standards aims to strike a balance between innovation and safety, ensuring that nanofertilizers contribute to sustainable agriculture without compromising human or environmental health (German Environment Agency (UBA) 2022 and Department of Biotechnology (DBT) Govt. of India 2020).

### *4.4 Sustainability Considerations*

Beyond regulatory compliance, the sustainability profile of nanofertilizers is critically important. This includes examining the full life cycle of nanofertilizers, from raw material extraction and production, through on-farm use, to end-of-life disposal or environmental fate. Life Cycle Assessment (LCA) has emerged as a key tool to evaluate these stages holistically. Recent studies have begun comparing the environmental impacts of nano-enabled fertilizers with conventional fertilizers using LCA. Notably, one such assessment reported that introducing nanofertilizers did *not* significantly increase overall energy use or environmental impacts across the fertilizer's life cycle, and in some aspects improved resource efficiency (Escribà-Gelonch et al. 2023). This suggests that, if designed and used properly, nanofertilizers can potentially offer a smaller environmental footprint per unit of nutrient delivered (thanks to higher nutrient use efficiency) than traditional fertilizers. However, LCA of nanofertilizers is still an evolving methodology, and it must account for some nano-specific factors (Romeo et al. 2022). For example, conventional LCAs are being updated to include nanoparticle-specific toxicity characterization and the potential for nanoparticles to behave differently in ecosystems (Romeo et al. 2022). These advancements in LCA tools – such as incorporating new impact categories for nanoparticle exposure or using in vitro toxicity data for nanomaterials – are enabling a more accurate assessment of nanofertilizer sustainability beyond just energy or greenhouse gas metrics (Romeo et al. 2022). Such refined LCA methodologies help identify any hidden trade-offs (e.g., if manufacturing a nanofertilizer consumes rare materials or extra energy) so that researchers can improve nano-formulations accordingly.

#### *4.4.1 Environmental Impact Assessments (EIA)*

They are also crucial before the large-scale deployment of nanofertilizers. An EIA focuses on the localized effects on soil health, water quality, and biota in the areas where nanofertilizers are used. Given that nanofertilizers introduce engineered nanoparticles into agroecosystems, EIAs try to predict how these particles will interact with the environment. Current evidence indicates a mix of potential benefits and concerns. On the one hand, nanofertilizers can significantly reduce the negative environmental impacts associated with conventional fertilizers. By achieving higher nutrient uptake efficiency, nanofertilizers can minimize nutrient leaching into soil and runoff into waterways (Ude 2024). This means less eutrophication risk for water bodies and less accumulation of excess chemicals in the soil. Indeed, studies have noted that due to improved nutrient delivery, nano-formulations can curtail nutrient runoff, thereby reducing soil degradation and water pollution compared to traditional fertilization (Ude 2024). Some nanofertilizers are designed as slow or controlled-release formulations, which not only feed crops over a longer period but also lower the emission of greenhouse gases like nitrous oxide (a byproduct of excessive fertilizer use) (Ude 2024). These attributes align with climate-smart agriculture goals, suggesting nanofertilizers might help shrink the carbon and nitrogen footprint of crop production. Additionally, certain nano-coatings or nano-carriers can be made from biodegradable materials (like polysaccharides or clay), which gradually break down in soil, further mitigating long-term residue buildup (Saurabh et al. 2024).

On the other hand, sustainability assessments also highlight areas of concern. Engineered nanoparticles introduced into the environment may persist or behave in unforeseen ways. A prominent concern is that many nanofertilizer components (e.g., metal oxide nanoparticles) are not readily biodegradable and could accumulate in soil and water, potentially leading to ecotoxic effects on soil microorganisms, insects, or aquatic life over time (Arora et al. 2024). If these particles enter food chains, there is a possibility of biomagnification, where nanoparticles could move up the food chain and concentrate in organisms (though research on this is still in early stages). Scientists stress that understanding the long-term fate of nanofertilizers requires studying their biodegradability and bioaccumulation potential in ecosystems (Arora et al. 2024). For instance, a nano formulation that greatly benefits crop growth might still pose sustainability issues if it leaves behind residual nanoparticles that harm beneficial soil microbes or earthworms. Recent experiments have shown some nanoparticles can indeed affect soil biota or plant root development negatively at high concentrations, underlining the need for cautious application rates and further ecotoxicological research (Arora et al. 2024, Saurabh et al. 2024). Thus, environmental monitoring programs are recommended whenever new nanofertilizers are introduced in the field to detect any adverse changes in soil microbial communities, soil fertility, or water quality. To address these sustainability challenges, several mitigation strategies and best practices are being proposed.

### *4.4.2 Optimal Dosing and Precision Application*

From a practical standpoint, using nanofertilizers in a precision agriculture framework enhances sustainability. Because nanofertilizers can be effective at much lower doses (often cited as requiring only 20–25% of the weight of conventional fertilizers for similar yields (Ude 2024), it is crucial to apply them at the right rate and time). Over-application not only wastes resources but could exacerbate any toxicity issues. Thus, farmers should be educated to use the minimum effective nano-dose and employ targeted delivery (e.g., foliar sprays directly on leaves or seed coatings) so that nanoparticles are absorbed by the plant and not left free in the soil in large quantities. This strategy reduces the environmental loading of nanomaterials and is intended to reduce excess nutrient runoff.

### *4.4.3 Life Cycle Assessment and Continuous Improvement*

Advanced LCA tools should be employed as decision support for nanofertilizer development. By identifying hotspots in the life cycle (for example, energy use in nanoparticle production or any end-of-life emissions), manufacturers can improve the design and process. Recent methodologies integrate LCA with risk assessment, meaning they evaluate not just the carbon footprint or energy, but also incorporate potential human and ecotoxicity impacts of nanoparticles (Romeo et al. 2022). This combined approach is sometimes referred to as "sustainable by design," where safety and sustainability are considered concurrently in R&D. As data on nanofertilizer impacts accumulates, LCA models can be refined with more realistic assumptions (such as actual field nano-release rates, degradation rates in soil, etc.), thereby guiding more sustainable practices or policy decisions (e.g., whether a certain nano-coating should be phased out due to persistence).

### *4.4.4 Environmental Monitoring and Remediation*

To ensure long-term sustainability, monitoring programs can be put in place in areas using nanofertilizers. Soil and water samples can be periodically tested for the presence of residual nanoparticles or changes in key indicators (like soil enzyme activities, microbial diversity, etc.). If buildup is detected, mitigation could involve phytoremediation (using plants known to uptake heavy metals or nanoparticles) or soil amendments that immobilize or degrade nanoparticles. Proactive guidelines suggest having a plan for the safe disposal of nanofertilizer packaging and leftovers (Saurabh et al. 2024). For example, since nanofertilizers are often applied in suspension or coated form, any unused suspension should be disposed of properly (not dumped into water bodies), and containers rinsed and recycled safely to prevent environmental release of concentrated nanoparticles. Some countries are considering whether residues from nano-agrochemicals could be treated as hazardous waste if they contain high levels of certain nanomaterials, thereby necessitating specialized disposal, all aimed at preventing environmental contamination (Saurabh et al. 2024).

In terms of recent advancements, tools and frameworks like the European Union's Safe and Sustainable by Design (SSbD) initiative explicitly incorporate life-cycle thinking for nano-enabled products, including fertilizers. This approach, gaining traction in the last two to three years, means that from the early design phase, developers evaluate material choices, production methods, and end-of-life scenarios to minimize environmental and health risks (German Environment Agency 2022). Additionally, updated computational models are being used to predict nanoparticle behavior in soil and aquatic environments, which helps in performing risk assessment alongside LCA. On the methodological front, researchers have been working on bridging nano-toxicology data with LCA impact categories (Romeo et al. 2022). For instance, if a certain nanofertilizer has a slight toxic effect on algae, that information can be translated into an LCA indicator for ecosystem quality – something not traditionally done in fertilizer LCAs. Such interdisciplinary progress in the past few years has been enhancing our ability to judge nanofertilizer sustainability with greater confidence.

In conclusion, the sustainability of nanofertilizers hinges on careful design, responsible use, and ongoing evaluation. When implemented thoughtfully, nanofertilizers offer promising environmental advantages: improving nutrient efficiency, reducing pollution and greenhouse emissions, and potentially lowering the resource intensity of agriculture (Ude 2024, Arora et al. 2024). However, it is equally important to acknowledge and manage the uncertainties – ensuring that these novel inputs do not introduce new ecological problems. Life cycle and environmental impact assessments, guided by the latest scientific tools, will play a pivotal role in steering nanofertilizer technology towards truly sustainable agriculture. By coupling innovation with precautionary measures and transparent science-based policies, stakeholders can maximize the benefits of nanofertilizers for food production while safeguarding soil health and environmental integrity for the future (Saurabh et al. 2024, Arora et al. 2024).

## 5. Conclusion

Nanofertilizers seem to better alternative to traditional fertilizers in improving soil fertility, hence improving the agricultural yield. They offer enhanced nutrient delivery to the plants compared to conventional fertilizers. However, they also have some limitations. They pose an environmental and health risk. During prolonged application in the field, they tend to get accumulate in the soil and hence pose potential harmful effects to plants and soil microflora. It is also felt that there is a lack of clear regulatory guidelines and approval processes for the development and commercialization of nanofertilizers. It is essential to address the limitations and harmful effects of nanofertilizers through continued research and development. There is a need to understand the long-term effects of nanofertilizers and undertake studies to overcome these challenges through enhanced nano-formulations. The upgradation of existing regulatory process or establishment of robust regulatory frameworks will ensure the safe and effective use of nanofertilizers in the future.

## References

Aboelghait, K. M., Walid, E., Abdallah, I., Abdelfattah, A. M. and El-Shamy. (2024). Green synthesis of silver nanoparticles by waste of *Murcott mandarin* peel as a sustainable approach for efficient heavy metal removal from metal industrial wastewater, Separation and Purification Technology 347: 127609.

Adhikari, T. (2025). Transformations of nanofertilizers in soil and their uptake by plants. *In*: Patra, A. K. (eds.). Nanofertilizers in Agriculture. Sustainability Sciences in Asia and Africa. Springer, Singapore. https://doi.org/10.1007/978-981-97-8112-6_3.

Ahmad, M. Z., Alasiri, A. S., Ahmad, J., Alqahtani, A. A., Abdullah, M. M., Abdel-Wahab, B. A. et al. (2022). Green synthesis of titanium dioxide nanoparticles using *Ocimum sanctum* leaf extract: *In vitro* characterization and its healing efficacy in diabetic wounds. Molecules 27(22): 7712.

Ahmed, B., Rizvi, A., Ali, K. Lee, J., Zaidi, A., Khan, M. S. et al. (2021). Nanoparticles in the soil–plant system: a review. Environ Chem Lett 19: 1545–1609. https://doi.org/10.1007/s10311-020-01138-y.

Alfalahi, A. O. and Abdulqahar, F. W. (2021). Nanonutrients: plant nutritive and possible antioxidant regulators. *In*: Nanobiotechnology: Mitigation of Abiotic Stress in Plants; Springer: Cham, Switzerland, p. 471.

Alsaiari, N. S., Alzahrani, F. M., Amari, A., Osman, H., Harharah, H. N., Elboughdiri, N. et al. (2023). Plant and microbial approaches as green methods for the synthesis of nanomaterials: synthesis, applications, and future perspectives. Molecules 28(1): 463.

Arora, P. K., Tripathi, S., Omar, R. A., Chauhan, P., Sinhal, V. K., Singh, A. et al. (2024). Next-generation fertilizers: the impact of bionanofertilizers on sustainable agriculture. Microbial Cell Factories 23: 254. https://doi.org/10.1186/s12934-024-02528-5.

Aswathi, V. P., Meera, S., Maria, C. G. A. and Nidhin, M. (2023). Green synthesis of nanoparticles from biodegradable waste extracts and their applications: a critical review. Nanotechnology for Environmental Engineering 8: 377–397. https://doi.org/10.1007/s41204-022-00276-8.

Avellan, A., Yun, J., Morais, B. P., Clement, E. T., Rodrigues, S. M. and Lowry, G. V. (2021). Critical review: role of inorganic nanoparticle properties on their foliar uptake and *in Planta* translocation. Environment Science and Technology 55(20): 13417–13431.

Capek, J. and Roušar, T. (2021). Detection of oxidative stress induced by nanomaterials in cells—the roles of reactive oxygen species and glutathione. Molecules 26(16): 4710. https://doi.org/10.3390/molecules26164710.

Channab, B. E., El Idrissi, A., Ammar, A., Dardari, O., Marrane, S. E., El Gharrak, A. et al. (2024). Recent advances in nano-fertilizers: synthesis, crop yield impact, and economic analysis. Nanoscale 2024 Feb 29; 16(9): 4484–4513.

Chhipa, H. (2017). Nanofertilizers and nanopesticides for agriculture. Environment Chemistry Letters 15: 15–22. doi: 10.1007/s10311-016-0600-4. [DOI] [Google Scholar].

Congreves, K. A. and Van Eerd, L. L. (2015). Nitrogen cycling and management in intensive horticultural systems. Nutrient Cycling in Agroecosystems 102: 299–318. doi: 10.1007/s10705-015-9704-7.

Deka, K., Nongbet, R. D., Das, K., Saikia, P., Kaur, S., Talukder, A. et al. (2025). Understanding the mechanism underlying the green synthesis of metallic nanoparticles using plant extract(s) with special reference to Silver, Gold, Copper and Zinc oxide nanoparticles, Hybrid Advances 9: 100399. https://doi.org/10.1016/j.hybadv.2025.100399.

Department of Biotechnology (DBT), Govt. of India. (2020). Guidelines for Evaluation of Nano-Based Agri-Input and Food Products in India. New Delhi: Ministry of Science & Technology, India dbtindia.gov.in.

El-Khawaga, A. M., Elsayed, M. A., Gobara, M., Suliman, A. A., Hashem, A. H., Zaher, A. A. et al. (2025). Green synthesized ZnO nanoparticles by *Saccharomyces cerevisiae* and their antibacterial activity and photocatalytic degradation. Biomass Conversion and Biorefinery 15(2): 2673–2684.

Escribà-Gelonch, M., Butler, G. D., Goswami, A., Tran, N. N. and Hessel, V. (2023). Definition of agronomic circular economy metrics and use for assessment for a nanofertilizer case study. Plant Physiology and Biochemistry 196: 917–924. https://doi.org/10.1016/j.plaphy.2023.02.042.

Fadzil, N. I., Anoam, S. R., Mohd Rosmi, M. N., Mohd Anuar, M. F. and Masdor, N. A. (2022). Toxicity assessment of colloidal nanofertilizers using zebrafish embryo model through acute toxicity assay. pp. 93–104. In Materials Science Forum (Vol. 1055). Trans Tech Publications, Ltd. https://doi.org/10.4028/p-swiwg6.

German Environment Agency (UBA). (2022, June 30). OECD-Guideline: Determination of particle size of nanomaterials (Press Release). Retrieved from https://www.umweltbundesamt.de/.

Gupta, I. R., Anderson, A. J. and Rai, M. (2015). Toxicity of fungal-generated silver nanoparticles to soil-inhabiting *Pseudomonas putida* KT2440, a rhizospheric bacterium responsible for plant protection and bioremediation. Journal of Hazardous Materials 286: 48–54.

Gupta, I., Duran, N. and Rai, M. (2012). Nano-silver toxicity: emerging concerns and consequences in human health. pp. 525–548. *In*: Rai, M. and Cioffi, N. (eds.). Nano-antimicrobials: Progress and Prospects Germany: Springer Verlag.

Habas, K., Demir, E., Guo, C., Brinkworth, M. H. and Anderson, D. (2021). Toxicity mechanisms of nanoparticles in the male reproductive system. Drug Metabolism Reviews 53(4): 604–617.

Hayat, K., Din, I. U., Alam, K., Khan, F. U., Khan, M. and Mohamed, H. I. (2025). Green synthesis of zinc oxide nanoparticles using plant extracts of *Fumaria officinalis* and *Peganum harmala* and their antioxidant and antibacterial activities. Biomass Conversion and Biorefinery 15: 9565–9579.

Hussien, N. A. (2023). Antimicrobial potential of biosynthesized zinc oxide nanoparticles using banana peel and date seeds extracts. Sustainability 15(11): 9048.

Idris, D. S., Roy, A., Malik, A., Khan, A. A.. Sharma, K. and Roy, A. (2025). Green synthesis of silver oxide-nickel oxide bimetallic nanoparticles using peels of *Citrus Sinensis* and their application. Journal of Inorganic and Organometallic Polymers and Materials 35: 594–606. https://doi.org/10.1007/s10904-024-03316-9.

Jabeen, I., Rashid, Z., Waheed, R., Zafar, S., Ahmad, K., Arooj, I. et al. (2025). Green synthesis and biological applications of *Peganum harmala* mediated copper oxide nanoparticles. Journal of Molecular Structure 140838, https://doi.org/10.1016/j.molstruc.2024.140838.

Jadoun, S., Chauhan, N. P. S., Zarrintaj, P., Barani, M., Varma, R. S., Chinnam, S. et al. (2022). Synthesis of nanoparticles using microorganisms and their applications: a review. Environment Chemistry Letters 20: 3153–3197. https://doi.org/10.1007/s10311-022-01444-7.

Khan, F., Ayesha Shahid, A., Zhu, H., Ning Wang, N., Javed, M. R., Ahmad, N. et al. (2022). Prospects of algae-based green synthesis of nanoparticles for environmental applications. Chemosphere 293: 133571, https://doi.org/10.1016/j.chemosphere.2022.133571.

Kumar, S., Sharma, Y., Khandelwal, V., Rawat, K. and Patil, A. (2025). Applications of nanotechnology in fertilizers: a review study. Sustainable Chemistry for the Environment 10: 100247, ISSN 2949-8392.

Kumar, Y., Tiwari, K., Singh, T. and Ramesh. R. (2021). Nanofertilizers and their role in sustainable agriculture. Annals of Plant and Soil Research 23(3): 238–255. 10.47815/apsr.2021.10067.

Kumari, R., Suman, K., Karmakar, S., Mishra, V., Lakra, S. G., Saurav, G. K. et al. (2023). Regulation and safety measures for nanotechnology-based agri-products. Frontiers in Genome Editing 5: 1200987 pmc.ncbi.nlm.nih.gov. https://doi.org/10.3389/fgeed.2023.1200987.

Mahesha, K. N., Singh, N. and Amarshettiwar, S. B. (2023). Entering a new agricultural era through the impact of nano-fertilizers on crop development: a review. 1: 10.9734/ijpss/2023/v35i203789.

Lopez-Moreno, M. L., de la Rosa, G., Hernández-Viezcas, J. A., Castillo-Michel, H., Botez, C. E., Peralta-Videa, J. R. et al. (2010). Evidence of the differential biotransformation and genotoxicity of ZnO and $CeO_2$ nanoparticles on soybean (*Glycine max*) plants. Environmental Science and Technology 44(19): 7315–20.

Mbanga, O., Cukrowska, E. and Gulumian, M. (2022). Dissolution of titanium dioxide nanoparticles in synthetic biological and environmental media to predict their biodurability and persistence. Toxicology *in Vitro* 84.

Munir, A., Salah, M. A., Ali, M., Baber Ali, Saleem, M. H., Samarasinghe, K. G. B. A. et al. (2024). Advancing agriculture: harnessing smart nanoparticles for precision fertilization. BioNanoSci. 14: 3846–3863. https://doi.org/10.1007/s12668-024-01597-5.

Nisha Raj, S., Anooj, E. S., Rajendran, K. and Vallinayagam, S. A. (2021). Comprehensive review on regulatory invention of nano pesticides in agricultural nano formulation and food system. Journal of Molecular Structure 1239: 130517. doi: 10.1016/j.molstruc.2021.130517.

Organisation for Economic Co-operation and Development (OECD). (n.d.). (2025). OECD Series on the Safety of Manufactured Nanomaterials and other Advanced Materials oecd.org. Retrieved April 30, 2025, from https://www.oecd.org/chemicalsafety/.

Pan, W., Zhang, H. J., Zhang, Y. F., Wang, M., Tsui, M. T., Yang, L. et al. (2023). Silica nanoparticle accumulation in plants: current state and future perspectives. Nanoscale 15(37): 15079–15091.

Prachi, Mirza, A. (2025). Sustainable nanomaterials: principles and practices. *In*: Rajan, R., Ahmad, F., Khoshariya, A. and Seth, C. S. (eds.). Sustainable Era of Nanomaterials. Sustainable Development and Biodiversity, vol 40. Springer, Singapore.

Qureshi, A., Singh, D. K. and Dwivedi, S. (2018). Nano-fertilizers: a novel way for enhancing nutrient use efficiency and crop productivity. International Journal of Current Microbiological and Applied Sciences 7(2): 3325–3335.

Rautela, I., Dheer, P., Thapliyal, P., Shah, D., Joshi, M., Upadhyay, S. et al. (2021). Current scenario and future perspectives of nanotechnology in sustainable agriculture and food production. Plant Cell Biotechnology and Molecular Biology 22: 99–121.

Romeo, D., Hischier, R., Nowack, B., Jolliet, O., Fantke, P. and Wick, P. (2022). *In vitro*-based human toxicity effect factors: challenges and opportunities for nanomaterial impact assessment. Environmental Science: Nano 9(6): 1913–1925. https://doi.org/10.1039/d1en01014j.

Saraiva, R., Ferreira, Q., Rodrigues, G. C. and Oliveira, M. (2023). Nanofertilizer use for adaptation and mitigation of the agriculture/climate change dichotomy effects. Climate 11(6): 129. https://doi.org/10.3390/cli11060129.

Sales, H. B. E., de S. Carolino, A., de A. Nunes, R. Z., Macalia, C. M. A., Ruzo, C. M., da C. Pinto et al. (2024). Advances in agricultural technology: a review of slow-release nanofertilizers and innovative carriers. Communications in Soil Science and Plant Analysis 55(12): 1849–1882. https://doi.org/10.1080/00103624.2024.2326145.

Samaddar, P., Ok, Y. s., Kim, K-H., Kwon, E. K. and Tsang, D. C. W. (2018). Synthesis of nanomaterials from various wastes and their new age applications. Journal of Cleaner Production 197(1): 1190–1209,

Saurabh, K., Prakash, V., Dubey, A. K. and Dubey, P. K. (2024). Enhancing sustainability in agriculture with nanofertilizers. Discover Applied Sciences 6: 559 link.springer.com. https://doi.org/10.1007/s42452-024-06267-5.

Saxena, R., Kotnala, S., Bhatt, S. C., Uniyal, M., Rawat, B. S., Negi, P. et al. (2025). A review on green synthesis of nanoparticles toward sustainable environment. Sustainable Chemistry for Climate Action 6: 100071, ISSN 2772-8269, https://doi.org/10.1016/j.scca.2025.100071.

Shah, G. M., Ali, H., Ahmad, I., Kamran, M., Hammad, M., Shah, G. A. et al. (2022). Nano agrochemical zinc oxide influences microbial activity, carbon, and nitrogen cycling of applied manures in the soil-plant system. Environment and Pollution 293: 118559.

Sharma, P. and Chauhan, N. S. (2023). Effect on nanoparticles on plant cell morphology, physiology, and metabolism. pp. 95–113. *In*: The Impact of Nanoparticles on Agriculture and Soil; Academic Press: Cambridge, MA, USA.

Sharma, R., Kumar, N., Sharma, P., Anita Yadav, A. and Aggarwal, N. K. (2025). Biosynthesis, characterisation and therapeutic potential of green coconut waste derived zinc oxide nanoparticles. Results in Surfaces and Interfaces 18: 100372, https://doi.org/10.1016/j.rsurfi.2024.100372.

Singh, M., Goswami, S. P., Ranjitha, G., Sachan, P., Sahu, D. K., Beese, S. et al. (2024). Nanotech for fertilizers and nutrients-improving nutrient use efficiency with nano-enabled fertilizers. Journal of Experimental Agriculture International 46(5): 220–247. https://doi.org/10.9734/jeai/2024/v46i52372.

Slman, D. K., Ihsan, I., Ahmed, H. K. and Jawad, A. A. (2025). Biosynthesis of silver nanoparticles by Piper nigrum (Black Pepper) extract and their application as antibacterial. Advanced Journal of Chemistry-Section A 8: 864–873.

Stoll, J. (2022). OECD-Guideline: Determination of particle size of nanomaterials. Retrieved April 30, 2025, from Umweltbundesamt website: https://www.umweltbundesamt.de/en/topics/oecd-guideline-determination-of-particle-size-of#:~:text=On%2030th%20June%202022%2C%20the,example%20for%20the%20REACH%20Regulation.

Suganthi, N., Kavitha, B., Priya, R., Yuva raj, M., Sharmila, R., Saranya, M. et al. (2025). Emerging tools and technologies in plant nanotechnology research. *In*: Al-Khayri, J. M., Anju, T. R. and Jain, S. M. (eds.). Plant Nanotechnology Fundamentals and Methodologies. Nanotechnology in Plant Sciences, vol 1. Springer, Cham. https://doi.org/10.1007/978-3-031-81896-7_18.

Susana, I. L. Gomes, Sekerani, B. Chidiamassamba, Tito Trindade, Janeck J., Scott-Fordsmand and Mónica J. B. Amorim. (2023). Environmental hazards of WELGRO® Cu+Zn: A nano-enabled fertilizer. Environmental Pollution 10.1016/j.envpol.2023.122469.

Ude, R., Kujiper, I., Takiguchi, T. and Bußman, J. P. (2024). Nanofertilizers: A green window of opportunity for food security in Sub-Saharan Africa (Science-Policy Brief)sdgs.un.org. United Nations Multistakeholder Forum on Science, Technology and Innovation for the SDGs. Retrieved from https://sdgs.un.org/resources/.

Verma, K. K., Song, X.-P., Joshi, A., Tian, D.-D., Rajput, V. D., Singh, M. et al. (2022). Recent trends in nano-fertilizers for sustainable agriculture under climate change for global food security. Nanomaterials 12(1): 173. https://doi.org/10.3390/nano12010173.

Verma, M. G., Das, R. S., Bhardwaj, A. K. and Kumar, A. (2024). Global status of biogenic and nonbiogenic waste production and their employability in nanomaterial production. pp. 1–15. *In*: Abhishek Kumar Bhardwaj, Arun Lal Srivastav, Kuldip Dwivedi and Mika Sillanpää (eds.). Green and Sustainable Approaches Using Wastes for the Production of Multifunctional Nanomaterials, Elsevier.

Wang, X., Xie, H., Wang, P. and Yin, H. (2023). Nanoparticles in plants: uptake, transport and physiological activity in leaf and root. Materials 16(8): 3097. https://doi.org/10.3390/ma16083097.

Yadav, A., Yadav, K. and Abd-Elsalam, K. A. (2023). Nanofertilizers: Types, delivery and advantages in agricultural sustainability. Agrochemicals 2(2): 296–336. https://doi.org/10.3390/agrochemicals2020019.

Zamani, A., Marjani, A. and Mousavi, Z. (2019). Agricultural waste biomass-assisted nanostructures: Synthesis and application. Green Processing and Synthesis 8(1): 421–429. https://doi.org/10.1515/gps-2019-0010.

Zhang, X., Davidson, E. A., Mauzerall, D. L., Searchinger, T. D., Dumas, P. and Shen, Y. (2015). Managing nitrogen for sustainable development. Nature 528(7580): 51–9.

Zulfiqar, F., Navarro, M., Ashraf, M., Akram, N. A. and Munne-Bosch, S. (2019). Nanofertilizer use for sustainable agriculture: Advantages and limitations. Plant Science 289: 110270.

# Chapter 6

# Next-Generation Farming

## Crop Monitoring with Nanosensor Technologies

*Sergei Tarasov, Yulia Plekhanova* and *Anatoly Reshetilov**

## 1. Introduction

Agriculture and the food industry can be called the foundation of modern society, because they play a vital role in ensuring food security, economic stability, and sustainable development of mankind. It is imperative that continued progress in these sectors is made in order to increase productivity, improve quality and meet the growing global demand for food. Agriculture, specifically, is a pivotal catalyst of economic growth, particularly within developing countries, where it facilitates livelihoods and contributes to rural development. Nevertheless, the global community continues to face significant challenges related to hunger and food insecurity. According to the report published by the Food and Agriculture Organization of the United Nations (FAO), approximately 725 million people worldwide, representing nine per cent of the global population, currently face food insecurity to some degree (FAO, IFAD, UNICEF, WFP and WHO 2024). In addition, the report estimates that 2.33 billion individuals – constituting almost one in three people globally are either moderately or severely food insecure, indicating a deficiency in their access to sufficient and nutritious food on a consistent basis. These staggering figures underscore the urgent need for innovative solutions and continued efforts to ensure equitable access to nutritious food for all. Achieving sustainable food production is significantly influenced by agricultural plants and their yields.

In order to address the challenges outlined above, it is imperative to have access to measurement techniques that provide inexpensive, reliable and continuous monitoring of aquatic and agricultural systems. Continuous monitoring facilitates the timely identification of contaminants or stressors such as pathogens and high/low temperature, etc., affecting agricultural plants, thereby enabling a rapid response to emerging problems (Li et al. 2024b). Furthermore, the utilization of monitoring data is instrumental in the development of prediction models for potential plant disease outbreaks, thereby facilitating the timely implementation of preventative measures. This, in turn, contributes to

G.K. Skryabin Institute of Biochemistry and Physiology of Microorganisms, FRC Pushchino Centre for Biological Research, Russian Academy of Sciences, 142290 Pushchino, Russian Federation.

Email: yu_plekhanova@pbcras.ru

* Corresponding author: anatol@pbcras.ru

DOI: 10.1201/9781003659907-8

the optimization of crop growth conditions, increasing crop resilience and increasing yields, which is key to ensuring food security.

Traditional crop monitoring, including assessment of crop health, growth and development, is based mainly on visual inspection and field observations (Lv et al. 2024). This approach requires the involvement of qualified specialists and regular field visits, which makes the process labor-intensive, costly, and its accuracy is extremely limited. Furthermore, such methods frequently lack the necessary rapidity to detect problems such as stress factors, nutrient deficiencies or early signs of disease in a timely manner. This reduces the effectiveness of measures to improve plant health. Modern laboratory techniques such as high-performance liquid chromatography (HPLC), gas chromatography (GC), etc., can provide highly accurate analysis of crop samples. However, these techniques are limited by high cost, labor intensity, the need for skilled technicians and specialized instruments (Bharti et al. 2024b). In addition, such equipment can only be operated in centralized laboratories, often far away from the farmland itself. It is imperative to acknowledge that the health and growth rates of agricultural plants are contingent on their interaction with the environment, which can exert a dual impact, both positive and negative. Consequently, agriculture requires apparatus capable of expeditious monitoring not only of the plants themselves, but also of the environment conducive to the farmland.

A promising alternative to traditional monitoring methods is the use of nanosensors, especially nano-biosensors. A biosensor is an integrated device that combines a biological sensing element and a physicochemical transducer. Analyte detection occurs by converting molecular interactions into a measurable and interpretable signal. These devices are compact, cost-effective and user-friendly, requiring minimal experience to operate. These devices can be used for real-time monitoring of both aquatic environments (e.g., to detect contaminants in water) and agricultural systems (e.g., to detect biotic and abiotic stresses in plants). This approach has the potential to enhance the efficiency and accuracy of environmental and crop monitoring, particularly for small-scale farmers who prioritize cost-effectiveness, expediency, and simplicity in analysis (He et al. 2023).

The present chapter does not aspire to provide a comprehensive review of all advancements in the domain of nanosensors for agriculture and food safety. Rather, its objective is to underscore the broad spectrum of potential applications in the realm of crop monitoring. These applications encompass the analysis of environmental factors that might be conducive to agricultural land, the monitoring of plant health, the identification of diseases and pathogens, and the analysis of crops intended for human consumption (Figure 1).

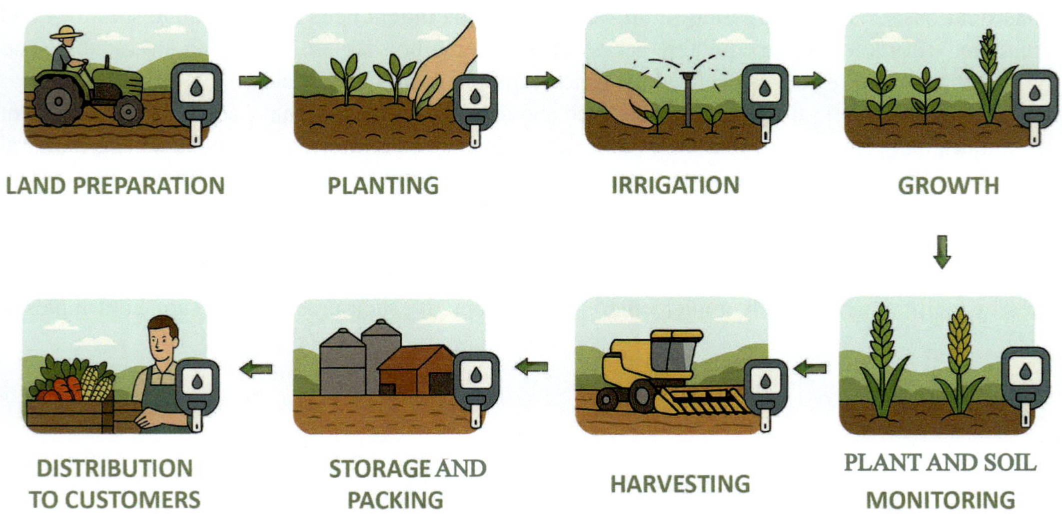


**Figure 1.** Application of biosensors at different stages of the agricultural cycle.

## 2. Environmental Monitoring to Improve Plant Yields

In order to ensure a successful harvest, the soil must first be prepared adequately, fertilizers must then be applied following the plant's needs, and the nutrient content (macro- and micronutrients necessary for plant growth) of the soil must be determined. Furthermore, assessments of the temperature, moisture and pH levels in the soil are essential, as is an evaluation of the soil's ability to support agricultural activity, taking into account the presence of toxins, biotoxins, chemical compounds and other potential contaminants that may harm plants during their growth cycle. Traditional testing methods are expensive, time-consuming, require highly trained personnel, and are often limited to laboratory settings. Portable devices, such as biosensors, are associated with simpler operations and are available to a wider range of users. They allow on-site analysis and have high level of accuracy and selectivity. The utilization of such portable soil assessment devices has the potential to optimize the use of fertilizers for crop yields, thereby facilitating budgetary savings and mitigating the risk of soil contamination from superfluous fertilizers.

### *2.1 Soil Moisture and Temperature Control*

In this area of research, a large amount of work is related not just to the development of individual sensors that detect soil moisture and temperature, but to the development of controllable systems that estimate moisture and decide whether to irrigate a garden or a field. For example, a group from the University of Pisa presented a wireless sensor network based on soil moisture to support field-scale irrigation management in the Tuscany region (Italy) (Hamouda et al. 2024). The application of such an intelligent system allowed maintaining soil moisture content within a predetermined optimal range, where the upper and lower limits corresponded respectively to the soil moisture capacity and the threshold value below which water stress occurs. The intelligent irrigation management employed during the study's growing seasons resulted in water savings of up to 50% of the total amount of water supplied under a normal schedule. Concurrently, the system is being trained on a real dataset. Data fusion from multiple Internet-of-Things sensors can predict moisture (Baldi et al. 2025), and thus optimize irrigation. Similar systems are also being studied in the USA, for example, a module for anomaly detection in soil moisture time series observations is presented in (Bandaru et al. 2024), where a bi-directional Long Short-Term Memory (LSTM) deep learning approach was used. The new system allows for more precise soil moisture management.

The development of moisture sensors is a complex process that involves a series of calculations to determine the optimal level of immersion in the soil, the required number of sensors per field, the distance between them, and the operating frequency if they are electromagnetic sensors (Sheng et al. 2024). Also, standard technologies can be used as sensors. In the paper (Ullah et al. 2024) authors presented an easy-to-use and highly sensitive screen-printed flexible soil moisture sensor array to monitor soil moisture remotely. The proposed sensor is based on a unique interdigitated electrode (IDE)-based capacitive structure with conductive carbon ink printed on a flexible polyethylene terephthalate (PET) substrate using the screen-printing technique. Before soil moisture monitoring, the sensor is tested for humidity monitoring in an environmental chamber. The response of the sensor is observed for a temperature range from 20 to 50°C. Typically, soil moisture sensors work on the principle of dielectric characteristics of soil and water and use conversion ratios to obtain soil moisture. Thus, these sensors require proper calibration for accurate measurements. The utilization of these devices in field studies is constrained by various factors, including the complexity of device calibration and the reliability of instrument calibration procedures. The review by Mane et al. (2024) evaluates calibration methods, including regression equations and neural networks, highlighting effectiveness and limitations, and provides practical recommendations for selecting a suitable calibration method/approach for dielectric soil moisture sensors. In addition, it is important to consider soil temperature when measuring soil moisture, as temperature will affect the measurement error (Kelley et al. 2025), and temperature correction should be considered in humidity control systems. The temperature varies significantly throughout the day, which affects the response of soil moisture sensors. Therefore, errors occurring in the response of moisture sensors are analyzed, simulated and then corrected when the sensors are operating under real conditions (Cao et al. 2021).

Thus, the integration of a humidity sensor, capable of detecting the presence of water vapor in the air, with a soil moisture sensor, which measures the water content of the soil, has the potential to result in a self-activating irrigation system. The system is then able to release the optimal amount of water required to facilitate healthy plant growth (Ammar et al. 2024).

## *2.2 Soil Composition Control*

Soil fertility is typically indicated by the presence of organic matter in the soil. The presence of pesticides, herbicides, heavy metals, and fertilizer residues in the soil indicates that it is contaminated. The presence of these compounds can lead to the accumulation of contaminants, which can eventually result in a decline in agricultural yields. Furthermore, the presence of these contaminants has the potential to enter the food chain, thereby causing health complications for the end consumer. Consequently, a comprehensive assessment of soil suitability for specific crops, taking into account these parameters, is imperative to ensure optimal harvest outcomes.

*Pesticide Detection*

The utilization of pesticides, which are chemical or biological preparations used for the control of pests and external parasites of plants, weeds and pests of stored agricultural products, on the one hand increases the gross yield of horticultural products, but on the other hand leads to the poisoning of the environment and, if absorbed by the human body, can have both acute and chronic toxic effects, depending on the dose and the way of absorption.

Highly sensitive nanosensors based on new conductive composites are being developed to detect residues of various pesticides. For example, a unique combination based on Au and zinc tungstate ($ZnWO_4$) composite materials is reported by Shad et al. (2023). The Au@$ZnWO_4$ material is applied for the detection of organophosphate pesticide (chlorpyrifos) residues with 1 pM LoD. The fabrication of a robust, nonenzymatic electrochemical-sensing electrode modified with electrochemically reduced graphene oxide (ERGO) to detect parathion residues (with a lower detection limit of 10.9 pM) in environmental samples (e.g., soil, water) as well as in vegetables and cereals is presented in (Sen et al. 2022). The utilization of nanoparticles for the detection of various compounds has become a prevalent practice in contemporary research. A wide variety of nanoparticles, including gold, silver, iron oxide, zirconium, titanium dioxide, silicon dioxide, carbon nanotubes, graphene nanoparticles, quantum dots, and others, have been employed to detect pesticide residues (Pestovsky and Martínez-Antonio 2017, Zhang et al. 2023). A variety of methods, including colorimetry, SERS, fluorescence, electrochemistry and chemiluminescence are employed to detect these particles with high sensitivity. Concurrently, inorganic nanoparticles have the capacity to function as catalysts, i.e., nanoenzymes, which in certain instances permit the substitution of the biocatalyst of the nanosensor with these particles (Bolat and Abaci 2018, Rahmani et al. 2018). Pesticide detection in this case is most often based on the fact that pesticides can inhibit both the activity of enzymes used as nanosensor bioreceptors (Rajangam et al. 2017) as well as the catalytic activity of nanoparticles (Zhang et al. 2023).

In addition, immunocomponents can become the receptor of the biosensor, such as in (Sharma et al. 2025). The authors synthesized a new nanocomposite of MXene and reduced graphene oxide (rGO), and used it to immobilize antibodies in a biosensor for the detection of endosulfan (organochlorine pesticide); the detection limit was 0.497 ppt with a linear range of 0.1 $ppt^{-1}$ ppm. In (Kou et al. 2025) a visual immunosensor with gold nanoparticles for rapid detection of fipronil (belongs to the class of phenylpyrazoles and is an active component of various pesticides) in complex conditions, including soil with a detection limit of 6.46 μg/kg, is presented. For the development of immunosensors, an important step is the need to synthesize highly specific antibodies. The majority of pesticides are low-molecular-weight compounds, which means that they are not capable of inducing an immunogenic response that would lead to the production of antibodies (Cui et al. 2018). To induce antibody production in the host, pesticide derivatives, haptens, must be synthesized and bound to carrier proteins, or altered in certain and specific positions so that a functional group can be inserted to make it more active (Khan et al. 2021). Methods of synthesis of such compounds often include several steps and involve complex reactions (Kim et al. 2003, Xu et al. 2023). Computer-aided immunoassay design can help to simulate all these critical steps such that the hypothetical immunogen with a high probability of producing required titres of antibodies in an animal model can be selected for synthesis *in vitro*. For example, in a recent paper from Kou et al. (2025), computer-simulated semi-antigens and then novel Fipronil (FPN)-immunogens were designed and synthesized by modifying carrier proteins to expose them to specific sites, so as to prepare monoclonal antibodies with high specificity and sensitivity to FPN.

*Heavy Metal Detection*

Heavy metal accumulation in the soil beyond permissible limits has a detrimental effect on the soil quality (Mahajan et al. 2024). The process of accumulation of heavy metals in soil results in alterations to its structure, water-absorbing and water-holding capacity, and pH. These changes in turn lead to alterations in soil microbiota, disturbing its properties and structure, and disrupting the balance of nutrients in the soil. Ultimately, these changes can result in a decrease in yield. The employment of biosensors as an analytical instrument facilitates the monitoring of heavy metal content in soil. The majority of nanosensors are based on the inhibitory effect of heavy metals on the activity of the receptor (Elsebai et al. 2017, Dalkıran 2020). For example, Ospina et al. (2023) showed the possibility of detection of $Pb^{2+}$ and $Cu^{2+}$ ions (detection limits were 0.8 μg/L and 0.3 μg/L, respectively) by an enzyme sensor. Heavy metal ions reduced the activity of the enzyme, in the composition of poly(propyleneimine) dendrimer as a bonding-layer between electrodeposited and colloidal gold nanoparticles, which form an electrocatalytic self-assembled layer to the covalent immobilization and direct electrochemistry of the enzyme HRP. A green fluorescent protein (GFP) based whole cell bacterial biosensor was prepared using a bacterial strain sensitive to several heavy metals in order to detect bioavailable heavy metals in soils (Rathnayake et al. 2021). The transformant, named as *Bacillus megaterium* VR1 was immobilized *in silica* matrix using sol–gel technology. The biosensor allowed detection of metal ions with lower detection limits of $1.42 \times 10^{-4}$, $3.16 \times 10^{-4}$, and $2.42 \times 10^{-4}$ mg/L for Cd, Cu and Zn, respectively.

Immunosensors based on monoclonal, polyclonal and recombinant antibodies can also be used to recognize heavy metals (Velusamy et al. 2022). To increase the sensitivity of the assay, antibodies or antigens are bound to nanoparticles. For example, in Ling et al. (2022) signal amplification probes incorporating multi-branched gold nanoflowers and latex microspheres were utilized to enhance the sensitivity of immunosensors, resulting in a twenty-fold decrease in the detection limit for mercury (II) ion (0.44 ng/mL) in comparison to the same sensor with conventional gold nanoparticles (8.92 ng/mL).

One of the most sensitive bioreceptors for the detections of heavy metals are DNA molecules. Heavy metal ions are known to cause DNA alterations and damage, and can disrupt their replication and transcription. The basic principle of biosensors based on this receptor is the hybridization of complementary DNA strands. DNA and DNAzymes (single-stranded DNA (ssDNA) molecules with catalytic activity) are biodegradable, highly selective, and easy to obtain through *in vitro* methods. The working surface of the electrode is modified with ssDNA; hybridization of the probe with the complementary target ssDNA or disruption of its structural integrity can be detected using, for example, electrochemical assays (Zhou et al. 2017b).

After the identification of a particular interaction between thymine (T)-rich DNA and mercury ions ($Hg^{2+}$) that is discernible within the UV-visible range, there has been a marked increase in the development of optical methodologies for the analysis of heavy metals (Katz 1952). The Hg-N3 bond in the T-$Hg^{2+}$-T hybrid was found to be less covalent but more ionic. The ionic nature allows $Hg^{2+}$ to act as an electron acceptor, which can quench fluorescence through electron transfer from the labeled fluorophore to the T-$Hg^{2+}$-T couple (Uchiyama et al. 2012). A fluorescence biosensor for the ultrasensitive detection of the available lead on the basis of DNAzyme biorecognition and hairpin DNA probes assembly was presented by Chen et al. (2023). In the presence of the available lead, the released trigger strand (T) can activate the hairpin assembly among H1, H2, and H3 to produce the Y-shaped sensing scaffold (H1-H2-H3) through toehold-mediated strand displacement reactions. In the H1-H2-H3 products, the formed active $Mg^{2+}$-DNAzymes at the three arms can cleave H4 to produce the fluorescence signal for the available lead detection. The detection limit of the fluorescence biosensor is 0.2 pM.

Microbial biosensors have also been shown to be valuable tools for the detection of heavy metal ions, both in mixtures and as individual ions (O Hara and Singh 2021, Zeng et al. 2021). However, selectivity, sensitivity, and maintenance of cell activity, especially under non-native environmental conditions, remain a challenge. To improve microbial biosensor parameters, microorganisms are genetically modified. The development of genetic circuits (such as toggle switches, logic gates, and amplification modules) for microbe-based biosensors has become more popular in recent years for heavy metal detection (Mathur et al. 2023). Genetic circuits are used to increase the sensitivity and specificity of biosensors, also allowing for rapid and multiple analyte detection at the same time. For example, an immobilized bioluminescent bacterium (*Escherichia coli* MC106) with embedded recombinant plasmid pmerRBPmerlux is used for mercury detection (Babapoor et al. 2020). In the presence of mercury-contaminated soil sample extract, a significant decrease in the luminescence level was observed in the mercury concentration range of 0–20 ppm. The selective mercury detection system was incorporated into *E. coli* using MerR regulatory protein and mer promoter (Branyikova et al. 2020). The detection limit for mercury was 0.035 mg/kg and was independent of the presence of other toxic heavy metal ions in the soil.

Recently, there have been many developments of biosensors based on microbial fuel cells (Zhou et al. 2017a, Liu et al. 2024). Such biosensors can perform several functions at once, as shown in (Wang et al. 2023). The microbial biosensor has been demonstrated to possess the capacity to detect residual metal ions, including $Cd^{2+}$, $Zn^{2+}$, $Pb^{2+}$, and $Hg^{2+}$, within the range of 0.5 to 30 mg/L over a period of four months, exhibiting minimal decline in performance. The biosensor generates an output voltage of approximately 400 mV when subjected to an external load of 1000 ohms, thereby demonstrating autonomy. Additionally, the microorganisms within the bioreceptor utilize heavy metal ions as electron acceptors, a property that facilitates soil purification.

*Detection of Other Soil Contaminants and its Assessment*

In order to determine the extent of soil contamination or the availability of nutrients, it is often necessary to undertake a comprehensive assessment. In this context, microbial fuel cells represent a particularly intriguing area of development, given their capacity to not only assess soil contamination but also generate energy and facilitate soil remediation. The soil microbial fuel cell (SMFC) is a distinct category of devices. These devices are characterized by their low cost, ease of use, in situ functionality, and long-term self-powering capabilities (Abbas et al. 2022). By varying the immersion depth of such sensors, the effect of toxic pollutants on microbial activity at different soil depths can be monitored, and the accumulation of toxic compounds in different soil layers can be assessed (Liu et al. 2020, Yang and Chen 2021).

Many laboratory models of microbial biosensors have been developed for monitoring contaminants in soil, like polychlorinated biphenyls, nitrogenous compounds, aromatic hydrocarbons, antibiotic residues, etc.

In the majority of cases, microorganisms with a constructed genome are utilized. In the first method, the lux operon is inserted into the required plasmid in the corresponding microorganism, and luminescence is observed in the presence of the analyte. In the second method, genes responsible for the production of certain proteins that catalyze the metabolism of the required analytes are inserted into microorganisms that have genes responsible for luminescence (Bazhenov et al. 2023, Rafeeq et al. 2023). Thus, *E. coli* based biosensor models for toluene detection were constructed (Bae et al.

2020) for monitoring soil contamination in areas of petrochemical industrial zones or gas stations; five new whole-cell biosensors based on five *E. coli* strains for direct detection of p-nitrophenol concentrations in soils near a chemical plant (Nanjing, China) (Ma et al. 2023); an autonomous microbial sensor based on genetically modified *Bacillus subtilis* bacteria for the detection of trinitrotoluene in farm soil (Essington et al. 2024).

A solute transport-soil microbial fuel cell was presented in Liu et al. (2025a). The degradation effect of benzo [a]pyrene (BaP) has been demonstrated to be positively correlated with sulfate reduction and organic matter consumption, and negatively correlated with distance from the anode. Water flow-induced solute transport promoted sulfur conversion and organic matter metabolism and altered the community characteristics of sulfate-reducing bacteria and degrading bacteria, thereby enhancing BaP degradation.

In Khmelevtsova et al. (2024), a battery of whole-cell bacterial lux-biosensors based on *E. coli* MG1655 strain with different inducible promoters and the natural luminescent *Vibrio aquamarinus* strain VKPM B-11245 was presented to evaluate the effects of agrochemical soil treatments. The results demonstrated that the joint application of pesticides and mineral fertilizers in soil led to an enhancement in the total toxicity of the soil, which was greater than the toxicity of the individual substances applied separately. Several probable mechanisms for the realization of agrochemical toxic effects have been identified. These include DNA damage caused by both the SOS response induction and alkylation, oxidative stress due to increased superoxide levels, and damage to cellular proteins and membranes. Consequently, the employment of biosensors facilitates the evaluation of the cumulative impacts of diverse toxicants on living organisms.

Quantification of antibiotics in soils is important to assess their bioavailability and mobility, and ultimately their ecotoxicological and medical risks, so the development and application of microbial biosensors in such cases is an important area of research. Ma et al. (2020) proposed a biosensor based on *E. coli* containing a reporter plasmid (pMTGFP or pMTmCherry) carrying fluorescent protein genes tightly controlled by tetracycline-responsive control region (tetRO). The biosensor was used to detect six tetracyclines in the range of 5.32–10.2 μg/kg soil, tested on soils from the most agriculturally productive areas in China. Lautenschläger et al. (2020) successfully designed and built a β-lactam biosensor in *B. subtilis* using the heterologous regulatory system BlaR1I from *S. aureus*.

The utilization of lux-biosensors facilitates the monitoring of not only toxic compounds, but also compounds that are essential for plant nutrition. For instance, biosensors based on a bioluminescent strain of *E. coli* have been developed to assess the concentration of phosphorus and nitrogen in the rhizosphere of the main part of the soil. The detection ranges of phosphate and ammonia were found to be 1–24 ppm and 0.1–1.6 ppm, respectively (Cardemil et al. 2010). And cyanobacteria (*Nostoc* sp. PCC 7120 and *Synechococcus elongatus* PCC 7942) were used to detect bioavailable nitrogen (Muñoz-Martín et al. 2014).

In a study by Siontorou and Georgopoulos (2016), the development of a soil nitrite biosensor is presented, detailing the design rationale, optimization of chemical and physical variables, and preliminary validation results using both simulated and real soil samples. The sensor exhibits high tolerance to interfering ions, a detection limit of 2.1 μg/L for soil extracts and a lowest detectable field nitrite ion concentration of 1 ppm. The biosensor is designed using a versatile lipid nano-platform that allows the easy customization of the selectivity and sensitivity of the sensor through the adjustment of macro-parameters (e.g., lipid composition, ionic strength, and pH).

A biosensor that, when directly placed on plant roots, indicates the concentration of proline (an important plant metabolite that is a plant defender against osmotic stress, its level is critical for assessing plant health) in real time is presented in Rubia et al. (2020). The proline biosensor was constructed by cloning the promoter region of pRL120553, a gene with high levels of induction in the presence of proline, in front of the lux cassette in *Rhizobium leguminosarum* bv. viciae. By using a biosensor based on the common soil bacterium *Pseudomonas putida* KT2440 as host of the plasmid pZKH2 containing a fusion of the strong constitutive promoter nptII and luxCDABE (encoding light-emitting proteins) from *Vibrio fischeri* (Herron et al. 2013), it is possible to monitor rhythms of change in carbon excretion related or unrelated to diurnal cycles in different plants. Harrison and her colleagues have developed a biosensor (Zhang et al. 2022) for use in mycorrhizal root cells, and this allowed the researchers to observe changes in the phosphate content among *Brachypodium distachyon* root cells and in cells of a nearby fungus at different stages of development. They were also able to detect the rate of phosphate transfer to nearby plant cells.

The utilization of plant microbial fuel cells (PMFCs) has enabled researchers to ascertain that plant roots are capable of exerting a significant influence on the production of bioelectricity. This influence stems from the ability of plant roots to modify unsaturated soil properties, such as soil breathability and soil pore structure (Chen et al. 2024). It is therefore possible to monitor a number of factors, including the effect of planting density on suction reactions in the topsoil, the effect of fertilizer application and competition between plants.

The development and implementation of intelligent irrigation systems and devices for tracking key soil parameters will therefore lead to more efficient farming and reduce the cost of production of final products. The utilization of versatile devices, which function as platforms for detecting various compounds, can be instrumental in the assessment of specific contaminants or nutrients. These devices are characterized by their ease of modification and customization, thereby ensuring their relevance in the context of soil property management.

## 3. Biosensors for Plant Disease Monitoring

To increase crop yields, plants must be healthy, which requires timely disease monitoring and disease management decisions. The majority of plant diseases are caused by fungal infections (Chelladurai et al. 2023), followed by diseases caused by viruses (Jablonski et al. 2021) and bacteria (Nazarov et al. 2020). In addition, plant diseases can be associated with insects and parasitic plants, with chemical compounds synthesized by humans, and caused by abiotic factors (Xu et al. 2017, Kumar and Arora 2020). Early detection of disease outbreaks and their pathogens is important not only for assessing plant health but also for minimizing the risk of spreading infections to other areas and other plants. Therefore, the development of methods for early detection of various diseases is one of the most important areas of laboratory analysis.

DNA-based biosensors are particularly sensitive because they can detect the pathogen even when single cells infect samples (Vatankhah et al. 2022), which allows the disease to be detected at an early stage. Immunoassay technologies offer high specificity and the potential to detect, for example, plant viruses *in situ*, but are limited by low virus concentrations and disposable sensors. They are typically used to confirm plant diseases after visual symptoms have occurred (Jablonski et al. 2021). Immunoassay is the most common method used for the identification of viruses. This is due to the fact that obtaining antibodies with the necessary specificity for the identification of fungi and bacteria is difficult. The antibodies obtained can be specific to individual life forms of microorganisms or to a group of species. The employment of bacteriophages as biocatalysts facilitates the detection of bacterial infestations in plants with high specificity, sensitivity, and efficiency (Xu et al. 2019). For instance, the utilization of bioluminescent phages enables the detection of nucleic acids exclusively from viable bacterial cells, thereby mitigating the occurrence of false positives. The primary constraint of bacterial phage-based biosensors is their inability to detect fungi and viruses. Table 1 presents a selection of laboratory models of biosensors for the detection of infectious plant diseases.

**Table 1.** Examples of biosensors for detection of plant diseases caused by viruses, bacteria and fungi.

| Pathogen | Crop | Bioreceptor type | Nanomaterial | Analysis time | Detection limits | Reference |
|---|---|---|---|---|---|---|
| **Viruses** | | | | | | |
| *Tomato yellow leaf curl China virus* and *papaya leaf curl China virus* | Tomato | DNA | Gold nanoparticles | 60 min | 10 copies/uL | Sun et al. 2025 |
| *Phalaenopsis virus*, *Odontoglossum ringspot virus* (ORSV) and *Cymbidium mosaic virus* (CymMV) | Orchid | Antibodies | Gold nanoparticles | 2 min | 0,5 ng/ml | Lin et al. 2022 |
| *Tobacco mosaic virus* (TMV) | Tobacco | Antibodies | Nanoparticles $Fe_3O_4$@TMV | 10–20 min | 4.8 fM | Gong et al. 2025 |
| *Tobacco mosaic virus* (TMV) | *Rehmannia glutinosa* | DNA | | 1 min | 0.17 pM | Liu et al. 2025b |
| *Chilli leaf curl virus* (ChiLCV) | Chilli seedlings | DNA | Gold nanoparticles | 10 min | 1,0 µg/ml | Das et al. 2021 |
| **Bacteria** | | | | | | |
| *Phytophthora infestans* | Potatoes | ssDNA | Gold nanoparticles | 20 min | 0.1 pg $\mu L^{-1}$ genomic DNA | Zhan et al. 2018 |
| *Acidovorax citrulli* | Watermelon, cantaloupe, cucumber, and zucchini | A monoclonal antibody | 4',4''',4''''',4'''''''-(ethene-1,1,2,2-tetrayl) tetrabiphenyl-4-carboxylic acid (TCBPE) carboxyl activation product (CA-TCBPE) | 35 min | $10^5$–$10^8$ КОЕ/мл | Li et al. 2024a |
| *Ralstonia solanacearum* | Eggplant | DNA | Gold and chromium nanoparticles | | 0.1 ng/µL | Patel et al. 2023 |
| *Pantoea stewartii* subsp. stewartii | Sweet corn | DNA | | 10 min | 0.0005 pg $\mu L^{-1}$ bacterial DNA | Cai et al. 2023 |
| *Xanthomonas oryzae pv. oryzae* | Rice | Antibodies | Graphene quantum dots and gold nanoparticles | 1 hour | 22 CFU $mL^{-1}$ | Awaludin et al. 2020 |
| *Clavibacter sepedonicus* | Potato | RNA | | 2 hours | 24 CFUs of *C. sepedonicus* per 1 g of tuber tissue | Khmeleva et al. 2024 |

*Table 1 contd. ...*

*...Table 1 contd.*

| Pathogen | Crop | Bioreceptor type | Nanomaterial | Analysis time | Detection limits | Reference |
|---|---|---|---|---|---|---|
| **Fungi** | | | | | | |
| *Alternaria* | Citruses | DNA | Gold nanoparticle, magnetic bead | | 15.6 pM of the artificial target DNA | He et al. 2022 |
| *Sclerotinia sclerotiorum* | Carrot | Polyclonal antibodies | Gold nanoparticle | | $7.8 \times 10^4$ ascospores/mL | Shoute et al. 2018 |
| *Puccinia striiformis f.* sp. *tritici* | Wheat | Nanozymes | Gold-nickel-activated carbon nanocomposite | 72 hours | | Hassan et al. 2022 |
| *Pseudocercospora fijiensis* | Banana | Polyclonal antibody | Self-assembled monolayer of gold and chromium | 15 min | 11.7 $\mu g\ mL^{-1}$ | Luna-Moreno et al. 2019 |
| *Leptosphaeria maculans* | Rapeseed | DNA | Gold nanoparticle-enhanced dynamic microcantilever | | 57 ppm | Lei et al. 2021 |
| *Ustilaginoidea virens* | Rice | ssDNA | Oxidized graphene | | 10 fM | Rana et al. 2021 |
| *Botrytis cinerea* | Temperate Legumes | DNA | Magnetic nanoparticles | | 100 fg/μL | Sambasivam et al. 2024 |

Additionally, biosensors have been developed to detect indirect signs of infection because during the infection process, plants initiate their defense reactions, resulting in changes in their primary carbon metabolism, respiration, increased $H_2S$ synthesis, increased emission of volatile organic compounds, etc. (Gan et al. 2023).

Additionally, smartphone-based sensors have become a special focus. An innovative Apple® and Android™ mobile application for citizen science has been developed, to enable real-time detection and identification of plant leaf diseases and pests (Christakakis et al. 2024). Such smartphone programs allow users to identify the name of a plant or pest, as well as plants affected by microorganisms, based on an image. They can also collect huge databases containing complex images of diseases and pests, including several interclass and extraclass variations, such as infection status and location on the plant. These applications leverage advanced technologies such as image processing techniques, Artificial Intelligence (AI), Machine Learning (ML), and Deep Learning (DL). For instance, convolutional neural networks (CNNs) have been employed to detect *Tuta absoluta* infestations at early stages, significantly improving the efficiency and speed of detection (Singh et al. 2016).

## 4. Monitoring of Crop Suitability for Humans

An equally intriguing area of biosensor development is their application for post-harvest quality assessment. Horticultural crops undergo numerous post-harvest processing stages, including cleaning, sorting, packing, and storage, culminating in delivery to the consumer. At any of these stages, the application of biosensors can facilitate rapid and non-invasive analysis of samples. These applications encompass a wide range of purposes, including the assessment of maturity, nutritional value, the presence of contaminants, and the estimation of shelf life (Hua et al. 2021). The utilization of biosensors by participants in the agricultural supply chain has the potential to enhance product quality, ensure food safety, and reduce waste. These benefits contribute to more efficient and sustainable production. Let us take a closer look at each step where biosensors can be applied to analyze the quality of the harvested product (Figure 2).

The most significant application of biosensors is their capacity to detect contaminants, including pesticides, herbicides, heavy metals, and pathogens, in agricultural products (Zhou et al. 2023). Biosensors can also detect microbial spoilage in vegetables, fruits, and cereals. This capability enhances food safety by preventing the distribution of substandard goods, as biosensors can sort and evaluate foods based on contamination level and quality. For example, Zhang et al. (2024) developed a WaveFlex biosensor based on localized surface plasmon resonance, which enabled accurate detection of the herbicide glyphosate in soybeans and corn. The receptor element contained a complex of gold nanoparticles, zinc oxide nanowires (ZnO-NWs), and molybdenum disulfide nanosheets, providing detection in the range of 0–80 μM with a sensitivity of 76.4 pm/μM. Biosensors have also been developed for the detection of other commonly used herbicides and insecticides, including formetanate (Ribeiro et al. 2014), triazofos (Guo et al. 2018), acetamiprid (Li et al. 2021b) and chlorpyrifos (Lin et al. 2021), as well as less selective biosensors for the total detection of organophosphorus pesticide residues in fruits and vegetables (Mukherjee et al. 2021).

Mycotoxins produced by some fungi can contaminate crops and pose serious health risks (Barac 2019). To date, more than 300 mycotoxins are known (Trucksess and Diaz-Amigo 2011), twenty of which have been detected in food in amounts sufficient to adversely affect humans or animals. Mycotoxins are highly toxic and have mutagenic, teratogenic, carcinogenic, immunosuppressive properties (Khan et al. 2024), as well as nephrotoxicity and genotoxicity (Ülger et al. 2020). The damage can occur both during maturation and storage (especially if its conditions are violated) of plant crops. The most prevalent mycotoxins include aflatoxin B1, predominantly present in peanuts, pistachios, nuts, and oilseeds; ochratoxin A, detected in rice, rye, wheat, buckwheat, and barley; and fumonisins (found in cereals. corn, rice), zearalenone (in wheat, cereals, rice, barley, oats), T-2 toxin (in rice, rye, wheat, barley, oats), deoxynivalenol (in corn, rice, rye, wheat, buckwheat). The

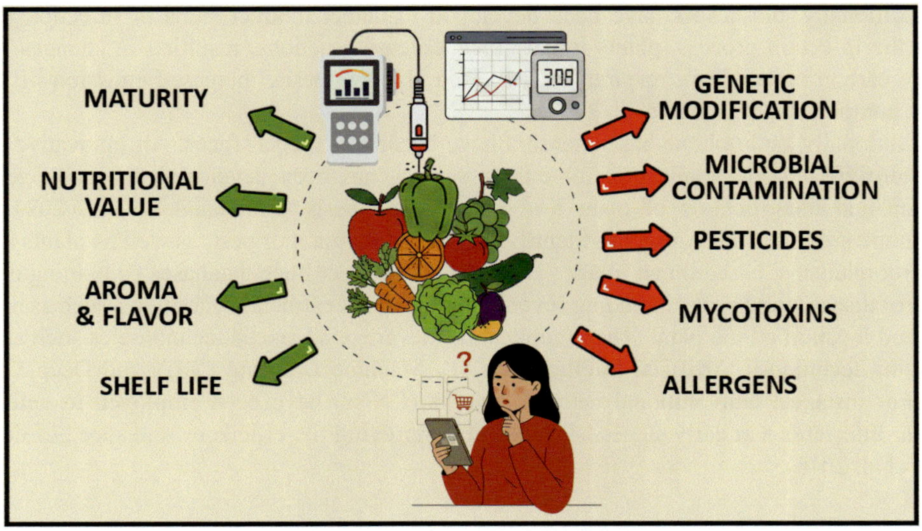


**Figure 2.** Application of biosensors for the evaluation of the harvested crops.

contaminant that poses the greatest risk to grain products is deoxynivalenol; for nuts and oilseeds, it is aflatoxin B1; for food grains and flour, and cereal products, it is ochratoxin A; and for corn and its products, it is fumonisin. Biosensors have been developed to detect these toxins at very low concentrations, enabling the detection of contaminated products and their removal from the supply chain. Li et al. (2021a) presented an electrochemiluminescent immunosensor capable of detecting ochratoxin A in coffee beans, with a detection limit of 5.7 pg/mL. In a subsequent study, Li et al. (2022b) reported an ultra-low detection limit of 8.5 fg/mL for aflatoxin B1 in corn oil samples using a MIP-ECL sensor based on MAPb QDs@$SiO_2$ encapsulated in silicon. Several papers have described the use of biosensors for the determination of heavy metals in food products. Wang et al. (2020) investigated a new type of DNA nanomachine based on dual-recognition aptazyme beacon (DRAB) for two-in-one detection of pesticides and heavy metal ions. The detection limit of this DNA nanomachine can reach 0.178 nM for chlorpyrifos and 0.034 nM for $Pb^{2+}$. Subsequently, Li et al. (2022a) developed a novel fluorescent sensor based on polyfluorene (CPs) for the rapid and accurate determination of $Cr_2O_7^{2-}$ in citrus fruit with a detection limit of 16.6 fM. Yuan et al. (2022) have developed an ultrasensitive electrochemical aptasensor for simultaneous detection of $Pb^{2+}$ and $Cd^{2+}$ ions in fruits and vegetables. The biosensor was tested on real samples of lettuce and oranges with recovery rates of 90.06% ~ 97.24%. The detection limits of $Cd^{2+}$ and $Pb^{2+}$ reached 89.31 and 16.44 pmol/L, respectively. In the field of food testing, most residual heavy metals in solid sources exist in the form of complexes, which leads to the complexity of detection. In addition, most of the measurements have to be carried out in the field to speed up the analysis process. To overcome these problems, in Tang et al. (2023), a glove-worn hydrogel sensor for monitoring trace $Cd^{2+}$ residues in rice, tea, and soil with a detection limit of up to 0.112 mg/kg and fast signal acquisition (< 5 min) was proposed.

Furthermore, biosensors can detect allergenic components in agricultural products, thereby facilitating the sorting and labeling of allergen-free or allergen-specific products. This is of particular significance for individuals afflicted with food allergies or sensitivities, as it ensures precise labeling and mitigates health risks. It is imperative to note that allergen detection does not occur during the harvesting and processing stage of agricultural products, but rather at the manufacturing and inspection stage of processed products. Notably, a highly sensitive sensor has been engineered for the detection of the allergen beta-lactoglobulin in rice porridge formulated for infants (Yang et al. 2023). This aptasensor provided a sensitivity of 96.91 μg/L and can be used for portable detection using a smartphone. Wu et al. (2024) reported the rapid detection of 40-kDa major mango allergens in processed foods using an electrochemical immunosensor based on immunomagnetic nanoparticles. Assay validation was performed on twenty-four commercial food products that were sequentially analyzed by Western blot to confirm the immunosensor results. The validation showed a high specificity of the assay of 100% and a high sensitivity of 91.7%.

Biosensors have been demonstrated to be effective in detecting microbial contamination on the surface of crops, including bacteria and fungi. This contamination can have a significant impact on both the quality and safety of the produce. This is particularly important for fresh products consumed raw, as it ensures that only microbiologically safe items reach the market. In particular, a detailed description of the amperometric detection of *Fusarium oxysporum* using nanomaterials in the receptor elements of biosensors is given by Beilinson et al. (2022). The biosensor was configured to detect the antigens of *F. oxysporum*. The authors developed immunoenzyme systems, employing L-cysteine desulfhydrase as a label. These immunosensors enabled the detection of *F. oxysporum* within a concentration range of $5 \times 10^{-10}$ to $5 \times 10^{-6}$ milligrams per milliliter. Cross-reactivity of the receptor element with other pathogenic fungi, *Alternaria alternata*, *Cladosporium herbarum*, *Aspergillus fumigatus*, and *Aspergillus niger* was shown to be absent. The proposed immunosensor method for detecting *F. oxysporum* was evaluated for its effectiveness in assessing the quality of wheat groats. *Penicillium digitatum* is a major post-harvest citrus pathogen causing losses in all citrus-growing countries. Chalupowicz et al. (2020), proposed the utilization of luminescent reactions of bacteria to modify the content of volatile organic compounds following pathogen infection in oranges. The study's findings, which were based on the reactions of four distinct bioluminescent strains on infected oranges, revealed the capacity to detect fungal activity on the third day of infection. This detection occurred prior to the manifestation of visible signs on the fruit surface, offering a novel approach for early and precise detection of fungal activity in citrus fruits.

Among bacterial pathogens dangerous to humans, *Listeria monocytogenes*, *Escherichia coli*, and various members of the genus *Salmonella* should be emphasized. There are many biosensors that can quickly and accurately detect their presence in various fruits and vegetables. Back in 2013, a phage-coated E2 magnetoelastic biosensor was presented for the detection of *Salmonella typhimurium* on the surface of tomatoes (Chai et al. 2013). In the study by El-Moghazy et al. (2022), an electrochemical biosensor based on the use of genetically engineered bacteriophage T7 encoding the phoA gene as biorecognition elements that can trigger overexpression of the alkaline phosphatase enzyme upon infection of target bacteria is presented. The use of screen-printed electrodes modified with carbon nanotubes for the electrochemical detection of *E. coli* in fresh spinach leaves has been demonstrated to be a rapid and effective method, with detection limits ranging from 1 to 104 colony-forming units (CFU) per milliliter within one hour. In conclusion, a novel tri-signal biosensor was developed in 2025 for the simultaneous and ultra-sensitive detection of live and dead *L. monocytogenes* cells (Zhang et al. 2025). In particular, multifunctional phage/Prussian blue nanoparticles (PBNPs) and phage/polyethylenimine@magnetic nanoparticles (PEI@MNPs) probes were obtained. In the presence of *L. monocytogenes*, a phage/PBNP- *L. monocytogenes*-phage/PEI@MNP sandwich complex was formed and separated by a magnetic field. The bioluminescent signal allows accurate detection of live *L. monocytogenes* with a limit of detection (LOD) of 1 CFU/mL, while the photothermal and colorimetric signals allow measurement of total bacterial counts *in situ* with a LOD of 5 CFU/mL. In this case, the parallel use of three signal transducers simultaneously allowed the increase of the accuracy of detection results.

Biosensors can accurately assess the maturity of fruits and vegetables by detecting specific biomarkers or changes in chemical composition. Biosensors can monitor specific biomarkers, such as ethylene gas or sugar content, to determine the optimal time to harvest fruits and vegetables (Medina-Plaza et al. 2016, Vong et al. 2019, Mallick et al. 2023). This ensures that the crop is harvested at the peak of ripeness, maximizing flavor, nutritional value and shelf life. Particularly in (Jin et al. 2023), a nanosensor based on ultrafine Pd nanoparticles decorated with porous ZnO nanosheets was shown. Using mango as an example, it is shown that the signal of the nanosensor on mangoes at different stages of maturity was quite different, with the detection limit of the sensor not exceeding 10 ppb.

In addition, biosensors are capable of measuring the nutrient content of agricultural products, including vitamins, minerals, proteins and carbohydrates (Bharti et al. 2024a). By quantifying these nutrients or using biomarkers related to nutritional quality, biosensors allow sorting foods based on their nutritional value, providing consumers with high-quality and nutrient-rich foods. In particular, screen-print electrode-based sensors have been proposed to monitor fresh fruits (e.g., watermelons and apples) with low vitamin C content (Spissu et al. 2021). In order to achieve this objective, the electrodes were modified with graphene nanomaterials, thereby augmenting the oxidative capacity of the functionalized SPEs and increasing their electroactive area. Consequently, the LOD of ascorbic acid was reduced from 0.55 to 0.15 and 0.28 μM.

It has been established that the content of sugars in fruit and vegetable products can be ascertained through the utilization of both conventional methodologies, such as high-performance liquid chromatography (HPLC), and biosensors. These methods primarily facilitate the estimation of the total sugar content of a product. However, biosensors have been developed that are designed to detect specific sugars, such as glucose, with a high degree of specificity. A microelectrode glucose biosensor based on a three-dimensional hybrid platinum/graphene oxide nanoporous nanostructure has been developed for the rapid determination of glucose in tomato and cucumber fruits (Wu et al. 2022). The glucose oxidase-based biosensor exhibited a wide range of glucose determination within 0.1–20.0 mmol/L and a rapid detection time of up to three seconds. The relative deviation of this method was less than 5% compared to the HPLC method, indicating the high accuracy of the biosensor in determining glucose levels in plants. Another indirect method for the determination of sugars in fruits is the use of biofuel cells, which are devices similar in nature to biosensors but are used to generate electricity. Consequently, in (MacVittie et al. 2015) the impact of conditionally "positive chemotaxis" was delineated. An enzyme fuel cell immersed in orange pulp yielded potential generation, thereby depleting substrates essential for microbial activity within the fruit. The buckypaper material, modified with PQQ-dependent glucose dehydrogenase and FAD-dependent fructose dehydrogenase at the anode and laccase at the cathode, used to activate a wireless information transfer system, was used as the anode. It can be hypothesized that the voltage level can be utilized to ascertain the nutrient content of a specific fruit. This suggests that biofuel cells can be regarded as an ancillary method for evaluating the quality of agricultural products. This principle can also work "in reverse". The decrease of the voltage generated by the MFC may indicate the presence of certain substances (e.g., antibiotics, pesticides) that negatively affect the catalytic activity of microorganisms and therefore can be unsafe for humans. Biosensors also play a key role in assessing quality attributes such as acidity (Rai et al. 2022), antioxidant levels (Ribeiro et al. 2021), flavor characteristics and additives in processed foods (Kumar et al. 2024a). Secondary metabolites such as polyphenols and flavonoids are not only antioxidants but also account for the taste, color and health benefits of crops (Lzaod and Dutta 2022). Biosensors can quantify these compounds, allowing sorting and grading of products based on their functional or nutraceutical properties. Mediavilla et al. (2022) presented a fluorescent biosensor based on laccase and carbon nanorods was demonstrated. This biosensor exhibited a detection limit of 7.4 μM, enabling quantitative analysis of polyphenols. Its performance was evaluated in both processed food products, such as wine and orange juice, and the direct detection of polyphenols in agricultural leaf extracts. Luteolin, a naturally occurring flavonoid compound found in various herbs, vegetables, and fruits, was also examined. Xu et al. (2024) developed a biosensor for the detection of luteolin in plants widely used in traditional Chinese medicine, such as *Scutellaria barbata* and *Dandelion*. The biosensor was constructed using a nanocomposite of multi-walled carbon nanotubes, bovine serum albumin, laccase, titanium dioxide, and chitosan. The square-wave voltammetry study demonstrated linear ranges of 80 nanomoles to 1 micromole and 1 to 6 micromoles, respectively, with a detection limit of 11 nanomoles for luteolin.

Biosensors can detect volatile organic compounds (VOCs) that contribute to the aroma and flavor of crops (Zhao et al. 2024). In addition, microbial contamination of fruits and vegetables can also be detected by the level of some VOCs (carbon dioxide, p-ethylphenol, methyl salicylate, etc.) (Ventura-Aguilar et al. 2024). This methodology can be employed to sort products based on their sensory qualities, ensuring that only products of the highest caliber reach consumers. By regulating these parameters, biosensors facilitate precise sorting according to the stipulated quality standards, thereby ensuring that only products meeting specific criteria reach the market.

In addition, when combined with traditional data analysis methods, biosensors can predict the shelf life of agricultural products (Kumar et al. 2024b). This necessitates the implementation of sensors that are capable of monitoring indicators of deterioration, including changes in pH, gas composition, and enzyme activity. Complementary physical sensors must also be employed to assess environmental conditions during storage, such as temperature, humidity, and gas composition. The integration of these sensors facilitates the optimization of inventory management, the reduction of waste, and the assurance of the prioritized distribution of products with the longest shelf life.

Finally, biosensors can be developed to detect specific genetic markers associated with desirable traits such as disease resistance, drought tolerance, or increased yield. This facilitates the identification and selection of high-quality crops that possess superior genetic characteristics. Another potential application for such devices is the detection of transgenic crops, given the regulatory framework that governs their use. Consequently, Zhu et al. (2024) developed a CRISPR/Cas12a-mediated entropy-driven electrochemiluminescence (ECL) biosensor. This biosensor was employed to detect MON810, the world's most widely used insect-resistant transgenic maize. The biosensor demonstrated sufficient accuracy and selectivity, with an ECL linear range of 1–106 fM and a detection limit of 3.3 fM. Nevertheless, the authors emphasize that the biosensor's efficacy is not yet optimal, and that false positive or false negative errors may occur due to environmental factors.

In summary, biosensors are facilitating the development of novel methodologies for the purpose of crop monitoring by offering rapid, accurate, and non-invasive methods for assessing crop health. These technologies have the potential to enhance various processes, including cultivation, storage, and distribution, by optimizing them. The integration of biosensors into agricultural practices has the potential to enhance crop yield, mitigate losses, and ensure food security, particularly in the context of escalating demand for resources. Despite current limitations, the potential for biosensors to transform agriculture remains enormous. Consequently, the further development and commercialization of these solutions will contribute to the creation of more sustainable and efficient agri-food systems.

## 5. Conclusions

This chapter emphasizes the key role of innovative nano- and biosensor technologies in addressing the pressing challenges of modern agriculture, directly related to global food security and the need for sustainable development. There is now a critical need for effective monitoring of agricultural systems, and nano- and biosensors show great potential to address this challenge, from assessing soil suitability for farming to monitoring the final product. By evaluating fundamental soil parameters such as moisture, temperature, pH, and nutrient availability, sensor technologies can optimize growing conditions and significantly improve water and fertilizer use efficiency, reducing both costs and environmental burdens. It is equally important to apply them for the timely detection of various soil contaminants such as pesticide residues, heavy metals and other toxins. The platforms utilized in this process encompass a wide range of sensors, including electrochemical and optical sensors based on nanoparticle catalysts or fluorescent transducers, as well as complex bioreceptors based on enzymes, antibodies, DNA, and whole microbial cells, including those genetically modified to enhance specificity. The utilization of nano- and biosensors supplants the conventional approach of time-consuming laboratory analyses with rapid on-site assessments. The capacity of sensor devices to deliver prompt and precise diagnoses of plant diseases contributes to the mitigation of crop losses and the reduction in the utilization of chemical control agents. At the postharvest stage, biosensors also play an indispensable role in quality and safety control of final horticultural products. They allow rapid assessment of maturity (e.g., ethylene levels), nutritional value (vitamins, sugars, antioxidants), and detection of hazardous contaminants including pesticide residues, mycotoxins, heavy metals and pathogens (such as *Salmonella*, *E. coli*, *Listeria*) down to trace concentrations. The ability to detect volatile organic compounds to assess taste or product quality, detect allergens, and identify genetic markers (including GMOs) highlights their versatility. The comprehensive assessment and prediction of shelf life contributes to reducing losses and enhancing the efficiency of the entire agri-food chain.

Consequently, the integration of nano- and biosensors enables the prospective capacity for uninterrupted, accelerated, and highly sensitive monitoring throughout all phases of agro-production. However, challenges related to field reliability, standardization, and cost must still be addressed. Nevertheless, ongoing development and integration of these sensors into agricultural practices offer genuine prospects for enhancing sustainability, resource efficiency, and productivity in global agriculture. Innovations in the domain of nano- and biosensors are, therefore, instrumental in addressing the pressing global challenges of ensuring food security and establishing efficient agro-systems for the future.

## References


Abbas, S. Z., Wang, J. Y., Wang, H., Wang, J. X., Wang, Y. T. and Yong, Y. C. (2022). Recent advances in soil microbial fuel cells based self-powered biosensor. Chemosphere 303(1): Article 135036. https://doi.org/10.1016/j.chemosphere.2022.135036.

Ammar, E. E., Sayed, A. A. S., Rabee, M. M., Keshta, A. E., Daher, M. G. and Ali, G. A. M. (2024). Environmental and agricultural applications of sensors. *In*: Ali, G. A. M., Chong, K. F. and Makhlouf, A. S. H. (eds.). Handbook of Nanosensors. Springer, Cham. https://doi.org/10.1007/978-3-031-47180-3_50.

Awaludin, N., Abdullah, J., Salam, F., Ramachandran, K., Yusof, N. A. and Wasoh, H. (2020). Fluorescence-based immunoassay for the detection of *Xanthomonas oryzae pv. oryzae* in rice leaf. Analytical Biochemistry 610: Article 113876. https://doi.org/10.1016/j.ab.2020.113876.

Babapoor, A., Hajimohammadi, R., Jokar, S. M. and Paar, M. (2020). Biosensor design for detection of mercury in contaminated soil using rhamnolipid biosurfactant and luminescent bacteria. Journal of Chemistry 2020: Article 9120959. https://doi.org/10.1155/2020/9120959.

Bae, J. W., Seo, H. B., Belkin, S. and Gu, M. B. (2020). An optical detection module-based biosensor using fortified bacterial beads for soil toxicity assessment. Anal. Bioanal. Chem. 412: 3373–3381. https://doi.org/10.1007/s00216-020-02469-z.

Baldi, A., Carnevali, L., Collodi, G., Lippi, M. and Manes, A. (2025). Multi-task neural networks for multi-step soil moisture forecasting in vineyards using Internet-of-Things sensors. Smart Agricultural Technology 10: Article 100769. https://doi.org/10.1016/j.atech.2025.100769.

Bandaru, L., Irigireddy, B. C., Koutilya, P. V. N. R. and Davis, B. (2024). DeepQC: A deep learning system for automatic quality control of *in-situ* soil moisture sensor time series data. Smart Agricultural Technology 8: Article 100514. https://doi.org/10.1016/j.atech.2024.100514.

Barac, A. M. (2019). Mycotoxins and human disease: A practical approach. pp. 213–225. *In*: Presterl, E. (ed.). Clinically Relevant Mycoses. Springer Cham. DOI:10.1007/978-3-319-92300-0_14.

Bazhenov, S. V., Novoyatlova, U. S., Scheglova, E. S., Prazdnova, E. V., Mazanko, M. S., Kessenikh, A. G. et al. (2023). Bacterial lux-biosensors: Constructing, applications, and prospects. Biosensors and Bioelectronics: X 13: Article 100323. https://doi.org/10.1016/j.biosx.2023.100323.

Beilinson, R. M., Medyantseva, E. P., Kirillova, E. A., Khaldeeva, E. V. and Budnikov, H. C. (2022). Amperometric enzyme immunoassay sensors in determination of *Fusarium oxysporum* antigens. Uchenye Zapiski Kazanskogo Universiteta. Seriya Estestvennye Nauki 164(1): 22–35. doi: 10.26907/2542-064X.2022.1.22-35 (In Russian).

Bharti, A., Jain, U. and Chauhan, N. (2024a). From lab to field: nano-biosensors for real-time plant nutrient tracking. Plant Nano Biology 9: Article 100079, doi:10.1016/j.plana.2024.100079.

Bharti, A., Jain, U. and Chauhan, N. (2024b). Progressive analytical techniques utilized for the detection of contaminants attributed to food safety and security. Talanta Open 10: Article 100368. https://doi.org/10.1016/j.talo.2024.100368.

Bolat, G. and Abaci, S. (2018). Non-enzymatic electrochemical sensing of malathion pesticide in tomato and apple samples based on gold nanoparticles-chitosan-ionic liquid hybrid nanocomposite. Sensors 18: Article 773.

Branyikova, I., Lucakova, S., Kuncova, G., Trogl, J., Synek, V., Rohovec, J. et al. (2020). Estimation of Hg(II) in soil samples by bioluminescent bacterial bioreporter *E. coli* ARL1, and the effect of humic acids and metal ions on the biosensor performance. Sensors 20: Article 3138. doi: 10.3390/s20113138.

Cai, L., Tian, Q., Meng, Q., Bao, X., Xu, P., Liu, J. et al. (2023). Recombinase polymerase amplification assay for rapid field diagnosis of stewart's wilt of corn pathogen *Pantoea stewartii* subsp. *stewartii*. Agriculture 13: Article 1982. https://doi.org/10.3390/agriculture13101982.

Cao, D.-F., Zhu, H.-H., Wu, B., Wang, J.-C. and Shukla, S. K. (2021). Investigating temperature and moisture profiles of seasonally frozen soil under different land covers using actively heated fiber Bragg grating sensors. Engineering Geology 290: Article 106197. https://doi.org/10.1016/j.enggeo.2021.106197.

Cardemil, C. V., Smulski, D. R., Larossa, R. A. and Vollmer, A. C. (2010). Bioluminescent *Escherichia coli* strains for the quantitative detection of phosphate and ammonia in coastal and suburban watersheds. DNA and Cell Biology 29(9): 519–531. https://doi.org/10.1089/dna.2009.0984.

Chai, Y., Horikawa, S., Li, S., Wikle, H. C. and Chin, B. A. (2013). A surface-scanning coil detector for real-time, *in-situ* detection of bacteria on fresh food surfaces. Biosensors and Bioelectronics 50: 311–317. https://doi.org/10.1016/j.bios.2013.06.056.

Chalupowicz, D., Veltman, B., Droby, S. and Eltzov, E. (2020). Evaluating the use of biosensors for monitoring of *Penicillium digitatum* infection in citrus fruit. Sensors and Actuators B: Chemical 311: Article 127896. doi:10.1016/j.snb.2020.127896.

Chelladuraia, K., Sujathac, N. and Valli, L. N. (2023). Fungal infection in plant leaves—A review. E3S Web of Conferences 399: 02010. https://doi.org/10.1051/e3sconf/202339902010.

Chen, B., Li, L., Cai, W. and Garg, A. (2024). Bioelectricity generation in plant microbial fuel cells: Influence of vegetation density and unsaturated soil properties. Biomass and Bioenergy 181: Article 107053. https://doi.org/10.1016/j.biombioe.2024.107053.

Chen, J., Chen, M., Tong, H., Wu, F., Liu, Y. and Liu, C. (2023). Fluorescence biosensor for ultrasensitive detection of the available lead based on target biorecognition-induced DNA cyclic assembly. Science of The Total Environment 905: Article 167253. https://doi.org/10.1016/j.scitotenv.2023.167253.

Christakakis, P., Papadopoulou, G., Mikos, G., Kalogiannidis, N., Ioannidis, D., Tzovaras, D. et al. (2024). Smartphone-based citizen science tool for plant disease and insect pest detection using artificial intelligence. Technologies 12: Article 101. https://doi.org/10.3390/technologies12070101.

Cui, X., Jin, M., Du, P., Chen, G., Zhang, C., Zhang, Y. et al. (2018). Development of immunoassays for multi-residue detection of small molecule compounds. Food Agric Immunol. 29(1): 638–652. https://doi.org/10.1080/09540105.2018.1428284.

Dalkıran, B. (2020). Amperometric determination of heavy metal using an HRP inhibition biosensor based on ITO nanoparticles-ruthenium (III) hexamine trichloride composite: central composite design optimization. Bioelectrochemistry 135: Article 107569.

Das, S., Agarwal, D. K., Mandal, B., Rao, V. R. and Kundu, T. (2021). Detection of the *Chilli leaf curl virus* using an attenuated total reflection-mediated localized surface-plasmon-resonance-based optical platform. ACS Omega 6(27): 17413–17423. DOI: 10.1021/acsomega.1c01702.

El-Moghazy, A. Y., Wisuthiphaet, N., Yang, X., Sun, G. and Nitin, N. (2022). Electrochemical biosensor based on genetically engineered bacteriophage T7 for rapid detection of *Escherichia coli* on fresh produce. Food Control 135: Article 108811. https://doi.org/10.1016/j.foodcont.2022.10881.

Elsebai, B., Ghica, M. E., Abbas, M. N. and Brett, C. M. A. (2017). Catalase based hydrogen peroxide biosensor for mercury determination by inhibition measurements. J. Hazard Mater 340: 344–350.

Essington, E. A., Vezeau, G. E., Cetnar, D. P., Grandinette, E., Bell, T. H. and Salis, H. M. (2024). An autonomous microbial sensor enables long-term detection of TNT explosive in natural soil. Nat Commun. 15: Article 10471. https://doi.org/10.1038/s41467-024-54866-y.

FAO, IFAD, UNICEF, WFP and WHO. (2024). The State of Food Security and Nutrition in the World 2024 – Financing to end hunger, food insecurity and malnutrition in all its forms. Rome. https://doi.org/10.4060/cd1254en.

Gan, Z., Zhou, Q., Zheng, C. and Wang, J. (2023). Challenges and applications of volatile organic compounds monitoring technology in plant disease diagnosis. Biosensors and Bioelectronics 237: Article 115540. https://doi.org/10.1016/j.bios.2023.115540.

Gong, H., Pang, T., Yang, X., Chen, F., Jiang, N., Li, Y. et al. (2025). Rapid visual detection of *Tobacco mosaic virus* using a portable paper-based molecularly imprinted sensor. Sensors and Actuators B: Chemical 424: Article 136909. https://doi.org/10.1016/j.snb.2024.136909.

Guo, Y., Liu, R., Liu, Y., Xiang, D., Liu, Y., Gui, W. et al. (2018). A non-competitive surface plasmon resonance immunosensor for rapid detection of triazophos residue in environmental and agricultural samples. Science of The Total Environment 613–614: 783–791. doi:10.1016/j.scitotenv.2017.09.157.

Hamouda, F., Puig-Sirera, À., Bonzi, L., Remorini, D., Massai, R. and Rallo, G. (2024). Design and validation of a soil moisture-based wireless sensors network for the smart irrigation of a pear orchard. Agricultural Water Management 305: Article 109138. https://doi.org/10.1016/j.agwat.2024.109138.

Hassan, M. H., Omar, A. M., Daskalakis, E., Mohamed, A. A., Boyd, L. A., Blanford, C. et al. (2022). Multi-layer biosensor for pre-symptomatic detection of *Puccinia strifformis*, the causal agent of yellow rust. Biosensors 12(10): Article 829. doi: 10.3390/bios12100829.

He, J. Y., Chen, K., Pan, X. B., Zhai, J. F. and Lin, X. M. (2023). Advanced biosensing technologies for monitoring of agriculture pests and diseases: A review. J. Semicond. 44(2): Article 023104. https://doi.org/10.1088/1674-4926/44/2/023104.

He, Y., Zhou, J., Fu, R., Liu, Y., Wang, Y., Liu, H. et al. (2022). The application of DNA-HRP functionalized AuNP probes in colorimetric detection of citrus-associated *Alternaria* genes. Talanta 237: Article 122917. https://doi.org/10.1016/j.talanta.2021.122917.

Herron, P. M., Gage, D. J., Pinedo, C. A., Haider, Z. K. and Cardon, Z. G. (2013). Better to light a candle than curse the darkness: illuminating spatial localization and temporal dynamics of rapid microbial growth in the rhizosphere. Front. Plant Sci. 4: Article 323. https://doi.org/10.3389/fpls.2013.00323.

Hua, Z., Yu, T., Liu, D. and Xianyu, Y. (2021). Recent advances in gold nanoparticles-based biosensors for food safety detection. Biosensors and Bioelectronics 179: Article 113076. doi:10.1016/j.bios.2021.113076.

Jablonski, M., Poghossian, A., Keusgen, M., Wege, C. and Schöning, M. J. (2021). Detection of plant virus particles with a capacitive field-effect sensor. Anal. Bioanal. Chem. 413(22): 5669–5678. doi:10.1007/s00216-021-03448-8.

Jin, Z., Wang, D.-C., Xie, W.-J., Ding, Y. and Li, J. (2023). Highly sensitive ethylene sensors based on ultrafine Pd nanoparticles-decorated porous ZnO nanosheets and their application in fruit ripeness detection. Processes 11: Article 1686. https://doi.org/10.3390/pr11061686.

Katz, S. (1952). The reversible reaction of sodium thymonucleate and mercuric chloride. J. Am. Chem. Soc. 74: 2238–2245. DOI: 10.1021/ja01129a023.

Kelley, B., Ali, N. and Dong, Y. (2025). Methods to correct temperature-induced changes of soil moisture sensors to improve accuracy. MethodsX 14: Article 103100. https://doi.org/10.1016/j.mex.2024.103100.

Khan, N. S., Pradhan, D., Choudhary, S., Saxena, P. and Poddar, N. K. (2021). Immunoassay-based approaches for development of screening of chlorpyrifos. J. Anal. Sci. Technol. 12: Article 32. https://doi.org/10.1186/s40543-021-00282-6.

Khan, R., Anwar, F. and Ghazali, F. M. (2024). A comprehensive review of mycotoxins: Toxicology, detection, and effective mitigation approaches. Heliyon 10(8): Article e28361. https://doi.org/10.1016/j.heliyon.2024.e28361.

Khmeleva, S. A., Kurbatov, L. K., Ptitsyn, K. G., Timoshenko, O. S., Morozova, D. D., Suprun, E. V. et al. (2024). Detection of potato pathogen *Clavibacter sepedonicus* by CRISPR/Cas13a analysis of NASBA Amplicons. Int. J. Mol. Sci. 25: Article 12218. https://doi.org/10.3390/ijms252212218.

Khmelevtsova, L., Klimova, M., Karchava, S., Azhogina, T., Polienko, E., Litsevich, A. et al. (2024). Biosensor-based assessment of pesticides and mineral fertilizers' influence on ecotoxicological parameters of soils under soya, sunflower and wheat. Chemosensors 12: Article 73. https://doi.org/10.3390/chemosensors12050073.

Kim, M. J., Lee, H.-S., Chung, D. H. and Lee, Y. T. (2003). Synthesis of haptens of organophosphorus pesticides and development of enzyme-linked immunosorbent assays for parathion-methyl. Analytica Chimica Acta 493(1): 47–62. https://doi.org/10.1016/S0003-2670(03)00793-1.

Kou, S., Qiu, G., Liu, L., Wu, A., Zhang, Z., Kuang, H. et al. (2025). Visual immunosensor assay with double T line for on-site sensitive fipronil pesticide detection in water, soil and honeysuckle. Journal of Hazardous Materials 489: Article 137634. https://doi.org/10.1016/j.jhazmat.2025.137634.

Kumar, A., Chandel, R., Kaur, R., Kumar, S., Gautam, N. and Kumar, V. (2024a). Biosensors: recent advancements in postharvest management of fruits. pp. 253–280. *In*: Irfan, M. and Kumar, P. (eds.). Bioengineered Fruit and Vegetables. Apple Academic Press. DOI:10.1201/9781003501954-12.

Kumar, A., Kulshreshtha, S., Shrivastava, A. and Saini, A. (2024b). Biosensors for food spoilage detection: a comprehensive review of current advances. J. Food Chem. Nanotechnol. 10(S1): S73–S82.

Kumar, V. and Arora, K. (2020). Trends in nano-inspired biosensors for plants. Mater. Sci. Technol. 3: 255–273. doi:10.1016/j.mset.2019.10.004.

Lautenschläger, N., Popp, P. F. and Mascher, T. (2020). Development of a novel heterologous β-lactam-specific whole-cell biosensor in *Bacillus subtilis*. J. Biol. Eng. 14: Article 21. https://doi.org/10.1186/s13036-020-00243-4.

Lei, R., Wu, P., Li, L., Huang, Q., Wang, J., Zhang, D. et al. (2021). Ultrasensitive isothermal detection of a plant pathogen by using a gold nanoparticle-enhanced microcantilever sensor. Sens. Actuators B Chem. 338: Article 129874.

Li, C., Wu, Y., Zeng, H., Wang, Y., Yu, Y., Jiao, J. et al. (2024a). Label free immunochromatographic sensor based on TCBPE for the detection of *Acidovorax citrulli* in Cucurbitaceae plants. Microchemical Journal 206: Article 111493. https://doi.org/10.1016/j.microc.2024.111493.

Li, F., Duan, X., Li, H., Zou, L., Liu, G., Liu, F. et al. (2022a). Dual effect of aminobutyric acid group and "molecular wire effect" of conjugated polymer enables ultra-trace detection of $Cr_2O_7^{2-}$ in fruits. Microchem. J. 178: Article 107426. 10.1016/j.microc.2022.107426.

Li, J., Wang, Q., Xiong, C., Deng, Q., Zhang, X., Wang, S. et al. (2022b). An ultrasensitive $CH_3NH_3PbBr_3$ quantum dots@ $SiO_2$-based electrochemiluminescence sensing platform using an organic electrolyte for aflatoxin B1 detection in corn oil. Food Chem. 390: Article 133200. https://doi.org/10.1016/j.foodchem.2022.133200.

Li, L., Liu, X., He, S., Cao, H., Su, B., Huang, T. et al. (2021a). Electrochemiluminescence immunosensor based on nanobody and Au/$CaCO_3$ synthesized using waste eggshells for ultrasensitive detection of ochratoxin A. ACS Omega 6(44): 30148–30156.

Li, X., Wu, F., Pang, Y., Nie, H., Huang, S. and Luo, Y. (2021b). Detection of pesticide residue acetamiprid by SERS based on aptamers regulating carbon dots catalytic reactions. Sci. Technol. Food Ind. 42: 236–244.

Li, X.-H., Li, M.-Z., Li, J.-Y., Gao, Y.-Y., Liu, C.-R. and Hao, G.-F. (2024b). Wearable sensor supports *in-situ* and continuous monitoring of plant health in precision agriculture era. Plant Biotechnol. J. 22: 1516–1535. https://doi.org/10.1111/pbi.14283.

Lin, W.-P., Wang, W.-J., Lee, C.-H., Jan, F.-J. and Wang, G.-J. (2022). A two-in-one immunoassay biosensor for the simultaneous detection of *Odontoglossum ringspot virus* and *Cymbidium mosaic virus*. Sensors and Actuators B: Chemical 350: Article 130875. https://doi.org/10.1016/j.snb.2021.130875.

Lin, Z., Liu, X., Li, Y., Li, C., Yang, L., Ma, K. et al. (2021). Electrochemical aptasensor based on $Mo_2C/Mo_2N$ and gold nanoparticles for determination of chlorpyrifos. Microchimica Acta 188(5): Article 170. DOI: 10.1007/s00604-021-04830-0.

Ling, S., Dong, M., Xu, Y., Xu, A., Lin, J., Lin, M. et al. (2022). Development of sensitive and portable immunosensors based on signal amplification probes for monitoring the mercury(II) ions. Biosens. Bioelectron. 217: Article 114676. https://doi.org/10.1016/j.bios.2022.114676.

Liu, B., Zhai, H., Huang, Y., Yuan, T., Li, J., Li, M. et al. (2025a). Enhanced bioremediation of benzo [a]pyrene-polluted soil using high-efficiency soil microbial fuel cells with artificial solute transport. Separation and Purification Technology 363(2): Article 132031. https://doi.org/10.1016/j.seppur.2025.132031.

Liu, C., Cheng, L. and Jia, H. (2024). Application of microbial fuel cell-based biosensor in environmental monitoring. Electroanalysis 36: Article e202400100. https://doi.org/10.1002/elan.202400100.

Liu, L., Lu, Y., Zhong, W., Meng, L. and Deng, H. (2020). On-line monitoring of repeated copper pollution using sediment microbial fuel cell based sensors in the field environment. Sci. Total Environ. 748: Article 141544.

Liu, Z., Wang, Z., Ma, L., Guo, L., Yang, H., Liu, Y. et al. (2025b). Programmable target-driven MNAzyme and DNA tetrahedron framework synergistically constructed label-free electrochemical sensing platform for TMV RNA detection. Sensors and Actuators B: Chemical 428: Article 137256. https://doi.org/10.1016/j.snb.2025.137256.

Luna-Moreno, D., Sánchez-Álvarez, A., Islas-Flores, I., Canto-Canche, B., Carrillo-Pech, M., Villarreal-Chiu, J. F. et al. (2019). Early detection of the fungal banana black sigatoka pathogen *Pseudocercospora fijiensis* by an SPR immunosensor method. Sensors 19: Article 465. https://doi.org/10.3390/s19030465.

Lv, X., Zhang, X., Gao, H., He, T., Lv, Z. and Zhangzhong, L. (2024). When crops meet machine vision: A review and development framework for a low-cost nondestructive online monitoring technology in agricultural production. Agriculture Communications 2(1): Article 100029. https://doi.org/10.1016/j.agrcom.2024.100029.

Lzaod, S. and Dutta, T. (2022). Recent advances in the development of oxidoreductase-based biosensors for detection of phenolic antioxidants in food and beverages. ACS Omega 7(51): 47434–47448, https://doi.org/10.1021/acsomega.2c05604.

Ma, Z., Li, Y., Lu, Z., Pan, J. and Li, M. (2023). A novel biosensor-based method for the detection of p-nitrophenol in agricultural soil. Chemosphere 313: Article 137306. https://doi.org/10.1016/j.chemosphere.2022.137306.

Ma, Z., Liu, J., Li, H., Zhang, W., Williams, M., Gao, Y. et al. (2020). A fast and easily parallelizable biosensor method for measuring extractable tetracyclines in soils. Environ. Sci. Technol. 54(2): 758–767. https://doi.org/10.1021/acs.est.9b04051.

MacVittie, K., Conlon, T. and Katz, E. (2015). A wireless transmission system powered by an enzyme biofuel cell implanted in an orange. Bioelectrochemistry 106(A): 28–33. http://dx.doi.org/10.1016/j.bioelechem.2014.10.005.

Mahajan, P., Khanna, V., Singh, A. and Singh, K. (2024). Advances in nanomaterial-based biosensors for heavy metal detection and remediation in soil. J. Electrochem. Soc. 171: Article 117527. DOI 10.1149/1945-7111/ad9413.

Mallick, M., Basu, D., Hossain, S. M. and Das, J. (2023). Ethylene sensor based on graphene oxide for fruit ripeness sensing application. Appl. Phys. A 129: Article 140. https://doi.org/10.1007/s00339-023-06413-7.

Mane, S., Das, N., Singh, G., Cosh, M. and Dong, Y. (2024). Advancements in dielectric soil moisture sensor Calibration: A comprehensive review of methods and techniques. Computers and Electronics in Agriculture 218: Article 108686. https://doi.org/10.1016/j.compag.2024.108686.

Mathur, S., Singh, D. and Ranjan, R. (2023). Genetic circuits in microbial biosensors for heavy metal detection in soil and water. Biochemical and Biophysical Research Communications 652: 131–137. https://doi.org/10.1016/j.bbrc.2023.02.031.

Mediavilla, M., Revenga-Parra, M., Gutiérrez-Sánchez, C., Hernández-Apaolaza, L., Pariente, F. and Lorenzo, E. (2022). Fluorescent enzymatic assay for direct total polyphenol determination in food-related samples. Talanta 247: Article 123576. doi:10.1016/j.talanta.2022.123576.

Medina-Plaza, C., de Saja, J. A., Fernández-Escudero, J. A., Barajas, E., Medrano, G. and Rodriguez-Mendez, M. L. (2016). Array of biosensors for discrimination of grapes according to grape variety, vintage and ripeness. Analytica Chimica Acta 947: 16–22. doi:10.1016/j.aca.2016.10.032.

Mukherjee, S., Pal, S., Paria, P., Bhattacharyya, S., Ghosh, K., Pal, A. et al. (2021). On-spot biosensing device for organophosphate pesticide residue detection in fruits and vegetables. Current Research in Biotechnology 3: 308–316. doi:10.1016/j.crbiot.2021.11.002.

Muñoz-Martín, M. A., Mateo, P., Leganés, F. and Fernández-Piñas, F. (2014). A battery of bioreporters of nitrogen bioavailability in aquatic ecosystems based on cyanobacteria. Sci. Total Environ. 475: 169–179. https://doi.org/10.1016/j.scitotenv.2013.07.015.

Nazarov, P. A., Baleev, D. N., Ivanova, M. I., Sokolova, L. M. and Karakozova, M. V. (2020). Infectious plant diseases: etiology, current status, problems and prospects in plant protection. Acta Naturae. Jul–Sep 12(3): 46–59. doi: 10.32607/actanaturae.11026.

O Hara, T. and Singh, B. (2021). Electrochemical biosensors for detection of pesticides and heavy metal toxicants in water: recent trends and progress. ACS EST Water 1(3): 462–478. https://doi.org/10.1021/acsestwater.0c00125.

Ospina, E., Alonso, B., Casado, C. M. and Armada, M. P. G. (2023). Electrocatalytic self-assembled multilayer structures based on thiolated Fc-DAB dendrimers: Determination of heavy metal ions by enzymatic inhibition. Catalysis Today 423: Article 114293. https://doi.org/10.1016/j.cattod.2023.114293.

Patel, R., Vinchurkar, M., Shaikh, A. M., Patkar, R., Adami, A., Giacomozzi, F. et al. (2023). Part II: Impedance-based DNA biosensor for detection of isolated strains of phytopathogen *Ralstonia solanacearum*. Bioelectrochemistry 153: Article 108500. doi: 10.1016/j.bioelechem.2023.108500.

Pestovsky, Y. S. and Martínez-Antonio, A. (2017). The use of nanoparticles and nanoformulations in agriculture. J. Nanosci. Nanotechno. 17: 8699–8730.

Rafeeq, H., Afsheen, N., Rafique, S., Arshad, A., Intisar, M., Hussain, A. et al. (2023). Genetically engineered microorganisms for environmental remediation. Chemosphere 310: Article 136751. https://doi.org/10.1016/j.chemosphere.2022.136751.

Rahmani, T., Hajian, A., Afkhami, A. and Bagheri, H. (2018). A novel and high performance enzyme-less sensing layer for electrochemical detection of methyl parathion based on BSA templated Au–Ag bimetallic nanoclusters. New J. Chem. 42: 7213–7222.

Rai, P., Mehrotra, S. and Sharma, S. K. (2022). Challenges in assessing the quality of fruit juices: intervening role of biosensors. Food Chemistry 386: Article 132825. doi:10.1016/j.foodchem.2022.132825.

Rajangam, B., Daniel, D. K. and Krastanov, A. I. (2017). Progress in enzyme inhibition based detection of pesticides. Eng Life Sci. 18(1): 4–19. doi: 10.1002/elsc.201700028.

Rana, K., Mittal, J., Narang, J., Mishra, A. and Pudake, R. N. (2021). Graphene based electrochemical DNA biosensor for detection of false smut of rice (*Ustilaginoidea virens*). Plant Pathol J. 37(3): 291–298. doi: 10.5423/PPJ.OA.11.2020.0207.

Rathnayake, I. V. N., Megharaj, M. and Naidu, R. (2021). Green fluorescent protein based whole cell bacterial biosensor for the detection of bioavailable heavy metals in soil environment. Environmental Technology & Innovation 23: Article 101785. https://doi.org/10.1016/j.eti.2021.101785.

Ribeiro, D. B., Santos Silva, G., dos Santos, D. R., Castro Costa, A. R., Braga Ribeiro, E., Badea, M. et al. (2021). Determination of the antioxidant activity of samples of tea and commercial sources of vitamin c, using an enzymatic biosensor. Antioxidants 10: Article 324. https://doi.org/10.3390/antiox10020324.

Ribeiro, F. W. P., Barroso, M. F., Morais, S., Viswanathan, S., de Lima-Neto, P., Correia, A. N. et al. (2014). Simple laccase-based biosensor for formetanate hydrochloride quantification in fruits. Bioelectrochemistry 95: 7–14. doi:10.1016/j.bioelechem.2013.09.005.

Rubia, M. I., Ramachandran, V. K., Arrese-Igor, C., Larrainzar, E. and Poole, P. S. (2020). A novel biosensor to monitor proline in pea root exudates and nodules under osmotic stress and recovery. Plant Soil 452: 413–1422. https://doi.org/10.1007/s11104-020-04577-2.

Sambasivam, P., Bilkiss, M., Soda, N., Bar, I., Shiddiky, M. J. A. and Ford, R. (2024). A rapid electrochemical biosensor diagnostic for *Botrytis* ssp. causing botrytis gray mold of temperate legumes. CS Agric. Sci. Technol. 4(11): 1184–1193. https://doi.org/10.1021/acsagscitech.4c00136.

Sen, S., Roy, A., Sanyal, A. and Devi, P. S. (2022). A nonenzymatic reduced graphene oxide-based nanosensor for parathion. Beilstein J. Nanotechnol. 13: 730–744. doi:10.3762/bjnano.13.65.

Shad, N. A., Munawar, A., Javed, Y., Rakha, A., Riaz, A., Din, S. U. et al. (2023). In-field deployable and facile nanosensor for the detection of pesticides residues. Analytica Chimica Acta 1259: Article 341204. https://doi.org/10.1016/j.aca.2023.341204.

Sharma, R., Bhat, G. P. and Gandhi, S. (2025). MXene-rGO nanocomposite based electrochemical immunosensor for detection of endosulfan – An organochlorine pesticide. Chemosphere 370: Article 143997. https://doi.org/10.1016/j.chemosphere.2024.143997.

Sheng, W., Wu, Y., Chang, C.-Y., González-Teruel, J. D., Jones, S. B. and Meng, F. (2024). Sampling volume evaluation of seven electromagnetic soil moisture sensors. Agricultural and Forest Meteorology 358: Article 110240. https://doi.org/10.1016/j.agrformet.2024.110240.

Shoute, L. C. T., Anwar, A., MacKay, S., Abdelrasoul, G. N., Lin, D., Yan, Z. et al. (2018). Immuno-impedimetric biosensor for onsite monitoring of ascospores and forecasting of sclerotinia stem rot of canola. Sci. Rep. 8: Article 12396. https://doi.org/10.1038/s41598-018-30167-5.

Singh, A., Ganapathysubramanian, B., Singh, A. K. and Sarkar, S. (2016). Machine learning for high-throughput stress phenotyping in plants. Trends Plant Sci. 21: 110–124.

Siontorou, C. G. and Georgopoulos, K. N. (2016). A biosensor platform for soil management: the case of nitrites. Journal of Cleaner Production 111(A): 133–142. https://doi.org/10.1016/j.jclepro.2015.07.038.

Spissu, Y., Barberis, A., Bazzu, G., D'hallewin, G., Rocchitta, G., Serra, P. A. et al. (2021). Functionalization of screen-printed sensors with a high reactivity carbonaceous material for ascorbic acid detection in fresh-cut fruit with low vitamin C content. Chemosensors 9: Article 354. https://doi.org/10.3390/chemosensors9120354.

Sun, K., Liu, Y., Tang, Q., Ma, B., Zhang, B., Fu, X. et al. (2025). CADLAB: A contamination-free, argonaute-enhanced, dual detection of plant viruses via lateral flow assay biosensors system. Sensors and Actuators B: Chemical 426: Article 137045. https://doi.org/10.1016/j.snb.2024.137045.

Tang, W., Gu, Z., Chu, Y., Lv, J., Fan, L., Liu, X. et al. (2023). Magnetically-oriented porous hydrogel advances wearable electrochemical solidoid sensing heavy metallic ions. Chemical Engineering Journal 453: Article 139902. doi:10.1016/j.cej.2022.139902.

Trucksess, M. W. and Diaz-Amigo, C. (2011). Mycotoxins in foods. pp. 888–897. *In*: Nriagu, J. O. (ed.). Encyclopedia of Environmental Health. Burlington: Elsevier. doi:10.1016/b978-0-444-52272-6.00700-5.

Uchiyama, T., Miura, T., Takeuchi, H., Dairaku, T., Komuro, T., Kawamura, T. et al. (2012). Raman spectroscopic detection of the T-HgII-T base pair and the ionic characteristics of mercury. Nucleic Acids Res. 40: 5766–5774. DOI: 10.1093/nar/gks208.

Ülger, T. G., Uçar, A., Çakıroğlu, F. P. and Yilmaz, S. (2020). Genotoxic effects of mycotoxins. Toxicon 185: 104–113. https://doi.org/10.1016/j.toxicon.2020.07.004.

Ullah, A., Zubair, M., Zulfiqar, M. H., Kamsong, W., Karuwan, C., Massoud, Y. et al. (2024). Highly sensitive screen-printed soil moisture sensor array as green solutions for sustainable precision agriculture. Sensors and Actuators A: Physical 371: Article 115297. https://doi.org/10.1016/j.sna.2024.115297.

Vatankhah, A., Reezi, S., Izadi, Z., Ghasemi-Varnamkhasti, M. and Motamedi, A. (2022). Development of an ultrasensitive electrochemical biosensor for detection of *Agrobacterium tumefaciens* in *Rosa hybrida* L. Measurement 187: Article 110320. https://doi.org/10.1016/j.measurement.2021.110320.

Velusamy, K., Periyasamy, S., Kumar, P. S., Rangasamy, G., Nisha Pauline, J. M., Ramaraju, P. et al. (2022). Biosensor for heavy metals detection in wastewater: A review. Food Chem. Toxicol. 168: Article 113307. https://doi.org/10.1016/j.fct.2022.113307.

Ventura-Aguilar, R. I., Lucas-Bautista, J. A., Arévalo-Galarza, M. d. L. and Bosquez-Molina, E. (2024). Volatile organic compounds as a diagnostic tool for detecting microbial contamination in fresh agricultural products: mechanism of action and analytical techniques. Processes 12: Article 1555. https://doi.org/10.3390/pr12081555.

Vong, K., Eda, S., Kadota, Y., Nasibullin, I., Wakatake, T., Yokoshima, S. et al. (2019). An artificial metalloenzyme biosensor can detect ethylene gas in fruits and *Arabidopsis* leaves. Nat Commun 10: Article 5746. https://doi.org/10.1038/s41467-019-13758-2.

Wang, S.-H., Wang, J.-W., Zhao, L.-T., Abbas, S. Z., Yang, Z. and Yong, Y.-C. (2023). Soil microbial fuel cell based self-powered cathodic biosensor for sensitive detection of heavy metals. Biosensors 13: Article 145. https://doi.org/10.3390/bios13010145.

Wang, W., Xu, Y., Cheng, N., Xie, Y., Huang, K. and Xu, W. (2020). Dual-recognition aptazyme-driven DNA nanomachine for two-in-one electrochemical detection of pesticides and heavy metal ions. Sensors and Actuators B: Chemical 321: Article 128598. doi:10.1016/j.snb.2020.128598.

Wu, B., Xu, H., Shi, Y., Yao, Z., Yu, J., Zhou, H. et al. (2022). Microelectrode glucose biosensor based on nanoporous platinum/graphene oxide nanostructure for rapid glucose detection of tomato and cucumber fruits. Food Quality and Safety 6: Article fyab030. https://doi.org/10.1093/fqsafe/fyab030.

Wu, C.-C., Lin, C.-C., Yin, H.-Y., Tsai, W.-C., Yang, P.-F., Liu, H.-J. et al. (2024). Rapid detection of mango allergen in processed foods using an immunomagnetic nanoparticle-based electrochemical immunosensor. Microchemical Journal 198: Article 110070. doi:10.1016/j.microc.2024.110070.

Xu, J., Chau, Y. and Lee, Y. (2019). Phage-based electrochemical sensors: A review. Micromachines 10(12): Article 855. doi:10.3390/mi10120855.

Xu, J., Tran, T., Marcia, C. S. P., Braun, D. M. and Goggin, F. L. (2017). Superoxide-responsive gene expression in *Arabidopsis thaliana* and *Zea mays*. Plant Physiol. Biochem. 117: 51–60. DOI: 10.1016/j.plaphy.2017.05.018.

Xu, K., Ma, L., Chen, Y., Li, Y., Abdalbagemohammedabdalsadeg, S., Moosavi-Movahedi, A. A. et al. (2024). A biosensor based on MWCNTs-BSA and $TiO_2$-laccase nanocomposite modified glassy carbon electrode for sensitive detection of luteolin in traditional Chinese medicine. Microchemical Journal 207: Article 112103. doi:10.1016/j.microc.2024.112103.

Xu, L., Abd El Aty, A. M., Cao, Z., Lei, X., Zhao, J., Li, J. et al. (2023). Generating monoclonal antibodies against buprofezin and developing immunoassays for its residue detection in tea samples. J. Agric Food Chem. 71(41): 14967–14978. doi: 10.1021/acs.jafc.3c03263.

Yang, J., Zhang, Y., Wu, L. and Lu, Y. (2023). A coffee-ring effect-based paper sensor chip for the determination of beta-lactoglobulin in foods via a smartphone. Sensors and Actuat B Chem. 374: Article 132807. doi: 10.1016/J.SNB.2022.132807.

Yang, X. and Chen, S. (2021). Microorganisms in sediment microbial fuel cells: ecological niche, microbial response, and environmental function. Sci. Total Environ. 756: Article 144145.

Yuan, M., Qian, S., Cao, H., Yu, J., Ye, T., Wu, X. et al. (2022). An ultra-sensitive electrochemical aptasensor for simultaneous quantitative detection of $Pb^{2+}$ and $Cd^{2+}$ in fruit and vegetable. Food Chemistry 382: Article 132173. https://doi.org/10.1016/j.foodchem.2022.1321.

Zeng, N., Wu, Y., Chen, W., Huang, Q. and Cai, P. (2021). Whole-cell microbial bioreporter for soil contaminants detection. Front. Bioeng. Biotechnol. 9: 622994. doi: 10.3389/fbioe.2021.622994.

Zhan, F., Wang, T., Iradukunda, L. and Zhan, J. (2018). A gold nanoparticle-based lateral flow biosensor for sensitive visual detection of the potato late blight pathogen, *Phytophthora infestans.* Analytica Chimica Acta 1036: 153–161. https://doi.org/10.1016/j.aca.2018.06.083.

Zhang, C., Qiu, M., Wang, J. and Liu, Y. (2023). Recent advances in nanoparticle-based optical sensors for detection of pesticide residues in soil. Biosensors 13: Article 415. https://doi.org/10.3390/bios13040415.

Zhang, M., Zhou, J., Li, Y., Luo, Q., Xiong, X., Shi, X. et al. (2025). An integrated three-signal biosensor based on phage multifunctional probe for simultaneous and ultrasensitive detection of live/dead *Listeria monocytogenes*. Sensors and Actuators B: Chemical 423: Article 136709, doi:10.1016/j.snb.2024.136709.

Zhang, Q., Gu, C., Singh, R., Zhang, B. and Kumar, S. (2024). Development of WaveFlex biosensor for rapid detection of glyphosate herbicide in real agricultural products. IEEE Sensors Journal 24(9): 14320–14327. doi: 10.1109/JSEN.2024.3380601.

Zhang, S., Daniels, D. A., Ivanov, S., Jurgensen, L., Müller, L. M., Versaw, W. K. et al. (2022). A genetically encoded biosensor reveals spatiotemporal variation in cellular phosphate content in Brachypodium distachyon mycorrhizal roots. New Phytol. 234: 1817–1831. https://doi.org/10.1111/nph.18081.

Zhao, M., You, Z., Chen, H., Wang, X., Ying, Y. and Wang, Y. (2024). Integrated fruit ripeness assessment system based on an artificial olfactory sensor and deep learning. Foods 13: Article 793. https://doi.org/10.3390/foods13050793.

Zhou, J., Liu, Y., Du, X., Gui, Y., He, J., Xie, F. et al. (2023). Recent advances in design and application of nanomaterials-based colorimetric biosensors for agri-food safety analysis. ACS Omega 8(49): 46309–47317. https://pubs.acs.org/doi/full/10.1021/acsomega.3c06409.

Zhou, T., Han, H., Liu, P., Xiong, J., Tian, F. and Li, X. (2017a). Microbial fuels cell-based biosensor for toxicity detection: a review. Sensors 17: Article 2230. https://doi.org/10.3390/s17102230.

Zhou, W., Saran, R. and Liu, J. (2017b). Metal sensing by DNA. Chemical Reviews 117(12): 8272–8325. DOI: 10.1021/acs.chemrev.7b00063.

Zhu, X., Zhang, J., Pan, R., Zhang, K. and Dai, H. (2024). CRISPR/Cas12a-mediated entropy-driven electrochemical biosensor for detection of genetically modified maize Mon810. Analytica Chimica Acta 1296: Article 342290. doi:10.1016/j.aca.2024.342290.

# Chapter 7

# Nanobiotechnology for the Management of Plant Diseases

*María de Jesús Briseño,*[1] *Juan Jose Reyes-Pérez,*[2] *Bernardo Murillo-Amador,*[1] *Graciela Avila-Quezada,*[3] *Roberto Gregorio Chiquito-Contreras*[4] and *Luis Guillermo Hernández-Montiel*[1,*]

## 1. Introduction

Nanobiotechnology is a discipline that focuses on the study and manipulation of matter at the nanometric scale (1–100 nm) to develop new devices, materials and applications in the fields of biology and biochemistry (Sevimli et al. 2024, Díaz-Marcos and Mendoza-Gonzálvez 2025). This area integrates principles of nanotechnology and biotechnology, which are based on a deep knowledge of molecular biology to design and build functional nanomaterials, which can act as machines or systems at the nanoscale, capable of interacting with specific cellular processes to diagnose, to intervene or to modify biological functions (Kishore et al. 2025). This field has enabled the development of innovative tools such as nanoparticles with antimicrobial properties, nanosystems for targeted release of bioactive compounds, biosensors and nanobots with biomedical and agricultural applications (Khan et al. 2025, Massoud et al. 2025). The uniqueness of nanobiotechnology resides in the fact that, by working at the molecular level, it takes advantage of the unique physicochemical properties of nanomaterials that are not present at a larger scale. Plant diseases represent a significant threat to global agriculture; the main causes of the spread of these pathologies include climatic factors, climate change, inadequate use of agricultural practices and transport of infected plant material. Among the most common pathogens are fungi, bacteria, viruses and nematodes, which can impair crop productivity, generate economic losses and compromise safety (Oraon et al. 2024, Parrado and Quintanilla 2024). In this context, technological innovation has become a fundamental tool to deal with this problem. The development of advanced technologies, such as biotechnology, artificial intelligence (AI) and remote monitoring systems, has enabled a better understanding of pathogenic mechanisms and early identification of diseases (Mehta et al. 2024). These innovations offer the potential to transform disease management strategies, providing faster, more accurate

[1] Centro de Investigaciones Biológicas del Noroeste, La Paz, Baja California Sur 23096, México.
[2] Universidad Técnica Estatal de Quevedo, Santo Domingo, Quevedo 120501, Ecuador.
[3] Facultad de Ciencias Agrotecnológicas, Universidad Autónoma de Chihuahua, Chihuahua 31310, México.
[4] Facultad de Ciencias Agrícolas, Universidad Veracruzana, Xalapa, Veracruz 91090, México.
* Corresponding author: lhernandez@cibnor.mx

DOI: 10.1201/9781003659907-9

and sustainable solutions. The application of these technologies not only improves the efficacy of treatments but also contributes to reducing the use of chemicals, promoting a greener and less environmentally damaging approach. Through these tools, farmers can adopt smarter and more adaptive practices to protect their crops efficiently and sustainably, which is key to ensuring food security in an ever-changing world.

As a result of new technologies, nanobiotechnology has established itself as an innovative and promising tool in the field of agriculture, especially in the management of plant diseases. This discipline combines advances in nanotechnology and biotechnology, making it possible to develop solutions on a molecular scale to address the challenges facing crops today. One of the main advantages of nanobiotechnology in plant disease control is its ability to design specific nano-bioformulas that act with high precision on pathogens, without affecting plants or the environment. Nanoparticles, for example, can be used to release antifungal, bactericidal or virucidal agents in a controlled and efficient manner, improving the effectiveness of treatments. In addition, these particles can be designed to penetrate deeper into plants, reaching areas where traditional methods are not effective (Ghormade et al. 2011). Another crucial aspect is the ability of nanobiotechnology to improve the natural resistance of plants to disease. Through the engineering of nanoparticles and biological systems, it is possible to stimulate the immune defense of plants, activating natural protection mechanisms and favoring their resistance to infections and pests. In addition, the use of nanomaterials in the early diagnosis of diseases makes it possible to identify infections in their initial stages, which facilitates faster and more efficient intervention (Camara et al. 2019, Markandan et al. 2024). The application of nanobiotechnology also reduces the use of conventional chemical products, which contributes to more sustainable crop management. By being more specific and efficient, nanotechnology-based treatments reduce the environmental impact and the risk to human health associated with the excessive use of pesticides and fungicides. Therefore, the objective of this chapter is to describe the properties, functions and types of nanoparticles that have been used to control the main phytopathogens of crops of global importance.

## 2. Nanomaterials: Properties and Key Mechanisms of Action in Disease Plant Management

There are various types of metallic nanoparticles used to control phytopathogens (Table 1). However, silver (Ag) nanoparticles (NPs) have gained considerable scientific attention due to their unique characteristics, such as their size and shape, which influence optical, electrical and antimicrobial properties. Various synthesis methods have been developed to obtain them, including techniques such as laser ablation, gamma or electron irradiation, chemical reduction, photochemical processes, microwave, and biological or green approaches (Sharma et al. 2018, Remita and Lampre 2024). Current scientific evidence supports three main mechanisms by which silver nanoparticles (AgNPs) exert their antibacterial action, either jointly or independently.

The first consists of direct interaction with the cell membrane: AgNPs cross the outer membrane and accumulate in the inner membrane, destabilizing it, increasing its permeability and causing leakage of cellular contents, which ultimately leads to bacterial death. They can also bind to sulfur-rich proteins of the cell wall, structurally weakening it to the point of rupture (Zhang et al. 2020). The second mechanism involves the intracellular penetration of nanoparticles, where they can interact with sulfur and phosphorus groups present in proteins and DNA, altering their structure and function. In addition, they interfere with the respiratory chain by binding to thiol groups on enzymes, which induces the generation of reactive oxygen species (ROS) and free radicals, causing damage to cellular machinery and initiating apoptosis processes (Gomma 2017, Quinteros et al. 2016). The third mechanism is based on the release of silver ions ($Ag^{+}$) from the NPs. These ions, due to their small size and high reactivity, can interact with key cellular components, affecting metabolic pathways, membranes and genetic material, and thus enhance the antibacterial effect of AgNPs (Saeb et al. 2014, Xu et al. 2025). Among the NPs most used in agriculture for disease management

**Table 1.** Nanoparticles (NPs) used for the control of phytopathogens.

| **Nanoparticle** | **Phytopahogen** | **Test** | **Reference** |
|---|---|---|---|
| ZnO-CuO | *Fusarium oxysporum* | *In vitro* | Gaber et al. 2024 |
| CMC-Cu-Zn-Fe | | *In silico* and *in vitro* | Bouqellah 2023 |
| ZnO | | *In vitro* and *in vivo* | Abdelaziz et al. 2022 |
| $CuSO_4$ | *Botrytis cinerea* | | Bikdeloo et al. 2021 |
| S-NPs and Cu-NPs | *Botrytis cinerea* and *Sclerotinia sclerotiorum* | | Sadek et al. 2022 |
| Copper-boron (Cu-B) NPs | *Phytophthora* species | | Yiğit et al. 2024 |
| MgO-NPs | *Phytophthora infestans* | *In vitro* | Wang et al. 2022 |
| Cu and $Cu_2O$ | *Colletotrichum* spp. | | Montejo-Alvaro et al. 2024 |
| ZV-AgNPs | *Colletotrichum falcatum* | | Ajaz et al. 2021 |
| Ag and CU NPs | *Lasiodiplodia theobromae* | | Linh et al. 2021 |

are copper oxide (CuO) NPs. In addition to the antibacterial activity of CuONPs against numerous bacteria and plant pathogenic fungi, CuNPs could act through a nutritional mechanism (El-Abeid et al. 2020, Ghdeeb and Abdalameer 2024). Copper is an essential micronutrient for plants, which is necessary for their health and their protection against diseases (Elmer et al. 2018, Yogi et al. 2024).

## *2.1 Carbon Nanotubes and Other Carbon Nanostructures*

Research for sustainable and effective alternatives for crop protection management, carbon-based nanomaterials such as carbon nanotubes (CNTs), graphene oxide (GO) and carbon-metal nanocomposites have demonstrated remarkable antimicrobial, antiviral and fungicidal potential (Neelan et al. 2024, Mogadem et al. 2025). These nanostructures, thanks to their unique physicochemical properties, can act both directly against pathogens and through the activation of plant defenses (Enyashin et al. 2007, Shoudho et al. 2024). However, it is crucial to continue researching and evaluating their impact on the environment and health, as well as to establish appropriate regulations for their use. CNTs functionalized with carboxyl (-COOH), hydroxyl (-OH) or amino ($-NH_2$) groups have shown high efficacy against plant pathogenic fungi such as *Fusarium graminearum* and *Fusarium poae* (Wang et al. 2018). Their mechanism of action includes destabilization of fungal cell membranes, reduction in spore germination and induction of structural damage. This effect is enhanced by increasing the surface functionalization of CNTs, which also reduces their toxicity in plants.

## *2.2 Polymers and Nanocomposites*

Polymer nanocomposites combine the properties of polymers with the advantages of nanomaterials (Maghimaa et al. 2024). These combinations allow the development of materials with greater mechanical resistance, antimicrobial capacity and controlled release of active compounds. Some of the natural polymers used are chitosan, alginate, and nanocrystalline cellulose, and within the synthetic polymers are polylactide, polylactide-co-glycolic, and polyacrylamide, while nanocomposites are polymers with metallic nanoparticles or clays (Gomes et al. 2018).

## *2.3 Nanostructured Materials of Biological Origin*

Nanostructured materials of biological origin are obtained from renewable sources or generated through the metabolic activity of living organisms, such as plants, bacteria and fungi (Abady et al. 2025). Their synthesis is generally carried out through sustainable methods known as "green chemistry", which minimize the use of toxic reagents and reduce environmental impact

(Iravani et al. 2011). These materials include chitosan in nanometric form, derived from chitin; metal nanoparticles biosynthesized by plant extracts or by microorganisms; and nanocellulose, obtained from plant biomass or bacterial processes (Ahmed et al. 2016). Nanomaterials represent an innovative and sustainable alternative to traditional agrochemicals in the management of plant diseases. Their application not only reduces production costs but also offers ecological benefits due to their biodegradability and lower toxicity. They have shown antimicrobial potential against various plant pathogens and biostimulant properties, promoting plant growth and improving plant resistance to adverse conditions (Singh et al. 2018).

## 3. Properties of Nanomaterials

Physicochemical properties of nanomaterials determine their biological behavior, toxicity, stability and effectiveness in agricultural and microbiological applications. The most relevant are size, shape, surface charge, specific surface area and solubility, all of which directly influence the mechanisms of interaction with pathogens and plants, modulating their antimicrobial or biostimulant activity. Size (1–100 nm) is crucial, as smaller sizes increase the surface-to-volume ratio, improving interaction with microorganisms and allowing them to cross cell membranes to release metal ions or active compounds (Kah et al. 2018, Sathyamoorthy et al. 2025). Shape also matters: prismatic or sharp-edged nanoparticles are often more effective at penetrating cells and generating reactive oxygen species (Zhang et al. 2024). Surface charge regulates electrostatic interactions with biological membranes. Positively charged nanoparticles tend to adhere to microbial surfaces, favoring membrane destabilization (Slavin et al. 2017). In turn, a larger specific surface area increases contact with pathogens or plant tissues, enhancing the release of antimicrobial ions or the adsorption of bioactive compounds. Solubility and dispersion in aqueous media determine their bioavailability, and can be enhanced by polymeric coatings such as chitosan or polyethylene glycol (Soltani-Nezhad et al. 2024).

### *3.1 Mechanisms of Action of Nanomaterials in Disease Control*

Ag ions can bind to cysteine-containing proteins, interact with DNA bases, cause structural changes in the cell membrane, and inhibit cell division and reproduction (Mikhailova 2020). However, it is worth adding that metal ions in environments could interact with organic matter and other toxic effects on biota can be mitigated, in addition to stabilizing nanoparticles (Ale et al. 2021, Wang et al. 2023). In addition, another well-known mechanism of toxicity involves the production of reactive oxygen species (ROS) on the surface of NPs that can severely damage organisms. The increase in ROS alters cellular homeostasis, which damages cells at various levels and, leading to apoptosis and death (Adisa et al. 2019).

## 4. Direct Interaction with Pathogens

Nanomaterials applied in plant protection and pathogen control act through direct and indirect mechanisms, involving both effects on pathogenic microorganisms and plant physiology. These interactions result from the physicochemical properties of the nanoparticles, and their appropriate design allows taking advantage of their antimicrobial and biostimulant benefits.

### *4.1 Direct Mechanisms on Pathogens*

*Damage to the Cell Membrane and Internal Structures*

One of the main mechanisms of action is structural damage to the cell membrane of bacteria, fungi and other pathogens. Nanoparticles, especially those with a positive charge, adhere to the cell surface by electrostatic interactions, disrupting the permeability of the membrane. This disruption leads to leakage of cytoplasmic contents, loss of membrane potential, and ultimately cell death (Slavin et al.

2017). Some nanoparticles can penetrate inside pathogenic cells, affecting vital structures such as DNA, proteins and ribosomes.

*Generation of Reactive Oxygen Species (ROS)*

Metallic nanoparticles such as silver (Ag), zinc oxide (ZnO) or copper (CuO) are capable of generating reactive oxygen species (ROS), such as peroxides and hydroxyl radicals. These molecules, due to their high reactivity, induce oxidative stress in pathogens, resulting in lipid, protein and nucleic acid damage and consequently causing cellular disorganization and metabolic collapse (Kah et al. 2018).

*Interference with essential metabolic processes*: Nanoparticles also act by interfering with critical metabolic processes, such as protein synthesis or DNA replication. In addition, the release of metal ions by nanoparticles inhibits enzymes essential for energy production and biosynthetic metabolism of microorganisms.

*Induced systemic resistance (ISR)*: This refers to immune responses triggered by internal signals, priming the plant to respond more rapidly and effectively to subsequent infections. Nanoparticles can induce the production of phytoalexins, antioxidant enzymes and defense-related proteins (Hirose et al. 2025, Joshi 2025).

## 5. Improved Nutrition and Plant Growth

Another indirect route of action is the general strengthening of plant vigor and development. Some nanoparticles act as controlled sources of nutrients or improve the absorption of essential minerals. Improved nutritional status results in more robust plants, with more efficient defense systems, increased production of antioxidant compounds, and a better ability to tolerate biotic and abiotic stress conditions (Wang et al. 2018). This improvement in plant physiology reduces susceptibility to infections, while optimizing crop yield and quality (Wang et al. 2023).

*Applications of nanobiotechnology in plant disease management*: Nanobiotechnology offers very promising applications for plant disease management, as it enhances the efficacy of treatments and diagnostics. Through nanoparticles, controlled release of agrochemicals and early detection of pathogens is facilitated, contributing to the sustainability of agricultural systems (Kah et al. 2018, Nizamani et al. 2024). These innovations allow for more precise disease control, decreasing the excessive use of chemicals and reducing environmental impact.

## 6. Antimicrobial Nanoparticles for Disease Control

Antimicrobial nanoparticles offer an innovative approach to plant disease control, as they act directly on pathogens and help prevent their spread. Usually composed of metals such as silver, copper or zinc, these nanoparticles possess unique properties that allow them to effectively interact with plant pathogenic microorganisms, inhibiting their growth and development (Mishra et al. 2017, Singh et al. 2018). In addition, their use contributes to reducing dependence on conventional agrochemicals and mitigating the risk of microbial resistance, thus favoring more sustainable and efficient agricultural practices.

### *6.1 Fungal Diseases*

Fungal diseases represent one of the main threats to global food security, affecting staple crops such as wheat, rice and maize, causing significant economic losses and a decrease in agricultural productivity (Gai and Wang 2024, Muvunyi et al. 2024). In response to this problem, nanobiotechnology has proven to be an effective tool, especially through the use of metallic nanoparticles (e.g., silver and zinc oxide), whose antimicrobial properties are based on the generation of reactive oxygen species (ROS) and the alteration of the integrity of fungal cell membranes (Ghormade et al. 2011).

These nanoparticles act in a selective and controlled manner against pathogens such as *Fusarium* spp., *Botrytis cinerea,* and *Alternaria solani*, significantly reducing the incidence of fungal diseases in different crops. In addition, their application reduces the use of conventional fungicides, thus contributing to a more sustainable agriculture with less environmental impact. Antifungal nanoparticles exert their action through several mechanisms that directly affect pathogenic fungi. One of the main ones is the generation of reactive oxygen species (ROS), such as superoxide radicals and peroxides, which damage lipids, proteins and nucleic acids in fungal cells (Alfei et al. 2024, Rauf et al. 2024). In addition, NPs can interact with cell membranes, causing their destabilization and increasing permeability, leading to loss of structural and functional integrity (Sondi et al. 2004). Another relevant mechanism is the release of metal ions (e.g., silver or zinc ions) that interfere with essential enzymes and metabolic processes of fungi, inhibiting their growth and sporulation. These multifaceted mechanisms confer on NPs a high antifungal efficacy, surpassing in many cases the action of conventional fungicides and minimizing the appearance of resistance.

### *6.2 Diseases Caused by Bacteria*

Metal and metal oxide nanoparticles (NPs) have established themselves as an innovative tool for controlling bacterial diseases in plants due to their unique antimicrobial properties. These properties derive from several mechanisms of action, both physical and biochemical, that directly affect bacterial viability (Kashyap et al. 2024). One of the most relevant mechanisms is the generation of reactive oxygen species (ROS), which cause oxidative stress and damage to essential cellular components such as lipids, proteins and DNA (Hatami and Ghorbanpour 2024). NPs can alter the structure of the bacterial cell membrane, increasing its permeability, causing loss of cell content and, finally, cell death. Another important mechanism is the release of metal ions, such as $Ag^{+}$ or $Zn^{2+}$, which interfere with fundamental processes such as DNA replication or protein synthesis. Likewise, some nanoparticles are capable of interfering with the internal metabolism of bacteria, affecting vital functions such as energy production or ribosomal activity (Mammari et al. 2024). These mechanisms have been successfully tested in the control of several bacterial diseases of agricultural importance. For example, silver nanoparticles (AgNPs) synthesized with proteins showed remarkable activity against *Xanthomonas campestris* pv. *campestris*, responsible for diseases in cruciferous plants, by generating cell membrane damage and activating antioxidant responses in plants (Usman et al. 2020). Zinc oxide nanoparticles (ZnO NPs) have been studied extensively for their ability to inhibit *Pseudomonas syringae* pv. tabaci in tobacco plants. These nanoparticles not only affect bacterial viability, but also inhibit biofilm formation, affect motility and promote defensive responses in plants (Siddiqui et al. 2020). ZnO NPs biosynthesized with *Matricaria chamomilla* extract showed significant activity against Ralstonia solanacearum, reducing bacterial populations and the incidence of bacterial wilt in tomato (Ahmed et al. 2021). Innovatively, DNA-directed silver nanoparticles on graphene have been developed and demonstrated strong activity against *Xanthomonas perforans*, a tomato pathogen, significantly reducing the bacterial load on treated leaves, even at low concentrations, without causing phytotoxicity (Elmer et al. 2018). These applications reinforce the potential of nanoparticles as a sustainable and effective strategy for the integrated management of bacterial diseases, offering viable alternatives to the use of traditional antibiotics or bactericides.

### *6.3 Diseases Caused by Viruses*

Viral diseases are a significant threat to agricultural production worldwide. The deficiency of curative treatments and the rapid propagation of plant viruses hinder the efficient management of these diseases. In this context, the use of nanoparticles (NPs), especially metallic and metal oxides, has emerged as an alternative for the control of viral infections in plants (Hamidzade et al. 2024).

Antiviral nanoparticles can interfere with different stages of the infectious cycle of viruses. The most relevant mechanisms include:

*Direct inactivation of viral particles*: Nanoparticles, such as silver nanoparticles (AgNPs), can bind to virus structural proteins or genetic material, preventing virus adsorption to plant cells or replication. For example, AgNPs have been shown to reduce the accumulation of Pepper mild mottle virus (PMMoV) in Capsicum annuum plants (Elbeshehy et al. 2015).

*Induction of defensive responses in plants*: Some nanoparticles stimulate innate plant defense pathways, including the expression of genes related to defense proteins (RP) and phytoalexins, thereby increasing host resistance (Warghane et al. 2024).

*Controlled release of antiviral agents*: By using nanoformulated systems, it is possible to encapsulate antiviral compounds that are released in a sustained manner at the site of action, increasing their efficacy and reducing toxicity.

### *6.4 Development Strategies for the Control of Viral Diseases in Crops*

The strategies developed so far consist in foliar applications of metallic nanoparticles (Ag, Cu, Zn), the use of hybrid nanocomposites that improve leaf stability and adherence, as well as nanoencapsulation systems of natural extracts with antiviral properties (Singh et al. 2021). In addition, progress has been made in the design of nanotechnology-based biosensors capable of detecting viral infections in early stages, even before symptoms are visible, allowing timely intervention to mitigate crop damage.

### *6.5 Nanosensors for Early and Accurate Detection of Diseases*

In recent years, several studies have demonstrated the successful use of nanosensors for the early detection of plant diseases, allowing the identification of infections before the appearance of visible symptoms, which represents a key advance for a more accurate and sustainable agriculture (Choudhary et al. 2019). One of the most prominent cases is the development of an electrochemical nanosensor based on gold nanoparticles to detect the DNA of *Xanthomonas axonopodis*, the cause of bacterial canker in citrus fruits. This sensor was able to identify the pathogen in asymptomatic leaves weeks before visible symptoms appeared. Similarly, an optical sensor based on silver nanoparticles has been developed to detect Tomato mosaic virus (ToMV) in tomato plants.

## 7. Conclusions

Nanobiotechnology represents one of the most promising areas for sustainable innovation in the agricultural sector. As a result of the development of nanomaterials (NMs), applications such as nano-pesticides, nanofertilizers and nanosensors have emerged to improve crop yields, reduce losses due to diseases and optimize the use of natural resources. However, the accelerated progress in this field also raises fundamental questions about environmental safety, interaction with non-target organisms and the ultimate fate of these compounds in ecosystems. Extrapolation of laboratory results to real field conditions is one of the main challenges. Most of the current studies on nanomaterials are performed in highly controlled environments, using model organisms with well-characterized metabolisms, and under concentrations often higher than those that would occur in real agricultural scenarios. This limitation prevents an accurate understanding of the ecological response to environmentally relevant doses and makes it difficult to predict impacts on soil microorganisms, beneficial insects, non-target plants and food chains.

Furthermore, the bioavailability, mobility, and transformation of nanomaterials in complex environmental matrices remain a largely unexplored field. The interaction of nanoparticles with soil components (organic matter, minerals, microorganisms) could modify their reactivity and toxicity, indirectly affecting their expected efficacy or even generating side effects. Consequently, studies that integrate disciplines such as ecotoxicology, environmental chemistry, and soil microbiology are needed to understand these processes in their complexity. From an applied perspective, the future of nanobiotechnology in agriculture will depend on its ability to be integrated within sound ethical, regulatory and scientific frameworks. It is essential to establish clear guidelines to ensure its safe, effective and environmentally responsible use. This includes the design of biodegradable nanoformulations, the development of controlled release technologies directed exclusively at target organisms, and the continuous evaluation of the life cycle of nanotechnology products.

Therefore, future perspectives point towards a more precise, safe and adaptive agricultural nanobiotechnology, supported by long-term field studies, integrated evaluation methodologies and multidisciplinary collaboration. Only in this way will it be possible to maximize the benefits of this revolutionary technology, minimizing its risks and ensuring its sustainability in agricultural production systems.

## References

Abady, M. M., Mohammed, D. M., Soliman, T. N., Shalaby, R. A. and Sakr, F. A. (2025). Sustainable synthesis of nanomaterials using different renewable sources. Bulletin of the National Research Centre 49(1): 24.

Abdelaziz, A. M., Salem, S. S., Khalil, A. M., El-Wakil, D. A., Fouda, H. M. and Hashem, A. H. (2022). Potential of biosynthesized zinc oxide nanoparticles to control *Fusarium* wilt disease in eggplant (Solanum melongena) and promote plant growth. BioMetals 35(3): 601–616.

Adisa, I. O., Pullagurala, V. L. R., Peralta Videa, J. R., Dimkpa, C. O., Elmer, W. H., Gardea Torresdey, J. L. et al. (2019). Recent advances in nano enabled fertilizers and pesticides: a critical review of mechanisms of action. Environ. Sci. Nano. 6(7): 2002–2030.

Alfei, S., Schito, G. C., Schito, A. M. and Zuccari, G. (2024). Reactive oxygen species (ROS)-mediated antibacterial oxidative therapies: available methods to generate ROS and a novel option proposal. International Journal of Molecular Sciences 25(13): 7182.

Ahmed, S., Saifullah, Ahmad, M., Swami, B. L. and Ikram, S. (2016). Green synthesis of silver nanoparticles using *Azadirachta indica* aqueous leaf extract. J. Radiat. Res. Appl. Sci. 9(1): 1–7.

Ahmed, S., Saifullah, Ahmad, M., Swami, B. L. and Ikram, S. (2021). Green synthesis of ZnO nanoparticles using *Matricaria chamomilla* extract and their antimicrobial efficacy against Ralstonia solanacearum. Mater. Sci. Semicond. Process. 121: 105372.

Ajaz, S., Ahmed, T., Shahid, M., Noman, M., Shah, A. A., Mehmood, M. A. and Li, B. (2021). Bioinspired green synthesis of silver nanoparticles by using a native *Bacillus* sp. strain AW1-2: characterization and antifungal activity against *Colletotrichum falcatum* Went. Enzyme and Microbial Technology 144: 109745.

Ale, A., Galdopórpora, J. M., Mora, M. C., de la Torre, F. R., Desimone, M. F. and Cazenave, J. (2021). Mitigación de la toxicidad de nanopartículas de plata mediante ácidos húmicos en branquias de peces *Piaractus mesopotamicus*. Environ. Sci. Pollut. Res. 28: 31659.

Bikdeloo, M., Ahsani Irvani, M., Roosta, H. R. and Ghanbari, D. (2021). Green synthesis of copper nanoparticles using rosemary extract to reduce postharvest decays caused by *Botrytis cinerea* in tomato. Journal of Nanostructures 11(4): 834–841.

Bouqellah, N. A. (2023). *In silico* and *in vitro* investigation of the antifungal activity of trimetallic Cu–Zn-magnetic nanoparticles against *Fusarium oxysporum* with stimulation of the tomato plant's drought stress tolerance response. Microbial Pathogenesis 178: 106060.

Camara, M. C., Campos, E. V. R., Monteiro, R. A., Pereira, A. E. S., Proença, P. L. F. and Fraceto, L. F. (2019). Development of stimuli responsive nano based pesticides: emerging opportunities for agriculture. J. Nanobiotechnol. 17(1): 100.

Choudhary, R. C., Kumaraswamy, R. V., Kumari, S., Sharma, S. S., Pal, A., Raliya, R. et al. (2019). Zinc encapsulated chitosan nanoparticle to promote maize crop yield. Int. J. Biol. Macromol. 127: 126–35.

Díaz-Marcos, J. and Mendoza Gonzálvez, J. (2025). Shaping the future of nanotechnology: a perspective on education, dissemination and ethics. Frontiers in Nanotechnology 7: 1524578.

El-Abeid, S. E., Ahmed, Y., Daròs, J.-A. and Mohamed, M. A. (2020). Reduced graphene oxide nanosheets decorated with copper oxide nanoparticles: A potent antifungal nanocomposite against root rot and Fusarium wilt in tomato and pepper plants. Nanomaterials 10(5): 1001.

Elbeshehy, E. K. F., Elazzazy, A. M. and Aggelis, G. (2015). Silver nanoparticles synthesized by the novel bacterial strain *Streptomyces aegyptia* have potent antiviral activity against Bean yellow mosaic virus. Mater Sci. Eng. C 57: 273–80.

Elmer, W. H., De La Torre Roche, R. and White, J. C. (2018). The use of nanoscale materials in the management of plant pathogenic bacteria. Agriculture 8(6): 78.

Enyashin, A. N. and Ivanovskii, A. L. (2007). Functionalization of carbon nanotubes by covalently bonded graphite nanoplatelets: a theoretical study. Mendeleev Commun. 17(4): 199–201.

Gaber, S. E., Hashem, A. H., El-Sayyad, G. S. and Attia, M. S. (2024). Antifungal activity of myco-synthesized bimetallic ZnO-CuO nanoparticles against fungal plant pathogen *Fusarium oxysporum*. Biomass Conversion and Biorefinery 14(20): 25395–25409.

Gai, Y. and Wang, H. (2024). Plant disease: A growing threat to global food security. Agronomy 14(8): 1615.

Ghdeeb, N. J. and Abdalameer, N. K. (2024). Synthesis, characterization, and antimicrobial activity of CuO nanoparticles and CuO/Ag nanocomposites. Applied Nanoscience 14(2): 401–409.

Ghormade, V., Deshpande, M. V. and Paknikar, K. M. (2011). Perspectives for nano biotechnology enabled protection and nutrition of plants. Biotechnol. Adv. 29(6): 792–803.

Gomes, M. J., Neves, J. and Sarmento, B. (2018). Nanoparticle-based drug delivery systems: Advances and challenges. Int. J. Nanomedicine 13: 2735–2771.

Gomma, E. Z. (2017). Silver nanoparticles as an antimicrobial agent: A case study on *Staphylococcus aureus* and *Escherichia coli* as models for Gram-positive and Gramnegative bacteria. The Journal of General and Applied Microbiology 63(1): 36–43.

Hamidzade, M., Motlaghzadeh, S., Khales, P., Aminpanah, D., Minaeian, S., Hosseini-Hosseinabad, S. M. et al. (2024). Metal and metal oxide nanoparticles as agents against human infectious viruses. Current Nanoscience 20(4): 510–529.

Hatami, M. and Ghorbanpour, M. (2024). Metal and metal oxide nanoparticles induced reactive oxygen species: Phytotoxicity and detoxification mechanisms in plant cell. Plant Physiol. Biochem. 213: 108847.

Hirose, S., Horiyama, S., Morikami, A., Fujiwara, K. and Tsukagoshi, H. (2025). Eugenol and basil essential oil as priming agents for enhancing *Arabidopsis* immune response. Bioscience, Biotechnology, and Biochemistry 89(1): 41–50.

Iravani, S., Korbekandi, H., Mirmohammadi, S. V. and Zolfaghari, B. (2011). Synthesis of silver nanoparticles: the significance of plant extracts and biotechnological approaches. Green Chem. 13(10): 2638–2650.

Joshi, H. (2025). Exploring the efficacy of green nanoparticles in enhancing plant defense: a mechanistic investigation into immune response activation. Journal of Nanoparticle Research 27(2): 29.

Kah, M., Kookana, R. S., Gogos, A. and Bucheli, T. D. (2018). Nanopesticides: State of knowledge, environmental fate, and exposure modeling. Crit. Rev. Environ. Sci Technol. 48(6): 436–68.

Kashyap, A. S., Manzar, N., Vishwakarma, S. K., Mahajan, C. and Dey, U. (2024). Tiny but mighty: metal nanoparticles as effective antimicrobial agents for plant pathogen control. World J. Microbiol. Biotechnol. 40(3): 104.

Khan, Z., Khan, N., Geetha, M., Veettil, R. P., Kasote, D. M., Hasan, A. et al. (2025). Therapeutic applications of nanobots and nanocarriers in cancer treatment. Analytical Sciences 1–20.

Kishore, A., Singh, C. and Kaur, G. (2025). Bio-nanomaterials: An introduction. Bio-Nanomaterials in Environmental Remediation: Industrial Applications 1–46.

Linh, N. V. M., Thao, P. T. H., Lien, N. T. H., Van, D. T. H., Van Hieu, N., Van, N. T. et al. (2021). Antifungal potential of silver and copper nanoparticles against *Lasiodiplodia theobromae* causing stem-end rot of mango. Vietnam Journal of Biotechnology 9(4): 735–740.

Maghimaa, M., Sagadevan, S., Boojhana, E., Fatimah, I., Lett, J. A., Moharana, S. et al. (2024). Enhancing biocompatibility and functionality: Carbon nanotube-polymer nanocomposites for improved biomedical applications. Journal of Drug Delivery Science and Technology 105958.

Mammari, N., Lamouroux, E., Boudier, A. and Duval, R. E. (2024). Current knowledge on the oxidative stress mediated antimicrobial properties of metal based nanoparticles. Microorganisms 10(2): 437.

Massoud, E. N., Hebert, M. K., Siddharthan, A., Ferreira, T., Neron, A., Goodrow, M. et al. (2025). Delivery vehicles for light-mediated drug delivery: microspheres, microbots, and nanoparticles: a review. Journal of Drug Targeting 33(5): 691–703.

Markandan, K., Tiong, Y. W., Sankaran, R., Subramanian, S., Markandan, U. D., Chaudhary, V. et al. (2024). Emergence of infectious diseases and role of advanced nanomaterials in point-of-care diagnostics: a review. Biotechnology and Genetic Engineering Reviews 40(4): 3438–3526.

Mehta, A., Niaz, M., Adetoro, A. and Nwagwu, U. (2024). Advancements in manufacturing technology for the biotechnology industry: the role of artificial intelligence and emerging trends. International Journal of Chemistry, Mathematics and Physics 8(2): 12–18.

Mikhailova, E. O. (2020). Nanopartículas de plata: mecanismo de acción y probable bioaplicación. Rev. Biomater. Funct. 11(4): 84.

Mishra, S., Singh, H. B. and Nautiyal, C. S. (2017). Antifungal activity of chitosan-based nanoparticles: a new approach for managing plant diseases. Environ Sustain. 1(2): 123–32.

Mogadem, A., Almotairy, A. R. Z., El-Khatib, M., Ismail, P. A. and Elwakil, B. H. (2025). Synthesized Mg@ C Fullerene nanocomposite using a plasma reactor and its biological activity. Journal of Cluster Science 36(4): 1–12.

Montejo-Alvaro, F., Vásquez-López, A., Pariona, N., Mtz-Enriquez, A. I., Hernández-Sánchez, L., Rojas-Chávez, H. et al. (2024). Antifungal activities of Cu and $Cu_2O$ nanoparticles against *Colletotrichum* spp. Materials Letters 365: 136399.

Muvunyi, C. M., Ngabonziza, J. C. S., Florence, M., Mukagatare, I., Twagirumukiza, M., Ahmed, A. et al. (2024). Diversity and distribution of fungal infections in Rwanda: High risk and gaps in knowledge, policy, and interventions. Journal of Fungi 10(9): 658.

Neelan, Y. D., Bakthavatchalam, S., Mahalingam, S., Yoganand, K. S., Ramalingam, S. V., Rajendran, U. et al. (2024). Nanoalchemy: Unveiling the power of carbon nanostructures and carbon–metal nanocomposites in synthesis and photocatalytic activity. Catalysts 14(9): 618.

Nizamani, M. M., Hughes, A. C., Zhang, H. L. and Wang, Y. (2024). Revolutionizing agriculture with nanotechnology: Innovative approaches in fungal disease management and plant health monitoring. Science of The Total Environment 172473.

Oraon, S., Padamini, R., Shahni, Y. S., Das, N., Sinha, D., Sujatha, G. et al. (2024). Impact of emerging pathogens in crop production. Microbiology Research Journal International 34(7): 80–92.

Parrado, L. M. and Quintanilla, M. (2024). Plant-parasitic nematode disease complexes as overlooked challenges to crop production. Frontiers in Plant Science 15: 1439951.

Quinteros, M. A., Cano Aristizábal, V., Dalmasso, P. R., Paraje, M. G. and Páez, P. L. (2016). Oxidative stress generation of silver nanoparticles in three bacterial genera and its relationship with the antimicrobial activity. Toxicol. *In Vitro* 36: 216–23.

Rauf, A., Khalil, A. A., Awadallah, S., Khan, S. A., Abu-Izneid, T., Kamran, M. et al. (2024). Reactive oxygen species in biological systems: Pathways, associated diseases, and potential inhibitors—A review. Food Science & Nutrition 12(2): 675–693.

Remita, H. and Lampre, I. (2024). Synthesis of metallic nanostructures using ionizing radiation and their applications. Materials 17(2): 364.

Sadek, M. E., Shabana, Y. M., Sayed-Ahmed, K. and Abou Tabl, A. H. (2022). Antifungal activities of sulfur and copper nanoparticles against cucumber postharvest diseases caused by *Botrytis cinerea* and *Sclerotinia sclerotiorum*. Journal of Fungi 8(4): 412.

Saeb, A. T. M., Alshammari, A. S., Al-Brahim, H. and Al-Rubeaan, K. A. (2014). Production of silver nanoparticles with strong and stable antimicrobial activity against highly pathogenic and multidrug-resistant bacteria. Sci. World J. 704283.

Sathyamoorthy, S., Venugopal, A., Lenin, M. J., Tirkey, S. and Sevanan, M. (2025). Nanomaterials and microbial compatibility: synergistic and antagonistic mechanisms. Nano-Bioremediation for Wastewater Treatment 35–72.

Sharma, J., Singh, V. K., Kumar, A., Shankarayan, R. and Mallubhotla, S. (2018). Role of silver nanoparticles in treatment of plant diseases. pp. 435–54. *In*: Patra, J., Das, G. and Shin, H. S. (eds.). Microbial Biotechnology. Singapore: Springer.

Siddiqui, H., Fatima, N., Naz, S. and Khan, K. A. (2020). Zinc oxide nanoparticles as a novel antibacterial agent against *Pseudomonas syringae* pv. tabaci. Front Microbiol. 11: 1425.

Singh, J., Dutta, T., Kim, K. H., Rawat, M., Samddar, P. and Kumar, P. (2018). 'Green' synthesis of metals and their oxide nanoparticles: Applications for environmental remediation. J. Nanobiotechnol. 16(1): 84.

Singh, P., Pandit, S., Beshay, M., Mokkapati, V. R. S. S., Garnaes, J., Sultan, A. et al. (2021). Green synthesis of metallic nanoparticles and their potential applications to combat virus infection. Nanomaterials 11(4): 1109.

Sevimli, B., İlgin, C. Y., Acar, E., Almusawi, M., Aydostan, Ç., Halvac, E. et al. (2024). Overview of nanobiotechnology: Nanotechnological perspectives and development of nano solutions in the fight against Cancer. International Journal of Boron Science and Nanotechnology (1): 49–85.

Shoudho, K. N., Uddin, S., Rumon, M. M. H. and Shakil, M. S. (2024). Influence of physicochemical properties of iron oxide nanoparticles on their antibacterial activity. ACS Omega 9(31): 33303–33334.

Soltani-Nezhad, S., Mashreghi, A., Hasani, S., Daneshfar, Z., Rezvan, M. T. and Emami, A. (2024). Optimization of the coating process in the PEGylated chitosan-based doped cobalt ferrite nanoparticles for hyperthermia applications using the Taguchi method. Materials Chemistry and Physics 323: 129625.

Sondi, I. and Salopek-Sondi, B. (2004). Silver nanoparticles as antimicrobial agent: a case study on *E. coli* as a model for Gram-negative bacteria. J. Colloid. Interface. Sci. 275(1): 177–182.

Slavin, Y. N., Asnis, J., Häfeli, U. O. and Bach, H. (2017). Metal nanoparticles: Understanding the mechanisms behind antibacterial activity. J. Nanobiotechnol. 15: 65.

Usman, M. S., Zowalaty, M. E. E., Shameli, K., Zainuddin, N., Salama, M. and Ibrahim, N. A. (2020). Synthesis, characterization, and antimicrobial properties of copper nanoparticles. Int. J. Nanomedicine 15: 4467–4479.

Wang, X. P., Li, Q. Q., Pei, Z. M. and Wang, S. C. (2018). Effects of zinc oxide nanoparticles on the growth, photosynthetic traits, and antioxidative enzymes in tomato plants. Biol. Plant. 62(4): 801–808.

Wang, Z. L., Zhang, X., Fan, G. J., Que, Y., Xue, F. and Liu, Y. H. (2022). Toxicity effects and mechanisms of MgO nanoparticles on the oomycete pathogen *Phytophthora infestans* and its host Solanum tuberosum. Toxics 10(10): 553.

Wang, Z., Yu, L. and Wang, D. G. (2023). Dissolved organic matter and lignin modulate aquatic toxicity and oxidative stress response activated by layered double hydroxide nanomaterials. Arch. Environ. Contam. Toxicol. 84(3): 413–425.

Wang, Z., Wang, S., Ma, T., Liang, Y., Huo, Z. and Yang, F. (2023). Synthesis of zinc oxide nanoparticles and their applications in enhancing plant stress resistance: a review. Agronomy 13(12): 3060.

Warghane, A., Saini, R., Shri, M., Andankar, I., Ghosh, D. K. and Chopade, B. A. (2024). Application of nanoparticles for management of plant viral pathogen: Current status and future prospects. Virology 592: 109998. https://doi.org/10.1016/j.virol.2024.109998.

Xu, L., Wang, Z., Zhang, H., Sun, A., Zhou, G., Jiang, R. et al. (2025). Chronic toxic effects of silver nanoparticles on Chlamydomonas reinhardtii: Photosynthesis, antioxidation and internalization. Journal of Environmental Management 377: 124428.

Yiğit, U., Türkkan, M., İlhan, H., Şimşek, T., Güler, Ö. and Derviş, S. (2024). Activity of nanosized copper-boron alloys against *Phytophthora* species. Journal of Plant Pathology 106(1): 175–190.

Yogi, L. N., Bhandari, S., Thalal, T., Bhattarai, M., Upadhyay, A. and Kharel, B. (2024). Enhancing rice yields through foliar application of essential micronutrients: A study on zinc, copper, and boron nutrition in context of Nepal. Archives of Agriculture and Environmental Science 9(2): 373–378.

Zhang, L., Jiang, H. and Wang, W. X. (2020). Subcellular imaging of localization and transformation of silver nanoparticles in the oyster larvae. Environmental Science & Technology 54(18): 11434–11442.

Zhang, P., Shen, Q., Yang, J., Zhao, Z., Gao, A., Chen, S. et al. (2024). Shuttle-like nanoassemblies by isomeric photosensitizers to enhance ROS generation and tumor penetration for photodynamic therapy. Matter. 7(12): 4342–4355.

# Chapter 8

# Nanotechnology for Plant Growth Promotion and Stress Tolerance

*Sudhir S. Shende,*[1] *Aniket Gade,*[1,2] *Joanna Trzcińska-Wencel,*[1] *Mahendra Rai*[3,4] and *Patrycja Golińska*[1,*]

## 1. Introduction

The ever-increasing world population, which is expected to increase and surpass 9.7 billion people by 2050 and 11 billion by the year 2100 (United Nations 2019), will require a double increase in agricultural production to meet food demand in the future (Ioannou et al. 2020, Janni et al. 2024). In addition, due to the decreasing area of cultivated areas resulting from increasing urbanization and industrialization, food security is one of the biggest concerns now and in the near future (Ioannou et al. 2020, Kang et al. 2023). Moreover, various biotic and abiotic stresses are major factors limiting plant productivity worldwide. The plant abiotic stressors include salinity due to high sodium ions, extremely high and low levels of temperature, light, radiation, depletion or excess of water and essential nutrients, and soil contamination (Seleiman et al. 2021, González Guzmán et al. 2022). The biotic stressors that affect the quantity and quality of crops include pathogenic microorganisms and pests (González Guzmán et al. 2022, Koza et al. 2022, Ahmad et al. 2024).

To date, different methodologies have been employed to enhance stress tolerance. In the past few decades, many efforts have been devoted to breeding crop cultivars with various stress-tolerant traits, including breeding through techniques like wide hybridization and mutation breeding, and genetic modification by introducing exogenous genes or changing the expression levels of endogenous genes to improve stress tolerance (Hu and Xiong 2014, Ioannou et al. 2020). These methods are particularly time-consuming and usually bring unpredictable results (Hu and Xiong 2014, Ioannou et al. 2020) or are limited globally due to the fact that genetically modified plant cultivation practice is currently unacceptable in many countries (Ioannou et al. 2020, Trzcińska-Wencel et al. 2025).

[1] Department of Microbiology, Faculty of Biological and Veterinary Sciences, Nicolaus Copernicus University in Toruń, Toruń, Poland.

[2] Department of Biological Sciences and Biotechnology, Institute of Chemical Technology, Mumbai, 400019, Maharashtra, India.

[3] Nanobiotechnology Laboratory, Department of Biotechnology, Sant Gadge Baba Amravati University, Amravati, India,

[4] Department of Chemistry, Federal University of Piaui (UFPI), Teresina, Piauí, Brazil.

Email: sudhir.shende@umk.pl

* Corresponding author: golinska@umk.pl

DOI: 10.1201/9781003659907-10

Therefore, to increase plant production and food security, modern and sustainable agriculture requires new technologies, such as nanotechnology, and solutions to protect crops from stress factors and to increase the efficiency of agrochemical use (Ahmad et al. 2024).

Nanotechnology is a rapidly developing field that has led to various innovative solutions in different sectors such as medicine, energy or agrifood sector. In agriculture, nanotechnology provides nanoproducts, which may increase plant productivity and protect plants from microbial pathogens and pests or act as nanosensors and nanoremediators (Ahmed et al. 2021, El-Saadony et al. 2022, Ahmad et al. 2024, Trzcińska-Wencel et al. 2025). Nanomaterials can be delivered to the plants by seed priming, foliar spray, hydroponics or supplemented to the soil to promote their growth and stress tolerance (González Guzmán et al. 2022, El-Saadony et al. 2022, Trzcińska-Wencel et al. 2025).

Nanoparticles (NPs), which can be produced by different physical, chemical and biological methods, are extremely small-sized materials 1–100 nm in size that possess unique and diverse physicochemical properties when compared with their larger-sized equivalents (Ahmad et al. 2024, Trzcińska-Wencel et al. 2025). Nanoparticles have a large surface area-to-volume ratio, high adsorption efficacy, and multiple binding sites that make them excellent nanocarriers for bioactive molecules, such as plasmids or RNA, for the delivery of functional genetic material to the chloroplasts or nucleus of cells of various plant species. Such an approach neither requires the use of the bacterium *Agrobacterium* sp. to infect plants in order to introduce functional genes, which is limited to a small number of plants, nor low efficiency and plant-damaging gene gun bombardment (Zhao et al. 2020, Wu et al. 2023, Shivashakarappa et al. 2025). The nanocarriers may be loaded or encapsulated with traditional agrochemicals, which will be slowly and in a controlled manner released into the environment to target plants for the promotion of their growth and development (Zhao et al. 2020, González Guzmán et al. 2022). Nanoparticles, e.g., iron, magnesium, or zinc NPs, can be themselves the source of micro- and macronutrients for plants to alleviate nutrient deficiency symptoms and supplement essential elements that are poorly soil-available nutrients (El-Saadony et al. 2022).

Moreover, NPs due to their unique physico-chemical and catalytic properties, can directly promote seed germination, plant growth, increase chlorophyll content and improve photosynthesis, and increase plant resistance to various biotic and abiotic stress factors (Trzcińska-Wencel et al. 2025).

Currently, one of the biggest problems of using nanomaterials in agriculture is explaining the fate of nanomaterials introduced into ecosystems and how they affect other organisms inhabiting given environments, including microorganisms, plants, and animals and their inclusion in the food chain.

In the context of the impact on plant growth and development, the mechanisms underlying the interaction of NPs with plants have not yet been fully explained (El-Saadony et al. 2022, Mgadi et al. 2024).

Therefore, in this chapter we highlight the current knowledge and potential uses of NPs in agriculture, for plant growth promotion and stress tolerance to improve crop productivity and food security.

## 2. Abiotic and Biotic Stresses and Their Effects on Plants

Stress includes all factors that adversely affect the rate of photosynthesis and reduce the capacity of the plant for converting energy to biomass (Parihar et al. 2015). A wide range of biotic and abiotic stresses (Figure 1), including heat, cold, drought, salinity, flooding and heavy metals, demonstrate an adverse effect on plant growth and development and are commonly responsible for limiting crop yields (El-Saadony et al. 2022). As sedentary organisms, plants are unable to relocate under adverse environmental conditions and have developed numerous strategies, incorporating specialized morphological structures and biochemical mechanisms, to ensure survival and reproduction under

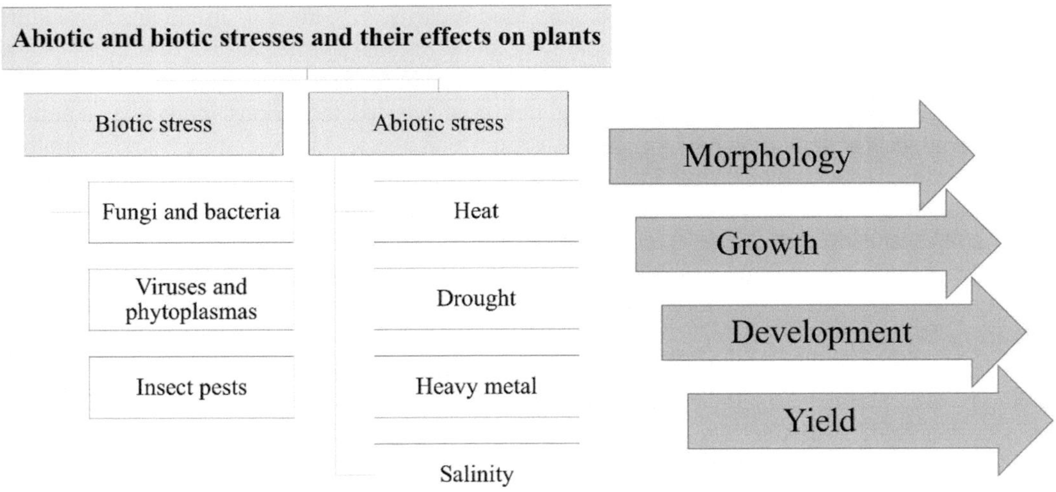


**Figure 1.** Showing the most prevalent biotic and abiotic stresses that negatively affect plant growth and development.

adverse conditions (Du et al. 2024). However, crops have not only to overcome stress, but also compensate for its consequences, to ensure the efficiency of agricultural production. Moreover, in the environment, stress factors very often occur simultaneously, therefore, the processes of recognizing and responding to particular combinations that can have a negative impact on plant growth and reproduction are crucial (Suzuki et al. 2014).

## *2.1 Abiotic Stress*

Abiotic stresses resulting from non-living environmental factors, include water stresses (drought or over-watering or flood), temperature stresses (cold, frost and heat), as well as salinity or toxicity related to the heavy metals (Gull et al. 2019). Reduced growth is a common consequence of stress, which is particularly concerning for crop plants (Du et al. 2024).

### *2.1.1 Heat*

Due to climate change, high temperatures are becoming a global concern in agricultural production. The exposure above optimal temperature results in heat stress in plants, which is the cause of reduced seed germination and decreased photosynthetic efficiency (Qu et al. 2013). In addition, during the reproductive growth period, excessively high temperatures are the cause of dysfunction of tapetal cells, decreased quantity and swelling of pollen grains during flowering, or inhibit differentiation of young ears, which negatively affects the fertility of spikelet and grain yields (Endo et al. 2009, Raja et al. 2019, Wang et al. 2020). Proteins and pathways constitute the basic network of the heat stress response, including a particularly important heat stress factor (Hsf) and heat shock proteins (Hsp) and catalase (CAT), which are signal transduction components and biomolecules responsible for neutralizing the negative effects, respectively (Qu et al. 2013).

### *2.1.2 Drought*

Agricultural drought caused by deficiencies in rainfall and simultaneous intensive evaporation and transpiration results in shortage of the moisture necessary for normal plant growth, development and completion of the life cycle (Farooq et al. 2012, Nyaupane et al. 2024). The major aspects of the effects of drought on plants include a decrease in the rate of cell division and expansion, reduced leaf size, effects on the functioning of stomatal apparatuses, and slowed stem and root elongation (Joshi et al. 2016). However, plants have developed several morphological and physiological adaptations that mitigate the negative effects of water shortage on plant growth and productivity

(Sheoran et al. 2022). For instance, transpiration is regulated by phytohormones, as shown in the case of minimizing transpiration losses through abscisic acid (ABA) synthesis (Saradadevi et al. 2017, Ullah et al. 2018).

#### *2.1.3 Heavy Metals*

Natural processes such as rock weathering and a wide range of anthropogenic activities, especially industrial ones, contribute to the contamination of agricultural soils with heavy metals (HMs) and metalloids such cadmium (Cd), mercury (Hg), arsenic (As), copper (Cu), lead (Pb), chromium (Cr), zinc (Zn) or nickel (Ni) (Hou et al. 2020). Heavy metals exhibit the potential to transfer gradually to plants, animals and humans, and can be transferred along the food chain, or undergo biotransformation and bioaccumulation in the environment (Srivastava et al. 2017). The impact of HMs on plants is determined by their bioavailability and uptake, which is strongly related to environmental conditions. The response of plants to HMs' toxicity comprises detoxification or immobilization strategies, among them retaining HMs in the cell wall, vascular sequestration and binding by phytochelators or organic acids (Kumar et al. 2022, Srivastava et al. 2017). The toxicity of HMs is emphasized in two aspects, firstly, they compete with nutrients in the soil and prevent their absorption, thereby impairing plant growth. Secondly, once taken up by plants, they disrupt basic metabolic processes and negatively affect cellular structures and enzyme activity (Ejaz et al. 2023).

#### *2.1.4 Salinity*

Many natural processes contribute to soil salinization, as well as agricultural intensification and climate change. Salt stress is defined as the degree of mineral saturation that determines the excessive level of $Na^+$, $Cl^-$ ions and ions of other allied salts in the soil solution. Salt stress in plants primarily consists of osmotic and ionic stresses as well as secondary stress (e.g., overproduction and excessive accumulation of reactive oxygen species (ROS)) or nutrient deficiency (Li et al. 2022). Salt stress alters the metabolic processes required for healthy plant growth and development. Prevention of water uptake is indicated as the main mechanism responsible for the negative effects of salt stress, for example, soil salinity generates a low osmotic potential around the seeds, which limits water uptake and consequently inhibits seed germination (Farooq et al. 2017). Although there is a certain level of tolerance in plants, the majority of crops are adversely affected by disruption of respiration, photosynthesis, mineral uptake and oxidative stress, which negatively regulates growth and yield (Wang et al. 2022).

### *2.2 Biotic Stress*

Biotic stress factors include living organisms, among which are microorganisms (viruses, bacteria, fungi), nematodes, insects, arachnids and weeds, that directly deprive their host of nutrients. Importantly, beyond the vegetative season, biotic stress can become a cause of crop loss during storage (Gull et al. 2019).

#### *2.2.1 Plant Pathogens Causing Diseases*

Microorganisms infect plants both in the above-ground parts (phyllosphere), the area of the root system (rhizosphere), and internally (endosphere) in order to utilize plant biomass as a substrate (Schulze et al. 2019). Among the most important microbial plant pathogens, bacteria are represented by *Agrobacterium tumefaciens, Pseudomonas syringae* pathovars, *Xanthomonas campestris* pathovars, *Xanthomonas oryzae* pv. *oryzae* (Mansfield et al. 2012) and fungi such as *Fusarium graminearum, Fusarium oxysporum, Botrytis cinereal* or *Blumeria graminis* (Ferreira et al. 2006). The infection process is dependent on the virulence of the pathogen and susceptibility of the host, as well as the stage of development or environmental conditions such as water availability, temperature and humidity. Only a limited number of pathogens can successfully attack a particular plant host

(Dutta et al. 2021, Ferreira et al. 2006). In order to control the diseases in plants, suitable cultivation practices are required that include crop rotation, high-quality seeds, use of resistant varieties, employment of early detection methods and use of biocontrol agents and chemicals (Dyussembayev et al. 2021).

#### *2.2.2 Viruses and Phytoplasmas*

Plant viruses consist of a protein coat that surrounds nucleic acids (DNA or RNA) required to synthesize essential proteins in the host cell. Following the formation of new particles, viruses move through plasmodesmata and attack neighboring cells (Schulze et al. 2019). Economically important viruses include potato virus Y (PVY), rice tungro spherical virus (RTSV), rice tungro bacilliform virus (RTBV), barley yellow dwarf virus (BYDV; genus Luteovirus; family Tombusviridae), maize streak virus (MSV), tomato spotted wilt virus (TSWV) (Tatineni and Hein 2023). Phytoplasmas are a group of obligate biotrophic pathogenic bacteria for monocotyledonous and dicotyledonous plants, and their replication is performed intracellularly in phloem sieve elements. Furthermore, the infection of plants triggers alterations in the developmental processes due to an imbalance of growth regulators, the symptoms becoming sterility, irregular internode elongation or virescence and flower leaf set (Bertaccini and Lee 2018, Miller et al. 2017).

#### *2.2.3 Insect Pests*

Insect pests consume living plant components, and they are classified as either monophagous (one or a few hosts) or polyphagous depending on the number of hosts (Abbas et al. 2024). The sucking type of feeding is particularly harmful to crop plants due to damage caused to the vascular system and disruption of transport of water and minerals. Aphids and whiteflies are examples of pests that are of a piercing-sucking feeding type (Kamatham et al. 2021). In contrast, mechanical damage is caused by the biting and chewing activities of insects like grasshoppers and beetles (Nakano et al. 2022). Plants feature a complex sensory system that enables them to perceive insect feeding and subsequently trigger defensive responses (Abbas et al. 2024).

## 3. Role of Nanoparticles in Stress Tolerance and Growth Promotion in Plants

In recent years, NPs have gained significant attention in agricultural biotechnology due to their potential in enhancing plant growth and stress tolerance. The unique physicochemical properties of NPs, such as high surface area, reactivity, and controlled-release capabilities, make them highly effective in modulating plant physiological and biochemical processes. NPs are known to tolerate various types of stress in plants like heat, drought, HMs, salinity as well as microbial and insect pests.

### *3.1 Tolerance of Heat Stress*

Due to global warming, the increase in temperature has become a key ecological limitation, threatening global food security. Heat or temperature stress can reduce crop yield at the morphological, physiological, biochemical, and cellular levels (Jahan et al. 2022). An increase of 1°C in annual temperature can decrease crop yields ranging from 2.5% to 16% (Jahan et al. 2021). To meet the rising demand for food due to population growth, it is crucial to enhance plant tolerance to temperature stress and employ eco-friendly technologies. Nanoparticles have gained considerable attention in recent years due to their environmentally favorable use in agriculture for enhancing the agriculture sector, since it introduces new techniques for imparting tolerance to temperature stresses and increasing crop yield (Alabdallah et al. 2024).

The exogenous application of titanium oxide ($TiO_2$) NPs to tomato (Qi et al. 2013) and the pretreatment of $TiO_2$ NPs to safflower cultivars (Dinler et al. 2024) were found to increase the plant tolerance to temperature stress by positively impacting the photosynthetic apparatus and enhancing photosynthesis in plants exposed to stressful conditions. The $TiO_2$ NPs improve photosynthetic efficiency under high temperatures by reducing photoinhibition and heat-induced oxidative stress. $TiO_2$ NPs enhance chlorophyll stability and increase the efficiency of electron transport chains, protecting plants from excessive heat exposure (Alabdallah et al. 2024).

Silicon (Si) is considered an important element for plant growth promotion and stress tolerance. The Si NPs application maintains Si level in plants, which in turn positively correlate with heat stress responses by improving photosynthesis, stabilizing plant cell membranes and reducing water loss through transpiration, and enhancing the expression of heat stress-responsive genes for the formation of Hsp, which help in protein folding and prevent thermal damage (Saha et al. 2021). The study conducted by Younis et al. (2020) assessed the ameliorative effects of Si NPs on wheat (*Triticum aestivum* L.) seedlings exposed to heat stress (45°C, 4 h). The Si NPs treatments significantly restored the heat stress-provoked ultrastructural distortions of cellular organelles. Further, Si NPs enhanced the photosynthetic capacity and membrane stability index. Cerium Oxide ($CeO_2$) NPs can act as antioxidants to neutralize ROS generated due to heat stress, protecting chloroplasts and mitochondria from oxidative damage. The $CeO_2$ NPs have redox properties that enable them to scavenge free radicals and reduce cellular stress (Wu et al. 2017). In addition, NPs such as silver (Ag), gold (Au), copper oxide (CuO), zinc oxide (ZnO), and graphene-based NPs can influence the physiological and biochemical responses of plants under heat stress. Sidhu et al. (2024) provided a detailed overview of the various mechanisms by which these NPs enhance plant resilience to heat stress.

### *3.2 Tolerance to Drought Stress*

Water availability to the plant can be affected by environmental factors and changing climatic conditions, such as rise in temperature and changing precipitation pattern, thereby affecting the plant's growth, metabolism, and survival resulting in drought stress in plants. NPs, due to their higher surface area to volume, can interact with plant cells and tissues in different ways, leading to stress mitigation (Kaplan et al. 2024).

Biomolecules like lipids, proteins and nucleic acids present in the plant can be damaged by the excessive generation of ROS such as superoxide radicals, hydrogen peroxide, and hydroxyl radicals during the draught stress. The NPs like ZnO (Shirvani-Naghani et al. 2024), $TiO_2$ (Karvar et al. 2022), $CeO_2$ (Soleymani et al. 2025), and $SiO_2$ (Nazim et al. 2024) are known to enhance the activity of antioxidant enzymes like superoxide dismutase (SOD), catalase (CAT), ascorbate peroxidase (APX), and peroxidase (POD). The NPs can neutralize these enzymes by scavenging ROS directly or indirectly by stimulating the plant's own defense pathways, reduce lipid peroxidation, and maintaining cell membrane integrity.

The Si, ZnO and $Fe_3O_4$ NPs promote accumulation of osmolytes like proline, glycine, betaine, and sugars under drought condition. Increase proline and soluble sugar content in plant tissues, enhance osmotic adjustment and allowing to retain water and sustain growth in plants (Shirvani-Naghani et al. 2024, Heidarian et al. 2025). The Si NPs are known to improve the uptake and retention of water in plants by improving the plant root architecture, reducing transpiration rate (inducing stomatal closure or forming a protective layer on the leaf surface) (Nazim et al. 2024), and thereby enhancing the relative water content of plants. Drought reduces chlorophyll content and photosynthetic capacity. The ZnO, $SiO_2$, and $TiO_2$ NPs preserve chlorophyll content and photosystem efficiency by protecting the photosynthetic apparatus from oxidative damage and maintaining higher rates of $CO_2$ assimilation and stomatal conductance under water-limited conditions (Daler et al. 2024, Shirvani-Naghani et al. 2024, Karvar et al. 2022).

Nanoparticles offer a multifaceted approach to drought stress mitigation in plants. By enhancing physiological, biochemical, and molecular responses, they act as stress modulators and performance boosters. However, the concentration, size, coating, and method of application of NPs are critical.

### *3.3 Tolerance to Heavy Metals Stress*

The HMs get accumulated in soil due to extensive anthropogenic activities, leaching of HM ions and the oxidation process, adversely affecting the plant growth. Nanoparticles play a significant role in enhancing plant tolerance to HM stress, a major environmental concern that affects crop productivity and food safety. There are several ways by which NPs, either singly or in combination with other molecules, help plants to tolerate the HM toxicity. The $Fe_3O_4$ NPs stimulated the root growth in tomato, cucumber, carrot and lettuce plants in the presence of Cd solution, the precipitation mainly contributed to phytotoxicity reduction by the NPs (Wang et al. 2012). The ZnO NPs in combination with sialic acid were responsible for modulation of oxidative stress in rice (*Oryza sativa* L.), plants under arsenic (As) stress by enhancing the enzymatic activities (CAT, POD, and SOD), proline and total soluble protein contents, indicating their role in the regulation of various transcriptional pathways participated in oxidative stress tolerance (Faizan et al. 2021). The $CeO_2$ NPs influences the uptake and subsequent translocation of HMs like Cu, Mn, Zn and Fe by sugar pea (*Pisum sativum* L.). The HM ions presumably transported via cytoplasm and cell to cell connections; the inner part of the plant cell, which enclosed by a plasma membrane containing the network of cytoplasm connected by plasmodesmata, also known as symplastic pathways, may compete with NPs for similar carriers (Skiba and Wolf 2019).

An elaborated account on recent progress concerning the role of NPs in HM stress tolerance by different NPs is given by Faizan et al. (2023). This review can be useful for understanding phytoremediation of contaminated soils and show how NPs, either singly or in combination with other molecules, help the soil withstand HM stress.

### *3.4 Tolerance to Salinity*

Salinity is a major abiotic stress that limits plant growth and agricultural productivity by disrupting water uptake, ionic balance, photosynthesis, and metabolic functions. Nanoparticles offer an innovative and eco-friendly approach to improving plant resilience under such stress (Muhammad et al. 2024). Salinity stress leads to excessive production of reactive oxygen species (ROS), which damages biomolecules and cell organelle. Nanoparticles like $TiO_2$ (Karvar et al. 2022), $CeO_2$ (Soleymani et al. 2025), help plants to detoxify ROS by the activation of antioxidant enzymes like SOD, CAT, APX and GR or by direct ROS scavenging or by upregulation of antioxidant gene expression, enhancing cellular protection mechanisms. High salt levels cause ion imbalance, particularly $Na^+$ toxicity and $K^+$ deficiency. Nanoparticles help maintain ion homeostasis by reducing $Na^+$ uptake and enhancing $K^+/Na^+$ selectivity, modulating ion transporter activity (e.g., HKT1, SOS1, NHX1), forming physical or biochemical barriers at the root surface to prevent $Na^+$ entry and compartmentalizing $Na^+$ ions into vacuoles, reducing cytosolic toxicity (Singh et al. 2024).

Salinity reduces water potential, making it difficult for plants to absorb water. Nanoparticles help counter this through enhancing aquaporin expression, stimulating the production of osmolytes, which aids in osmotic adjustment and maintaining relative water content in plant tissues (Soni et al. 2024). The disruption of photosynthesis by damaging chloroplasts and photosynthetic pigments during the salinity stress can be tolerated by ZnO, $TiO_2$, and $SiO_2$ NPs by preserving the chlorophyll content, stabilizing photosynthetic enzymes, enhancing light-harvesting efficiency and improving gas exchange and stomatal conductance, leading to the sustained energy production under stress conditions (Rehman et al. 2024). Nanoparticles also influence phytohormones and stress signaling pathways, contributing to salinity tolerance by upregulation of stress hormones like ABA and salicylic acid (SA), activate signaling molecules such as calcium ions ($Ca^{2+}$), nitric oxide (NO), and hydrogen sulfide ($H_2S$) and modulating the gene expression related to transcription factors like DREB, NAC, and WRKY, which regulate stress-responsive genes (Singh et al. 2024). The impaired nutrient uptake and metabolism during the salinity stress can be restored by NPs; ZnO NPs supply zinc, crucial for enzyme activity and hormone biosynthesis (Ahmed et al. 2022); $Fe_3O_4$ NPs enhance iron availability, supporting chlorophyll synthesis and respiration (Kreslavski et al. 2023) and nano-fertilizers improve nutrient use efficiency and reduce the need for excess chemical fertilizers.

### *3.5 Tolerance to Microbial Stress*

Adverse effects caused by pathogenic bacteria, fungi, and viruses can infect roots, stems, leaves, or vascular tissues, leading to reduced growth, yield loss, and in severe cases, death of a plant. Nanoparticles offer an eco-friendly alternative as compared to chemical pesticides by providing multiple modes of action to enhance plant defense systems, inhibiting the growth of pathogenic bacteria, fungi and viruses, and strengthening the plant health.

The intrinsic antimicrobial property of NPs that directly inhibit or kill plant pathogenic bacteria, fungi and viruses, NPs like Ag (Shende et al. 2024a, 2024b), ZnO (Ingle et al. 2024a), $TiO_2$ (Rathore et al. 2023), sulphur (Ingle et al. 2024a), chitosan (Ingle et al. 2022), and Cu and CuO NPs (Shende et al. 2015, Rai et al. 2018, Ghaywat et al. 2023, Ingle et al. 2024b) have demonstrated antimicrobial potential against the plant pathogenic microbes. The Cu NPs could attach to the fungal cell wall and release the $Cu^{++}$ into cells or attach to microbial membranes causing physical disruption of the membrane and leakage of cell content. In the fungal cell they interact with the enzymes inhibiting vital metabolic pathways or can generate ROS causing oxidative damage to cell components and biomolecules (Shende et al. 2021).

Plant innate immunity can be elicited by nanoparticles similar to a regulating molecule. They activate defensive responses by upregulating pathogenesis-related proteins such as chitinases or β-1,3-glucanases or by activation of defense signaling pathways such as salicylic acid pathway for systemic acquired resistance and jasmonic acid or ethylene pathways for induced systemic resistance (Zhang et al. 2024). Apart from that NPs also activate anti-oxidative defense systems (Singh et al. 2024).

Nanoparticles improve the physical structural defense of plants by enhancing the lignin biosynthesis for strengthening cell walls, cuticle reinforcement by stimulating wax and callose deposition, making it harder for pathogens to penetrate (Lamalakshmi et al. 2017). They are used as carriers for antimicrobial agents by nanoencapsulation of biocontrol agents for controlled-release and targeted delivery, ensuring higher effectiveness with lower doses.

### *3.6 Tolerance to Insect Pests*

Insect pests pose a significant challenge in agriculture as they can cause extensive damage to crops by feeding on plant tissues, transmitting pathogens, and disrupting physiological functions. Nanoparticles offer promising potential in pest management through targeted molecule delivery, enhanced control of release mechanisms, and direct toxic effects on insect pests (Wang et al. 2022).

Nanoparticles enter the insect body either by penetration through the cuticle due to small size, or by feeding on treated surfaces of plants, or by inhalation or tracheal exposure. Once inside the body, NPs can accumulate in tissues, interact with cellular membranes, and interfere with metabolic pathways. The $SiO_2$ NPs have demonstrated its entomotoxic effect on *Spodoptera litura* (Sushila et al. 2020) and *Plutella xylostella* (Shoaib et al. 2018). Nanoparticles can influence the stress-responsive gene expression profiles of Hsp, metallothioneins, antioxidant and detoxification pathways and cytochrome P450, in insect pests. Nair et al. (2013) have evaluated the effect of Ag NPs on stress-responsive gene expression in *Chironomus riparius*. Nanoparticles can interfere with endocrine signaling by disrupting ecdysteroids and juvenile hormones responsible for molting and metamorphosis, and neuropeptides and diuretic hormones, altering water balance and osmotic stress management, which leads to developmental delays, abnormal milting and increase mortality under environmental stress conditions in insect pests (Mossa et al. 2018). For sustainable and effective pest control, it is essential to adopt a balanced and environmentally responsible approach to NP use, taking into account appropriate dosage, precise delivery methods, and target specificity.

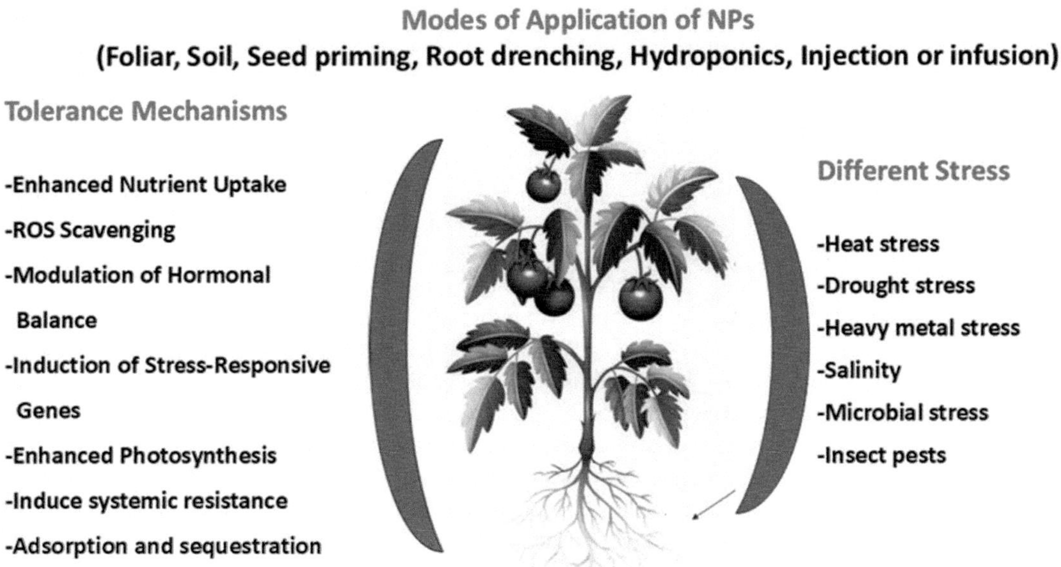


**Figure 2.** Role of NPs in tolerance of different types of stress in plants and tolerance mechanism.

Nanoparticles offer a multifaceted approach to improving plant stress resilience and promoting growth. An overview of different modes of application of NPs, different stress and their tolerance mechanism is given in the Figure 2. While considerable progress has been made in laboratory and greenhouse settings, translating these findings to field applications requires further investigation and regulatory oversight.

## 4. Mechanisms Involved in Enhancing Stress Tolerance in Plants by Nanoparticles

Plants are constantly exposed to various abiotic stresses, including drought, flood, salinity, temperature extremes (i.e., heat and cold), heavy metal toxicity, and oxidative stress, as well as biotic stresses, such as microbial stress, insect pests, viruses, and phytoplasmas, all of which negatively impact their growth, productivity, and survival. To mitigate these challenges, plants have developed various stress-responsive mechanisms, ranging from biochemical adjustments to morphological changes. Recent advances in nanobiotechnology have opened new avenues for enhancing stress tolerance in plants through the application of NPs. Nanoparticles, because of their unique shape, size, and surface properties, can interact with plant systems in ways that bulk materials cannot. The mechanisms by which NPs enhance stress tolerance in plants, focusing on biochemical, physiological, and molecular responses, are given in the sections below.

### *4.1 Enhancement in Osmotic Stress Tolerance*

Osmotic stress arises because of water scarcity or salinity and is one of the most common forms of abiotic stress confronted by plants. Nanoparticles, particularly metallic (e.g., Ag, Au, and ZnO) NPs, have been shown to play a significant role in improving water uptake and maintaining osmotic balance in plants. The hydration of cells in plant is one of the mechanisms that may be transformed by NPs which eventually improve the plants' water retention capacity by enhancing cellular hydration. Several studies are showing that metal NPs improve the roots' ability to absorb water under drought environments, possibly due to the increased reactivity and surface area of NPs (Zaman et al. 2025). For example, Ag NPs have been documented for stimulating root growth, leading to better water absorption in plants (Abasi et al. 2022, Ansari et al. 2023). The regulation of osmolytes by NPs is another mechanism that can affect the accumulation of osmoprotectants, like betaine, mannitol, proline, glycine, and sugars, which help to maintain the cellular osmotic potential. ZnO NPs, for instance, have been revealed to enhance the accumulation of proline and other osmolytes under salt stress, therefore reducing the negative impacts of high salinity (Gupta et al. 2024).

### *4.2 Regulation of Reactive Oxygen Species (ROS) and Antioxidant Defense*

Abiotic stresses frequently bring oxidative stress in plants, leading to the generation of ROS, including hydrogen peroxide ($H_2O_2$), superoxide radicals ($O^{2-}$), and hydroxyl radicals ($OH^-$), etc. Extreme ROS can impair cellular structures, lipids, and proteins, damaging plant function. Nanoparticles may improve the antioxidant defense systems of plant, thus mitigating oxidative damage. The NPs have been revealed to influence the antioxidant enzymes that can activate the plants' antioxidant defense system, particularly enzymes such as catalase (CAT), superoxide dismutase (SOD), peroxidase (POD), and glutathione reductase (GR). For example, $TiO_2$ NPs can stimulate the activity of SOD and CAT, help in reducing ROS levels in plants exposed to salinity and drought stress (Farahi et al. 2023). The ROS scavenging capacities of specific NPs themselves show antioxidant properties. For instance, NPs like $CeO_2$ NPs have inherent ROS scavenging capabilities, which benefit neutralize oxidative stress in plant cells (Mukherjee et al. 2023, Haghmadad Milani et al. 2024). These NPs can directly interact with ROS and prevent their accumulation, dropping oxidative damage to cellular membranes, DNA, and proteins.

### *4.3 Regulation of Plant Hormones*

Nanoparticles have the ability to modulate hormonal levels in plants, which play a crucial role in stress responses. Hormones like auxins, gibberellins, ABA, and cytokinins are involved in regulating stress tolerance mechanisms such as root growth, stomatal closure, and antioxidant responses. The ABA is an important regulator of plant responses in the water stress condition, and NPs have been revealed to enhance the ABA production. Especially, Au NPs have been found to rise ABA levels in plants exposed to drought stress, helping stomatal closure to save water (Al-Khayri et al. 2023, Sharma et al. 2024). Ag NPs might affect the levels of auxins, which are integrated in regulating root development and growth. Under stress conditions, Ag NPs may improve root branching and elongation, consequently enhancing nutrient and water uptake. Similarly, ZnO NPs have been reported to affect levels of cytokinin, contributing to mitigation of stress by endorsing cell division and elongation (El-Badri et al. 2021).

### *4.4 Augmentation of Nutrient Uptake and Homeostasis*

In plants, NPs may enhance nutrient uptake and homeostasis, which is crucial to maintain optimal metabolic functions under stress environments. The usage of NPs as nano-fertilizers has been revealed to increase essential nutrients such as nitrogen, phosphorus, potassium, and micronutrients availability and uptake. For instance, ZnO NPs can improve zinc availability to plants, leading to enhanced growth and tolerance to drought and salinity stress (Nongbet et al. 2022). Likewise, Si NPs have been employed to improve nutrients uptake like phosphorus, contributing to boosted stress resilience (Rakesh et al. 2024). Nanoparticles can also participate in the ion-channel regulation. Within plant cells, NPs may control ion transport and homeostasis. For example, Au NPs may regulate ion channels in root cells, improving the ability of plants to manage with high salinity by refining ion partitioning and minimalizing ion toxicity (Cao et al. 2025).

### *4.5 Molecular and Genetic Regulations of Stress Responses*

The application of NPs may affect molecular signaling pathways and gene expression involved in stress tolerance. Nanoparticles can control genes expression involved in stress-responsive proteins, antioxidant defenses, and transporters. Nanoparticles have been revealed to impact the genes expression associated to stress tolerance. For example, ZnO NPs bring the genes expression involved in stress-responsive transcription factors (e.g., dehydration-responsive element-binding protein 2 (DREB2), NAC), antioxidant defense, and detoxification pathways (Nuruzzaman et al. 2013, Al-Khayri et al. 2023). Besides this, emergent evidence proposes that NPs such as ZnO, CuO, and $\gamma$-$Fe_3O_4$ with diverse and unique physicochemical properties may also trigger epigenetic changes in plants, such as DNA methylation and histone modification, which control the stress-responsive genes expression. These NPs also displayed genotoxic effects on wheat (*Triticum aestivum* L.) and mature embryo cultures, when complemented in MS medium (Haliloğlu et al. 2022). These alterations can deliberate long-term stress tolerance even in successive generations. Some examples of molecular responses in plants under the effect of various NPs at different abiotic stresses have been given in Table 1.

From the above discussion, we can conclude that the NPs signify a promising tool for increasing stress tolerance in plants, offering a multifaceted method to modifying the adverse effects of different stresses. By influencing nutrient availability, water uptake, hormonal regulation, antioxidant defense, and gene expression, NPs provide an exceptional mechanism of action that improves plant resilience. However, further investigate is essential to fully comprehend the long-term impacts of NPs on plant health and their potential risks to the surroundings. Understanding the interaction between plant systems and NPs will be crucial to developing effective and sustainable nanotechnological applications in agriculture.

**Table 1.** Molecular responses in plants under the influence of various nanoparticles at different abiotic stresses (Al-Khayri et al. 2023).

| **Plant used in study** | **Applied nanoparticles** | **Condition** | **Plant response** | **Reference** |
|---|---|---|---|---|
| *Cicer arietinum* L. | $TiO_2$ | Cold stress | Upregulation of RUBISCO and PEPC to upsurge photosynthesis and decrease electrolyte leakage index | Amini et al. 2017 |
| *Arabidopsis thaliana* L. | $TiO_2$, Ag, MWCNTs | Drought stress | Noncoding RNA-mediated translational repression of physiological target genes | García-Sánchez et al. 2015 |
| *Solanum lycopersicum* L. | Ag, Si | Salinity and Osmotic stress | ABA upregulation and its dependent stress alleviation by NCED3, TAS14 and SnRK2 upregulation; osmoregulation and oxidative stress alleviation by AREB1-mediated P5CS1 overexpression and proline biosynthesis; SOS, ROS and MAPK-signalling activation | Almutairi 2016 |
| *Glycine max* (L.) Merr. | Ag, $Al_2O_3$ | Flood stress | Protein, lipid and energy metabolism perturbation by PAB2 and BKR1 upregulation and PDC downregulation, respectively, and antioxidant defense by glyoxalase II 3 downregulation | Mustafa et al. 2015 |

## 5. Risk Assessment

Nowadays, nanobiotechnology is swiftly progressing in many sectors, including agriculture, where it holds potential for improving plant growth and enhancing tolerance under environmental challenges. The application of NPs in agriculture, mainly for plant growth promotion, stress mitigation, and nutrient delivery, has revealed promising results (Shende et al. 2017, Jahagirdar et al. 2020). However, the exceptional physicochemical properties of NPs, like their high surface area, small size, and reactivity, raise concerns concerning their ecological impact, safety, and probable risks to plants, animals, and humans. In line with this, a comprehensive risk assessment of

NPs applications in plant growth promotion and stress tolerance, directing on the probable hazards, regulatory frameworks, and strategies for modifying risks, is briefly given in the subsequent section.

### *5.1 Phytotoxicity of Nanoparticles and Potential Risks to Plants*

Nanobiotechnology has the capability to upsurge agricultural output and productivity, protect crops from conservational threats, and decrease chemical emissions into the atmosphere. Several aspects, including their size, the cultivar or genotype, composition, dosage, surface area, redox state, surface coatings, and application practices of the NPs, significantly impact their unique effects. The potential toxicity of NPs to plants is one of the greatest noteworthy concerns. The interaction between plant tissues and NPs may lead to undesirable results, including inhibited growth, oxidative stress, and decreased physiological processes. Many plant species have established toxicity to NPs like $CeO_2$, Se, $TiO_2$, and ZnO, including broad bean, barley, soybean, sorghum, tomato, wheat, sweet basil, and onion (Vasseghian et al. 2022). While NPs have been revealed to recover plant health and resilience, their different properties also raise questions about their impacts on plant physiology and development. Nanoparticles cause numerous potential adverse effects on plants, including decreased seed germination and root elongation, reduced biomass, limited mineral uptake, growth inhibition, electron-hole pair generation, damage of cellular structures, ROS, impairment to the cell wall and cell membrane, and NP aggregation, which increases ROS levels and causes toxicity in tissues (Ghori et al. 2019, Ranjan et al. 2021). Metal-based NPs, like Ag and Cu NPs, can persuade oxidative stress in plant cells by producing ROS (Dhiman et al. 2021). Excessive ROS may impair plant cell membranes, DNA, and proteins, leading to cell death or reduced growth. However, research demonstrated that the NPs toxicity depends on aspects like the species and genotype of plants involved, the size and concentration of NPs, and the duration of plant exposure to them (Zhou et al. 2023). Some studies have revealed that high concentrations of NPs, mainly metal-based NPs, may interfere with root morphology by disturbing root branching and elongation. This may decrease nutrient and water uptake, ultimately delaying plant growth (Cao et al. 2025). Nanoparticles may accumulate in plant tissues, possibly leading to toxicity in the plant and successive entry into the food chain. For example, ZnO NPs have been reported to accumulate in plant leaves and roots, with possible long-term impacts on plant health and consumer protection (Strekalovskaya et al. 2024).

The function of protein corona in plants has just recently been explicated, notwithstanding noteworthy advances in medicine and ecology. The absorption, impacts, transport, and the ultimate fate of NPs in plants are obstructed by several factors, involving the protein corona surrounding the particles. The 'protein corona' is a dynamic coating of proteins and other biomolecules that is identified to adsorb onto the surface of NPs as they go in biological surroundings, like plant systems (Huang et al. 2017, Önal Acet et al. 2024, Mayordomo et al. 2025). The development of the protein corona is critical since it alters the physicochemical characteristics of NPs, including their size, charge, and surface topographies, which may affect how they are recognized and transported through plant tissues (Wang et al. 2020). Precisely, the protein corona changes the interaction of NPs with cell walls and membranes, henceforth persuading their uptake, translocation, and bioavailability in plants (Yu et al. 2023). To influence the eventual site of NPs inside plant tissues, for instance, specific proteins in the corona can interact with membrane receptors or transporters to either prevent or promote NPs entry in the plant cell (Kumar et al. 2022). Additionally, the protein corona may impact the biocompatibility and probable toxicity of NPs in plants, as certain proteins may reduce or increase the stress response of plants to foreign particles (Rampado et al. 2020). Thus, knowing how the protein corona forms and is composed in plant systems offers significant evidence on how NPs behave and end up in agricultural and environmental surroundings.

### *5.2 Environmental Risks of Nanoparticles in Agriculture*

The extensive use of NPs in agriculture also increases concerns about their possible influences on the environment, particularly water and soil systems. Soil microorganisms play a vital role in organic matter decomposition, nutrient cycling, and plant health. The application of NP-based products in soil can lead to NPs' accumulation, affecting soil microbial diversity and health, which may pollute the soil and disturb the microbiota (Shende et al. 2023). Some NPs may disturb and demonstrate toxic effects on soil microbial communities, particularly beneficial microorganisms such as nitrogen-fixing bacteria like species of *Bacillus* sp., *Azotobacter* sp., *Klebsiella* sp., and *Clostridium* sp. genus, mycorrhizal fungi like arbuscular mycorrhizal fungi (AMF), Mucoromycotina, etc., and those involved in decomposition of organic matters (Rajput et al. 2020, Shende et al. 2021, Balusamy et al. 2023). For instance, Ag NPs have been revealed to decrease the richness and diversity of soil bacteria, possibly disrupting nutrient cycling and soil health (Soltani Toularoud et al. 2025, Tończyk et al. 2025). There is also an opportunity of discharging of NPs from the soil into surface or ground water, where they may persist and affect aquatic ecosystems. The movement and bioavailability of NPs in the soil are affected by their surface charge, size, and coating materials (Atanda et al. 2025). For example, ZnO and $TiO_2$ NPs could be transported through the soil profile, disturbing soil chemistry and aquatic organisms (Dube and Okuthe 2023). The NPs used in plant or soil can be wash away into water bodies through runoff or irrigation, leading to water pollution. The accumulation of NPs in aquatic environments poses probable hazards to aquatic fauna, including amphibians, fish, and aquatic plants. They may display toxicity to aquatic organisms. Ag and Cu NPs, especially, have been found to show toxicity to aquatic species, producing oxidative stress, behavioral changes, and mortality in invertebrates and fish (Dube and Okuthe 2023). The leaching of NPs into water bodies can therefore have cascading impacts on aquatic ecosystems.

### *5.3 Animal and Human Health Risks*

There are increasing evidences that NPs might influences humans and other higher-level consumers. Nanoparticles not only pose jeopardies to plants and the environment but also to animal and human health. The opportunity of NPs entering the food chain via water or crops sources needs vigilant risk assessment. To examine the toxicity of NPs to animals and humans, investigators are also looking into their wider consequences (Rajpal et al. 2023). It was demonstrated that a food additive based on $TiO_2$ NPs may increase the spread and growth of gastrointestinal tumors in a mouse model by altering systems such as cancer signaling, inflammation, immunological responses, and the cell cycle (Li et al. 2021). Plants that absorb NPs from water or soil can accumulate these particles in eatable parts like leaves and fruits. If these NPs are toxic, they might pose a menace to human health when consumed. There are reports that some NPs, such as Ag and $TiO_2$ NPs, can accumulate in edible plant tissues and pose dangers to human health upon ingestion (Maharramov et al. 2019, Shende et al. 2020). The probable jeopardies comprise neurotoxicity, inflammation, and modifications in gut microbiota. The long-term acquaintance to NPs through food ingestion can lead to bioaccumulation in human tissues, triggering potential health issues like organ damage, cancer, or respiratory problems (Gupta et al. 2024). Nevertheless, further studies are required to launch safe exposure levels and the probable cumulative impacts. Besides this, NPs might also be injurious to animals, predominantly those that depend on plants as a food source. For example, herbivores consuming polluted plants might be at risk of NPs ingestion. The herbivorous animals feeding on plants preserved with NPs can ingest NPs, possibly leading to gastrointestinal toxicity, decreased reproductive success, or other health issues. Studies have shown that Ag NPs may originate toxicity in earthworms, which play an important role in soil aeration and decomposition of organic matter (Jesmer et al. 2017).

In order to assess the advantages and risks of NPs in agricultural practices, it is critical to have a better understanding of how they interrelate with plants at the cellular and molecular levels. Since cells respond differently to different characteristics, involving their concentration, introduction, variability, uptake, translocation, accumulation, surface activity, and, most significantly, the NPs association through cellular organelles and their influence on metabolism, extensive research is needed in the nanobiotechnology field before the technology can be commercially applied (Ali et al. 2021, Shende et al. 2023).

Given these limitations, future investigation efforts must look into conducts to make agricultural NP usage more productive and safer. To limit the opportunity of ecotoxicological threats to plants and humans, more study into nano-bio interactions and effect evaluations is essential. Another important question that the scientific community should address is the safe removal of NPs. If not burned, NPs must be discarded in closed vessels in a hazardous waste facility. Unfortunately, NPs are being used in several institutions and laboratories worldwide that lack access to suitable technologies. The usage and dumping of NPs requires rigorous safety procedures and guidelines (Donia and Carbone 2019).

### *5.4 Risk Mitigation Strategies*

For minimizing the risks related with the usage of NPs in agriculture, numerous strategies might be employed, including (a) development of eco-friendly NPs, (b) controlled-release systems, (c) comprehensive risk assessment protocols, and (d) regulatory frameworks. The development of environmentally friendly NPs may be possible by designing and synthesizing biocompatible, eco-friendly NPs that degrade speedily in the environment and may decrease long-term environmental and health risks. For instance, using plant-based or biodegradable polymers to coat NPs may decrease their toxicity and surge their environmental security (Vinzant et al. 2023). Alternatively, controlled-release NP formulation systems can support limit the environmental effect by guaranteeing that NPs are only available to plants when required. This may decrease discharge into the water and soil, minimalizing the risk of pollution (Fincheira et al. 2023). Moreover, the formation of rigorous, standardized risk assessment protocols for evaluating the environmental fate and toxicity of NPs must be established for their safe application in agriculture, which might be a critical step. These protocols must include testing for bioaccumulation, environmental persistence, phytotoxicity, and human and animal health effects (Oomen et al. 2018, Kumari et al. 2023). The regulatory frameworks must create clear guiding principles and protocols leading the application of NPs in agriculture by regulatory and governments bodies. These regulations should be based on systematic risk assessment and guarantee that NPs employed in agricultural applications encounter safety ethics (Amenta et al. 2015, Subramanian and Rajkishore 2018, Kumari et al. 2023, Ijaz et al. 2024).

Thus, from the above discussion, it can be concluded that while nanobiotechnology holds potential for increasing plant growth and stress tolerance, the usage of NPs in agriculture poses many risks to plant health, the environment, and human and animal protection. A careful and comprehensive risk evaluation is important to ensure that these technologies are employed sustainably. Mitigation approaches, including the development of environmentally friendly NPs, controlled release systems, and regulatory frameworks, are important to lessen probable adversative impacts. As research develops, more specific risk assessment models will be required to guarantee the safe incorporation of nanobiotechnology in agricultural practices.

## 6. Nanosensors and Their Application to Mitigate Various Stresses

To uphold sustainable agricultural development, novel technological progressions, for example, nanosensors, are essential. Nanosensors are economic, eco-friendly, real-time, powerful devices with a dimension of approx. 1–100 nm (Amjad 2022, Bhagat et al. 2023, Mondal et al. 2023). The addition of engineered nanomaterials and nanostructures such as NPs, quantum dots, nanotubes, nanocomposites, etc., into sensors curiously upgrades the efficiency of devices in terms of accuracy, reactivity, selectivity, transferability, and manifold recognition capabilities (Kumar et al. 2021, Sharma et al. 2021, Hameed et al. 2023). The use of NPs in agricultural fields as nanopesticides and

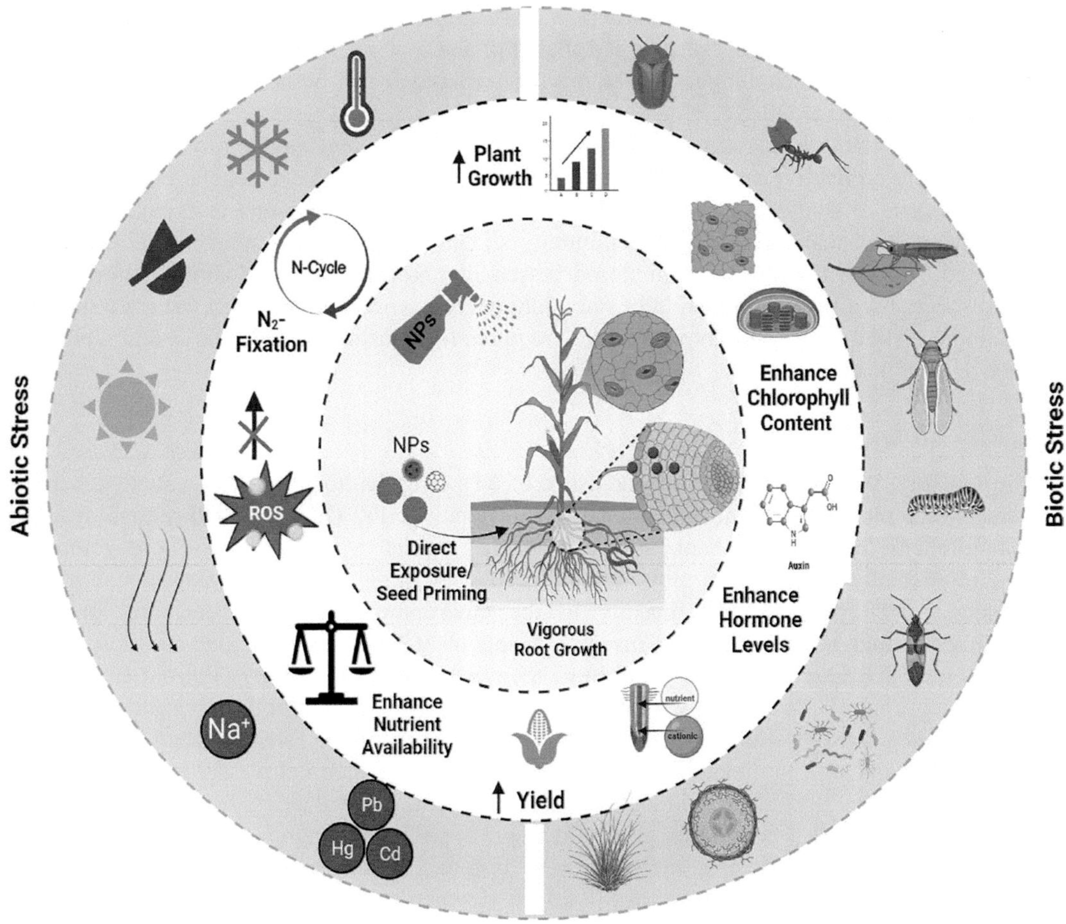


**Figure 3.** Schematic representation on response of plants to NPs under different abiotic and biotic stress conditions (Zain et al. 2023, reprinted under http://creativecommons.org/licenses/bync-nd/4.0/).

nanofertilizers has gained virtuous thrust in terms of increased crop production (Shen et al. 2023). Nanoparticles have high benefits concerning improved growth and yield, particularly under abiotic and biotic stress conditions (Figure 3).

Nanosensor-based technologies have already established their importance in several disciplines. Abiotic environmental stresses such as salinity, drought, waterlogging, temperature (heat and cold), and HMs are the foremost impediment to global agricultural crop production and food security. Every year, abiotic environmental stresses are responsible for 52 to 82% of losses in crop yield worldwide (Kajla et al. 2015, Li et al. 2021, Oshunsanya et al. 2019). In recent years, unembellished changes in the environment have had an adverse effect on the global agriculture sector. Climate change has affected the abiotic surroundings of plants, changing the development of plants, their physiology, and ultimately their productivity (Al-Khayri et al. 2023). To decrease crop loss produced by abiotic stress, it is critical to detect and retort to it quickly. There are several inbuilt specialized adaptation mechanisms that occur in plants and are used by them to mitigate abiotic stress, which includes (a) osmoprotectants, (b) change in physiological response, (c) stress-related proteins, (d) antioxidant defense, (e) gene expression, (f) cellular homeostasis restoration, (g) detoxification signaling for repairing damages caused by stress, etc. (Kocabay et al. 2013, Mashabela et al. 2023, Singh and Chauhan 2017). The environmental abiotic stress accelerates ROS production and produces oxidative stress in plants (Hasanuzzaman and Fujita 2022). The plants under abiotic stressed conditions produce certain harmful chemical entities, such as diatomic oxygen, which include of $H_2O_2$, $O^{2-}$, $OH^-$, etc. (Choudhury et al. 2013, Hasanuzzaman et al. 2020, Hasanuzzaman and Fujita 2022). Moreover, stressed plants show modified phenotypes.

The airborne sounds produced by stressed plants can be recorded, which can then be used to identify the condition of the plants (Khait et al. 2023). Therefore, investigators may assess health in response to different stress by identifying chemical composition, phenotypes, phytohormones, etc. (Ansari et al. 2020). Plants also emit chemicals during stress conditions to activate their defense system, and identifying these stress-related chemicals can open the door for finding the various stress conditions. A surface-enhanced Raman scattering (SERS)-based nanosensor was found by Son et al. (2023) and was then used for the real-time detection of several stress-related chemicals in living plants. Alova et al. (2020) have recognized oxygen ($O_2$) nanosensors using quartz micropipettes with platinum- and carbon-coated tips to measure the decreased oxygen level in *Chara corallina* plants. Lew et al. (2020) developed a non-destructive optical NS-probe for the detection of in planta hydrogen peroxide signaling molecules under stress conditions in different plant species such as *Arabidopsis thaliana*, *Eruca sativa*, and *Spinacia oleracea*. This sensor utilized a single-stranded (GT)15 enfolded CNT with single-walled that precisely binds to $H_2O_2$. In another study, genetically encoded Förster resonance energy transfer (FRET) nanosensors were used for the measurement of cytosolic glucose instability in rice (*Oryza sativa* L.) plants developed under abiotic stressed conditions by Zhu et al. (2017). The authors confirmed in rice plants that under different stress conditions, FRET-based nanosensors may be exercised for sensing increases in cellular glucose ($C_6H_{12}O_6$) up to the physiological limit of 0.2 and 20 μM. In addition, the electrochemistry-based immunobiosensor was designed by Wang et al. (2019) for sensing glycoprotein proteins, for example, vitronectin. The glycoprotein vitronectin is a biological marker found on the exterior of plant cells in response to HMs and metalloids. This method has a detection threshold of 25.6 nmol $L^{-1}$ and 18.5 nmol $L^{-1}$ for lead (Pb) and cadmium (Cd), respectively. Heavy metal ion detection has been possible with quantum dots-based nanomaterials-nanosensors (Wang et al. 2022a). The surface-enhanced Raman spectroscopy nano-biosensor has been employed for the detection of Cd by the AuNPs-trimercaptotriazine-based ratiometric sensing process with an identification threshold of 8 μg $kg^{-1}$ (Zuo et al. 2018). Several other Au nanomaterial-based nanosensors, like lateral flow and colorimetric assay, etc., are capable of detecting HM stress situations (Anwar et al. 2018, Gao et al. 2018, Xiao et al. 2018). Some examples of nanosensors used to sense abiotic environmental stresses have been given in Table 2.

An early detection of pests and diseases in plants may efficiently reduce the development of disease and pest spread, which also guarantees agricultural productivity. Nanosensor-based remote sensing is an exceptional method for predicting, detecting, and managing insect pests and plant diseases in several fruit orchards and crops (Abd El-Ghany et al. 2020). Particularly, in plants, nanosensors are employed for disease diagnosis, amino acid detection, nutrient assessment, plant hormones and other biomolecule identification. Real-time recognition and accurate management of the plant could be possible by these data. A number of nanosensors have been used to study different biological processes of plants: for example, optical nanosensors based on FRET have been used to discover cell compounds, amino acid interactions, and biophysical characteristics, while electrochemical nanosensors have been used to detect chemical redox (reduction-oxidation) reactions in plants (Zhu et al. 2017).

Furthermore, nanosensors are probable solutions for pests and insect management. Because of their great sensitivity and specificity, nanosensors may be exercised to immediately detect insects in agricultural fields (Martinazzo et al. 2022). A gas sensor was developed using micro-electro-mechanical systems (MEMS) by Brezolin et al. (2020), which was then used to sense insect sex pheromone ($C_{17}H_{34}O_2$) of stink bugs (*Euschistus heros*) at the concentrations of 0.005–0.03 μg $mL^{-1}$. In addition to this, researchers also developed biosensor-based nano-tuning for the detection of phytoplasmas, nematodes, and viroids (Bind et al. 2023, Dutta et al. 2023, Sharma and Dhadly 2023). Besides this, in pest control programs, insect pheromones and semiochemicals

**Table 2.** Some examples of nanosensors used to sense abiotic and biotic stresses in the surrounding (Zain et al. 2023).

| Nanosensors | Applied nanomaterial | Sensing purpose | Detection limit | Sensing process | References |
|---|---|---|---|---|---|
| **Detection of Heavy Metals** | | | | | |
| Colorimetric NSs | Au | Pd (II) | 4.23 μM | Pd (II) selectively induced the aggregation of APP-AuNPs as com- pared to other metals, resulting into the complete significant disappearance of surface plasmon resonance | Anwar et al. 2018 |
| | Silver-coated AuNBPs | Hg | 0.8 μM | Etching silver-coated AuNBPs results in a shift in color that allows for $Hg^{2+}$ detection | Qi et al. 2018 |
| | Mesoporous silica NPs | Hg | 60 pM | Hg (II)'s strong affinity for sulfur, it can interact with solid $S_2$ and produce new $Hg(S\text{-}R)_2$, which can disrupt the linear dithioacetal connections and cause the release of cargo | Jimenez-Falcao et al. 2020 |
| Lateral flow assay | AuNSs | Cd | 0.18 ng $mL^{-1}$ | Competitive immune-chromatographic strips are built on AuNSs' capacity to quench QD signal | Xiao et al. 2018 |
| | Core-shell UCNPs | Hg | 5 ppb | UCNPs-aptamer integrated mobile sensing system using smartphones | Jin et al. 2018 |
| Electrochemical | $Fc\text{-}NH_2$-UiO-66 | Cd (II); Pb (II) | 8.5 Nm; 0.6 nM | Cyclic voltammetry (CV), electrochemical impedance spectroscopy (EIS) | Wang et al. 2020 |
| | Ln-MOF/ ZJU-27 | Cd (II) | 1.66 Nm | Square wave anodic stripping voltammetry (SWASV) | Ye et al. 2020 |
| | 3DGO-Py10 | $Cd^{2+}$ | 3.6 μg $L^{-1}$ | SWASV | Guo et al. 2020 |
| | $MnCo_2O_4$ nanoparticles | $Pb^{2+}$; $Cd^{2+}$ | 8.06 nmol $dm^{-3}$ | Linear sweep anodic stripping voltammetry (LSASV) | Antunović et al. 2019 |
| | SWCNTs | Mercury | 3 f M | Vertical-SWCNTs based cyclic voltametric assay | Shi et al. 2017 |
| Single plasmonic Nanosensors | Gold nano-rods (GNRs) | $Pb^{2+}$; $Hg^{2+}$ | 5 nM; 1 nM | Position encoded aptamer functionalized GNRs sensing device | Ye et al. 2022 |
| Triboelectric Nanosensors | Ultra-thin MXene | $Cu^{2+}$; $Cr^{3+}$; $Zn^{2+}$ | 10–300 μM | Simple vacuum filtration method | Wang et al. 2022b |
| SERS | AuNPs | Cd | 8 μg $kg^{-1}$ | AuNPs-trimercapto triazine based radiometric sensing | Zuo et al. 2018 |
| Multimodal NS | Superparamagnetic $Fe_2O_3$ NPs | Hg | 0.49 nM | Fluorescence quenching as $Hg^{2+}$ concentrations increase | Satapathi et al. 2018 |

*Table 2 contd. ...*

*...Table 2 contd.*

| Nanosensors | Applied nanomaterial | Sensing purpose | Detection limit | Sensing process | References |
|---|---|---|---|---|---|
| **Detection of Pathogen** | | | | | |
| Chemiresistive biosensors | Polypyrrole (PPY) nanoribbons on Au microelectrode array (Au-MEAs) | Mosaic virus (Cucumber) | 10 ng $mL^{-1}$ | Amperometry | Chartuprayoon et al. 2015 |
| Electrochemical Biosensor | Colloidal Gold Nanoparticles (GNPs) | Pseudomonas syringae | 214 pM | Uses differential pulse voltammetry to evaluate electrochemical changes | Lau et al. 2017 |
| | AuNPs microelectrode | Avian influenza virus (H5N1) | 0.367 HAU $mL^{-1}$ | CV | Zhang et al. 2017 |
| | Multiwalled carbon nanotube (MWNTs) | Ganoderma boninense | 0.0414 mg $L^{-1}$ | Electrocatalytic actions of a modified electrode on G. boninense-infected oil palm leaves and healthy leaves | Isha et al. 2019 |
| Fluorescent NS | Carbon dots (CDs) and terbium ions ($Tb^{3+}$) | Citrus tristeza | 220 ng $mL^{-1}$ | Energy transfer via fluorescence resonance | Chen et al. 2018 |
| Surface plasmon resonance nanosensor (SPRNS) | Au | Afatoxin B1 (*Aspergillus flavus*) | 1.04 pg $mL^{-1}$ | The concentration of aflatoxin affected the reflectivity index of the gold electrode surface | Akgönüllü et al. 2020 |
| Optical NS | Selective single-walled carbon nanotube (SWCN) | Wound signaling | – | Selective SWCNs are superior fluorescent probes can monitor $H_2O_2$ created as a result of mechanical damage in plants on real-time | Lew et al. 2020 |
| Cu (I) tris(pyrazolyl) borate-SWCNT | Cu | Necrosis | 0.5 ppm | Chemiresistor | Cellini et al. 2017 |
| Pd (II)-pH dye-$SiO_2$ microspheres | Pd | Fusarium wilt | 0.17 ppm | Colorimetry | Cellini et al. 2017 |
| CdSe-PEI-QD | Cd | Bacterial wilt | 10 μg $mL^{-1}$ | Fluorescence | Gao et al. 2018 |
| $TiO_2$-NP-dsDNA | $TiO_2$ | Phytophthora spp. | 35 nM | Electrochemistry | Li 2018 |
| Au-NP-ssDNA | Au | Bacterial wilt | 7.5 ng | Colorimetry | Farber et al. 2019 |
| Ag NP-ssDNA | Ag | *Phytophthora infestans* | 0.2 μM | Nanoparticle functionalization and DNA hybridization | Li et al. 2020 |
| Anti-PPV IgG polyclonal antibody | Au | Pox virus (Plum) | 10 pg $mL^{-1}$ | Electrochemical impedance spectroscopy-based detection | Dyussembayev et al. 2021 |
| Pt-NP-IgG | Pt | Rot of potato | $10^2$ CFU | MALDI-TOF MS | Li et al. 2021 |

Note: SWASV: square wave anodic stripping voltammetry, CV: cyclic voltammetry.

are usually employed to detect pests and their monitoring and for determining the control of pesticide spray programs (Martinazzo et al. 2022). An $SiO_2$-based MEMS device that may effectively differentiate the sex pheromones in female lepidopteran insects (e.g., cotton bollworm) across several host plants, including chickpea and rice, in the air (Moitra et al. 2016). Some examples of the use of different nanosensors in the detection of diseases and pests in plants have been given in Table 2 (Zain et al. 2023).

## 7. Conclusions and Future Prospects

As the global population continues to grow rapidly, the demand for food is escalating at an extraordinary pace. Simultaneously, climate change, soil degradation, and the intensifying impact of biotic and abiotic stresses have created an urgent need for sustainable, innovative solutions to enhance agricultural productivity. In this context, nanotechnology emerges as a promising tool for addressing these challenges through its unique physicochemical properties, such as small size, high surface-area-to-volume ratio, and tunable surface charge.

Nanoparticles have demonstrated significant potential in improving plant tolerance to various abiotic stresses, including heat, drought, heavy metals, and salinity, as well as biotic stresses caused by microbial pathogens, viruses, phytoplasmas, and insect pests. Their multifunctional role includes early disease detection through nanobiosensors, targeted delivery of agrochemicals, modulation of stress-responsive genes, enhancement of photosynthetic efficiency, and fortification of the antioxidant defense system of the plant, all contributing to improved plant health and productivity under stress conditions.

Traditional agrochemicals such as synthetic fertilizers, pesticides, and fungicides, while effective, are associated with serious environmental concerns, including the emission of greenhouse gases, contamination of soil and water ecosystems, and adverse effects on non-target organisms, including humans. In contrast, nanofertilizers, nanoformulated pesticides, and nanofungicides offer advantages such as reduced input requirements, site-specific action, minimal environmental toxicity, and better compatibility with soil microflora and aquatic ecosystems. This paradigm shift toward nano-enabled agriculture holds the potential for high-yield, eco-friendly farming practices.

However, despite these promising advancements, the application of nanotechnology in agriculture remains in its early stages. There is a critical need for extensive field evaluations, long-term ecotoxicological studies on aquatic and terrestrial organisms (e.g., earthworms, beneficial soil microbes), and assessments of potential impacts on air quality and human health. The adoption of green synthesis approaches and biodegradable nanomaterials is essential to minimize toxicity and ensure environmental compatibility.

Moreover, interdisciplinary research is needed to deepen our understanding of plant–nanoparticle interactions at the cellular and molecular levels. Regulatory frameworks must also be established or strengthened to ensure the safe deployment of nanomaterials in agricultural systems.

To sum up, nanotechnology holds immense promise to transform conventional agriculture into a more precise, efficient, and sustainable system. The development of early disease detection systems, smart delivery platforms for agro-inputs, and stress-mitigating nanoformulations can significantly contribute to global food security. However, the success of nano-enabled agriculture depends on responsible innovation, robust safety assessments, regulatory oversight, and public acceptance to ensure long-term sustainability.

## Acknowledgement

SSS and PG acknowledge the YUFE4Postdocs (Y4P) programme, financially supported by European Union funds as a part of the Marie Sklodowska-Curie Actions Cofund programme (Agreement No. 101081327), Ministry of Science and Higher Education under the “International

Co-financed Projects" programme (Agreement No. 5678/HE/2023/2024/2). AKG acknowledges the National Science Centre and the European Union Framework Programme (Project No. UMO-2022/45/P/NZ9/01571), co-funded by for Research and Innovation Horizon 2020 under the Marie Skłodowska-Curie grant agreement No.945339. JTW acknowledge the National Science Centre, Poland for financial support under grant no. 2022/45/N/NZ9/01483. MR thankfully acknowledges Federal University of Piaui (UFPI), Teresina, for the Visiting Professorship.

## References

Abasi, F., Raja, N. I., Mashwani, Z. U. R., Amjad, M. S., Ehsan, M., Mustafa, N. et al. (2022). Biogenic silver nanoparticles as a stress alleviator in plants: A mechanistic overview. Molecules 27: 3378. https://doi.org/10.3390/molecules27113378.

Abbas, A., Dara, M. Z. N., Ullah, F., Saddam, B., Abbas, S., Alam, A. et al. (2024). Unraveling insect feeding patterns and their ecological impacts on plant defense mechanisms. Journal of Plant Diseases and Protection 132(1): 21. https://doi.org/10.1007/s41348-024-01002-z.

Abd El-Ghany, N. M., Abd El-Aziz, S. E. and Marei, S. S. (2020). A review: application of remote sensing as a promising strategy for insect pests and diseases management. Environ Sci. Pollut. Res. 27: 33503–33515. https://doi.org/10.1007/s11356-020-09517-2.

Ahmad, Z., Niyazi, S., Firdoos, A., Wang, C., Manzoor, M. A., Ramakrishnan, M. et al. (2024). Enhancing plant resilience: Nanotech solutions for sustainable agriculture. Heliyon 10: e40735. https://doi.org/10.1016/j.heliyon.2024.e40735.

Ahmed, M., Decsi, K. and Tóth, Z. (2022). Different tactics of synthesized zinc oxide nanoparticles, homeostasis ions, and phytohormones as regulators and adaptatively parameters to alleviate the adverse effects of salinity stress on plants. Life 13(1): 73. https://doi.org/10.3390/life13010073.

Ahmed, T., Noman, M., Ijaz, M., Ali, S., Rizwan, M., Ijaz, U. et al. (2021). Current trends and future prospective in nanoremediation of heavy metals contaminated soils: A way forward towards sustainable agriculture. Ecotoxicology and Environmental Safety 227: 112888. https://doi.org/10.1016/j.ecoenv.2021.112888.

Akgönüllü, S., Yavuz, H. and Denizli, A. (2020). SPR nanosensor based on molecularly imprinted polymer film with gold nanoparticles for sensitive detection of aflatoxin B1. Talanta 219: 121219. https://doi.org/10.1016/j.talanta.2020.121219.

Alabdallah, N. M., Alluqmani, S. M., Almarri, H. M. and Al-Zahrani, A. A. (2024). Physical, chemical, and biological routes of synthetic titanium dioxide nanoparticles and their crucial role in temperature stress tolerance in plants. Heliyon 10(4): e26537. https://doi.org/10.1016/j.heliyon.2024.e26537.

Ali, S., Mehmood, A. and Khan, N. (2021). Uptake, translocation, and consequences of nanomaterials on plant growth and stress adaptation. Journal of Nanomaterials 2021: 1–17. Doi: 10.1155/2021/6677616.

Al-Khayri, J. M., Rashmi, R., Surya Ulhas, R., Sudheer, W. N., Banadka, A., Nagella, P. et al. (2023). The role of nanoparticles in response of plants to abiotic stress at physiological, biochemical, and molecular levels. Plants 12(2): 292. https://doi.org/10.3390/plants12020292.

Almutairi, Z. M. (2016). Influence of silver nano-particles on the salt resistance of tomato (*Solanum lycopersicum*) during germination. International Journal of Agriculture and Biology 18: 449–457. Doi: 10.17957/IJAB/15.0114.

Alova, A., Erofeev, A., Gorelkin, P., Bibikova, T., Korchev, Y., Majouga, A. et al. (2020). Prolonged oxygen depletion in microwounded cells of Chara corallina detected with novel oxygen nanosensors. Journal of Experimental Botany 71(1): 386–398. https://doi.org/10.1093/jxb/erz433.

Amenta, V., Aschberger, K., Arena, M., Bouwmeester, H., Moniz, F. B., Brandhoff, P. et al. (2015). Regulatory aspects of nanotechnology in the agri/feed/food sector in EU and non-EU countries. Regulatory Toxic Pharma 73(1): 463–476. https://doi.org/10.1016/j.yrtph.2015.06.016.

Amini, S., Maali-Amiri, R., Mohammadi, R. and Shahandashti, S.-S. K. (2017). cDNA-AFLP analysis of transcripts induced in chickpea plants by $TiO_2$ nanoparticles during cold stress. Plant Physiology and Biochemistry 111: 39–49. https://doi.org/10.1016/j.plaphy.2016.11.011.

Amjad, S. and Serajuddin, M. (2022). Role of nanosensors and bionanosensors in crop abiotic stress. *In*: Ansari, S. A., Ansari, M. I. and Husen, A. (eds.). Augmenting Crop Productivity in Stress Environment. Springer, Singapore. https://doi.org/10.1007/978-981-16-6361-1_1.

Ansari, M., Ahmed, S., Abbasi, A., Khan, M. T., Subhan, M., Bukhari, N. A. et al. (2023). Plant mediated fabrication of silver nanoparticles, process optimization, and impact on tomato plant. Scientific Reports 13: 18048. https://doi.org/10.1038/s41598-023-45038-x.

Ansari, M. H. D., Lavhale, S., Kalunke, R. M., Srivastava, P. L., Pandit, V., Gade, S. et al. (2020). Recent advances in plant nanobionics and nanobiosensors for toxicology applications. Current Nanoscience 16(1): 27–41. DOI: 10.2174/1573413715666190409101305.

Antunović, V., Ilić, M., Baošić, R., Jelić, D. and Lolić, A. (2019). Synthesis of $MnCo_2O_4$ nanoparticles as modifiers for simultaneous determination of Pb(II) and Cd(II). PLoS ONE 14(2): e0210904. https://doi.org/10.1371/journal.pone.0210904.

Anwar, A., Minhaz, A., Khan, N. A., Kalantari, K., Afifi, A. B. M. and Shah, M. R. (2018). Synthesis of gold nanoparticles stabilized by a pyrazinium thioacetate ligand: A new colorimetric nanosensor for detection of heavy metal Pd(II). Sensors and Actuators B: Chemical 257: 875–881. https://doi.org/10.1016/j.snb.2017.11.040.

Atanda, S. A., Shaibu, R. O. and Agunbiade, F. O. (2025). Nanoparticles in agriculture: balancing food security and environmental sustainability. Discover Agriculture 3: 26. https://doi.org/10.1007/s44279-025-00159-x.

Balusamy, S. R., Joshi, A. S., Perumalsamy, H., Mijakovic, I. and Singh, P. (2023). Advancing sustainable agriculture: a critical review of smart and eco-friendly nanomaterial applications. Journal of Nanobiotechnology 21: 372. https://doi.org/10.1186/s12951-023-02135-3.

Bertaccini, A. and Lee, I.-M. (2018). Phytoplasmas: An update. pp. 1–29. *In*: Rao, G. P. A., Fiore, B. N. and Liefting, L. W. (eds.). Phytoplasmas: Plant Pathogenic Bacteria—I: Characterisation and Epidemiology of Phytoplasma—Associated Diseases. Springer. https://doi.org/10.1007/978-981-13-0119-3_1.

Bhagat, R., Ingle, A. P. and Chen, H. (2023). Nanosensors and nanobiosensors for sustainable agriculture. pp. 93–112. *In*: Avinash P. Ingle (ed.). Micro and Nano Technologies, Nanotechnology in Agriculture and Agroecosystems, Elsevier, https://doi.org/10.1016/B978-0-323-99446-0.00014-3.

Bind, S., Bind, S. and Chandra, D. (2023). Chapter 10 - Microbial nanotechnology: A green approach towards sustainable agriculture. pp. 195–211. *In*: Dinesh Chandra and Pankaj Bhatt (eds.). Developments in Applied Microbiology and Biotechnology, Unravelling Plant-Microbe Synergy, Academic Press, https://doi.org/10.1016/B978-0-323-99896-3.00008-4.

Brezolin, A. N., Martinazzo, J., Steffens, J. and Steffens, C. (2020). Nanostructured cantilever sensor using with Pani/MWCNT-COOH nanocomposites applied in the detection of pheromone. J. Mater Sci: Mater Electron. 31: 6008–6016. https://doi.org/10.1007/s10854-020-03152-w.

Cao, Y., Turk, K., Bibi, N., Ghafoor, A., Ahmed, N., Azmat, M. et al. (2025). Nanoparticles as catalysts of agricultural revolution: enhancing crop tolerance to abiotic stress: a review. Frontiers in Plant Science 15: 1510482. Doi: 10.3389/fpls.2024.1510482.

Cellini, A., Blasioli, S., Biondi, E., Bertaccini, A., Braschi, I. and Spinelli, F. (2017). Potential applications and limitations of electronic nose devices for plant disease diagnosis. Sensors 17(11): 2596. https://doi.org/10.3390/s17112596.

Chartuprayoon, N., Zhang, M., Bosze, W., Choa, Y.-H. and Myung, N. V. (2015). One-dimensional nanostructures based bio-detection. Biosensors and Bioelectronics 63: 432–443. https://doi.org/10.1016/j.bios.2014.07.043.

Chen, B. B., Liu, M. L., Zhan, L., Li, C. M. and Huang, C. Z. (2018). Terbium(III) modified fluorescent carbon dots for highly selective and sensitive ratiometry of stringent. Analytical Chemistry 90(6): 4003–4009. https://doi.org/10.1021/acs.analchem.7b05149.

Choudhury, S., Panda, P., Sahoo, L. and Panda, S. K. (2013). Reactive oxygen species signaling in plants under abiotic stress. Plant Signaling & Behavior 8(4). https://doi.org/10.4161/psb.23681.

Daler, S., Kaya, O., Canturk, S., Korkmaz, N., Kılıç, T., Karadağ, A. et al. (2024). Silicon nanoparticles ($SiO_2$ NPs) boost drought tolerance in grapevines by enhancing some morphological, physiological, and biochemical traits. Plant Molecular Biology Reporter 2024: 1–19. https://doi.org/10.1007/s11105-024-01520-y.

Dhiman, S., Bakshi, P., Kapoor, N., Sharma, P., Kohli, S. K., Mir, B. A. et al. (2021). Nanoparticle-induced oxidative stress in plant. pp. 269–313. *In*: Singh, V. P., Singh, S., Tripathi, D. K., Prasad, S. M. and Chauhan, D. K. (eds.). Plant Responses to Nanomaterials. Nanotechnology in the Life Sciences. Springer, Cham. Doi: 10.1007/978-3-030-36740-4_12.

Dinler, B. S., Cetinkaya, H., Koc, F. N., Gül, V. and Sefaoğlu, F. (2024). Effects of titanium dioxide nanoparticles against salt and heat stress in safflower cultivars. Acta Botanica Brasilica 38: e20230136.

Donia, D. and Carbone, M. (2019). Fate of the nanoparticles in environmental cycles. International Journal of Environmental Science and Technology 16: 583–600. Doi: 10.1007/s13762-018-1960-z.

Du, B., Haensch, R., Alfarraj, S. and Rennenberg, H. (2024). Strategies of plants to overcome abiotic and biotic stresses. Biological Reviews 99(4): 1524–1536. https://doi.org/10.1111/brv.13079.

Dube, E. and Okuthe, G. E. (2023). Engineered nanoparticles in aquatic systems: Toxicity and mechanism of toxicity in fish. Emerging Contaminants 9(2): 100212. https://doi.org/10.1016/j.emcon.2023.100212.

Dutta, A., Croll, D., McDonald, B. A. and Barrett, L. G. (2021). Maintenance of variation in virulence and reproduction in populations of an agricultural plant pathogen. Evolutionary Applications 14(2): 335–347. https://doi.org/10.1111/eva.13117.

Dutta, P., Bhattacharyya, A. and Kumari, A. (2023). Innovative integrated pest management paradigm for sustainable crop production with special reference to North East India Integrated Pest Management in Diverse Cropping Systems, Apple Academic Press, pp. 61–90.

Dyussembayev, K., Sambasivam, P., Bar, I., Brownlie, J. C., Shiddiky, M. J. A. and Ford, R. (2021). Biosensor technologies for early detection and quantification of plant pathogens. Frontiers in Chemistry 9(636245): 1–13. https://doi.org/10.3389/fchem.2021.636245.

Ejaz, U., Khan, S. M., Khalid, N., Ahmad, Z., Jehangir, S., Fatima Rizvi, Z. et al. (2023). Detoxifying the heavy metals: A multipronged study of tolerance strategies against heavy metals toxicity in plants. Frontiers in Plant Science 14. https://doi.org/10.3389/fpls.2023.1154571.

El-Badri, A. M., Batool, M., Wang, C., Hashem, A. M., Tabl, K. M., Nishawy, E. et al. (2021). Selenium and zinc oxide nanoparticles modulate the molecular and morpho-physiological processes during seed germination of *Brassica napus* under salt stress. Ecotoxicology and Environmental Safety 225: 112695. https://doi.org/10.1016/j.ecoenv.2021.112695.

El-Saadony, M. T., Saad, A. M., Soliman, S. M., Salem, H. M., Desoky, E.-S. M., Babalghith, A. O. et al. (2022). Role of nanoparticles in enhancing crop tolerance to abiotic stress: A comprehensive review. Frontiers in Plant Science 13: 946717. Doi: 10.3389/fpls.2022.946717.

Endo, M., Tsuchiya, T., Hamada, K., Kawamura, S., Yano, K., Ohshima, M. et al. (2009). High temperatures cause male sterility in rice plants with transcriptional alterations during pollen development. Plant and Cell Physiology 50(11): 1911–1922. https://doi.org/10.1093/pcp/pcp135.

Faizan, M., Alam, P., Rajput, V. D., Faraz, A., Afzal, S., Ahmed, S. M. et al. (2023). Nanoparticle mediated plant tolerance to heavy metal stress: what we know? Sustainability 15(2): 1446. https://doi.org/10.3390/su15021446.

Faizan, M., Sehar, S., Rajput, V. D., Faraz, A., Afzal, S., Minkina, T. et al. (2021). Modulation of cellular redox status and antioxidant defense system after synergistic application of zinc oxide nanoparticles and salicylic acid in rice (*Oryza sativa*) plant under arsenic stress. Plants 10(11): 2254. https://doi.org/10.3390/plants10112254.

Farahi, S. M. M., Yazdi, M. E. T., Einafshar, E., Akhondi, M., Ebadi, M., Azimipour, S. et al. (2023). The effects of titanium dioxide ($TiO_2$) nanoparticles on physiological, biochemical, and antioxidant properties of Vitex plant (*Vitex agnus* - Castus L). Heliyon 9(11): e22144. https://doi.org/10.1016/j.heliyon.2023.e22144.

Farber, C., Mahnke, M., Sanchez, L. and Kurouski, D. (2019). Advanced spectroscopic techniques for plant disease diagnostics. A review. TrAC Trends in Analytical Chemistry 118: 43–49. https://doi.org/10.1016/j.trac.2019.05.022.

Farooq, M., Gogoi, N., Hussain, M., Barthakur, S., Paul, S., Bharadwaj, N. et al. (2017). Effects, tolerance mechanisms and management of salt stress in grain legumes. Plant Physiology and Biochemistry 118: 199–217. https://doi.org/10.1016/j.plaphy.2017.06.020.

Farooq, M., Hussain, M., Wahid, A. and Siddique, K. H. M. (2012). Drought stress in plants: An overview. pp. 1–33. *In*: Aroca, R. (ed.). Plant Responses to Drought Stress: From Morphological to Molecular Features. Springer, Cham. https://doi.org/10.1007/978-3-642-32653-0_1.

Ferreira, R. B., Monteiro, S., Freitas, R., Santos, C. N., Chen, Z., Batista, L. M. et al. (2006). Fungal pathogens: The battle for plant infection. Critical Reviews in Plant Sciences 25(6): 505–524. https://doi.org/10.1080/07352680601054610.

Fincheira, P., Hoffmann, N., Tortella, G., Ruiz, A., Cornejo, P., Diez, M. C. et al. (2023). Eco-efficient systems based on nanocarriers for the controlled release of fertilizers and pesticides: Toward smart agriculture. Nanomaterials 13(13): 1978. https://doi.org/10.3390/nano13131978.

Gao, G., Jiang, Y.-W., Sun, W. and Wu, F.-G. (2018). Fluorescent quantum dots for microbial imaging. Chinese Chemical Letters 29(10): 1475–1485. https://doi.org/10.1016/j.cclet.2018.07.004.

García-Sánchez, S., Bernales, I. and Cristobal, S. (2015). Early response to nanoparticles in the *Arabidopsis* transcriptome compromises plant defence and root-hair development through salicylic acid signalling. BMC Genomics 16: 341. https://doi.org/10.1186/s12864-015-1530-4.

Ghaywat, S., Ingle, P., Shende, S., Hande, D., Rai, M., Shingote, P. et al. (2023). Fungus mediated copper oxide nanoparticles against fungi isolated from soft-rot infected ginger. Kavaka 59(4): 56–61. DOI: 10.36460/Kavaka/59/4/2023/56-61.

Ghori, N.-U.-H., Ghori, T., Hayat, M., Imadi, S. R., Gul, A., Altay, V. et al. (2019). Heavy metal stress and responses in plants. International Journal Environmental Science and Technology 16: 1807–1828. Doi: 10.1007/s13762-019-02215-8.

González Guzmán, M., Cellini, F., Fotopoulos, V., Balestrini, R. and Arbona, V. (2022). New approaches to improve crop tolerance to biotic and abiotic stresses. Physiologia Plantarium 174: e13547. Doi: 10.1111/ppl.13547.

Gull, A., Lone, A. A., Wani, N. U. I., Gull, A., Lone, A. A. and Wani, N. U. I. (2019). Biotic and abiotic stresses in plants. In abiotic and biotic stress in plants. IntechOpen. https://doi.org/10.5772/intechopen.85832.

Guo, X., Cui, R., Huang, H., Li, Y., Liu, B., Wang, J. et al. (2020). Insights into the role of pyrrole doped in three-dimensional graphene aerogels for electrochemical sensing Cd(II). Journal of Electroanalytical Chemistry 871: 114323. https://doi.org/10.1016/j.jelechem.2020.114323.

Gupta, A., Bharati, R., Kubes, J., Popelkova, D., Praus, L., Yang, X. et al. (2024). Zinc oxide nanoparticles application alleviates salinity stress by modulating plant growth, biochemical attributes and nutrient homeostasis in *Phaseolus vulgaris* L. Frontiers in Plant Science 15: 1432258. Doi:10.3389/fpls.2024.1432258.

Gupta, R. K., Guha, P. and Srivastav, P. P. (2024). Investigating the toxicological effects of nanomaterials in food packaging associated with human health and the environment. Journal of Hazardous Materials Letters 5: 100125. https://doi.org/10.1016/j.hazl.2024.100125.

Haghmadad Milani, M., Mohammadi, A., Panahirad, S., Farhadi, H., Labib, P., Kulak, M. et al. (2024). Cerium oxide nanoparticles ($CeO_2$ NPs) enhance salt tolerance in Spearmint (*Mentha spicata* L.) by boosting the antioxidant system and increasing essential oil composition. Plants 13(20): 2934. https://doi.org/10.3390/plants13202934.

Haliloğlu, K., Türkoğlu, A., Balpınar, Ö., Nadaroğlu, H., Alaylı, A. and Poczai, P. (2022). Effects of zinc, copper and iron oxide nanoparticles on induced DNA methylation, genomic instability and LTR retrotransposon polymorphism in wheat (*Triticum aestivum* L.). Plants (Basel) 11(17): 2193. Doi: 10.3390/plants11172193.

Hameed, A., Saif, M. J., Qayyum, M. A., Khalid, T. and Farooq, T. (2023). Chapter 16 - Nanomaterial-based sensors for real-time monitoring of crop plants growth, development, production, and protection. pp. 357–385. *In*: Azamal Husen (ed.).

Plant Biology, Sustainability and Climate Change, Engineered Nanomaterials for Sustainable Agricultural Production, Soil Improvement and Stress Management, Academic Press. https://doi.org/10.1016/B978-0-323-91933-3.00007-6.

Hasanuzzama, M. and Fujita, M. (2022). Plant oxidative stress: Biology, physiology and mitigation. Plants 11(9): 1185. https://doi.org/10.3390/plants11091185.

Hasanuzzaman, M., Bhuyan, M. H. M. B., Zulfiqar, F., Raza, A., Mohsin, S. M., Mahmud, J. A. et al. (2020). Reactive oxygen species and antioxidant defense in plants under abiotic stress: revisiting the crucial role of a universal defense regulator. Antioxidants 9(8): 681. https://doi.org/10.3390/antiox9080681.

Heidarian, F., Fallah, S., Pokhrel, L. R. and Rostamnejadi, A. (2025). Magnetite nanoparticles ($Fe_3O_4$NPs) promote drought tolerance and improve plant health, grain quality and yield in kidney bean (*Phaseolus vulgaris* L.). Science of The Total Environment 964: 178544. https://doi.org/10.1016/j.scitotenv.2025.178544.

Hou, D., O'Connor, D., Igalavithana, A. D., Alessi, D. S., Luo, J., Tsang, D. C. W. et al. (2020). Metal contamination and bioremediation of agricultural soils for food safety and sustainability. Nature Reviews Earth & Environment 1(7): 366–381. https://doi.org/10.1038/s43017-020-0061-y.

Hu, H. and Xiong, L. (2014). Genetic engineering and breeding of drought-resistant crops. Annual Review of Plant Biology 65: 715–741. https://doi.org/10.1146/annurev-arplant-050213-040000.

Huang, K., Boerhan, R., Liu, C. and Jiang, G. (2017). Nanoparticles penetrate into the multicellular spheroid-on-chip: effect of surface charge, protein corona, and exterior flow. Molecular Pharmaceutics 14: 4618–4627. Doi: 10.1021/acs.molpharmaceut.7b00726.

Ijaz, S., Iqbal, J., Abbasi, B. A., Ullah, Z., Ijaz, N., Yaseen, T. et al. (2024). Regulatory and ethical concerns of nanotechnology in agriculture. pp. 395–427. *In*: Shahzad, R., Fiaz, S., Qayyum, A., Ul Islam, M. and Lee, I. J. (eds.). Revolutionizing Agriculture: A Comprehensive Exploration of Agri-nanotechnology, Nanotechnology in the Life Sciences, Springer, Cham. Doi:10.1007/978-3-031-76000-6_18.

Ingle, P. U., Rai, M., Golińska, P. and Gade, A. K. (2024a). Phytomediated zinc oxide and sulfur nanoparticles for management of soft-rot causing pathogenic fungi in ginger. Biocatalysis and Agricultural Biotechnology 58: 103229. https://doi.org/10.1016/j.bcab.2024.103229.

Ingle, P. U., Shende, S. S., Hande, D., Rai, M., Golinska, P. and Gade, A. K. (2024b). Mycogenic copper oxide nanoparticles for fungal infection management in agricultural crop plants. Bionanoscience 14(1): 359–367. https://doi.org/10.1007/s12668-023-01255-2.

Ingle, P. U., Shende, S. S., Shingote, P. R., Mishra, S. S., Sarda, V., Wasule, D. L. et al. (2022). Chitosan nanoparticles (ChNPs): A versatile growth promoter in modern agricultural production. Heliyon 8(11): e11893. https://doi.org/10.1016/j.heliyon.2022.e11893.

Ioannou, A., Gohari, G., Papaphilippou, P., Panahirad, S., Akbari, A., Dadpour, M. R. et al. (2020). Advanced nanomaterials in agriculture under a changing climate: The way to the future? Environmental and Experimental Botany 176: 104048. https://doi.org/10.1016/j.envexpbot.2020.104048.

Isha, A., Akanbi, F. S., Yusof, N. A., Osman, R., Mui-Yun, W. and Abdullah, S. N. A. (2019). An NMR metabolomics approach and detection of *Ganoderma boninense*-infected oil palm leaves using MWCNT-based electrochemical sensor. Journal of Nanomaterials 4729706: 12. https://doi.org/10.1155/2019/4729706.

Jahagirdar, A. S., Shende, S., Gade, A. and Rai, M. (2020). Bioinspired synthesis of copper nanoparticles and its efficacy on seed viability and seedling growth in Mungbean (*Vigna radiata* L.). Current Nanoscience 16(2): 246–252(7). https://doi.org/10.2174/1573413715666190325170054.

Jahan, M. S., Guo, S., Sun, J., Shu, S., Wang, Y., Abou El-Yazied, A. et al. (2021). Melatonin-mediated photosynthetic performance of tomato seedlings under high-temperature stress. Plant Physiology and Biochemistry 167: 309–320. https://doi.org/10.1016/j.plaphy.2021.08.002.

Jahan, M. S., Hasan, M. M., Alotaibi, F. S., Alabdallah, N. M., Alharbi, B. M., Ramadan, K. M. et al. (2022). Exogenous putrescine increases heat tolerance in tomato seedlings by regulating chlorophyll metabolism and enhancing antioxidant defense efficiency. Plants (Basel) 11(8): 1038. Doi: 10.3390/plants11081038.

Janni, M., Maestri, E., Gullì, M., Marmiroli, M. and Marmiroli, N. (2024). Plant responses to climate change, how global warming may impact on food security: a critical review. Frontiers in Plant Science 14: 1297569. Doi: 10.3389/fpls.2023.1297569.

Jesmer, A. H., Velicogna, J. R., Schwertfeger, D. M., Scroggins, R. P. and Princz, J. I. (2017). The toxicity of silver to soil organisms exposed to silver nanoparticles and silver nitrate in biosolids-amended field soil. Environmental Toxicology and Chemistry 36(10): 2756–2765. doi: 10.1002/etc.3834.

Jimenez-Falcao, S., Villalonga, A., Parra-Nieto, J., Llopis-Lorente, A., Martinez-Ruiz, P., Martinez-Mañez, R. et al. (2020). Dithioacetal-mechanized mesoporous nanosensor for Hg(II) determination. Microporous and Mesoporous Materials 297: 110054. https://doi.org/10.1016/j.micromeso.2020.110054.

Jin, B., Yang, Y., He, R., Park, Y. I., Lee, A., Bai, D. et al. (2018). Lateral flow aptamer assay integrated smartphone-based portable device for simultaneous detection of multiple targets using upconversion nanoparticles. Sensors and Actuators B: Chemical 276: 48–56. https://doi.org/10.1016/j.snb.2018.08.074.

Joshi, R., Wani, S. H., Singh, B., Bohra, A., Dar, Z. A., Lone, A. A. et al. (2016). Transcription factors and plants response to drought stress: Current understanding and future directions. Frontiers in Plant Science 7: 1029. https://doi.org/10.3389/fpls.2016.01029.

Kajla, M., Yadav, V. K., Khokhar, J., Singh, S., Chhokar, R. S., Meena, R. P. et al. (2015). Increase in wheat production through management of abiotic stresses: A review. Journal of Applied and Natural Science 7(2): 1070–1080. https://doi.org/10.31018/jans.v7i2.733.

Kamatham, S., Munagapati, S., Manikanta, K. N., Vulchi, R., Chadipiralla, K., Indla, S. H. et al. (2021). Recent advances in engineering crop plants for resistance to insect pests. Egyptian Journal of Biological Pest Control 31(1): 120. https://doi.org/10.1186/s41938-021-00465-8.

Kang, J., Duan, X. and Yun, R. (2023). The impact of urbanization on food security: A case study of Jiangsu Province. Land. 12: 1681. https://doi.org/10.3390/land12091681.

Kaplan, A., Khan, M. N., Hayat, K., Iqbal, M., Ali, B., Wahab, S. et al. (2024). Coupling environmental factors and climate change: Impacts on plants and vegetation growth patterns in ecologically sensitive regions. *In*: Fahad, S., Saud, S., Nawaz, T., Gu, L., Ahmad, M. and Zhou, R. (eds.). Environment, Climate, Plant and Vegetation Growth. Springer, Cham. https://doi.org/10.1007/978-3-031-69417-2_11.

Karvar, M., Azari, A., Rahimi, A., Maddah-Hosseini, S. and Ahmadi-Lahijani, M. J. (2022). Titanium dioxide nanoparticles ($TiO_2$-NPs) enhance drought tolerance and grain yield of sweet corn (*Zea mays* L.) under deficit irrigation regimes. Acta Physiologiae Plantarum 44(2): 14. https://doi.org/10.1007/s11738-021-03349-4.

Khait, I., Lewin-Epstein, O., Sharon, R., Saban, K., Goldstein, R., Anikster, Y. et al. (2023). Sounds emitted by plants under stress are airborne and informative. Cell 186(7): 1328–1336.e10. DOI: 10.1016/j.cell.2023.03.009.

Kocabay, O., Emregul, E., Aydın, S. S. and Aras, S. (2013). Detection of superoxide radicals in tomato plants exposed to salinity, drought, cold and heavy metal stress using CMC-G-SOD biosensor. Artificial Cells, Nanomedicine, and Biotechnology 41(5): 352–358. https://doi.org/10.3109/10731199.2012.717944.

Koza, N. A., Adedayo, A. A., Babalola, O. O. and Kappo, A. P. (2022). Microorganisms in plant growth and development: Roles in abiotic stress tolerance and secondary metabolites secretion. Microorganisms 10: 1528. Doi: 10.3390/microorganisms10081528.

Kreslavski, V. D., Shmarev, A. N., Ivanov, A. A., Zharmukhamedov, S. K., Strokina, V., Kosobryukhov, A. et al. (2023). Effects of iron oxide nanoparticles ($Fe_3O_4$) and salinity on growth, photosynthesis, antioxidant activity and distribution of mineral elements in wheat (*Triticum aestivum*). Functional Plant Biology 50(11): 932–940. DOI: 10.1071/FP23085.

Kumar, A., Choudhary, A., Kaur, H., Guha, S., Mehta, S. and Husen, A. (2022). Potential applications of engineered nanoparticles in plant disease management: A critical update. Chemosphere 295: 133798. Doi: 10.1016/j.chemosphere.2022.133798.

Kumar, A., Choudhary, A., Kaur, H., Mehta, S. and Husen, A. (2021). Metal-based nanoparticles, sensors, and their multifaceted application in food packaging. Journal of Nanobiotechnology 19: 256. https://doi.org/10.1186/s12951-021-00996-0.

Kumar, S., Shah, S. H., Vimala, Y., Jatav, H. S., Ahmad, P., Chen, Y. et al. (2022). Abscisic acid: Metabolism, transport, crosstalk with other plant growth regulators, and its role in heavy metal stress mitigation. Frontiers in Plant Science 13. https://doi.org/10.3389/fpls.2022.972856.

Kumari, R., Suman, K., Karmakar, S., Mishra, V., Lakra, S. G., Saurav, G. K. et al. (2023). Regulation and safety measures for nanotechnology-based agri-products. Frontiers in Genome Editing 5: 1200987. https://doi.org/10.3389/fgeed.2023.1200987.

Lamalakshmi Devi, E., Kumar, S., Singh, T. B., Sharma, S. K., Beemrote, A., Devi, C. P. et al. (2017). Adaptation strategies and defence mechanisms of plants during environmental stress. pp. 359–413. *In*: Ghorbanpour, M. and Varma, A. (eds.). Medicinal Plants and Environmental Challenges. Springer, Cham. https://doi.org/10.1007/978-3-319-68717-9_20.

Lau, H., Wu, H., Wee, E., Trau, M., Wang, Y. and Botella, J. R. (2017). Specific and sensitive isothermal electrochemical biosensor for plant pathogen DNA detection with colloidal gold nanoparticles as probes. Scientific Reports 7: 38896. https://doi.org/10.1038/srep38896.

Lew, T. T. S., Koman, V. B., Silmore, K. S., Seo, J. S., Gordiichuk, P., Kwak, S.-Y. et al. (2020). Real-time detection of wound-induced $H_2O_2$ signalling waves in plants with optical nanosensors. Nature Plants 6: 404–415. https://doi.org/10.1038/s41477-020-0632-4.

Li, L. L. (2018). Self-assembled nanomaterials for bacterial infection diagnosis and therapy. *In*: Wang, H. and Li, L. L. (eds.). *In Vivo* Self-Assembly Nanotechnology for Biomedical Applications. Nanomedicine and Nanotoxicology. Springer, Singapore. https://doi.org/10.1007/978-981-10-6913-0_3.

Li, Y., Xie, J., Um, W., You, D. G., Kwon, S., Zhang, L. et al. (2021). Sono/photodynamic nanomedicine-elicited cancer immunotherapy. Advanced Functional Materials 31: 2008061. Doi: 10.1002/adfm.202008061.

Li, Z., Liu, Y., Hossain, O., Paul, R., Yao, S., Wu, S. et al. (2021). Real-time monitoring of plant stresses via chemiresistive profiling of leaf volatiles by a wearable sensor. Matter 4(7): 2553–2570. https://doi.org/10.1016/j.matt.2021.06.009.

Li, Z., Yu, T., Paul, R., Fan, J., Yang, Y. and Wei, Q. (2020). Agricultural nanodiagnostics for plant diseases: recent advances and challenges. Nanoscale Advances 2(8): 3083–3094. https://doi.org/10.1039/c9na00724e.

Li, Z., Zhu, L., Zhao, F., Li, J., Zhang, X., Kong, X. et al. (2022). Plant salinity stress response and nano-enabled plant salt tolerance. Frontiers in Plant Science 13. https://doi.org/10.3389/fpls.2022.843994.

Maharramov, A. M., Hasanova, U. A., Suleymanova, I. A., Osmanova, G. E. and Hajiyeva, N. E. (2019). The engineered nanoparticles in food chain: potential toxicity and effects. SN Applied Science 1: 1362. Doi:10.1007/s42452-019-1412-5.

Mansfield, J., Genin, S., Magori, S., Citovsky, V., Sriariyanum, M., Ronald, P. et al. (2012). Top 10 plant pathogenic bacteria in molecular plant pathology. Molecular Plant Pathology 13(6): 614–629. https://doi.org/10.1111/j.1364-3703.2012.00804.x.

Martinazzo, J., Ballen, S. C., Steffens, J. and Steffens, C. (2022). Sensing of pheromones from *Euschistus heros* (F.) stink bugs by nanosensors, Sensors and Actuators Reports 4: 100071. https://doi.org/10.1016/j.snr.2021.100071.

Mashabela, M. D., Masamba, P. and Kappo, A. P. (2023). Applications of metabolomics for the elucidation of abiotic stress tolerance in plants: A special focus on osmotic stress and heavy metal toxicity. Plants 12(2): 269. https://doi.org/10.3390/plants12020269.

Mayordomo, N. M., Zatarain-Beraza, A., Valerio, F., Álvarez-Méndez, V., Turegano, P., Herranz-García, L. et al. (2025). The protein corona paradox: challenges in achieving true biomimetics in nanomedicines. Biomimetics 10(5): 276. https://doi.org/10.3390/biomimetics10050276.

Mgadi, K., Ndaba, B., Roopnarain, A., Rama, H. and Adeleke, R. (2024). Nanoparticle applications in agriculture: Overview and response of plant-associated microorganisms. Frontiers in Microbiology 15: 1354440. Doi: 10.3389/fmicb.2024.1354440.

Miller, R. N. G., Costa Alves, G. S. and Van Sluys, M.-A. (2017). Plant immunity: Unravelling the complexity of plant responses to biotic stresses. Annals of Botany 119(5): 681–687. https://doi.org/10.1093/aob/mcw284.

Moitra, P., Bhagat, D., Pratap, R. and Bhattacharya, S. (2016). A novel bio-engineering approach to generate an eminent surface-functionalized template for selective detection of female sex pheromone of *Helicoverpa armigera*. Scientific Reports 6: 37355. https://doi.org/10.1038/srep37355.

Mondal, A., Kang, J. and Kim, D. (2023). Recent progress in fluorescent probes for real-time monitoring of Glioblastoma. ACS Applied Bio Materials 6(9): 3484–3503. https://doi.org/10.1021/acsabm.3c00052.

Mossa, A. T. H., Afia, S. I., Mohafrash, S. M. and Abou-Awad, B. A. (2018). Formulation and characterization of garlic (*Allium sativum* L.) essential oil nanoemulsion and its acaricidal activity on eriophyid olive mites (Acari: Eriophyidae). Environmental Science and Pollution Research 25: 10526–10537. https://doi.org/10.1007/s11356-017-0752-1.

Muhammad, M., Waheed, A., Wahab, A., Majeed, M., Nazim, M., Liu, Y. H. et al. (2024). Soil salinity and drought tolerance: An evaluation of plant growth, productivity, microbial diversity, and amelioration strategies. Plant Stress 11: 100319. https://doi.org/10.1016/j.stress.2023.100319.

Mukherjee, S., Krishnamoorthy, S. B., Subrayan, R., Goswami, A. and Mitra, S. (2023). A brief study on the role of cerium oxide nanoparticles in growth and alleviation of mercury-induced stress in *Vigna radiata* and soil bacteria *Bacillus coagulans*. Environmental Science and Pollution Research International 30(29): 73952–73963. https://doi.org/10.1007/s11356-023-27496-y.

Mustafa, G., Sakata, K. and Komatsu, S. (2015). Proteomic analysis of flooded soybean root exposed to aluminum oxide nanoparticles. Journal of Proteomics 128: 280–297. Doi: 10.1016/j.jprot.2015.08.010.

Nair, P. M. G., Park, S. Y. and Choi, J. (2013). Evaluation of the effect of silver nanoparticles and silver ions using stress responsive gene expression in *Chironomus riparius*. Chemosphere 92(5): 592–599. Doi: 10.1016/j.chemosphere.2013.03.060.

Nakano, M., Morgan-Richards, M., Trewick, S. A. and Clavijo-McCormick, A. (2022). Chemical ecology and olfaction in short-horned grasshoppers (Orthoptera: Acrididae). Journal of Chemical Ecology 48(2): 121–140. https://doi.org/10.1007/s10886-021-01333-3.

Nazim, M., Li, X., Anjum, S., Ahmad, F., Ali, M., Muhammad, M. et al. (2024). Silicon nanoparticles: a novel approach in plant physiology to combat drought stress in arid environments. Biocatalysis and Agricultural Biotechnology 58: 103190. https://doi.org/10.1016/j.bcab.2024.103190.

Nongbet, A., Mishra, A. K., Mohanta, Y. K., Mahanta, S., Ray, M. K., Khan, M. et al. (2022). Nanofertilizers: A smart and sustainable attribute to modern agriculture. Plants (Basel) 11(19): 2587. https://doi.org/10.3390/plants11192587.

Nuruzzaman, M., Sharoni, A. M. and Kikuchi, S. (2013). Roles of NAC transcription factors in the regulation of biotic and abiotic stress responses in plants. Frontiers in Microbiology 4. Doi:10.3389/fmicb.2013.00248.

Nyaupane, S., Poudel, M. R., Panthi, B., Dhakal, A., Paudel, H. and Bhandari, R. (2024). Drought stress effect, tolerance, and management in wheat – a review. Cogent Food & Agriculture 10(1): 2296094. https://doi.org/10.1080/23311932.2023.2296094.

Önal Acet, B., Gül, D., Stauber, R. H., Odabaşı, M. and Acet, Ö. (2024). A review for uncovering the "protein-nanoparticle alliance": implications of the protein corona for biomedical applications. Nanomaterials 14(10): 823. https://doi.org/10.3390/nano14100823.

Oomen, A. G., Steinhäuser, K. G., Bleeker, E. A. J., van Broekhuizen, F., Sips, A., Dekkers, S. et al. (2018). Risk assessment frameworks for nanomaterials: Scope, link to regulations, applicability, and outline for future directions in view of needed increase in efficiency. NanoImpact 9: 1–13. https://doi.org/10.1016/j.impact.2017.09.001.

Oshunsanya, S. O., Nwosu, N. J. and Li, Y. (2019). Abiotic stress in agricultural crops under climatic conditions. *In*: Jhariya, M., Banerjee, A., Meena, R. and Yadav, D. (eds.). Sustainable Agriculture, Forest and Environmental Management. Springer, Singapore. https://doi.org/10.1007/978-981-13-6830-1_3.

Parihar, P., Singh, S., Singh, R., Singh, V. P. and Prasad, S. M. (2015). Effect of salinity stress on plants and its tolerance strategies: A review. Environmental Science and Pollution Research 22(6): 4056–4075. https://doi.org/10.1007/s11356-014-3739-1.

Qi, M., Liu, Y. and Li, T. (2013). Nano-TiO(2) improve the photosynthesis of tomato leaves under mild heat stress. Biological Trace Element Research 156(1-3): 323–328. https://doi.org/10.1007/s12011-013-9833-2.

Qi, Y., Zhao, J., Weng, G., Li, J., Zhu, J. and Zhao, J. (2018). Modification-free colorimetric and visual detection of $Hg^{2+}$ based on the etching from core-shell structural Au-Ag nanorods to nanorices. Sensors and Actuators B: Chemical 267: 181–190. https://doi.org/10.1016/j.snb.2018.04.042.

Qu, A.-L., Ding, Y.-F., Jiang, Q. and Zhu, C. (2013). Molecular mechanisms of the plant heat stress response. Biochemical and Biophysical Research Communications 432(2): 203–207. https://doi.org/10.1016/j.bbrc.2013.01.104.

Rai, M., Ingle, A., Pandit, R., Paralikar, P., Shende, S., Gupta, I. et al. (2018). Copper and copper nanoparticles: role in management of insect-pests and pathogenic microbes. Nanotechnology Reviews 7(4): 303–315. https://doi.org/10.1515/ntrev-2018-0031.

Raja, M. M., Vijayalakshmi, G., Naik, M. L., Basha, P. O., Sergeant, K., Hausman, J. F. et al. (2019). Pollen development and function under heat stress: From effects to responses. Acta Physiologiae Plantarum 41(4): 47. https://doi.org/10.1007/s11738-019-2835-8.

Rajpal, V., Chaurasia, S., Mehta, S., Minkina, T., Rajput, V. and Deswal, R. (2023). A comprehensive review on mitigating abiotic stresses in plants by metallic nanomaterials: prospects and concerns. Clean Technology and Environmental Policy 25: 1–39. Doi: 10.1007/s10098-023-02561-9.

Rajput, V., Minkina, T., Mazarji, M., Shende, S., Sushkova, S., Mandzhieva, S. et al. (2020). Accumulation of nanoparticles in the soil-plant systems and their effects on human health. Annals of Agriculture Science 65(2): 137–143. https://doi.org/10.1016/j.aoas.2020.08.001.

Rakesh, B., Chitdeshwari, T. and Mohanapriya, G. (2024). Fascinating role of nanosilica in mitigating drought and nutrient stress—A review. Plant Stress 14: 100672, https://doi.org/10.1016/j.stress.2024.100672.

Rampado, R., Crotti, S., Caliceti, P., Pucciarelli, S. and Agostini, M. (2020). Recent advances in understanding the protein corona of nanoparticles and in the formulation of "Stealthy" nanomaterials. Frontiers in Bioengineering and Biotechnology 8: 166. Doi: 10.3389/fbioe.2020.00166.

Ranjan, A., Rajput, V. D., Minkina, T., Bauer, T., Chauhan, A. and Jindal, T. (2021). Nanoparticles induced stress and toxicity in plants. Environmental Nanotechnology, Monitoring & Management 15: 100457. https://doi.org/10.1016/j.enmm.2021.100457.

Rathore, C., Yadav, V. K., Gacem, A., AbdelRahim, S. K., Verma, R. K., Chundawat, R. S. et al. (2023). Microbial synthesis of titanium dioxide nanoparticles and their importance in wastewater treatment and antimicrobial activities: a review. Frontiers in Microbiology 14. DOI:10.3389/fmicb.2023.1270245.

Rehman, A., Khan, S., Sun, F., Peng, Z., Feng, K., Wang, N. et al. (2024). Exploring the nano-wonders: unveiling the role of Nanoparticles in enhancing salinity and drought tolerance in plants. Frontiers in Plant Science 14: 1324176. https://doi.org/10.3389/fpls.2023.1324176.

Saha, G., Mostofa, M. G., Rahman, M. M. and Tran, L. S. P. (2021). Silicon-mediated heat tolerance in higher plants: A mechanistic outlook. Plant Physiology and Biochemistry 166: 341–347. DOI: 10.1016/j.plaphy.2021.05.051.

Saradadevi, R., Palta, J. A. and Siddique, K. H. M. (2017). ABA-mediated stomatal response in regulating water use during the development of terminal drought in wheat. Frontiers in Plant Science 8. https://doi.org/10.3389/fpls.2017.01251.

Satapathi, S., Kumar, V., Chini, M. K., Bera, R., Halder, K. K. and Patra, A. (2018). Highly sensitive detection and removal of mercury ion using a multimodal nanosensor. Nano-Structures & Nano-Objects 16: 120–126, https://doi.org/10.1016/j.nanoso.2018.05.006.

Schulze, E.-D., Beck, E., Buchmann, N., Clemens, S., Müller-Hohenstein, K. and Scherer-Lorenzen, M. (2019). Biotic stress. pp. 257–299. *In*: Schulze, E.-D., Beck, E., Buchmann, N., Clemens, S., Müller-Hohenstein, K. and Scherer-Lorenzen, M. (eds.). Plant Ecology, Springer, Berlin, Heidelberg. https://doi.org/10.1007/978-3-662-56233-8_8.

Seleiman, M. F., Al-Suhaibani, N., Ali, N., Akmal, M., Alotaibi, M., Refay, Y. et al. (2021). Drought stress impacts on plants and different approaches to alleviate its adverse effects. Plants (Basel) 10: 259. Doi: 10.3390/plants10020259.

Sharma, M. M. M., Kapoor, D., Loyal, A., Kumar, R., Sharma, P. and Husen, A. (2024). Mitigation of drought stress in plants using gold nanoparticles. pp. 193–209. *In*: Husen, A. (eds.). Plant Response to Gold Nanoparticles. Smart Nanomaterials Technology. Springer, Singapore. https://doi.org/10.1007/978-981-97-7174-5_10.

Sharma, P., Pandey, V., Sharma, M. M. M., Patra, A., Singh, B., Mehta, S. et al. (2021). A review on biosensors and nanosensors application in agroecosystems. Nanoscale Research Letters 16: 136. https://doi.org/10.1186/s11671-021-03593-0.

Sharma, S. and Dhadly, D. K. (2023). Chapter 5 - Nano-inspired biosensors and plant diseases: recent advances and challenges. pp. 135–162. *In*: Sahadevan Seena, Akhilesh Rai and Santosh Kumar (eds.). Nanomaterial-Plant Interactions, Nanoparticles and Plant-Microbe Interactions, Academic Press. https://doi.org/10.1016/B978-0-323-90619-7.00002-3.

Shen, M., Liu, S., Jiang, C., Zhang, T. and Chen, W. (2023). Recent advances in stimuli-response mechanisms of nano-enabled controlled-release fertilizers and pesticides. Eco-Environment & Health 2(3): 161–175. https://doi.org/10.1016/j.eehl.2023.07.005.

Shende, S., Ingle, A. P., Gade, A. and Rai, M. (2015). Green synthesis of copper nanoparticles by *Citrus medica* Linn. (Idilimbu) juice and its antimicrobial activity. World Journal of Microbiology and Biotechnology 31: 865–873. https://doi.org/10.1007/s11274-015-1840-3.

Shende, S., Rathod, D., Gade, A. and Rai, M. (2017). Biogenic copper nanoparticles promote the growth of pigeon pea (*Cajanus cajan* L.). IET Nanobiotechnology 11: 773–781. https://doi.org/10.1049/iet-nbt.2016.0179.

Shende, S., Rajput, V., Gade, A., Minkina, T., Rai, M. and Singh, S. P. (2020). Impact of nanomaterials on human health through/via food chain. pp. 211–245. *In*: Keswani, C. (ed.). Intellectual Property Issues in Nanotechnology, Edition 1st CRC Press, Boca Raton.

Shende, S., Bhagat, R., Raut, R., Rai, M. and Gade, A. (2021). Myco-fabrication of copper nanoparticles and its effect on crop pathogenic fungi. IEEE Transactions on Nanobioscience 20(2): 146–153. DOI: 10.1109/TNB.2021.3056100.

Shende, S. (S), Rajput, V. (D), Minkina, T., Sushkova, S. and Mandzhieva, S. (2023). Strategic role of nanotechnology in plant growth improvement and crop production. pp. 25–49. *In*: Ingle, A. P. (eds.). In Micro and Nano Technologies, Nanotechnology in Agriculture and Agroecosystems, Elsevier. https://doi.org/10.1016/B978-0-323-99446-0.00015-5.

Shende, S. S., Rajput, V. D., Gorovtsov, A. V., Harish, Saxena, P., Minkina, T. M., Chokheli, V. A. et al. (2021). Interaction of nanoparticles with microbes. pp. 175–188. *In*: Singh, P., Singh, R., Verma, P., Bhadouria, R., Kumar, A. and Kaushik, M. (eds.). Plant-Microbes-Engineered Nano-particles (PM-ENPs) Nexus in Agro-Ecosystems. Advances in Science, Technology & Innovation. Springer, Cham. https://doi.org/10.1007/978-3-030-66956-0_12.

Shende, S. S., Gade, A. K., Minkina, T. M., Ingle, P. U., Rajput, V. D., Sushkova, S. N. et al. (2024a). Exploring sustainable management by using green nano-silver to combat three post-harvest pathogenic fungi in crops. Discover Nano 19(1): 53. Doi: 10.1186/s11671-024-03986-x.

Shende, S. S., Hande, D. V., Ingle, P., Bhagat, R., Golinska, P., Rai, M. et al. (2024b). Myco-fabrication of silver nanoparticles from endophytic fungus *Epicoccum nigrum* Ehrenb. Ex Schlecht: A novel approach for sustainable plant disease management. Kavaka 60: 32–38. DOI:10.36460/Kavaka/60/1/2024/32-38.

Sheoran, S., Kaur, Y., Kumar, S., Shukla, S., Rakshit, S. and Kumar, R. (2022). Recent advances for drought stress tolerance in maize (*Zea mays* L.): Present status and future prospects. Frontiers in Plant Science 13. https://doi.org/10.3389/fpls.2022.872566.

Shi, L., Wang, Y., Chu, Z., Yin, Y., Jiang, D., Luo, J. et al. (2017). A highly sensitive and reusable electrochemical mercury biosensor based on tunable vertical single-walled carbon nanotubes and a target recycling strategy. Journal of Materials Chemistry B 5(5): 1073–1080. DOI: 10.1039/C6TB02658C.

Shirvani-Naghani, S., Fallah, S., Pokhrel, L. R. and Rostamnejadi, A. (2024). Drought stress mitigation and improved yield in Glycine max through foliar application of zinc oxide nanoparticles. Scientific Reports 14(1): 27898. https://doi.org/10.1038/s41598-024-78504-1.

Shivashakarappa, K., Marriboina, S., Dumenyo, K., Taheri, A. and Yadegari, Z. (2025). Nanoparticle-mediated gene delivery techniques in plant systems. Frontiers in Nanotechnology 7: 1516180. Doi: 10.3389/fnano.2025.1516180.

Shoaib, A., Elabasy, A., Waqas, M., Lin, L., Cheng, X., Zhang, Q. et al. (2018). Entomotoxic effect of silicon dioxide nanoparticles on *Plutella xylostella* (L.) (Lepidoptera: Plutellidae) under laboratory conditions. Toxicological & Environmental Chemistry 100(1): 80–91. https://doi.org/10.1080/02772248.2017.1387786.

Sidhu, A. K., Sharma, M., Bhickchand Agrawal, S., Pradip Bhavsar, P. and Samota, M. K. (2024). Nanomaterial strategies for enhancing plant resilience in the face of temperature stress. CABI Agriculture and Bioscience 5(1): 60. https://doi.org/10.1186/s43170-024-00255-w.

Singh, A., Agrawal, S., Rajput, V. D., Ghazaryan, K., Yesayan, A., Minkina, T. et al. (2024). Nanoparticles in revolutionizing crop production and agriculture to address salinity stress challenges for a sustainable future. Discover Applied Sciences 6(6): 317. https://doi.org/10.1007/s42452-024-06009-7.

Singh, A. and Chauhan, P. S. (2017). Chapter 16 Ecological significance of soil-associated plant growth–promoting biofilm-forming microbes for stress management. pp. 291–326. Book Editor(s): Ahmad, I. and Husain, F. M. John Wiley & Sons Ltd, Hoboken, NJ, USA. ISBN:9781119246329. https://doi.org/10.1002/9781119246329.ch16.

Singh, R., Choudhary, P., Kumar, S. and Daima, H. K. (2024). Mechanistic approaches for crosstalk between nanomaterials and plants: plant immunomodulation, defense mechanisms, stress resilience, toxicity, and perspectives. Environmental Science: Nano 11: 2324. https://doi.org/10.1039/D4EN00053F.

Skiba, E. and Wolf, W. M. (2019). Cerium oxide nanoparticles affect heavy metals uptake by pea in a divergent way than their ionic and bulk counterparts. Water, Air, & Soil Pollution 230(10): 248. https://doi.org/10.1007/s11270-019-4296-5.

Soleymani, S., Piri, S., Aazami, M. A. and Salehi, B. (2025). Cerium oxide nanoparticles alleviate drought stress in apple seedlings by regulating ion homeostasis, antioxidant defense, gene expression, and phytohormone balance. Scientific Reports 15(1): 11805. https://doi.org/10.1038/s41598-025-96250-w.

Soltani Toularoud, A. A., Nasrollahi, S., Goli Kalanpa, E., Rouhi Kelarlou, T. and Nematollahzadeh, A. (2025). Toxicological impact of silver nanoparticles on soil microbial indicators in contaminated soil (pot experiment). Nanotoxicology 19(2): 216–231. https://doi.org/10.1080/17435390.2025.2454967.

Son, W. K., Choi, Y. S., Han, Y. W., Shin, D. W., Min, K., Shin, J. et al. (2023). *In vivo* surface-enhanced Raman scattering nanosensor for the real-time monitoring of multiple stress signalling molecules in plants. Nature Nanotechnology 18: 205–216. https://doi.org/10.1038/s41565-022-01274-2.

Soni, S., Jha, A. B., Dubey, R. S. and Sharma, P. (2024). Nanowonders in agriculture: Unveiling the potential of nanoparticles to boost crop resilience to salinity stress. Science of the Total Environment 925: 171433. https://doi.org/10.1016/j.scitotenv.2024.171433.

Srivastava, V., Sarkar, A., Singh, S., Singh, P., de Araujo, A. S. F. and Singh, R. P. (2017). Agroecological responses of heavy metal pollution with special emphasis on soil health and plant performances. Frontiers in Environmental Science 5. https://doi.org/10.3389/fenvs.2017.00064.

Strekalovskaya, E. I., Perfileva, A. I. and Krutovsky, K. V. (2024). Zinc oxide nanoparticles in the "soil–bacterial community-plant" system: impact on the stability of soil ecosystems. Agronomy 14(7): 1588. https://doi.org/10.3390/agronomy14071588.

Subramanian, K. S. and Rajkishore, S. K. (2018). Regulatory framework for nanomaterials in agri-food systems. pp. 319–342. *In*: Rai, M. and Biswas, J. (eds.). Nanomaterials: Ecotoxicity, Safety, and Public Perception. Springer, Cham. Doi:10.1007/978-3-030-05144-0_16.

Sushila, N., Pavitra, S. A. G., Ashoka, J. and Sharanagouda, H. (2020). Biosynthesis and effect of green silica nanoparticles on tobacco caterpillar, *Spodoptera litura* on cotton. Journal of Entomology and Zoology Studies 8: 1564–1570.

Suzuki, N., Rivero, R. M., Shulaev, V., Blumwald, E. and Mittler, R. (2014). Abiotic and biotic stress combinations. New Phytologist 203(1): 32–43. https://doi.org/10.1111/nph.12797.

Tatineni, S. and Hein, G. L. (2023). Plant viruses of agricultural importance: Current and future perspectives of virus disease management strategies. Phytopathology 113(2): 117–141. https://doi.org/10.1094/PHYTO-05-22-0167-RVW.

Tończyk, A., Niedziałkowska, K. and Lisowska, K. (2025). Ecotoxic effect of mycogenic silver nanoparticles in water and soil environment. Scientific Reports 15: 10815. https://doi.org/10.1038/s41598-025-95485-x.

Trzcińska-Wencel, J., Mucha, N., Rai, M., Tyburski, J. and Golińska, P. (2025). Pre-sowing grain treatment with bio-AgNPs stimulates plant growth and affects redox homeostasis in maize. Frontiers in Plant Science 16: 1494741. Doi: 10.3389/fpls.2025.1494741.

Ullah, A., Manghwar, H., Shaban, M., Khan, A. H., Akbar, A., Ali, U. et al. (2018). Phytohormones enhanced drought tolerance in plants: A coping strategy. Environmental Science and Pollution Research 25(33): 33103–33118. https://doi.org/10.1007/s11356-018-3364-5.

United Nations. (2019). Department of economic and social affairs, population division. World Population Prospects 2019: Highlights (ST/ESA/SER.A/423).

Vasseghian, Y., Arunkumar, P., Joo, S.-W., Gnanasekaran, L., Kamyab, H., Rajendran, S. et al. (2022). Metal-organic framework-enabled pesticides are an emerging tool for sustainable cleaner production and environmental hazard reduction. Journal of Cleaner Production 373: 133966. Doi: 10.1016/j.jclepro.2022.

Vinzant, K., Rashid, M. and Khodakovskaya, M. V. (2023). Advanced applications of sustainable and biological nano-polymers in agricultural production. Frontiers in Plant Science 13: 2022. Doi: 10.3389/fpls.2022.1081165.

Wang, C.-F., Han, G.-L., Yang, Z.-R., Li, Y.-X. and Wang, B.-S. (2022). Plant salinity sensors: Current understanding and future directions. Frontiers in Plant Science 13. https://doi.org/10.3389/fpls.2022.859224.

Wang, E., Cao, Y., Bai, Y., Gai, Y., Shan, Y., Li, Q. et al. (2022). A triboelectric nanosensor based on ultra-thin MXene composite paper for heavy metal ion detection. Journal of Micromechanics and Microengineering 32(4): 044003. https://dx.doi.org/10.1088/1361-6439/ac542b.

Wang, H.-Q., Liu, P., Zhang, J.-W., Zhao, B. and Ren, B.-Z. (2020). Endogenous hormones inhibit differentiation of young ears in maize (*Zea mays* L.) under heat stress. Frontiers in Plant Science 11. https://doi.org/10.3389/fpls.2020.533046.

Wang, L., Ning, C., Pan, T. and Cai, K. (2022). Role of silica nanoparticles in abiotic and biotic stress tolerance in plants: a review. International Journal of Molecular Sciences 23(4): 1947. DOI: 10.3390/ijms23041947.

Wang, M., Chen, L., Chen, S. and Ma, Y. (2012). Alleviation of cadmium-induced root growth inhibition in crop seedlings by nanoparticles. Ecotoxicology and Environmental Safety 79: 48–54. DOI: 10.1016/j.ecoenv.2011.11.044.

Wang, X., Cheng, M., Yang, Q., Wei, H., Xia, A., Wang, L. et al. (2019). A living plant cell-based biosensor for real-time monitoring invisible damage of plant cells under heavy metal stress. Science of The Total Environment 697: 134097. https://doi.org/10.1016/j.scitotenv.2019.134097.

Wang, X., Kong, L., Zhou, S., Ma, C., Lin, W., Sun, X. et al. (2022a). Development of QDs-based nanosensors for heavy metal detection: A review on transducer principles and *in-situ* detection. Talanta 239: 122903. https://doi.org/10.1016/j.talanta.2021.122903.

Wang, X., Qi, Y., Shen, Y., Yuan, Y., Zhang, L., Zhang, C. et al. (2020). A ratiometric electrochemical sensor for simultaneous detection of multiple heavy metal ions based on ferrocene-functionalized metal-organic framework. Sensors and Actuators B: Chemical 310: 127756. https://doi.org/10.1016/j.snb.2020.127756.

Wang, Z., Yue, L., Dhankher, O. P. and Xing, B. (2020). Nano-enabled improvements of growth and nutritional quality in food plants driven by rhizosphere processes. Environment International 142: 105831. Doi: 10.1016/j.envint.2020.105831.

Wu, H., Tito, N. and Giraldo, J. P. (2017). Anionic cerium oxide nanoparticles protect plant photosynthesis from abiotic stress by scavenging reactive oxygen species. ACS Nano 11(11): 11283–11297. https://doi.org/10.1021/acsnano.7b05723.

Wu, K., Xu, C., Li, T., Ma, H., Gong, J., Li, X. et al. (2023). Application of nanotechnology in plant genetic engineering. International Journal of Molecular Sciences 24: 14836. Doi: 10.3390/ijms241914836.

Xiao, M., Fu, Q., Shen, H., Chen, Y., Xiao, W., Yan, D. et al. (2018). A turn-on competitive immunochromatographic strips integrated with quantum dots and gold nano-stars for cadmium ion detection. Talanta 178: 644–649. https://doi.org/10.1016/j.talanta.2017.10.002.

Ye, W., Li, Y., Wang, J., Li, B., Cui, Y., Yang, Y. et al. (2020). Electrochemical detection of trace heavy metal ions using a Ln-MOF modified glass carbon electrode. Journal of Solid State Chemistry 281: 121032. https://doi.org/10.1016/j.jssc.2019.121032.

Ye, W., Yu, M., Wang, F., Li, Y. and Wang, C. (2022). Multiplexed detection of heavy metal ions by single plasmonic nanosensors. Biosensors and Bioelectronic 196: 113688. https://doi.org/10.1016/j.bios.2021.113688.

Younis, A. A., Khattab, H. and Emam, M. M. (2020). Impacts of silicon and silicon nanoparticles on leaf ultrastructure and TaPIP1 and TaNIP2 gene expressions in heat stressed wheat seedlings. Biologia Plantarum 64(1): 343–352. DOI: 10.32615/bp.2020.030.

Yu, Y., Dai, W. and Luan, Y. (2023). Bio- and eco-corona related to plants: Understanding the formation and biological effects of plant protein coatings on nanoparticles. Environmental Pollution 317: 120784, https://doi.org/10.1016/j.envpol.2022.120784.

Zain, M., Ma, H., Nuruzzaman, Md., Chaudhary, S., Nadeem, M., Shakoor, N. et al. (2023). Nanotechnology based precision agriculture for alleviating biotic and abiotic stress in plants, Plant Stress 10: 100239. https://doi.org/10.1016/j.stress.2023.100239.

Zaman, W., Ayaz, A. and Park, S. (2025). Nanomaterials in agriculture: A pathway to enhanced plant growth and abiotic stress resistance. Plants 14: 716. https://doi.org/ 10.3390/plants14050716.

Zhang, P., Jiang, Y., Schwab, F., Monikh, F. A., Grillo, R., White, J. C. et al. (2024). Strategies for enhancing plant immunity and resilience using nanomaterials for sustainable agriculture. Environmental Science & Technology 58(21): 9051–9060. https://doi.org/10.1021/acs.est.4c03522.

Zhang, Q., Li, L., Qiao, Z., Lei, C., Fu, Y., Xie, Q. et al. (2017). Electrochemical conversion of $Fe_3O_4$ magnetic nanoparticles to electroactive prussian blue analogues for self-sacrificial label biosensing of avian influenza virus H5N1. Analytical Chemistry 89(22): 12145–12151. https://doi.org/10.1021/acs.analchem.7b02784.

Zhao, L., Lu, L., Wang, A., Zhang, H., Huang, M., Wu, H. et al. (2020). Nano-biotechnology in agriculture: Use of nanomaterials to promote plant growth and stress tolerance. Journal of Agriculture and Food Chemistry 68: 1935–1947. Doi: 10.1021/acs.jafc.9b06615.

Zhou, P., Wang, L., Gao, J., Jiang, Y., Adeel, M. and Hou, D. (2023). Nanoplastic–plant interaction and implications for soil health. Soil Use Manage 39: 13–42. Doi: 10.1111/sum.12868.

Zhu, Q., Wang, L., Dong, Q., Chang, S., Wen, K., Jia, S. et al. (2017). FRET-based glucose imaging identifies glucose signalling in response to biotic and abiotic stresses in rice roots, Journal of Plant Physiology 215: 65–72. https://doi.org/10.1016/j.jplph.2017.05.007.

Zuo, Q., Chen, Y., Chen, Z.-P. and Yu, R. Q. (2018). Quantification of cadmium in rice by surface-enhanced raman spectroscopy based on a ratiometric indicator and conical holed enhancing substrates. Analytical Sciences 34: 1405–1410. https://doi.org/10.2116/analsci.18P342.

# Section III
# CRISPR-Cas9 Technology

# Chapter 9

# CRISPR-Cas9 Editing Technology

## Concept, Applications, and Prospects for Agriculture

*Gleice Ribeiro Orasmo** and *Lidiane de Lima Feitoza*

## 1. Introduction

A modern technology capable of making specific modifications to the genome of living beings has revolutionized different fields of science, including agriculture. The changes made to DNA by CRISPR/Cas9 technology seek to correct genetic mutations or resolve disease issues by developing new therapies (Agunbiade and Babalola 2024). For editing, the Cas9 enzyme is guided by an RNA molecule to a specific location in the genome, performing changes such as deletion, insertion, or editing of genes (Asmamaw et al. 2021).

Genome editing of an organism is based on a set of techniques developed to alter the genetic composition of an organism (Zaidi and Mansoor 2017). The CRISPR-Cas9 system, although simpler, has advantages over previously developed methods such as meganuclease, Zinc Finger Nucleases (ZFNs), and Transcription Activator-Like Effector Nucleases (TALENs) (Montecillo et al. 2020).

Transgenic techniques have been widely used in recent years to understand plant biology for application in breeding. The CRISPR/Cas system (clustered regularly spaced short palindromic repeats with associated proteins) is a tool that has produced positive results in plant genetic and epigenetic characteristics. CRISPR/Cas genome editing is fast and accurate, resulting in crops that are resistant to biotic and abiotic stresses, in contrast to the usual techniques that take many years to promote resilience (Mutezo et al. 2021). In fact, Jaganathan and colleagues (2018) state that first-generation genome editing tools are expensive and time-consuming, as they involve complicated steps that require protein engineering. On the other hand, CRISPR/Cas9 genome editing technology involves simple editing methods, using the same Cas9 and different guide RNAs to target different sites in the genome.

Several modifications using the CRISPR/Cas9 editing system have been established and adopted for plant varieties due to the highly satisfactory results obtained, as well as its versatility, capable of adapting to different crops and promoting the creation of improved varieties without the need for trangenics in economically important crops (Montecillo et al. 2020). The application of this

Department of Biology, Federal University of Piauí, Teresina, Piaui, Brazil.
* Corresponding author: gleice@ufpi.edu.br

DOI: 10.1201/9781003659907-12

technology in model plant genetics has resulted in greater precision and genome editing capacity (Schaeffer and Nakata 2015).

Several studies have emerged in recent years reviewing work using CRISPR/Cas9 technology, varying the focus of discussions and confirming its potential with the results presented. A study reviewed by Gan and Ling (2022) discussed the use of CRISPR/Cas9 technology applied to a large number of plant species for crop improvement, as well as the mechanisms of genetically edited plants compared to transgenic organisms. Thus, the CRISPR/Cas9 system has been of primary use to other methods due to its good results (Gan and Ling 2022).

Agunbiade and Babalola (2024) analyzed the adverse effects of water stress on cultivated plants, the management measures adopted, and the role of genome editing technologies such as CRISPR-Cas9. To mitigate the adverse effects of drought on plants, the authors state that it is necessary to analyze the causes of water stress, the mechanisms of action in the plant, and the long-term genetic effects. They also report that various strategies are needed, such as the use of chemicals, growth hormones, biostimulants, and genetic modifications, among others. In addition, the authors state that through gene editing, using CRISPR/Cas9 technology, drought-tolerant products can be developed, improving understanding of the relationship between soil microorganisms and drought tolerance. Plant recovery has corroborated the improving drought resistance in agricultural production.

Jin and his collaborators (2024) reviewed the mechanisms related to plant heat tolerance, highlighting the pivotal role of epigenetic regulation in facilitating adaptive responses to heat stress. Emerging genome-editing technologies, particularly CRISPR/Cas9, have significantly advanced the ability to introduce targeted modifications in epigenetic landscapes, thereby expanding the potential for developing climate-resilient crop varieties. The integration of these innovative biotechnological, such as CRISPR/Cas9, with traditional breeding methods holds significant promise for the development of more thermotolerant crops. The authors state that, based on the knowledge gained from epigenetic studies, it is possible to develop strategy for enhancing agricultural sustainability under increasingly variable environmental conditions (Jin et al. 2024).

According to Kamali and Singh (2023), recent advances in the use of genomic and transcriptomic tools have increased crop resistance to stress factors. Genomic studies help identify genes that respond to stress, revealing regulatory mechanisms and pointing to genetic characteristics related to stress resistance, while transcriptomics reveals the complex expression of genes when subjected to stress, indicating new genes and signaling pathways that respond to it. All this knowledge helps in the creation of stress-tolerant cultivars through traditional genetic improvement combined with genetic engineering techniques and cutting-edge gene editing, such as CRISPR-Cas9.

This review does not aim to provide a comprehensive overview of CRISPR/Cas9 technology, but simply to reveal, albeit partially, its potential and effectiveness, and to present some initiatives using it. In addition, this chapter presents the future prospects and possibilities pointed out by authors involved in the world of genome editing techniques applied to agriculture.

## 2. How CRISPR-Cas9 Biotechnology Works

CRISPR technology was initially described by Doudna and Charpentier (2014) and has become a fundamental tool in molecular biology and genetics, enabling the specific editing of DNA sequences. The fundamental principles of CRISPR technology lie in its ability to use guide RNA sequences (sgRNA) to direct the Cas9 enzyme to specific locations in the genome (Andrade et al. 2024). Simply put, CRISPR-Cas9 works like a kind of "genetic scissors", with two main parts: the guide RNA and the Cas9 protein.

In general terms, the guide RNA is responsible for finding the exact DNA sequence to be edited, while Cas9 cuts the DNA in that specific region. Once the cut is made, scientists can edit the genetic sequence, either by removing, adding or correcting parts of the DNA. After creating a guide RNA

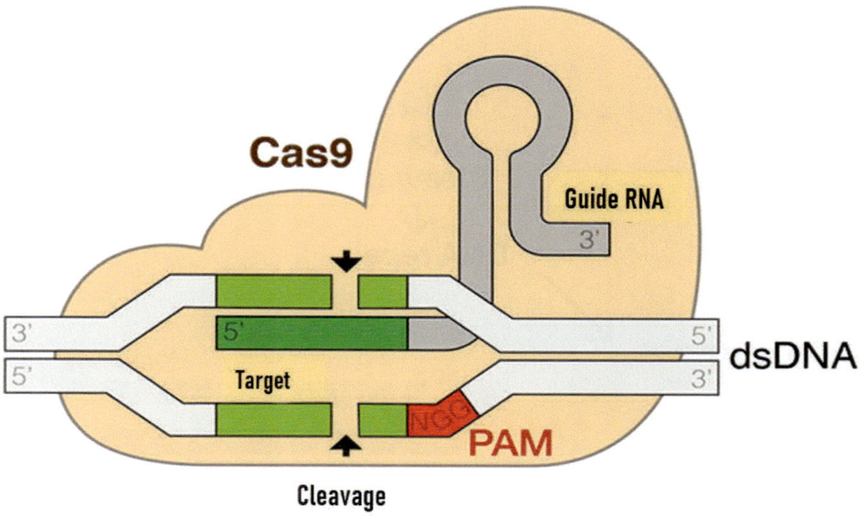


**Figure 1.** Representative diagram of the CRISPRS/Cas9 system.

(sgRNA) specific to the target DNA sequence, a complex is formed with the Cas9 protein, thereby initiating specific double-strand breaks in the target DNA (Costa et al. 2017) (Figure 1).

In a more detailed explanation, Jiang and Doudna (2017) describe that a typical CRISPR locus comprises a set of repetitive sequences (repeats) interspersed with short stretches of non-repetitive sequences (spacers), as well as a set of CRISPR-associated genes (cas). The CRISPR RNA (tracrRNA) gene encodes a non-coding RNA that is homologous to the repeated sequences. After phage incorporation, a new spacer is incorporated into the CRISPR matrix. Following integration, into the CRISPR array, the new spacer is transcribed alongside the existing spacers, producing a long precursor CRISPR RNA (pre-crRNA) composed of both repeat and spacer sequences. Separately, the trans-activating CRISPR RNA (tracrRNA) is transcribed and subsequently hybridizes with the repeat regions of the pre-crRNA. The additional cut at the 5' end of the crRNA reduces the length of the guide sequence to 20 nt (nucleotides). During interference of the crRNA-tracrRNA complex, the Cas9 endonuclease binds and directs it to cleave the foreign DNA, which contains a complementary 20 nt sequence of crRNA, which precedes the PAM sequence. Thus, the CRISPR-Cas9 system was developed from a bacterial defense mechanism against phage infection, and the plasmid is reused as a powerful RNA-guided DNA tool for genome editing, transcriptional interruption, or epigenetic modulation (Jiang and Doudna 2017) (Figure 2).

According to Kamburova and colleagues (2017), this method utilizes the adaptive immunity of bacteria and archaea, relying on specialized genomic sites termed CRISPR loci. Within these loci, the Cas9 protein is encoded alongside repeated arrays of spacer sequences. These spacers are short segments derived from exogenous DNA, such as viral or plasmid sequences, which are integrated into the bacterial genome following recombination events.

CRISPR-Cas9 is one of the most widely used gene editing technologies. Since the discovery of the first CRISPR locus by Ishino et al. (1987) and the pioneering study by Jansen et al. (2002), it has emerged as the most widely researched method for genome editing in diverse organisms. CRISPR-Cas9 is one of the most widely used gene editing technologies. This system functions as an efficient and adaptable method, using a guide RNA to guide the Cas9 enzyme to a defined DNA sequence, where it can perform accurate cuts or facilitate desired genetic alterations (Agunbiade and Babalola 2024). The Cas9 enzyme plays a central role in CRISPR technology; it works like molecular scissors, cutting the DNA at the site designated by the sgRNA. The simplicity and efficiency of this

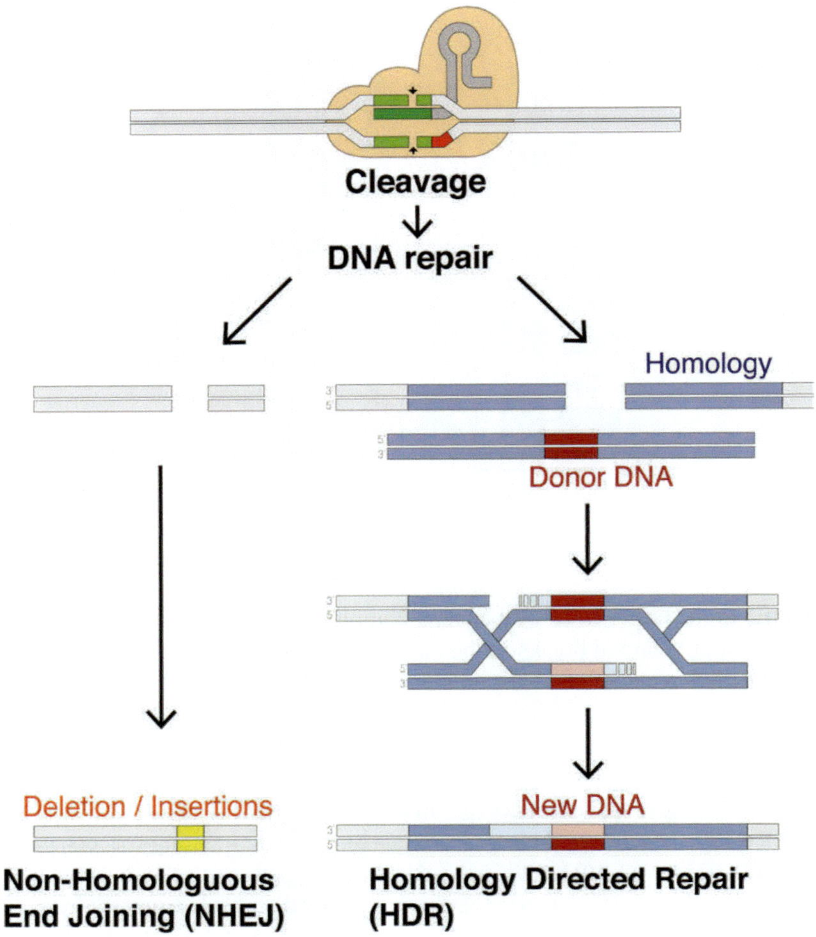


**Figure 2.** DNA repair after double-stranded break in a schematic overview of CRISPR-Cas9 gene editing.

system has enabled groundbreaking advances in gene editing experiments (Doudna 2020). However, a thorough understanding of Cas9 is essential to improve precision and minimize off-target effects.

## 3. Application of the CRISPR/Cas9 System in Agriculture

The use of gene-editing technologies has revolutionized agriculture by allowing the creation of crops with superior characteristics, including higher yields, higher nutriente content and increased resistance to pests, diseases, and environmental stresses (Zaidi et al. 2020), in addition, enabling a faster breeding process and shortening the period necessary to create new crop varieties (Begna 2022). Andrade et al. (2024) reviewed the technologies used in plant breeding, the significant advances made in agriculture, sustainability and efficiency in food production, including the use of multifunctional microorganisms. Among the technologies covered in the review, the authors place CRISPR technology as a revolutionary technique, which involves gene editing and enables precise modifications to plant DNA to improve desired characteristics.

CRISPR/Cas9-guided editing has made it possible to deepen our understanding of fundamental biological systems while also enhancing crop productivity and quality. This technology enables the development of plants with valuable compositional traits and characteristics that confer resistance to a wide range of biotic and abiotic stresses (Kamburova et al. 2017). In their review, Kamburova et al. (2017) provided an overview of genome editing tools in plants, with particular emphasis on the CRISPR/Cas9 system and its diverse applications.

The CRISPR/Cas9 technology has recently been applied to various crops, including bread wheat (Shan et al. 2014, Liang et al. 2017, Zhang et al. 2019, Okada et al. 2019), potato (*Solanum tuberosum*) (Wang et al. 2015, Andersson et al. 2017), rice (*Oryza sativa*) (Shan et al. 2013, Fang et al. 2015, Shen et al. 2017, Wang et al. 2017, Jaganathan et al. 2018, Makarova et al. 2020), oilseed rape (*Brassica napus*) (Okuzaki et al. 2018); grape (*Vitis vinífera* L.) (Cong et al. 2013, Foria et al. 2019, Villano et al. 2020, Fedorina et al. 2022), ramie (*Boehmeria nivea*) (Rasheed et al. 2023), cassava (*Manihot esculenta*) (Gomez et al. 2019), grapefruit (*Citrus paradisi*) (Jia et al. 2017), corn (*Zea mays*) (Liu et al. 2020b), cucumber (*Cucumis sativus* L.) (Chandrasekaran et al. 2016); tomato (*Solanum lycopersicum*) (Li et al. 2018a, Li et al. 2018b, Li et al. 2018c, Nekrasov et al. 2017, Wang et al. 2017b), among other crops, and the technology has also been applied to model plants such as the *Arabidopsis thaliana* (Ryder et al. 2017).

The grape (*Vitis vinifera* L.) is affected by many fungal diseases, significantly reducing the yield and quality of the berries. A study by Orasmo et al. (2015) revealed that traditional cultivation, which uses a huge amount of fungicides, has caused alterations at the functional level in carboxylesterases and acetylesterases. The authors indicate that this evidence represents a warning to grape producers about the dangers inherent in the indiscriminate use of powerful, modern fungicides widely used in agriculture today. As an alternative to the frequent use of fungicides in grape cultivation, novel approaches such as marker-assisted selection, rhizosphere bioengineering, and genetic engineering have been employed to develop disease-resistant grape cultivars. However, these methods are associated with several limitations and challenges (Fedorina et al. 2022).

On the other hand, CRISPR/Cas9-mediated genome editing is recognized as a promising strategy for improving disease resistance in grapevine cultivars. Fedorina and colleagues (2022) reviewed the primary mechanisms governing plant–pathogen interactions, the defense responses mounted by grapevine, in addition to the resistance and susceptibility genes associated with the pathogens *Erysiphe necator* and *Botrytis cinerea*. According to the authors, recent research is looking for new target genes in the grapevine that can be modified using CRISPR/Cas9 technology to improve pathogen resistance (Cong et al. 2013, Foria et al. 2019, Villano et al. 2020). Furthermore, modern genetic research in grapevine has enabled the identification of resistance and susceptibility loci associated with major diseases such as powdery mildew and gray mold.

Genome editing using the CRISPR/Cas9 system has been successfully tested on the most important vegetatively propagated fruit and berry crops (apple, pear, orange, kumquat, grapefruit, banana, strawberry and kiwi) and grapes (Fizikova et al. 2021). The authors discussed the existing limitations of genome editing in order to improve fruit and berry crop varieties.

Fizikova et al. (2021) reviewed the results of genome editing by CRISPR/Cas9 the studies involved different groups of target genes in fruit and berry crops with the aim of: (1) decreasing susceptibility to biotic and abiotic stress (Malnoy et al. 2016, Wang et al. 2021), (2) alter flowering time and plant architecture (Ren et al. 2020), (3) increase shelf life (Hu et al. 2021), (4) study gene functions (Zhou et al. 2018, Wang et al. 2021) and (5) optimize CRISPR/Cas9 editing of fruit and berry cultures (Varkonyi-Gasic et al. 2019, Malabarba et al. 2020, Zhang et al. 2021). Inactivation of the disease resistance gene MdDIPM4 was achieved using the high-efficiency CRISPR/Cas9 system in apple (*Malus domestica*) (Pompili et al. 2020).

Ramie (*Boehmeria nivea*), a member of the Urticaceae family, is native to East Asia and represents one of the most important fiber crops in China due to its excellent fiber quality and immense industrial importance. Planting cadmium-tolerant ramie cultivars can prevent yield losses in contaminated soils. However, the mechanism of cadmium tolerance is still, not fully understood and requires high-level molecular studies. Research initiatives, including QTL mapping, transcriptomic analyses, the identification of transcription factors, and genetic engineering approaches, have been undertaken to enhance cadmium tolerance in ramie; however, studies are still limited. In this context,

CRISPR/Cas9 biotechnology is considered a highly promising tool, as it enables precise and targeted gene editing. Candidate genes or QTLs associated with cadmium tolerance can be introduced or modified in sensitive cultivars through genetic engineering or CRISPR/Cas9-mediated approaches (Rasheed et al. 2023).

Rice (*Oryza sativa*) is an important crop for more than half of the world's population. Due to the small size of its genome, this plant has been studied extensively and has been a model among monocots. While the application of CRISPR-mediated genome editing in rice has been well documented, studies addressing its role in the genetic improvement of rice for biotic and abiotic stress tolerance are still scarce (Jaganathan et al. 2018). For example, Wang et al. (2017) generated mutagenic rice lines using CRISPR/Cas9 technology, which exhibited significantly enhanced resistance to blast compared with wild-type rice (Makarova et al. 2020).

Shan et al. (2013) reported the first sequence-specific CRISPR/Cas9-mediated genome modifications in a crop plant by targeting three rice genes: phytoene desaturase (*OsPDS*), betaine aldehyde dehydrogenase (*OsBADH2*), and mitogen-activated protein kinase (*OsMPK2*). These genes, which are involved in responses to various abiotic stress stimuli, were successfully edited using both protoplast systems and particle-bombarded rice calli. Plant annexins are $Ca^{2+}$-dependent phospholipid-binding proteins implicated in the regulation of plant development and in protection against environmental stress. In rice, the annexin gene *OsAnn3* was subsequently modified through CRISPR/Cas9-mediated genome editing (Shen et al. 2017).

Gan and Ling (2022) discussed recent progress in the use of CRISPR/Cas9-mediated genome editing aimed at improving key agronomic traits in major crops such as rice (*Oryza sativa*), soybean (*Glycine max* L.), and oilseed rape (*Brassica napus*). Precise base editing in defined genomic regions mediated through CRISPR-Cas9 has enabled functional evaluation of the OsACC gene, which encodes a protein in which amino acids determine herbicide resistance in the ACCase variety of rice (Liu et al. 2020a). CRISPR/Cas9-mediated mutagenesis in the GmFT2a and GmFT5a genes (Cai et al. 2020) and mutation in the GmPRR37 gene (Wang et al. 2020), which affected flowering time, contributed to expanding the capacity for regional adaptability in soybean.

The development of powdery mildew-resistant wheat (*Triticum aestivum* L.) varieties was achieved by mutating through CRISPR/Cas9-mediated mutagenesis in tree genes in the wheat genome (Shan et al. 2014). The bread wheat genome was efficiently edited using CRISPR/Cas9 ribonucleoprotein complexes (Liang et al. 2017), in another study, an Agrobacterium-mediated CRISPR/Cas9 system was developed to facilitate stable genome editing in wheat (Zhang et al. 2019a). The modification of the *Ms1* gene via CRISPR/Cas9 enabled the swift generation of male-sterile hexaploid wheat lines for heat-sensitive hybrid seed production (Okada et al. 2019).

Modifications CRISPR/Cas9-based in the elF4E, nCBP-1 and nCBP-2 genes in cassava (*Manihot esculenta*) resulted in increased resistance to double-strand break repair (Gomez et al. 2019). Similarly, CRISPR/Cas9-mediated alterations has been applied to modify genes associated with citrus canker tolerance in grapefruit, including the Duncan cultivar (Jia et al. 2017). Chandrasekaran et al. (2016) also opted for gene alteration using CRISPR/Cas9, the eIF4E gene was modified to improve virus resistance in cucumber (*Cucumis sativus* L.).

High-throughput targeted mutagenesis using multiplex CRISPR/Cas9, integrated with genetic mapping and genomic approaches, has been successfully applied to identify 743 candidate genes associated with agronomic and nutritional traits in maize (*Zea mays*). Following barcode-based sequencing, 412 edited sequences corresponding to 118 genes from individuals exhibiting clear phenotypic changes were accurately characterized (Liu et al. 2020b). Additionally, CRISPR/Cas9 has been employed to knock out or overexpress the *ZmPHYC1* and *ZmPHYC2* genes in maize to investigate their effects on flowering time and plant height (Li et al. 2020b).

Several studies have applied CRISPR/Cas9 technology have been carried out on the genetic improvement of tomatoes (*Solanum lycopersicum*) for various purposes, such as alterations to the *SGR1*, *LCY-E*, *Blc*, *LCY-B1*, and *LCY-B2* genes in order to obtain a greater accumulation of lycopene, taking into account its positive effects on the visual and functional properties of the fruit (Li et al. 2018a); modifications in several genes involved in the metabolism of aminobutyric acid (GABA) to obtain elevate content (Li et al. 2018b); modification to the *Mlo* gene involved in powdery mildew resistance (Nekrasov et al. 2017); and modification to the *SlMAPK3* gene involved in drought tolerance (Wang et al. 2017).

According to Jin et al. (2024) thermotolerance the ability of plants to withstand high temperatures, is a critical trait for plant survival, this complex trait involves a combination of physiological, biochemical, and molecular mechanisms. Recent studies highlight the significant impact of epigenetics on the development of climate-resilient crops and the transformation of plant breeding practices. Understanding epigenetic pathways can help to enhance stress tolerance, increase crop yield and improve plant adaptability to changing environmental conditions (Popova et al. 2013).

The advent of genome editing technologies, particularly CRISPR-Cas9, has transformed the capacity to precisely manipulate epigenetic traits. This innovation, referred to as epigenome editing, enables researchers to selectively regulate the expression of stress-responsive genes. Through targeted modifications of DNA methylation patterns or histone marks at stress-associated loci, plant tolerance to environmental stresses can be substantially enhanced (Samanta and Thakur 2015, Sable et al. 2018).

Epigenetic information can define complex traits, such as fruit maturity in tomatoes (*Solanum lycopersicum*) (Shen et al. 2014) and somaclonal variation in palm trees (Shen et al. 2019). Technologies like CRISPR-Cas9 enable precise epigenetic manipulation, supporting the targeted control of stress-related genes and strengthening plant adaptability (Jin et al. 2024).

A number of genomic techniques and approaches applied to crop improvement can help improve tolerance to abiotic stress. The functional roles of multiple genes involved in abiotic stress response in plant species have been discovered using RNAi techniques (Kamali and Singh 2023). Xu et al. (2022) reported that overexpression of the tubby-like protein gene *GmTLP8* in soybean (*Glycine max* L.) improved plant tolerance to drought and salinity, while its downregulation diminished stress resistance. Silencing of the AP2/ERF gene *NtRAV-4* in *Nicotiana tabacum* resulted in improved root architecture, enhanced photosynthetic efficiency, and greater tolerance to drought stress (Gao et al. 2020). The functional analysis of plant genes can be achieved through genome editing approaches. Although TALEN- and ZFN-based technologies have shown potential, their application in enhancing abiotic stress tolerance in crops remains limited. Conversely, CRISPR-Cas9 has emerged as the predominant genome editing tool in crop research during the past decade (Kamali and Singh 2023).

CRISPR/Cas9-mediated miRNA knockout was used to generate *OsMIR408* and *OsMIR528* rice lines, which exhibited increased sensitivity to salt stress, indicating that these genes act as positive regulators of salt tolerance (Zhou et al. 2017). CRISPR/Cas9-mediated editing also resulted in increased salt tolerance, resulting in a loss-of-function mutation in the *OsRR22* gene (Zhang et al. 2019b). In tomato (*Solanum lycopersicum*), precise deletion of the negative response domain(s) of the *SlHyPRP1* gene via CRISPR/Cas9 conferred high salinity tolerance during both germination and vegetative stages (Tran et al. 2023).

CRISPR-Cas–mediated genome editing has been employed to elucidate the functional roles of key genes in plant drought tolerance. For instance, editing of the *DST* gene induced a 366-bp deletion, resulting in a loss-of-function mutation; the corresponding mutant line demonstrated enhanced leaf water retention under water-deficit conditions in rice (Santosh-Kumar et al. 2020). Similarly, the *ARGOS8* gene, a negative regulator of the ethylene signaling pathway, has been associated with the activation of drought-responsive genes in maize. Owing to its inherently low expression, CRISPR-Cas9 was utilized in combination with the *GOS2* promoter to achieve elevated expression of *ARGOS8*, thereby enhancing drought tolerance (Jain and Garg 2021). Furthermore, targeted CRISPR-Cas9 editing of genes encoding the enzymes poly ADP-ribose polymerase (PARP) and Nudix hydrolase conferred increased resistance to both oxidative drought stress and

**Table 1.** Application of the CRISPR/Cas9 system in various crops. Source: the authors.

| Crop/species | Goals | References |
|---|---|---|
| Rice (*Oryza sativa*) | Abiotic and biotic stress tolerance.<br><br>Mutagenic strains.<br><br>Gene modification (OsAnn3);<br><br>Mutation in the OsRR22 gene to increase salt tolerance;<br><br>Alteration in the OsDST gene for drought and salt tolerance;<br><br>Mutagenesis in the OsACC gene for herbicide resistance in the ACCase variety; | Fang et al. 2015<br>Jaganathan et al. 2018<br>Wang et al. 2017a<br><br>Shan et al. 2013<br>Shen et al. 2017<br>Zhang et al. 2019b<br><br>Santosh-Kumar et al. 2020<br>Liu et al. 2020a<br>Li et al. 2020a |
| Potato (*Solanum tuberosum*) | Efficient multiallelic mutagenesis guided by CRISPR/Cas9;<br><br>Targeted multiallelic mutagenesis in tetraploid potatoes; | Wang et al. 2015<br><br>Andersson et al. 2017 |
| Oilseed rape (*Brassica napus*) | Increased oleic acid content | Okuzaki et al. 2018 |
| Grapefruit ou Grapefruit Duncan (*Citrus paradisi*) | Alteration in canker tolerance genes | Jia et al. 2017 |
| Apple (*Malus domestica*) | Inactivation of the disease resistance MdDIPM4 gene | Pompili et al. 2020 |
| Cassava (*Manihot esculenta*) | Increased resistance to repair of double chain breaks | Gomez et al. 2019 |
| Corn (*Zea mays*) | Identification of candidate genes corresponding to agronomic characteristics and nutrition<br><br>Alteration of the GOS2 promoter to increase expression of the ARGOS8 gene<br><br>Tolerance to oxidative stress, drought and genotoxic stress<br><br>Knockout and overexpression of the ZmPHYC1 and ZmPHYC2 genes targeting flowering time/plant height | Liu et al. 2020b<br><br>Jain and Garg 2021<br><br>Jain and Garg 2021<br><br>Li et al. 2020b |
| Cucumber (*Cucumis sativus* L.) | Modification of the eIF4E virus resistance gene | Chandrasekaran et al. 2016 |
| Ramie (*Boehmeria nivea*) | Tolerance to cadmium | Rasheed et al. 2023 |
| Soy or Soybean (*Glycine max* L.) | Mutagenesis in the GmFT2a/5a genes for flowering time<br><br>Mutagenesis in the GmPRR37 gene contributing to regional adaptability | Cai et al. 2020<br><br>Wang et al 2020 |
| Sorghum (*Sorghum bicolor* (L.) Moench) | Targeted mutagenesis in two SbFT and SbGA2ox5 genes mediated by Agrobacterium; | Char et al. 2020 |
| Tomato (*Solanum lycopersicum*) | Changes in five genes to increase lycopene content<br><br>Multiplex system aimed at accumulating GABA content<br><br>Alteration to the SICBF gene for cold tolerance<br><br>Alteration to the Mlo gene for resistance to powdery mildew<br><br>Alteration in the SIMAPK3 gene for drought tolerance<br><br>Mutagenesis in the SlNPR1 gene to reduce drought tolerance<br><br>Alteration in the SlHyPRP1 gene for salinity tolerance | Li et al. 2018a<br><br>Li et al. 2018b<br><br>Li et al. 2018c<br><br>Nekrasov et al. 2017<br><br>Wang et al. 2017b<br><br>Li et al. 2019<br><br>Tran et al. 2023 |

*Table 1 contd. ...*

*...Table 1 contd.*

| Crop/species | Goals | References |
|---|---|---|
| Wheat (*Triticum aestivum* L.) | Use of Agrobacterium and ribonucleoproteins to edit the wheat genome | Zhang et al. 2019a<br>Liang et al. 2017 |
| | Alteration of several genes for resistance to powdery mildew | Shan et al. 2014 |
| | Alteration of the Ms1 gene to develop heat-sensitive male sterile lines | Okada et al. 2019 |
| Grape (*Vitis vinífera* L.) | Tolerance to pathogens (fungi) | Fedorina et al. 2022<br>Foria et al. 2019<br>Villano et al. 2020<br>Cong et al. 2013 |

genotoxic stress in maize (Jain and Garg 2021). In contrast, mutagenesis of the *SlNPR1* gene in tomato via CRISPR-Cas9 led to reduced drought tolerance, concomitant with downregulation of key drought-responsive genes, including *SlGST*, *SlDHN*, and *SlDREB* (Li et al. 2019). Collectively, these studies underscore the potential of CRISPR-Cas technology as a precise tool for modulating gene function to enhance abiotic stress resilience in crops.

Sorghum (*Sorghum bicolor* L. Moench) ranks as the fifth most important cereal crop worldwide. However, the genome editing platform in sorghum was lacking compared to other plants due to its low efficiency of stable transformation (Char et al. 2020). Cas9/sgRNA activity was demonstrated for mutagenesis in sorghum callus cells derived from immature embryos using the reporter gene encoding red fluorescence protein (Jiang et al. 2013). Char and co-workers (2020) reported the development of a CRISPR/Cas9 system adapted for Agrobacterium-mediated transfer and the validation of its efficiency for targeted mutagenesis in two endogenous genes SbFT (Sb10G045100) and SbGA2ox5 (Sb09G230800) in sorghum.

A study by Agunbiade and Babalola (2024) evaluated the potential of leveraging the plant-associated microbiome to enhance drought resistance, as well as the use of various agronomic biotechnologies and advanced gene editing, to develop plant varieties with enhanced drought tolerance. According to the authors, water deficit in plants is caused by the expression of sensitive genes, causing hormonal imbalance, increased formation of ROS (reactive oxygen species) and decreased antioxidant activity.

In *Arabidopsis*, the promoter activity of the transcription factor *AREB1* was enhanced using a modified Cas9 (dCas9) fused to an acetyltransferase domain. This epigenome editing approach upregulated drought-responsive genes, resulting in increased *AREB1* transcription and improved drought tolerance in the edited lines (Paixão et al. 2019). Increased expression of multiple transcription factors (TFs) and stress-responsive genes under drought conditions facilitates the accumulation of diverse metabolites, signaling molecules, and osmolytes, thereby contributing to increased tolerance to drought stress (Agunbiade and Babalola 2024).

The initial applications of the CRISPR-Cas9 genome editing system in plants highlighted its capacity to introduce precise genetic mutations and corrections across a variety of species, ranging from model organisms to economically significant crops. Rapidly established as a versatile tool for genome manipulation, CRISPR-Cas9 has been employed to enhance crop quality and yield, modify metabolic pathways, confer stress tolerance, and elucidate gene functions. Following its successful establishment as a programmable RNA-guided editing system, numerous studies have validated the efficiency of CRISPR/Cas9 in a variety of plant species, as summarized in Table 1.

## 4. The Expected Future Prospects of This Technology

CRISPR/Cas9 technology represents one of the most promising genome editing approaches, enabling precise modification of specific genes, targeted suppression or activation of gene expression, functional alteration of genes, and epigenome editing (Fedorina et al. 2022). The work of Chen et al. (2019) highlights the potential of this technology for the achievement of precise genome editing, offering an optimistic outlook for plant breeding. The authors revealed the future prospects in plant genetic transformation, including emerging technologies such as CRISPR-Cas9 in the approach.

According to Mutezo et al. (2021), the CRISPR/Cas9 system is a powerful technology for genetically modifying crops. The development of technologies that contribute to crop improvement has the potential to enhance agricultural productivity, albeit within certain limits. In recent decades, genetic manipulation approaches, including the insertion of transposons or T-DNA into plant genomes, have played a pivotal role in elucidating gene functions and the underlying biological mechanisms that drive the improvement of cultivated species (Ma et al. 2016). Modern approaches that integrate physiological, molecular, and genetic methodologies offer significant advantages for mitigating plant diseases and abiotic stresses, including water deficit in rice (*Oryza sativa*) (Agunbiade and Babalola 2024).

CRISPR/Cas9-guided mutagenesis enables the development of non-transgenic plants with specific, targeted mutations that can be stably inherited across generations (Fedorina et al. 2022). Cadmium tolerance in ramie (*Boehmeria nivea*) is governed by complex genetic mechanisms; nevertheless, CRISPR/Cas9 is highly recommended for its precise gene-editing capabilities. Although its application in ramie remains limited, the technology has the potential to fill existing research gaps and drive significant advancements in agricultural biotechnology (Rasheed et al. 2023).

According to Mutezo et al. (2021), CRISPR/Cas technology represents a valuable tool for promoting sustainable agricultural production, having been successfully employed to manipulate a range of plant traits. While the technical challenges associated with translating CRISPR/Cas9-engineered crops from the laboratory to the field remain significant, additional hurdles include regulatory uncertainty, intellectual property considerations, and public acceptance. Nevertheless, a variety of Cas9-based methodologies have gained prominence, enabling researchers to efficiently and precisely enhance plant characteristics.

CRISPR/Cas9 technology has been successfully applied to modify canker tolerance genes in citrus, demonstrating its effectiveness as a tool for inducing disease resistance in plants (Jia et al. 2017). Building on functional characterization of genes in various crops, Liu and colleagues (2020b) demonstrated that combining forward and reverse genetics via a targeted mutagenesis library enables rapid validation of key agronomic genes, even in crops with complex genomes. Beyond these specific results, their study provides valuable guidance for optimizing high-throughput CRISPR experiments in plants.

According to Montecillo and collaborators (2020) the optimizing and combining known parameters that influence CRISPR/Cas9 editing efficiency, fidelity, and portability will pave the way for new technological advancements, enhancing the precision of targeted gene editing in plants.

Despite the potential of CRISPR-Cas9 technology for the development of thermotolerance in plants, some challenges must be considered, such as understanding the stability and heritability of epigenetic markers, as well as the interaction between different epigenetic modifications. Jin et al. (2024) state that future investigations should prioritize the comprehensive elucidation of these interactions and the identification of viable epigenetic markers for selection. The authors further emphasize the significance of epigenetic regulation in plant adaptation to heat stress and highlight its potential to enhance crop resilience. This approach represents an innovative strategy to safeguard food production and promote agricultural sustainability under increasingly variable climatic conditions.

Abiotic stress, which involves rising temperatures, droughts, soil salinity, and heavy metal contamination, is a continuing concern in agriculture. These adverse conditions impair crop growth and yield, as well as global food security (Kamali and Singh 2023). The authors emphasize that integrating genomics and transcriptomics offers significant potential for elucidating the molecular mechanisms underlying crop stress tolerance, providing a pathway toward sustainable agriculture and enhanced food security under changing environmental conditions. The results obtained by CRISPR-Cas9 technology in crop improvement, particularly for traits related to abiotic stress tolerance characteristics, demonstrate the potential of the technique. The information generated contributes to the development of transgenic or mutant crop lines oriented towards abiotic stress tolerance (Kamali and Singh 2023).

Despite advances, important challenges remain, including precision in editing, possible off-target effects, and ethical issues related to genetic alteration. Lopes Filho et al. (2021) highlight the need for ethical regulations to ensure the responsible use of CRISPR technology in various applications. According to Fizikova and collaborators (2021), involving more crops in multi-target gene editing projects will optimize existing protocols and thus facilitate the integration of the resulting lines and varieties of fruit and berry crops, as well as grapes.

The new CRISPR-Cas9 technology has demonstrated the countless possibilities of genome editing in a wide variety of living organisms. In this context, numerous ethical questions arise, which require the protection of legal precepts to control the risks of this revolutionary technology. Hupffer and Berwing (2020) reviewed ethical, legal, and governance issues, aiming to better understand and deal with the challenges. The authors concluded that discussions need to take place mainly between the legal and scientific systems, thus allowing for the expansion of knowledge about the risks and opportunities offered by this technology.

CRISPR technology, which makes genome sequences available in various crops, as well as advances in genome editing approaches, has opened the possibility of creating breeds with any desirable trait (Jaganathan et al. 2018). The creation of grape cultivars modified by mutagenesis through CRISPR/Cas9 technology, resistant to pathogens, has been one of the most promising methods (Fedorina et al. 2022).

The rice genome contains a high density of potential protospacer adjacent motif (PAM) sites, occurring at an estimated frequency of approximately one site per ten base pairs (Xie and Yang 2013). This abundance provides extensive opportunities for precise CRISPR-mediated targeting of genes associated with traits of interest in rice. In cassava (*Manihot esculenta*), successful CRISPR/Cas9-induced mutagenesis of the phytoene desaturase gene has further validated the versatility and effectiveness of this system as a genetic transformation tool for the species (Gomez et al. 2019, Mutezo et al. 2021). Collectively, the application of these genome editing approaches is expected to facilitate the development of non-transgenic (non-GMO) crops carrying desirable agronomic traits, thereby enhancing yield potential under both biotic and abiotic stress conditions (Jaganathan et al. 2018).

Genome editing tools have emerged as powerful molecular approaches in functional genomics, significantly contributing to crop improvement. In recent years, numerous applications of these editing systems has been used to improve the understanding of plant biology and contribute to improvements in crop yields through rapid and targeted mutagenesis with associated breeding (Kamburova et al. 2017). Due to its several advantageous features, such as simplicity, efficiency, high specificity and multiplexing capability, genome editing technology, such as the CRISPR system, has revolutionized the way crop breeding is done and is paving the way for next-generation breeding.

## 5. Conclusions

The CRISPR/Cas9 system is a powerful gene-editing tool that has been widely adopted as an innovative strategy in plant biotechnology, opening new avenues in agriculture. Advances in genome editing technologies, particularly CRISPR/Cas9, have enabled the precise targeting of specific genes in plants, facilitating the development of cultivated species with enhanced resistance to pathogens and improved productivity.

CRISPR-Cas9 has revolutionized agriculture and is therefore one of the most widely used gene editing biotechnologies today. It is an auxiliary tool in the genetic improvement of cultivated plants, resulting in the increase of desirable characteristics and can be applied both to increase productivity and to combat the diseases that affect crops. The studies carried out, some of which have been presented here, provide clues to help future researchers gain a comprehensive understanding of CRISPR/Cas9 biotechnology and could help solve problems in agriculture.

## References

Agunbiade, V. F. and Babalola, O. O. (2024). Drought stress amelioration attributes of plant-associated microbiome on agricultural plants. Bioinformatics and Biology Insights 18: 1–12. https://doi.org/10.1177/11779322241233442.

Andersson, M., Turesson, H., Nicolia, A., Fält, A. S., Samuelsson, M. and Hofvander, P. (2017). Efficient targeted multiallelic mutagenesis in tetraploid potato (*Solanum tuberosum*) by transient CRISPR-Cas9 expression in protoplasts. Plant Cell Reports 36(1): 117–128. https://doi.org/10.1007/s00299-016-2062-3.

Andrade, A. G. L., Menechini, W., Ortiz, T. A., de Souza, S. G. H. and da Silva, E. C. (2024). Challenges and opportunities in the incorporation of biotechnological technologies in agriculture. DELOS: Desarrollo Local Sostenible 17(55): e1444. https://doi.org/10.55905/rdelosv17.n55-013.

Asmamaw, M. and Zawdie, B. (2021). Mechanism and applications of CRISPR/Cas-9-mediated genome editing. Biologics: Targets and Therapy 15: 353–361. https://doi.org/10.2147/BTT.S326422.

Begna, T. (2022). Speed breeding to accelerate crop improvement. International Journal of Agricultural Science and Food Technology 8(2): 178–186. https://doi.org/10.17352/2455-815X.000161.

Cai, Y., Wang, L., Chen, L., Wu, T., Liu, L., Sun, S. et al. (2020). Mutagenesis of *GmFT2a* and *GmFT5a* mediated by CRISPR/Cas9 contributes for expanding the regional adaptability of soybean. Plant Biotechnology Journal 18(1): 298–309. https://doi.org/10.1111/pbi.13179.

Chandrasekaran, J., Brumin, M., Wolf, D., Leibman, D., Klap, C., Pearlsman, M. et al. (2016). Development of broad virus resistance in non-transgenic cucumber using CRISPR/Cas9 technology. Molecular Plant Pathology 17(7): 1140–1153. https://doi.org/10.1111/mpp.12385.

Char, S. N., Wei, J., Mu, O., Li, X., Zhang, Z. J., Yu, J. et al. (2020). An Agrobacterium-delivered CRISPR/Cas9 system for targeted mutagenesis in sorghum. Plant Biotechnology Journal 18(2): 319–321. https://doi.org/10.1111/pbi.13217.

Chen, K., Wang, Y., Zhang, R., Zhang, H. and Gao, C. (2019). CRISPR/Cas genome editing and precision plant breeding in agriculture. Annual Review of Plant Biology 70: 667–697. https://doi.org/10.1146/annurev-arplant-050718-100049.

Cong, L., Ran, F. A., Cox, D., Lin, S., Barretto, R., Habib, N. et al. (2013). Multiplex genome engineering using CRISPR/Cas systems. Science 339(6121): 819–823. https://doi.org/10.1126/science.1231143.

Costa, J. R., Bejcek, B. E., McGee, J. E., Fogel, A. I., Brimacombe, K. R. and Ketteler, R. (2017). Genome editing using engineered nucleases and their use in genomic screening. *In*: Markossian, S., Grossman, A. and Brimacombe, K. (eds.). Assay Guidance Manual. Bethesda, MD: Eli Lilly & Company and the National Center for Advancing Translational Sciences. Recuperado de https://www.ncbi.nlm.nih.gov/books/NBK464635/.

Doudna, J. A. and Charpentier, E. (2014). The new frontier of genome engineering with CRISPR Cas9. Science 346(6213): 1258096. https://doi.org/10.1126/science.1258096.

Doudna, J. A. (2020). The promise and challenge of therapeutic genome editing. Nature 578(7794): 229–236. https://doi.org/10.1038/s41586-020-1978-5.

Fang, H., Meng, Q., Xu, J., Li, X., Hu, J., Li, Y. et al. (2015). Knock-down of stress inducible OsSRFP1 encoding an E3 ubiquitin ligase with transcriptional activation activity confers abiotic stress tolerance through enhancing antioxidant protection in rice. Plant Molecular Biology 87(5-6): 441–458. https://doi.org/10.1007/s11103-015-0336-9.

Fedorina, J., Tikhonova, N., Ukhatova, Y., Ivanov, R. and Khlestkina, E. (2022). Grapevine gene systems for resistance to gray mold *Botrytis cinerea* and powdery mildew *Erysiphe necator*. Agronomy 12(3): 499. https://doi.org/10.3390/agronomy12030499.

Fizikova, A., Tikhonova, N., Ukhatova, Y., Ivanov, R. and Khlestkina, E. (2021). Applications of CRISPR/Cas9 system in vegetatively propagated fruit and berry crops. Agronomy 11(9): 1849. https://doi.org/10.3390/agronomy11091849.

Foria, S., Monte, C., Testolin, R., Di Gaspero, G. and Cipriani, G. (2019). Pyramidizing resistance genes in grape: A breeding program for the selection of elite cultivars. Acta Horticulturae 1248: 549–554. https://doi.org/10.17660/ActaHortic.2019.1248.70.

Gan, W. C. and Ling, A. P. K. (2022). CRISPR/Cas9 in plant biotechnology: Applications and challenges. BioTechnologia 103(1): 81–93. https://doi.org/10.5114/bta.2022.113919.

Gao, Y., Han, D., Jia, W., Ma, X., Yang, Y. and Xu, Z. (2020). Molecular characterization and systematic analysis of NtAP2/ERF in tobacco and functional determination of NtRAV-4 under drought stress. Plant Physiology and Biochemistry 156: 420–435. https://doi.org/10.1016/j.plaphy.2020.07.029.

Gomez, M. A., Lin, Z. D., Moll, T., Chauhan, R. D., Hayden, L., Renninger, K. et al. (2019). Simultaneous CRISPR/Cas9-mediated editing of cassava eIF4E isoforms nCBP-1 and nCBP-2 reduces cassava brown streak disease symptom severity and incidence. Plant Biotechnology Journal 17(2): 421–434. https://doi.org/10.1111/pbi.12990.

Hu, C., Sheng, O., Deng, G., He, W., Dong, T., Yang, Q. et al. (2021). CRISPR/Cas9-mediated genome editing of MaACO1 (aminocyclopropane-1-carboxylate oxidase1) promotes the shelf life of banana fruit. Plant Biotechnology Journal 19(5): 654–656. https://doi.org/10.1111/pbi.13519.

Hupffer, H. M. and Berwig, J. A. (2020). A tecnologia CRISPR-CAS9: da sua compreensão aos desafios éticos, jurídicos e de governança. Pensar. Advance Online Publication. https://doi.org/10.5216/pensar.v0i0.63757.

Ishino, Y., Shinagawa, H., Makino, K., Amemura, M. and Nakata, A. (1987). Nucleotide sequence of the iap gene, responsible for alkaline phosphatase isoenzyme conversion in *Escherichia coli*, and identification of the gene product. Journal of Bacteriology 169(12): 5429–5433. https://doi.org/10.1128/jb.169.12.5429-5433.1987.

Jaganathan, D., Ramasamy, K., Sellamuthu, G., Jayabalan, S. and Venkataraman, G. (2018). CRISPR for crop improvement: An update review. Frontiers in Plant Science 9: 985. https://doi.org/10.3389/fpls.2018.00985.

Jain, M. and Garg, R. (2021). Enhancers as potential targets for engineering salinity stress tolerance in crop plants. Physiologia Plantarum 173(4): 1382–1391. https://doi.org/10.1111/ppl.13384.

Jansen, R., van Embden, J. D. A., Gaastra, W. and Schouls, L. M. (2002). Identification of genes that are associated with DNA repeats in prokaryotes. Molecular Microbiology 43(6): 1565–1575. https://doi.org/10.1046/j.1365-2958.2002.02839.x.

Jia, H., Zhang, Y., Orbovic, V., Xu, J., White, F. F., Jones, J. B. et al. (2017). Genome editing of the disease susceptibility gene CsLOB1 in citrus confers resistance to citrus canker. Plant Biotechnology Journal 15(7): 817–823. https://doi.org/10.1111/pbi.12677.

Jiang, F. and Doudna, J. A. (2017). CRISPR–Cas9 structures and mechanisms. Annual Review of Biophysics 46: 505–529. https://doi.org/10.1146/annurev-biophys-062215-010822.

Jiang, W., Zhou, H., Bi, H., Fromm, M., Yang, B. and Weeks, D. P. (2013). Demonstration of CRISPR/Cas9/sgRNA-mediated targeted gene modification in *Arabidopsis*, tobacco, sorghum and rice. Nucleic Acids Research 41(20): e188. https://doi.org/10.1093/nar/gkt780.

Jin, Q., Chachar, M., Ali, A., Chachar, Z., Zhang, P., Riaz, A. et al. (2024). Epigenetic regulation for heat stress adaptation in plants: New horizons for crop improvement under climate change. Agronomy 14(8): 2105. https://doi.org/10.3390/agronomy14082105.

Kamali, S. and Singh, A. (2023). Genomic and transcriptomic approaches to developing abiotic stress-resilient crops. Agronomy 13(13): 2903. https://doi.org/10.3390/agronomy13132903.

Kamburova, V. S., Nikitina, E. V., Shermatov, S. E., Buriev, Z. T., Kumpatla, S. P., Emani, C. et al. (2017). Genome editing in plants: An overview of tools and applications. International Journal of Agronomy 2017: Article ID 7315351, 15 pages. https://doi.org/10.1155/2017/7315351.

Li, X., Wang, Y., Chen, S., Tian, H., Fu, D., Zhu, B. et al. (2018). Lycopene is enriched in tomato fruit by CRISPR/Cas9-mediated multiplex genome editing. Frontiers in Plant Science 9: 559. https://doi.org/10.3389/fpls.2018.00559.

Li, R., Li, R., Li, X., Fu, D., Zhu, B., Tian, H. et al. (2018). Multiplexed CRISPR/Cas9-mediated metabolic engineering of γ-aminobutyric acid levels in *Solanum lycopersicum*. Plant Biotechnology Journal 16(2): 415–427. https://doi.org/10.1111/pbi.12724.

Li, R., Zhang, L., Wang, L., Chen, L., Zhao, R., Sheng, J. et al. (2018). Reduction of tomato-plant chilling tolerance by CRISPR–Cas9-mediated SlCBF1 mutagenesis. Journal of Agricultural and Food Chemistry 66(36): 9042–9051. https://doi.org/10.1021/acs.jafc.8b03024.

Li, Q., Lu, L., Liu, H., Bai, X., Zhou, X., Wu, B. et al. (2020). A minor QTL, SG3, encoding an R2R3-MYB protein, negatively controls grain length in rice. Theoretical and Applied Genetics 133(8): 2387–2399. https://doi.org/10.1007/s00122-020-03641-9.

Liang, Z., Chen, K., Li, T., Zhang, Y., Wang, Y., Zhao, Q. et al. (2017). Efficient DNA-free genome editing of bread wheat using CRISPR/Cas9 ribonucleoprotein complexes. Nature Communications 8: Article 14261. https://doi.org/10.1038/ncomms14261.

Liu, X., Qin, R., Li, J., Liao, S., Shan, T., Xu, R. et al. (2020). A CRISPR-Cas9-mediated domain-specific base-editing screen enables functional assessment of ACCase variants in rice. Plant Biotechnology Journal 18(8): 1845–1847. https://doi.org/10.1111/pbi.13329.

Liu, H. J., Jian, L., Xu, J., Zhang, Q., Zhang, M., Jin, M. et al. (2020). High-throughput CRISPR/Cas9 mutagenesis streamlines trait gene identification in maize. The Plant Cell 32(5): 1397–1413. https://doi.org/10.1105/tpc.19.00740.

Lopes Filho, J. H., Silva, V. C. H. da, dos Santos, J. C., Dante, R. A., Gerhardt, I. R., Yassitepe, J. E. C. T. et al. (2020). Introdução à edição genômica de plantas: desafios da agricultura moderna para o presente e o futuro. pp. 11–48. *In*: Molinari, H. B. C., Vieira, L. R., Silva, N. V., Prado, G. S. and Lopes Filho, J. H. (eds.). Tecnologia CRISPR na edição genômica de plantas: biotecnologia aplicada à agricultura. Brasília, DF: Embrapa. http://www.alice.cnptia.embrapa.br/alice/handle/doc/1126481.

Ma, X., Zhu, Q., Chen, Y. and Liu, Y. G. (2016). CRISPR/Cas9 platforms for genome editing in plants: developments and applications. Molecular Plant 9(7): 961–974. https://doi.org/10.1016/j.molp.2016.04.003.

Makarova, K. S., Wolf, Y. I., Iranzo, J., Shmakov, S. A., Alkhnbashi, O. S. and Brouns, S. J. J. (2020). Evolutionary classification of CRISPR–Cas systems: a burst of class 2 and derived variants. Nature Reviews Microbiology 18(2): 67–83. https://doi.org/10.1038/s41579-019-0299-x.

Malabarba, J., Chevreau, E., Dousset, N., Veillet, F., Moizan, J. and Vergne, E. (2020). New strategies to overcome present CRISPR/Cas9 limitations in apple and pear: Efficient dechimerization and base editing. International Journal of Molecular Sciences 22(1): 319. https://doi.org/10.3390/ijms22010319.

Malnoy, M., Viola, R., Jung, M. H., Koo, O. J., Kim, S., Kim, J. S. et al. (2016). DNA-free genetically edited grapevine and apple protoplast using CRISPR/Cas9 ribonucleoproteins. Frontiers in Plant Science 7: Article 1904. https://doi.org/10.3389/fpls.2016.01904.

Montecillo, J. A. V., Chu, L. L. and Bae, H. (2020). CRISPR-Cas9 system for plant genome editing: Current approaches and emerging developments. Agronomy 10(7): 1033. https://doi.org/10.3390/agronomy10071033.

Mutezo, W., Sedibe, M. M., Mofokeng, A., Shargie, N. and Soko, T. (2021). The application of CRISPR/Cas9 technology in the management of genetic and nongenetic plant traits. International Journal of Agronomy 2021: Article 9993784. https://doi.org/10.1155/2021/9993784.

Nekrasov, V., Wang, C., Win, J., Lanz, C., Weigel, D. and Kamoun, S. (2017). Rapid generation of a transgene-free powdery mildew resistant tomato by genome deletion. Scientific Reports 7(1): 482. https://doi.org/10.1038/s41598-017-00578-x.

Okada, A., Arndell, T., Borisjuk, N., Sharma, N., WatsonHaigh, N. S., Tucker, E. J. et al. (2019). CRISPR/Cas9-mediated knockout of Ms1 enables the rapid generation of male-sterile hexaploid wheat lines for use in hybrid seed production. Plant Biotechnology Journal 17(10): 1905–1913. https://doi.org/10.1111/pbi.13106.

Orasmo, G. R., Oliveira-Collet, S. A. de, Mangolin, C. A., Lapenta, A. S. and Machado, M. F. P. S. (2015). Esterase isozymes patterns of grape vine (*Vitis vinifera* L.) are altered in response to fungicide exposure. Acta Scientiarum. Biological Sciences 37(4): 463–469. https://doi.org/10.4025/actascibiolsci.v37i4.23400.

Paixao, J., Gillet, F. and Ribeiro, T. (2019). Improved drought stress tolerance in Arabidopsis by CRISPR/dCas9 fusion with a histone acetyl transferase. Scientific Reports 9: 8080. https://doi.org/10.1038/s41598-019-44571-6.

Pompili, V., Dalla Costa, L., Piazza, S., Pindo, M. and Malnoy, M. (2020). Reduced fire blight susceptibility in apple cultivars using a high-efficiency CRISPR/Cas9-FLP/FRT-based gene editing system. Plant Biotechnology Journal 18(3): 845–858. https://doi.org/10.1111/pbi.13208.

Popova, O. V., Dinh, H. Q., Aufsatz, W. and Jonak, C. (2013). The RdDM pathway is required for basal heat tolerance in Arabidopsis. Molecular Plant 6(2): 396–410. https://doi.org/10.1093/mp/sss139.

Rasheed, A., Jie, H., He, P., Lv, X., Ali, B., Ma, Y. et al. (2023). A systematic review on the improvement of Cd stress tolerance in ramie crop, limitations and future prospective. Agronomy 13: 1793. https://doi.org/10.3390/agronomy13081793.

Ren, C., Guo, Y., Kong, J., Lecourieux, F., Dai, Z., Li, S. et al. (2020). Knockout of VvCCD8 gene in grapevine affects shoot branching. BMC Plant Biology 20: 47. https://doi.org/10.1186/s12870-020-2267-7.

Ryder, P., McHale, M., Fort, A. and Spillane, C. (2017). Generation of stable nulliplex autopolyploid lines of Arabidopsis thaliana using CRISPR/Cas9 genome editing. Plant Cell Reports 36(6): 1005–1008. https://doi.org/10.1007/s00299-017-2116-1.

Sable, A., Rai, K. M., Choudhary, A., Yadav, V. K., Agarwal, S. K. and Sawant, S. V. (2018). Inhibition of heat shock proteins HSP90 and HSP70 induce oxidative stress, suppressing cotton fiber development. Scientific Reports 8: 3620. https://doi.org/10.1038/s41598-018-21970-3.

Samanta, S. and Thakur, J. K. (2015). Importance of mediator complex in the regulation and integration of diverse signaling pathways in plants. Frontiers in Plant Science 6: 757. https://doi.org/10.3389/fpls.2015.00757.

Santosh Kumar, V. V., Verma, R. K., Yadav, S. K., Yadav, P., Watts, A., Rao, M. V. et al. (2020). CRISPR-Cas9 mediated genome editing of drought and salt tolerance (OsDST) gene in indica mega rice cultivar MTU1010. Physiology and Molecular Biology of Plants 26(6): 1099–1110. https://doi.org/10.1007/s12298-020-00838-w.

Schaeffer, S. M. and Nakata, P. A. (2015). CRISPR/Cas9-mediated genome editing and gene replacement in plants: Transitioning from lab to field. Plant Science 240: 130–142. https://doi.org/10.1016/j.plantsci.2015.08.012.

Shan, Q., Wang, Y., Li, J., Zhang, Y., Chen, K., Liang, Z. et al. (2013). Targeted genome modification of crop plants using a CRISPR-Cas system. Nature Biotechnology 31(8): 686–688. https://doi.org/10.1038/nbt.2650.

Shan, Q., Wang, Y., Li, J. and Gao, C. (2014). Genome editing in rice and wheat using the CRISPR/Cas system. Nature Protocols 9(10): 2395–2410. https://doi.org/10.1038/nprot.2014.157.

Shen, X., De Jonge, J., Forsberg, S. K. G., Pettersson, M. E., Sheng, Z., Hennig, L. et al. (2014). Natural CMT2 variation is associated with genome-wide methylation changes and temperature seasonality. PLOS Genetics 10(10): e1004842. https://doi.org/10.1371/journal.pgen.1004842.

Shen, C., Que, Z., Xia, Y., Tang, N., Li, D., He, R. et al. (2017). Knockout of the annexin gene OsAnn3 via CRISPR/Cas9-mediated genome editing decreased cold tolerance in rice. Journal of Plant Biology 60(6): 539–547. https://doi.org/10.1007/s12374-017-0470-7.

Shen, Y., Lei, T., Cui, X., Liu, X., Zhou, S., Zheng, Y. et al. (2019). Arabidopsis histone deacetylase HDA15 directly represses plant response to elevated ambient temperature. The Plant Journal 100(5): 991–1006. https://doi.org/10.1111/tpj.14452.

Tran, M. T., Son, G. H., Song, Y. J., Nguyen, N. T., Park, S., Thach, T. V. et al. (2023). CRISPR Cas9-based precise engineering of SlHyPRP1 protein towards multi-stress tolerance in tomato. Frontiers in Plant Science 14: 1186932. https://doi.org/10.3389/fpls.2023.1186932.

Varkonyi-Gasic, E., Wang, T., Voogd, C., Jeon, S., Drummond, R. S. M., Gleave, A. P. et al. (2019). Mutagenesis of kiwifruit CENTRORADIALIS-like genes transforms a climbing woody perennial with long juvenility and axillary flowering into a compact plant with rapid terminal flowering. Plant Biotechnology Journal 17(5): 869–880. https://doi.org/10.1111/pbi.13020.

Villano, C. and Aversano, R. (2020). Towards grapevine (*Vitis vinifera* L.) mildews resistance: Molecular defence mechanisms and new breeding technologies. Italus Hortus 27: 1–17. https://doi.org/10.26353/j.itahort/2020.1.1712.

Wang, S., Zhang, S., Wang, W., Xiong, X., Meng, F. and Cui, X. (2015). Efficient targeted mutagenesis in potato by the CRISPR/Cas9 system. Plant Cell Reports 34(9): 1473–1476. https://doi.org/10.1007/s00299-015-1784-2.

Wang, M., Lu, Y., Botella, J. R., Mao, Y., Hua, K. and Zhu, J.-K. (2017). Gene targeting by homology-directed repair in rice using a geminivirus-based CRISPR/Cas9 system. Molecular Plant 10(7): 1007–1010. https://doi.org/10.1016/j.molp.2017.04.009.

Wang, L., Chen, L., Li, R., Zhao, R., Yang, M., Sheng, J. et al. (2017). Reduced drought tolerance by CRISPR/Cas9-mediated SlMAPK3 mutagenesis in tomato plants. Journal of Agricultural and Food Chemistry 65(37): 8674–8682. https://doi.org/10.1021/acs.jafc.7b03838.

Wang, X., Tu, M., Wang, D., Liu, J., Li, Y., Li, Z. et al. (2018). CRISPR/Cas9-mediated efficient targeted mutagenesis in grape in the first generation. Plant Biotechnology Journal 16(5): 844–855. https://doi.org/10.1111/pbi.12831.

Wang, L., Sun, S., Wu, T., Liu, L., Sun, X., Cai, Y. et al. (2020). Natural variation and CRISPR/Cas9-mediated mutation in GmPRR37 affect photoperiodic flowering and contribute to regional adaptation of soybean. Plant Biotechnology Journal 18(9): 1869–1881. https://doi.org/10.1111/pbi.13348.

Wang, Z., Wong, D. C. J., Wang, Y., Xu, G., Ren, C., Liu, Y. et al. (2021). GRAS-domain transcription factor PAT1 regulates jasmonic acid biosynthesis in grape cold stress response. Plant Physiology 186(4): 1660–1678. https://doi.org/10.1093/plphys/kiab074.

Xie, K. and Yang, Y. (2013). RNA-guided genome editing in plants using a CRISPR–Cas system. Molecular Plant 6(6): 1975–1983. https://doi.org/10.1093/mp/sst119.

Xu, H.-R., Liu, Y., Yu, T.-F., Hou, Z.-H., Zheng, J.-C., Chen, J. et al. (2022). Comprehensive profiling of tubby-like proteins in soybean and roles of the gmtlp8 gene in abiotic stress responses. Frontiers in Plant Science. https://doi.org/10.3389.

Zaidi, S. S.-A. and Mansoor, S. (2017). Viral vectors for plant genome engineering. Frontiers in Plant Science 8: 539. https://doi.org/10.3389/fpls.2017.00539.

Zaidi, S. S.-E.-A., Mahas, A., Vanderschuren, H. and Mahfouz, M. M. (2020). Engineering crops of the future: CRISPR approaches to develop climate-resilient and disease-resistant plants. Genome Biology 21: 1–19. https://doi.org/10.1186/s13059-020-02097-y.

Zhang, Z., Hua, L., Gupta, A., Tricoli, D., Edwards, K. J., Yang, B. et al. (2019). Development of an Agrobacterium-delivered CRISPR/Cas9 system for wheat genome editing. Plant Biotechnology Journal 17(9): 1623–1635. https://doi.org/10.1111/pbi.13112.

Zhang, A., Liu, Y., Wang, F., Li, T., Chen, Z., Kong, D. et al. (2019). Enhanced rice salinity tolerance via CRISPR/Cas9-targeted mutagenesis of the OsRR22 gene. Molecular Breeding 39: 47. https://doi.org/10.1007/s11032-019-0943-8.

Zhang, Y., Zhou, P., Bozorov, T. A. and Zhang, D. (2021). Application of CRISPR/Cas9 technology in wild apple (*Malus sieversii*) for paired sites gene editing. Plant Methods 17: 79. https://doi.org/10.1186/s13007-021-00764-4.

Zhou, S., Jiang, W., Long, F., Cheng, S., Yang, W., Zhao, Y. et al. (2017). Rice homeodomain protein WOX11 recruits a histone acetyltransferase complex to establish programs of cell proliferation of crown root meristem. The Plant Cell 29(5): 1088–1104. https://doi.org/10.1105/tpc.17.00070.

Zhou, J., Wang, G. and Liu, Z. (2018). Efficient genome-editing of wild strawberry genes, vector development, and validation. Plant Biotechnology Journal 16(10): 1868–1877. https://doi.org/10.1111/pbi.12929.

# Chapter 10

# Genome Editing in Cereal Crops Using CRISPR-Cas9 Technology

*Madhvi Sharma,*[1] *Khaja Mohinuddin,*[1] *Priya Shah*[1] and *Amanpreet K. Sidhu*[2,*]

## 1. Introduction

Cereals have been consumed since the beginning of agriculture due to their nutritional value, favorable impact on human health, and relative ease of production. The three most important crop species in the world today are rice (*Oryza sativa*), wheat (*Triticum aestivum*), and maize (*Zea mays*), which together account for more than 42% of the calories consumed by all people (Neupane et al. 2022, http://ricepedia.org/rice-as-food/the-global-staple-rice-consumers). By 2050, there will be 9.8 billion people on the planet, according to a United Nations forecast. However, because of the continued climate change and the scarcity of fertile land, global agricultural production has not increased in recent years, in fact, it has decreased in some areas. In order to feed 9.8 billion people in the current arable lands, agricultural plant yields must be increased by 60% (Springmann et al. 2018). Targeted genome editing has recently advanced, speeding up genetic research and creating new opportunities to enhance crops for higher yields and quality. Any gene of interest can be targeted by genome editing methods such as zinc finger nucleases (ZFN) and transcription activator-like effector nucleases (TALENs). Along with their labor-intensive protein construction methodology and high cost, these systems do have certain limitations. A new generation of genome editing technology includes an easy-to-use tool called Clustered Regularly Interspaced Short Palindromic Repeats (CRISPR)-CRISPR-associated protein9 (Cas9). It is an RNA-based method that involves a straightforward and economical construct construction. CRISPR-Cas9 has revolutionized genome editing, offering precise and efficient modifications in plant genomes, including major cereal crops like rice, wheat, and maize. This powerful tool enables targeted gene modifications, facilitating trait improvement for higher yields, stress tolerance, and disease resistance (Hillary et al. 2019). Unlike traditional breeding or transgenic approaches, CRISPR-Cas9 allows precise edits without introducing foreign DNA, making it a more acceptable and sustainable technology. In cereals, CRISPR-Cas9 has been successfully applied to enhance drought and salinity tolerance, improve grain quality, and increase resistance to fungal, bacterial, and viral diseases. The technology has

[1] Center of Excellence in Genomics & Systems Biology (CEGSB), and Center for Pre-Breeding Research (CPBR), International Crop Research Institute for the Semi-Arid Tropics, (ICRISAT), Patancheru-502324, Hyderabad, India.
[2] Department of Sciences, Khalsa College for Women, G.T. Road, Amritsar, 143002, India.
* Corresponding author: aman.preet1807@gmail.com

DOI: 10.1201/9781003659907-13

also been used to modify genes controlling flowering time, nutrient content, and plant architecture, directly benefiting food security and agricultural sustainability. Despite its potential, challenges such as off-target effects, regulatory concerns, and genotype-dependent efficiency remain. Ongoing research aims to refine the technology, improve delivery systems, and enhance editing precision. As regulatory frameworks evolve, CRISPR-Cas9 is poised to play a crucial role in accelerating cereal crop improvement, addressing global food demands, and mitigating climate change impacts in agriculture. The CRISPR/Cas9 technology has been effectively used to modify a number of cereal crops, including rice, wheat, maize, and barley (Lee et al. 2021, Liu et al. 2021, Raffan et al. 2021). Additionally, this technique could be used to enhance these species' yield-related phenotypic features. For instance, altering the genes for barley for row-type or tillering could boost wheat output (Liu et al. 2017, Zhang et al. 2018a, Zhang et al. 2018b, Zhou et al. 2019). The mechanism of CRISPR/Cas9-mediated GE has been demonstrated by numerous published studies, recently summarized in our earlier works (Ahmar et al. 2020, Ahmar et al. 2021, Ahmar et al. 2022). Moreover, CRISPR is far easier and less expensive than previous gene-editing methods like zinc finger nucleases (ZFNs) and transcription activator-like effector nucleases (TALENs). Compared to protein-based targeting systems, it is simpler to develop and execute because it uses a guide RNA to precisely target the Cas9 protein in the genome. Additionally, multiplexing—the simultaneous editing of many genes—is made possible by CRISPR, whereas ZFNs and TALENs are technically challenging. CRISPR-Cas9 is the perfect technology for quick, precise, and affordable genetic modification in agriculture because of these benefits.

For instance, rice serves as a model plant for CRISPR/Cas functional gene investigations (Feng et al. 2013, Xu et al. 2016). The creation of rice mutant library collections has become simpler since the advent of CRISPR/Cas9 technology. Since the full rice genome has been sequenced, analyzing and determining the roles of each predicted gene in the genome is currently the main concern (Bao et al. 2019). For the past five years, scientists have produced a sizable, superior CRISPR/Cas9 library of rice mutants in order to study gene functions. More rice mutant libraries are anticipated to be created shortly, which might be crucial for determining gene function and accelerating the genetic advancement of rice (Meng et al. 2017). In the first report on the successful use of the CRISPR/Cas9 technique in rice, the OsPDS and OsBADH2 genes were edited. Nonetheless, only a small percentage of mutant plants were reported: 7.1% (from 98 transgenic plants) for OsBADH2 and 9.4% (from 96 transgenic plants overall) for OsPDS (Banakar et al. 2020).

Remarkably, the mildew-resistance locus (MLO) gene in wheat has been successfully knocked out by the CRISPR/Cas9 system, leading to a mutation frequency of 28.5% and 26.5–38.0% (Shan et al. 2013). The CRISPR/Cas9 system is a promising tool for increasing desirable features in wheat, according to the first research of knockout mutation of MLO in wheat. Moreover, the maize genome has been successfully altered using the CRISPR/Cas9 technique. Three hundred and forty-eight studies about maize and CRISPR/Cas9 have been published so far, according to the Web of Science (Shi et al. 2017). In 2014, Liang et al. published the first study on the use of CRISPR/Cas9 in maize. Therefore, CRISPR and other emerging GE techniques show promise as a workable way to expedite the creation of better cultivars of this allotetraploid plant. Quinoa seeds have many healthy qualities that make it a nutrient-dense food, including being gluten-free, having a high protein, fat, and carbohydrate ratio, and having more essential amino acids (Pereira et al. 2024). Therefore, to boost cereal production, all of the new GE techniques should be used in the less studied crops.

## 2. Mechanism of CRISPR-Cas9 in Cereal Crops

The CRISPR-Cas9 genome editing system has transformed plant biotechnology by allowing for precise, efficient, and targeted changes in the genomes of various crop species. In cereal crops such as rice, wheat, maize, and barley, CRISPR-Cas9 provides unprecedented opportunities for improving agronomic traits, increasing stress tolerance, and speeding up breeding programs (Razzaq et al. 2019). Its mechanism is based on a simple but powerful idea: create a controlled break in the DNA at a specific location, which the plant's own repair systems then modify in a predictable way. This section delves into the key components of the CRISPR-Cas9 system, the cellular processes involved in gene targeting and repair, and the overall efficiency and precision of this tool for modifying cereal crop genomes.

## *2.1 Components of the CRISPR-Cas9 System*

The CRISPR-Cas9 genome editing system consists of two main components: the Cas9 endonuclease and the guide RNA (gRNA). The Cas9 protein, derived from Streptococcus pyogenes, is an RNA-guided DNA endonuclease that cleaves double-stranded DNA at specific sites (Nadakuduti and Enciso-Rodríguez 2021). For successful binding and cleavage, Cas9 requires a short DNA sequence known as the protospacer adjacent motif (PAM), commonly abbreviated as "NGG", located immediately downstream of the target site (Nip 2019)

The guide RNA is a synthetic fusion of two naturally occurring RNAs: CRISPR RNA (crRNA), which provides specificity by binding to the target DNA and trans-activating crRNA (tracrRNA), which promotes Cas9 binding. They work together to form a single guide RNA (sgRNA) in engineered systems (Filippova et al. 2019). The sgRNA directs the Cas9 enzyme to a specific genomic location via complementary base pairing with a 20-nucleotide target sequence. This specificity is critical to the precision of genome editing with CRISPR-Cas9 (Tsai and Joung 2016).

Cas9 and gRNA components can be delivered in cereal crops using a variety of methods, including Agrobacterium-mediated transformation, biolistics (gene gun), and transient expression systems. These delivery methods influence efficiency, expression level, and whether modifications are stably inherited (Munir et al. 2023).

## *2.2 Process of Gene Targeting, Double-strand Breaks, and Repair Mechanisms*

Once the Cas9-sgRNA complex binds to the DNA target site, the Cas9 enzyme causes a site-specific double-strand break (DSB). This break is a critical event that activates the plant cell's natural DNA repair mechanisms, which ultimately determine the outcome of the genome editing process (Liu et al. 2022).

The two main DNA repair mechanisms include Non-homologous End Joining (NHEJ) which is the primary repair mechanism in plant cells, including cereal crops. NHEJ repairs DSBs by directly ligating the broken DNA ends, eliminating the need for a homologous template. However, this process is error-prone and frequently causes insertions or deletions (indels) at the repair site. These indels can cause frameshift mutations that result in gene knockout or loss of function. This feature makes NHEJ particularly useful for gene inactivation and functional genomics research in cereals (Gehrke et al. 2022, Shen and Li 2022).

On the other hand, Homology-Directed Repair (HDR) is a high-fidelity repair pathway that uses a homologous DNA template to accurately repair breaks. Researchers can use an exogenous donor template with the desired genetic change flanked by sequences homologous to the break site. HDR allows for precise gene editing, such as base substitution or gene insertion. However, HDR is relatively inefficient in plant systems, particularly in non-dividing cells, which limits its use in cereal crops. Current research focuses on strategies to improve HDR efficiency, such as synchronizing cell cycle stages or suppressing NHEJ components (Haider and Mussolino 2025).

## *2.3 Efficiency and Precision in Cereal Crop Genome Modifications*

CRISPR-Cas9 is a more efficient and precise tool for modifying cereal crop genomes than previous genome editing platforms such as ZFNs and TALENs (Shamshirgaran et al. 2022). The high specificity of sgRNA binding minimizes off-target effects, especially when combined with improved Cas9 variants or high-fidelity Cas9 enzymes. The ability to design gRNAs quickly and cost-effectively adds to the system's versatility (Kleinstiver et al. 2016).

The CRISPR-Cas9 system has been used successfully to modify key agronomic traits in cereal crops including rice, wheat, maize, barley, and sorghum. These include improvements in abiotic stress tolerance (e.g., drought, salinity, and temperature extremes), biotic stress resistance (e.g., fungal, bacterial, and viral diseases), nutritional quality (e.g., increased grain protein content or micronutrient levels), and yield-related traits (e.g., tiller number, panicle architecture, grain size) (Fiaz et al. 2019, Zafar et al. 2020, Zhang et al. 2021, Li et al. 2022, Sharma et al. 2024).

Furthermore, transgene-free genome editing can be achieved by delivering pre-assembled Cas9-gRNA ribonucleoproteins (RNPs) or transient expression systems. This method prevents the incorporation of foreign DNA into the plant genome, which is beneficial for regulatory approval and consumer acceptance, particularly for food crops (Poddar et al. 2023). Multiplex genome editing, or targeting multiple genes at the same time with multiple sgRNAs, is also possible and has been demonstrated in several cereal species. This allows for simultaneous manipulation of gene families or pathways, which accelerates trait stacking and complex trait improvement (Minkenberg et al. 2017).

## 3. Applications of CRISPR-Cas9 in Cereal Crop Improvement

CRISPR-Cas9 has emerged as a transformative tool in cereal crop improvement, enabling precise genome editing to enhance yield, stress tolerance, disease resistance, and nutritional quality. Its application in major cereals such as rice, wheat, and maize has accelerated the development of improved varieties by allowing targeted modifications of key genes associated with important agronomic traits. By overcoming the limitations of conventional breeding and earlier molecular tools, CRISPR-Cas9 offers a powerful and efficient approach to meet the growing demands of food security and climate-resilient agriculture. The broad applications of CRISPR-Cas9 in cereal crops—including abiotic and biotic stress tolerance, yield enhancement, nutritional improvement, and resistance to pests and herbicides—are illustrated in Figure 1.

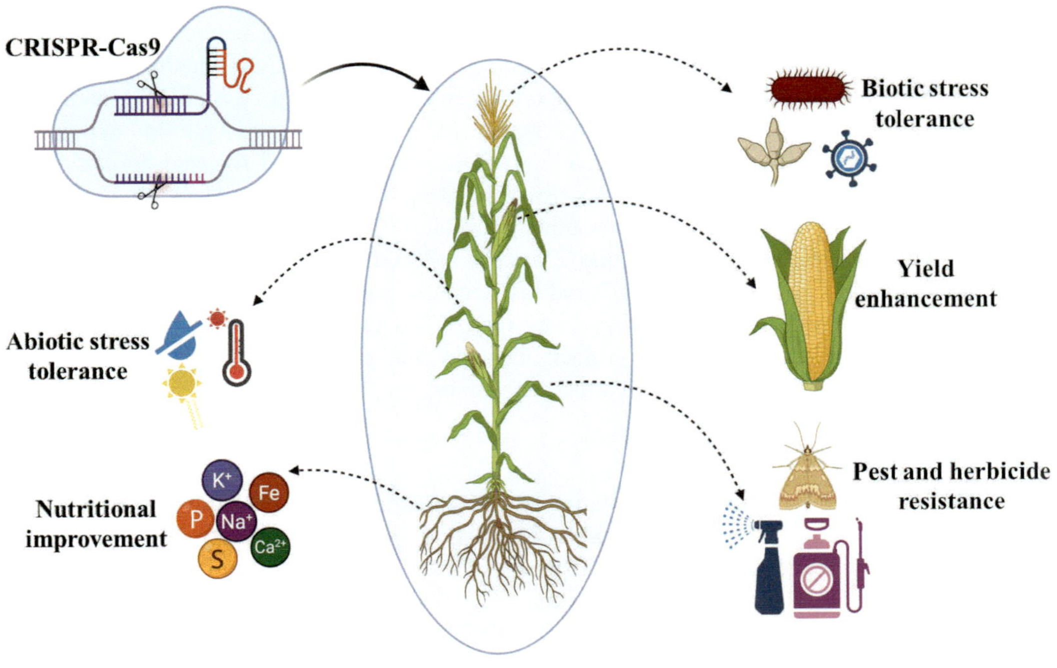


**Figure 1.** Applications of CRISPR-Cas9 gene editing technology for targeted improvement: Enhancing abiotic and biotic stress tolerance, nutritional quality, yield, and pest/herbicide resistance.

### *3.1 Abiotic Stress Tolerance*

CRISPR/Cas9 technology has enabled significant advances in crop tolerance to abiotic stresses such as drought, salinity, and extreme temperatures. In rice, editing the OsRR22 gene, which encodes a cytokinin response regulator, has improved salt tolerance by modulating hormone signaling (Zhang et al. 2019a). Modifications to OsSAPK2, a gene involved in ABA signaling, have enhanced drought resistance by activating stress-responsive genes (Lou et al. 2023). The OsPUB7 gene, encoding a U-box E3 ligase, is implicated in stress-induced protein degradation during drought (Kim et al. 2023), while OsTPP3 regulates trehalose metabolism, aiding osmoprotection under salt stress (Ye et al. 2023). OsCAT2, which encodes catalase, helps mitigate oxidative stress by detoxifying hydrogen peroxide. Genes like OsDEP1 and OsROCs are associated with heat stress tolerance via protein folding and developmental regulation. The transcription factor bHLH57 contributes to cold resistance and also boosts yield by influencing flowering time. In wheat, editing TaRR12, a type-B response regulator, has been linked to improved drought resistance (Li et al. 2024a). The TaHKT1;5 gene, encoding a sodium transporter, is crucial for maintaining ion homeostasis under saline conditions. TaPGK, a phosphoglycerate kinase, supports cold stress adaptation through energy metabolism. In maize, ZmHDT103, a histone deacetylase, modulates ABA signaling to regulate drought-responsive gene expression (Wang et al. 2024), while ZmG6PDH1 enhances the antioxidant system under cold stress. In barley, HvGSK1.1, a GSK-3-like kinase, modulates stress signal transduction under salinity. In potato, the StDRO2 gene supports drought tolerance by improving root development, and in soybean, GmNHL1 enhances salt resistance via a non-race-specific disease resistance-like mechanism.

### *3.2 Biotic Stress Resistance*

CRISPR/Cas9 has strengthened plant defenses against a variety of biotic agents. In rice, editing OsMPK5, a MAP kinase, has improved resistance to fungal pathogens by regulating downstream resistance genes (Ma et al. 2015). The blast resistance genes Pi-d2 and Pi21 have been edited to increase resistance through R gene-mediated signaling. OsCPR5.1, homologous to a nuclear pore complex protein, broad-spectrum viral resistance by regulating immune responses. OsPUB9, a ubiquitin ligase, was modified to improve resistance to bacterial leaf blight by modulating defense protein stability. OsSULTR3;6, a sulfate transporter, contributes to blast resistance by regulating sulfur-mediated defense. In wheat, knockout of the TaMLO gene disrupts susceptibility to powdery mildew by impairing host-pathogen interaction. TaEDR1 and TaCIPK14 regulate hormonal and calcium signaling to enhance resistance to stripe rust. In maize, editing ZmAGO18b, part of the Argonaute family, improves resistance to southern leaf blight. The Cry1F gene, derived from Bt, was introduced for pest resistance. Editing Zmpdrp1 reduced susceptibility to viral infections. In potatoes, targeting viral genomes conferred resistance to PVY, PVX, and PLRV. In soybean, Glyma05g29080 is linked to fungal defense, and GmUGT, a UDP-glycosyltransferase, enhances pest resistance by modifying secondary metabolites. In rice, OsWRKY71 and Bph15 regulate hormone and defense signaling, conferring resistance to brown planthopper.

### *3.3 Yield and Quality Improvement*

CRISPR has been used to improve yield by targeting genes associated with plant growth, architecture, and nutrient utilization. In rice, editing OsSPL14 has led to ideal plant architecture with increased tillering and panicle branching (Hua et al. 2018). OsGS2 and GRF4 regulate nitrogen assimilation and growth, while OsAPL enhances carbohydrate metabolism for yield improvement. In wheat, TaGW2 is a negative regulator of grain size, and TaRPK1 influences spike traits, both contributing to yield. In maize, RZ2MS9 regulates growth hormones for biomass accumulation. In barley, GW2.1 is associated with grain filling and seed setting. For quality traits, barley's Hina gene modifies starch granule size, and FtsZ1 in potato is involved in plastid division, affecting tuber quality.

### 3.4 Nutritional Enhancement

Nutritional improvements have been achieved by editing genes that enhance micronutrient accumulation and macronutrient composition. In rice, editing OsSBEIIb alters amylopectin synthesis, improving starch digestibility (Li et al. 2017). OsPDS influences carotenoid biosynthesis, and OsNAS2, a nicotianamine synthase gene, increases zinc content in grains. OsAUX5 and OsWRKY78 regulate amino acid metabolism, enhancing protein quality. In maize, PSY1, Crtl, and LCYB have been targeted to increase pro-vitamin A (β-carotene) through the carotenoid biosynthetic pathway. In soybean, editing GmIPK1 reduces phytic acid, improving mineral bioavailability.

### 3.5 Herbicide and Pest Resistance

CRISPR-mediated herbicide resistance is achieved by modifying key herbicide target genes. In rice, OsACC-T1 and OsALS were edited for tolerance to herbicides like imazamox and bispyribac-sodium (Yu et al. 2024, Zafar et al. 2020). In maize, editing EPSPS, ALS, and ACCase genes allowed plants to survive glyphosate and other common herbicides by altering the herbicide binding sites. The applications of CRISPR/Cas are given the table:

**Table 1.** CRISPR/Cas9 applications in cereal crop improvement.

| Category | Crop | Target Gene | Application/Improvement | Reference |
|---|---|---|---|---|
| Abiotic Stress | Rice | *OsRR22* | Salt tolerance | Zhang et al. (2019a) |
| Abiotic Stress | Rice | *OsSAPK2* | Drought tolerance | Liu et al. (2019) |
| Abiotic Stress | Maize | *ARGOS8* | Drought tolerance | Shi et al. (2017) |
| Abiotic Stress | Maize | *ZmHDT103* | Enhanced drought resistance | Wang et al. 2024 |
| Abiotic Stress | Wheat | *TaRR12* | Drought resistance | Li et al. 2024a |
| Abiotic Stress | Rice | *OsPUB7* | Drought resistance | Kim et al. 2023 |
| Abiotic Stress | Rice | *OsDEP1, OsROCs* | Heat resistance | Wanikar et al. (2025) |
| Abiotic Stress | Soybean | *GmNHL1* | Salt tolerance | Liu et al. 2023 |
| Abiotic Stress | Barley | *HvGSK1.1* | Salt tolerance | Kloc et al. 2024 |
| Abiotic Stress | Rice | *OsTPP3* | Salt resistance | Ye et al. 2023 |
| Abiotic Stress | Rice | *bHLH57* | Cold resistance and yield | Zhang et al. 2023 |
| Biotic Stress | Wheat | *TaMLO* | Powdery mildew resistance | Shan et al. (2014) |
| Biotic Stress | Rice | *OsMPK5* | Disease resistance | Ma et al. (2015) |
| Biotic Stress | Rice | *Pi-d2* | Blast resistance | Ren et al. (2018) |
| Fungal | Wheat | *TaEDR1* | Powdery mildew resistance | Char and Yang (2020) |
| Bacterial | Rice | *OsCERK1* | Defense response | Ren et al. (2018) |
| Virus | Barley | *WDV genes* | Virus resistance | Kis et al. (2019) |
| Pest | Maize | *ZmGL2* | Epicuticular wax (pest resistance) | Char et al. (2015) |
| Biotic Stress – Virus | Maize | *Zmpdrp1* | Reduced virus impact | Xie et al. 2024 |
| Biotic Stress – Virus | Rice | *OsCPR5.1* | Yellow mottle virus resistance | Arra et al. 2024 |
| Biotic Stress - Bacteria | Rice | *OsPUB9* | Bacterial leaf blight resistance | Kim et al. 2024 |
| Biotic Stress – Fungi | Soybean | *Glyma05g29080* | White mold resistance | Zhang et al. 2023 |
| Biotic Stress – Fungi | Wheat | *TaCIPK14* | Stripe rust resistance | He et al. 2023 |
| Biotic Stress – Fungi | Maize | *ZmAGO18b* | Southern leaf blight resistance | Dai et al. 2023 |
| Biotic Stress –Fungi | Rice | *Pi21, OsSULTR3;6* | Blast resistance | Yang et al. 2023 |
| Biotic Stress – Pest | Maize | *Cry1F* | Improved pest resistance | Kumari et al. 2024 |

*Table 1 contd. ...*

*...Table 1 contd.*

| Category | Crop | Target Gene | Application/Improvement | Reference |
|---|---|---|---|---|
| Biotic Stress – Pest | Rice | *OsWRKY71, Bph15* | Brown plant hopper resistance | Li et al. 2023 |
| Nutritional | Rice | *OsSBEIIb* | Starch quality | Li et al. (2017) |
| Nutritional | Barley | *HvPAPhy-a* | Phytase activity | Wendt et al. (2013) |
| Nutritional | Rice | *OsNAS2* | Increased Zn uptake | Ludwig et al. 2024 |
| Nutritional | Maize | *PSY1, Crtl, LCYB* | Increased pro-vitamin A | Sobrino-Mengual et al. 2024 |
| Nutritional | Soybean | *GmIPK1* | Reduced phytic acid | Song et al. 2022 |
| Yield & Quality | Rice | *OsSPL14* | Grain yield | Hua et al. (2018) |
| Yield & Quality | Wheat | *TaGW2* | Grain weight | Li et al. (2020) |
| Yield & Quality | Rice | *OsAPL* | Increased yield | Zhang et al. 2024 |
| Yield & Quality | Rice | *OsGS2/GRF4* | Increased grain size and yield | Wang et al. 2022 |
| Herbicide | Maize | *ZmIPK1* | Herbicide tolerance | Shukla et al. (2009) |
| Herbicide | Rice | *OsACC-T1* | Herbicide resistance | Yu et al. 2024 |
| Herbicide | Rice | *OsALS* | Herbicide resistance (imazamox) | Zafar et al. 2023 |
| Herbicide | Maize | *EPSPS, ALS, ACCase* | Herbicide resistance | Sang et al. 2024 |

## 4. Challenges and Future Perspectives

CRISPR/Cas9 gene editing can increase cereal yield, nutritional value, pest and disease resistance, and tolerance to a range of abiotic stressors. Before cereal farmers can fully utilize the promise of this technology, a number of obstacles must be addressed. In addition, CRISPR can modify existing elite kinds, carry out intricate editing, and produce new genetic variations. A variety of genes have recently been knocked out and activated using MGE systems, which may result in better multigenic features including disease resistance without any growth penalties (Zheng et al. 2024). The existing green revolution varieties were transformed by Song et al. (2023) into more contemporary ones created using an MGE method (Ahmar and Gruszka 2023). However, the editing efficiency of CRISPR-mediated T-DNA-free cereals remains a significant hurdle. Optimizing Cas9 transport into cells, employing numerous sgRNAs, and altering DNA with BE without causing a typical double break could all lead to better editing efficiency.

Another major issue facing GE in cereals is regulatory and legal barriers. Regulation of GE crops is a dynamic and complicated topic. It is not possible to create a universally applicable regulatory framework for these products because every nation regulates them differently. Strict laws governing genetically modified foods are required, particularly those about their manufacture and international distribution. In Japan, for instance, consumers may already buy GABA tomatoes, but they still need to be handled carefully and safely. This technology was first approved in China in April 2023 when a gene-edited soybean was approved for commercial use (https://www.reuters.com/science/china-approves-safety-first-gene-edited-crop-2023-05-04/). CRISPR-edited crops are largely deregulated in the US if they mimic natural mutations, while in the European Commission (EC), they are still regulated as GMOs, though reforms are underway to exempt certain gene edits that resemble conventional breeding (Ahmad et al. 2021). Furthermore, one of the main nations that carries out field tests of plants with altered genomes is China.

There are still a lot of barriers to overcome before CRISPR-enhanced grains may be sold. To overcome this obstacle, more details regarding the effective use of this technology and its advantages for the plant breeding community must be made available. Even though CRISPR/Cas can generate extremely specific and targeted mutations in the genome, it can also have off-target effects, like adding more, occasionally unfavorable, modifications to the genome (Guo et al. 2023). Numerous factors, including the quantity and locations of mismatches between the sgRNA and the targeted DNA, the GC levels, and the delivery technique, can result in off-target consequences. Nevertheless, little is currently known about the genomes of these grains, which makes it difficult to pinpoint the precise genotypes required to produce the desired characteristics.

## 5. Conclusion

Cereal genomes may be precisely modified to provide desired agronomic features thanks to the CRISPR/Cas9 GE technologies. Scientists have already produced rice, wheat, maize, and barley mutants with improved features, including higher yield, increased nutritional value, improved disease resistance, and tolerance to abiotic environmental challenges, using CRISPR/Cas9 alone or in conjunction with other cutting-edge technologies. Numerous minor cereal crops, such as rye, quinoa, millet, and sorghum, are growing in popularity as daily food sources, and their output could be greatly increased through gene editing with CRISPR/Cas9. Various researchers are now working to optimize the CRISPR/Cas9 methods for minor cereals, as we outlined in the section of our review above. Nevertheless, little is currently known about the genomes of these grains, which makes it difficult to pinpoint the precise genotypes required to produce the desired characteristics. In subsequent breeding initiatives, a large number of the mutants created in these cereal crops might be regarded as promising substitutes. Additionally, we propose that these GE methods might be used in conjunction with innovative breeding approaches to expedite the development of new cereal types.

## References


Ahmad, A., Ghouri, M. Z., Munawar, N., Ismail, M., Ashraf, S. and Aftab, S. O. (2021). Regulatory, ethical, and social aspects of CRISPR crops. CRISPR Crops: The Future of Food Security 261–287.

Ahmar, S., Saeed, S., Khan, M. H. U., Ullah Khan, S., Mora-Poblete, F., Kamran, M. et al. (2020). A revolution toward gene-editing technology and its application to crop improvement. International Journal of Molecular Sciences 21(16): 5665.

Ahmar, S., Ballesta, P., Ali, M. and Mora-Poblete, F. (2021). Achievements and challenges of genomics-assisted breeding in forest trees: From marker-assisted selection to genome editing. International Journal of Molecular Sciences 22(19): 10583.

Ahmar, S., Zhai, Y., Huang, H., Yu, K., Khan, M. H. U., Shahid, M. et al. (2022). Development of mutants with varying flowering times by targeted editing of multiple SVP gene copies in *Brassica napus* L. The Crop Journal 10(1): 67–74.

Ahmar, S. and Gruszka, D. (2023). CRISPR/Cas9 boosts wheat yield by reducing brassinosteroidsignaling. Trends in Biochemical Sciences 48(11): 917–919.

Ansari, W. A., Chandanshive, S. U., Bhatt, V., Nadaf, A. B., Vats, S., Katara, J. L. et al. (2020). Genome editing in cereals: approaches, applications and challenges. International Journal of Molecular Sciences 21(11): 4040.

Arra, Y., Auguy, F., Stiebner, M., Chéron, S., Wudick, M. M., Miras, M. et al. (2024). Rice Yellow Mottle Virus resistance by genome editing of the *Oryza sativa* L. ssp. japonica nucleoporin gene OsCPR5. 1 but not OsCPR5. 2. Plant Biotechnology Journal 22(5): 1299–1311.

Banakar, R., Schubert, M., Collingwood, M., Vakulskas, C., Eggenberger, A. L. and Wang, K. (2020). Comparison of CRISPR-Cas9/Cas12a ribonucleoprotein complexes for genome editing efficiency in the rice phytoene desaturase (OsPDS) gene. Rice 13(1): 4.

Bao, A., Burritt, D. J., Chen, H., Zhou, X., Cao, D. and Tran, L. S. P. (2019). The CRISPR/Cas9 system and its applications in crop genome editing. Critical Reviews in Biotechnology 39(3): 321–336.

Char, S. N. and Yang, B. (2020). Genome editing in grass plants. aBIOTECH 1(1): 41–57. doi:10.1007/s42994-019-00005-x.

Char, S. N., Unger-Wallace, E., Frame, B., Briggs, S. A., Main, M., Spalding, M. H. et al. (2015). Heritable site-specific mutagenesis using TALEN s in maize. Plant Biotechnol. J. 13(7): 1002–1010. doi:10.1111/pbi.12344.

Dai, Z., Yang, Q., Chen, D., Li, B., Que, J., Hu, L. et al. (2023). ZmAGO18b negatively regulates maize resistance against southern leaf blight. Theoretical and Applied Genetics 136(7): 158.

Feng, Z., Zhang, B., Ding, W., Liu, X., Yang, D. L., Wei, P. et al. (2013). Efficient genome editing in plants using a CRISPR/Cas system. Cell Research 23(10): 1229–1232.

Fiaz, S., Ahmad, S., Noor, M. A., Wang, X., Younas, A., Riaz, A. et al. (2019). Applications of the CRISPR/Cas9 system for rice grain quality improvement: perspectives and opportunities. International Journal of Molecular Sciences 20(4): 888.

Filippova, J., Matveeva, A., Zhuravlev, E. and Stepanov, G. (2019). Guide RNA modification as a way to improve CRISPR/Cas9-based genome-editing systems. Biochimie 167: 49–60.

Gehrke, F., Schindele, A. and Puchta, H. (2022). Nonhomologous end joining as key to CRISPR/Cas-mediated plant chromosome engineering. Plant Physiology 188(4): 1769–1779.

Guo, C., Ma, X., Gao, F. and Guo, Y. (2023). Off-target effects in CRISPR/Cas9 gene editing. Frontiers in Bioengineering and Biotechnology 11: 1143157.

Haider, S. and Mussolino, C. (2025). Fine-tuning homology-directed repair (HDR) for precision genome editing: current strategies and future directions. International Journal of Molecular Sciences 26(9): 4067.

Hillary, V. E. and Ceasar, S. A. (2019). Application of CRISPR/Cas9 genome editing system in cereal crops. The Open Biotechnology Journal 13(1).

Hua, K., Tao, X., Yuan, F., Wang, D. and Zhu, J. K. (2018). Precise A·TtoG·C base editing in the rice genome. Mol. Plant 11(4): 627–630. doi:10.1016/j.molp. 2018.02.007.

Kim, M.-S., Ko, S.-R., Jung, Y. J., Kang, K.-K., Lee, Y.-J. and Cho, Y.-G. (2023). Knockout mutants of OsPUB7 generated using CRISPR/Cas9 revealed abiotic stress tolerance in rice. Int. J. Mol. Sci. 24: 5338. doi: 10.3390/ijms24065338.

Kim, M. S., Le, V. T., Jung, Y. J., Kang, K. K. and Cho, Y. G. (2024). OsPUB9 gene edited by CRISPR/Cas9 enhanced resistance to bacterial leaf blight in rice (*Oryza sativa* L.). International Journal of Molecular Sciences 25(13): 7145.

Kleinstiver, B. P., Pattanayak, V., Prew, M. S., Tsai, S. Q., Nguyen, N. T., Zheng, Z. et al. (2016). High-fidelity CRISPR–Cas9 nucleases with no detectable genome-wide off-target effects. Nature 529(7587): 490–495.

Kloc, Y., Dmochowska-Boguta, M., Żebrowska-Różańska, P., Łaczmański, Ł., Nadolska-Orczyk, A. and Orczyk, W. (2024). HvGSK1. 1 controls salt tolerance and yield through the brassinosteroid signaling pathway in barley. International Journal of Molecular Sciences 25(2): 998.

Kumari, R., Saha, T., Kumar, P. and Singh, A. K. (2024). CRISPR/Cas9-mediated genome editing technique to control fall armyworm (*Spodoptera frugiperda*) in crop plants with special reference to maize. Physiology and Molecular Biology of Plants 30(7): 1161–1173.

Lee, J. H., Won, H. J., Hoang Nguyen Tran, P., Lee, S. M., Kim, H. Y. and Jung, J. H. (2021). Improving lignocellulosic biofuel production by CRISPR/Cas9 mediated lignin modification in barley. GCB Bioenergy 13(4): 742–752.

Li, H., Yang, Y., Hong, W., Huang, M., Wu, M. and Zhao, X. (2020). Applications of genome editing technology in the targeted therapy of human diseases: Mechanisms, advances and prospects. Signal Transduct. Target. Ther. 5(1): 1–23. doi:10.1038/s41392-019-0089-y.

Li, J., Sun, Y., Du, J., Zhao, Y. and Xia, L. (2017). Generation of targeted point mutations in rice by a modified CRISPR/Cas9 system. Mol. Plant 10(3): 526–529. doi:10.1016/j.molp.2016.12.001.

Li, S., Lin, D., Zhang, Y., Deng, M., Chen, Y., Lv, B. et al. (2022). Genome-edited powdery mildew resistance in wheat without growth penalties. Nature 602(7897): 455–460.

Li, X., Zhang, J., Shangguan, X., Yin, J., Zhu, L., Hu, J. et al. (2023). Knockout of OsWRKY71 impairs Bph15-mediated resistance against brown planthopper in rice. Frontiers in Plant Science 14: 1260526.

Li, S., Zhang, Y., Liu, Y., Zhang, P., Wang, X., Chen, B. et al. (2024a). The E3 ligase TaGW2 mediates transcription factor TaARR12 degradation to promote drought resistance in wheat. Plant Cell 36: 605–625. doi: 10.1093/plcell/koad307.

Li, Y., Wu, X., Zhang, Y. and Zhang, Q. (2022). CRISPR/Cas genome editing improves abiotic and biotic stress tolerance of crops. Frontiers in Genome Editing 4: 987817.

Liang, Z., Zhang, K., Chen, K. and Gao, C. (2014). Targeted mutagenesis in *Zea mays* using TALENs and the CRISPR/Cas system. Journal of Genetics and Genomics 41(2): 63–68.

Liu, G., Li, J. and Godwin, I. D. (2019). Genome editing by CRISPR/Cas9 in sorghum through biolistic bombardment. pp. 169–183. In Sorghum (New York, NY: Humana Press).

Liu, J., Chen, J., Zheng, X., Wu, F., Lin, Q., Heng, Y. et al. (2017). GW5 acts in the brassinosteroid signalling pathway to regulate grain width and weight in rice. Nature Plants 3(5): 1–7.

Liu, L., Gallagher, J., Arevalo, E. D., Chen, R., Skopelitis, T., Wu, Q. et al. (2021). Enhancing grain-yield-related traits by CRISPR–Cas9 promoter editing of maize CLE genes. Nature Plants 7(3): 287–294.

Liu, S. C., Feng, Y. L., Sun, X. N., Chen, R. D., Liu, Q., Xiao, J. J. et al. (2022). Target residence of Cas9-sgRNA influences DNA double-strand break repair pathway choices in CRISPR/Cas9 genome editing. Genome Biology 23(1): 165.

Liu, L., Wang, J., Zhang, Q., Sun, T. and Wang, P. (2023). Cloning of the soybean GmNHL1 gene and functional analysis under salt stress. Plants 12(22): 3869.

Lou, D., Lu, S., Chen, Z., Lin, Y., Yu, D. and Yang, X. (2023). Molecular characterization reveals that OsSAPK3 improves drought tolerance and grain yield in rice. BMC Plant Biology 23(1): 53.

Ludwig, Y., Dueñas Jr, C., Arcillas, E., Macalalad-Cabral, R. J., Kohli, A., Reinke, R. et al. (2024). CRISPR-mediated promoter editing of a cis-regulatory element of OsNAS2 increases Zn uptake/translocation and plant yield in rice. Frontiers in Genome Editing 5: 1308228.

Ma, X., Zhang, Q., Zhu, Q., Liu, W., Chen, Y., Qiu, R. et al. (2015). A robust CRISPR/Cas9 system for convenient, high-efficiency multiplex genome editing in monocot and dicot plants. CRISPR/Cas9 system for convenient, high-efficiency multiplex genome editing in monocotand dicot plants. Mol. Plant 8(8): 1274–1284. doi:10.1016/j.molp.2015.04.007.

Meng, X., Yu, H., Zhang, Y., Zhuang, F., Song, X., Gao, S. et al. (2017). Construction of a genome-wide mutant library in rice using CRISPR/Cas9. Molecular Plant 10(9): 1238–1241.

Minkenberg, B., Wheatley, M. and Yang, Y. (2017). CRISPR/Cas9-enabled multiplex genome editing and its application. Progress in Molecular Biology and Translational Science 149: 111–132.

Munir, F., Azeem, S., Amir, R., Gillani, M., Gul, A., Mustafa, A. et al. (2023). Genetic transformation methods in cereal crops. pp. 273–294. In Cereal Crops. CRC Press.

Nadakuduti, S. S. and Enciso-Rodríguez, F. (2021). Advances in genome editing with CRISPR systems and transformation technologies for plant DNA manipulation. Frontiers in Plant Science 11: 637159.

Neupane, D., Adhikari, P., Bhattarai, D., Rana, B., Ahmed, Z., Sharma, U. et al. (2022). Does climate change affect the yield of the top three cereals and food security in the world? Earth 3(1): 45–71.

Nip, L. (2019). Mechanistic insight on a chimeric Cas9 protein's specificity for DNA target with 5'-NAA-3'PAM (Doctoral dissertation, Massachusetts Institute of Technology).

Pereira, L., Cotas, J. and Gonçalves, A. M. (2024). Seaweed proteins: A step towards sustainability?. Nutrients 16(8): 1123.

Poddar, S., Tanaka, J., Running, K. L., Kariyawasam, G. K., Faris, J. D., Friesen, T. L. et al. (2023). Optimization of highly efficient exogenous-DNA-free Cas9-ribonucleoprotein mediated gene editing in disease susceptibility loci in wheat (*Triticum aestivum* L.). Frontiers in Plant Science 13: 1084700.

Raffan, S., Sparks, C., Huttly, A., Hyde, L., Martignago, D., Mead, A. et al. (2021). Wheat with greatly reduced accumulation of free asparagine in the grain, produced by CRISPR/Cas9 editing of asparagine synthetase gene TaASN2. Plant Biotechnology Journal 19(8): 1602–1613.

Razzaq, A., Saleem, F., Kanwal, M., Mustafa, G., Yousaf, S., Imran Arshad, H. M. et al. (2019). Modern trends in plant genome editing: an inclusive review of the CRISPR/Cas9 toolbox. International Journal of Molecular Sciences 20(16): 4045.

Ren, B., Yan, F., Kuang, Y., Li, N., Zhang, D., Zhou, X. et al. (2018). Improved base editor for efficiently inducing genetic variations in rice with CRISPR/Cas9 guided hyperactive hAID mutant. Mol. Plant 11(4): 623–626. doi:10.1016/j.molp. 2018.01.005.

Sang, S., Wang, Y., Yao, G., Ma, T., Sun, X., Zhang, Y. et al. (2024). A critical review of conventional and modern approaches to develop herbicide-resistance in rice. Physiologia Plantarum 176(2): e14254.

Shamshirgaran, Y., Liu, J., Sumer, H., Verma, P. J. and Taheri-Ghahfarokhi, A. (2022). Tools for efficient genome editing; ZFN, TALEN, and CRISPR. Applications of Genome Modulation and Editing 29–46.

Shan, Q., Wang, Y., Li, J. and Gao, C. (2014). Genome editing in rice and wheat using the CRISPR/Cas system. Nat. Protoc. 9(10): 2395–2410. doi:10.1038/nprot. 2014.157.

Shan, Q., Wang, Y., Li, J., Zhang, Y., Chen, K., Liang, Z. et al. (2013). Targeted genome modification of crop plants using a CRISPR-Cas system. Nature Biotechnology 31(8): 686–688.

Sharma, M., Sidhu, A. K., Samota, M. K., Shah, P., Pandey, M. K. and Gangurde, S. S. (2024). Technological advancements in the CRISPR toolbox for improving plant salt tolerance. Discover Agriculture 2(1): 1–31.

Shen, H. and Li, Z. (2022). DNA double-strand break repairs and their application in plant DNA integration. Genes 13(2): 322.

Shi, J., Gao, H., Wang, H., Lafitte, H. R., Archibald, R. L., Yang, M. et al. (2017). ARGOS 8 variants generated by CRISPR-Cas9 improve maize grain yield under field drought stress conditions. Plant Biotechnol. J. 15(2): 207–216. doi:10.1111/pbi. 12603.

Shukla, V. K., Doyon, Y., Miller, J. C., DeKelver, R. C., Moehle, E. A., Worden, S. E. et al. (2009). Precise genome modification in the crop species *Zea mays* using zinc-finger nucleases. Nature 459(7245): 437–441. doi:10.1038/nature07992.

Sobrino-Mengual, G., Alvarez, D., Twyman, R. M., Gerrish, C., Fraser, P. D., Capell, T. et al. (2024). Activation of the native PHYTOENE SYNTHASE 1 promoter by modifying near-miss cis-acting elements induces carotenoid biosynthesis in embryogenic rice callus. Plant Cell Reports 43(5): 118.

Song, J. H., Shin, G., Kim, H. J., Lee, S. B., Moon, J. Y., Jeong, J. C. et al. (2022). Mutation of GmIPK1 gene using CRISPR/Cas9 reduced phytic acid content in soybean seeds. International Journal of Molecular Sciences 23(18): 10583.

Song, L., Liu, J., Cao, B., Liu, B., Zhang, X., Chen, Z. et al. (2023). Reducing brassinosteroid signalling enhances grain yield in semi-dwarf wheat. Nature 617(7959): 118–124.

Springmann, M., Clark, M., Mason-D'Croz, D., Wiebe, K., Bodirsky, B. L., Lassaletta, L. et al. (2018). Options for keeping the food system within environmental limits. Nature 562(7728): 519–525.

Tsai, S. Q. and Joung, J. K. (2016). Defining and improving the genome-wide specificities of CRISPR–Cas9 nucleases. Nature Reviews Genetics 17(5): 300–312.

Wang, Wenshu, Weipeng Wang, Yanlin Pan, Chao Tan, Hongjing Li, Ya Chen, Xingdan Liu et al. (2022). A new gain-of-function OsGS2/GRF4 allele generated by CRISPR/Cas9 genome editing increases rice grain size and yield. The Crop Journal 10(4): 1207–1212. Accessed January 26, 2026. https://doi.org/10.1016/j.cj.2022.01.004.

Wang, X., Guo, Y., Wang, Y., Peng, Y., Zhang, H. and Zheng, J. (2024). ZmHDT103 Negatively Regulates Drought Stress Tolerance in Maize Seedlings. Agronomy 14: 134. doi: 10.3390/agronomy14010134.

Wanikar, R., Dixit, P., Khater, M., Dange, M., Yevale, R. and Mandal, S. (2025). Prime editing through CRISPR-Cas and its applications in plants. pp. 122–141. In Gene-Edited Crops. CRC Press.

Wendt, T., Holm, P. B., Starker, C. G., Christian, M., Voytas, D. F., Brinch Pedersen, H. et al. (2013). TAL effector nucleases induce mutations at a pre selected location in the genome of primary barley transformants. Plant Mol. Biol. 83(3): 279–285. doi:10.1007/s11103-013-0078-4.

Xie, J., Fei, X., Yan, Q., Jiang, T., Li, Z., Chen, H. et al. (2024). The C4 photosynthesis bifunctional enzymes, PDRPs, of maize are co-opted to cytoplasmic viral replication complexes to promote infection of a prevalent potyvirus sugarcane mosaic virus. Plant Biotechnology Journal 22(7): 1812–1832.

Xu, R., Yang, Y., Qin, R., Li, H., Qiu, C., Li, L. et al. (2016). Rapid improvement of grain weight via highly efficient CRISPR/Cas9-mediated multiplex genome editing in rice. Journal of Genetics and Genomics= Yi chuanxue bao 43(8): 529–532.

Yang, J., Fang, Y., Wu, H., Zhao, N., Guo, X., Mackon, E. et al. (2023). Improvement of resistance to rice blast and bacterial leaf streak by CRISPR/Cas9-mediated mutagenesis of Pi21 and OsSULTR3; 6 in rice (*Oryza sativa* L.). Frontiers in Plant Science 14: 1209384.

Ye, N., Wang, Y., Yu, H., Qin, Z., Zhang, J., Duan, M. et al. (2023). Abscisic acid enhances trehalose content via OsTPP3 to improve salt tolerance in rice seedlings. Plants 12(14): 2665.

Yu, M., Kuang, Y., Wang, C., Wu, X., Li, S., Zhang, D. et al. (2024). Diverse nucleotide substitutions in rice base editing mediated by novel TadA variants. Plant Communications 5(8).

Zafar, S. A., Zaidi, S. S. E. A., Gaba, Y., Singla-Pareek, S. L., Dhankher, O. P., Li, X. et al. (2020). Engineering abiotic stress tolerance via CRISPR/Cas-mediated genome editing. Journal of Experimental Botany 71(2): 470–479.

Zafar, K., Khan, M. Z., Amin, I., Mukhtar, Z., Zafar, M. and Mansoor, S. (2023). Employing template-directed CRISPR-based editing of the OsALS gene to create herbicide tolerance in Basmati rice. AoB Plants 15(2): plac059.

Zhang, A., Liu, Y., Wang, F., Li, T., Chen, Z., Kong, D. et al. (2019a). Enhanced rice salinity tolerance via CRISPR/Cas9-targeted mutagenesis of the OsRR22 gene. Mol. Breed. 39(3): 47. doi:10.1007/s11032-019-0954-y.

Zhang, J., Zhang, H., Li, S., Li, J., Yan, L. and Xia, L. (2021). Increasing yield potential through manipulating of an ARE1 ortholog related to nitrogen use efficiency in wheat by CRISPR/Cas9. Journal of Integrative Plant Biology 63(9): 1649–1663.

Zhang, M., Cao, Y., Wang, Z., Wang, Z. Q., Shi, J., Liang, X. et al. (2018a). A retrotransposon in an HKT1 family sodium transporter causes variation of leaf Na+ exclusion and salt tolerance in maize. New Phytologist 217(3): 1161–1176.

Zhang, Y., Li, D., Zhang, D., Zhao, X., Cao, X., Dong, L. et al. (2018b). Analysis of the functions of Ta GW 2 homoeologs in wheat grain weight and protein content traits. The Plant Journal 94(5): 857–866.

Zhang, L., Xiang, Z., Li, J., Wang, S., Chen, Y., Liu, Y. et al. (2023). bHLH57 confers chilling tolerance and grain yield improvement in rice. Plant, Cell & Environment 46(4): 1402–1418.

Zhang, Y., Blahut-Beatty, L., Zheng, S., Clough, S. J. and Simmonds, D. H. (2023). The role of a soybean 14-3-3 gene (Glyma05g29080) on white mold resistance and nodulation investigations using CRISPR-Cas9 editing and RNA silencing. Molecular Plant-Microbe Interactions 36(3): 159–164.

Zhang, H., Liu, M., Yin, K., Liu, H., Liu, J. and Yan, Z. (2024). A novel OsHB5-OsAPL-OsMADS27/OsWRKY102 regulatory module regulates grain size in rice. Journal of Plant Physiology 295: 154210.

Zheng, W., Wang, Y., Cui, J., Guo, G., Li, Y., Hou, J. et al. (2024). ReaL-MGE is a tool for enhanced multiplex genome engineering and application to malonyl-CoA anabolism. Nature Communications 15(1): 9790.

Zhou, J., Xin, X., He, Y., Chen, H., Li, Q., Tang, X. et al. (2019). Multiplex QTL editing of grain-related genes improves yield in elite rice varieties. Plant Cell Reports 38: 475–485.

# Chapter 11

# Genome Editing of Horticultural Crops Using CRISPR/Cas for Enhanced Improvement

*Shivangi Rahangdale,*[1] *Himanshu Pandey,*[1] *Suhas Gorakh Karkute,*[2,]* 
*Ram Krishna*[1] and *Achuit Kumar Singh*[1]

## 1. Introduction

Horticultural crops, including a wide range of fruits and vegetables, are vital for human health and well-being (Knez et al. 2024). Horticultural crops also hold significant economic importance globally, contributing to both agricultural economies and food security (Dong et al. 2022). The production of horticultural crops varies widely across regions, reflecting differences in climate, soil, and agricultural practices (Bisbis et al. 2018). For instance, China and India are major producers of vegetables, with vast areas dedicated to crops such as tomatoes, potatoes, and onions, while countries like the United States, Brazil, and Spain are leading producers of fruits like apples, oranges, and grapes. They are a vital component of the agricultural sector, with millions of farmers relying on vegetable cultivation for their livelihoods. Vegetables, such as tomatoes, potatoes, onions, and leafy greens, are grown in diverse climates and regions, making them a key export commodity for many countries. The vegetable industry supports numerous sectors, including processing, retail, and distribution, creating jobs across the supply chain, from farm to table (Barrett et al. 2022). In addition, vegetables play a crucial role in improving public health by providing essential vitamins, minerals, and fibre, which are integral to a balanced diet (Gupta et al. 2024). This has led to increased demand for fresh produce and processed vegetable products, further driving economic activity. The global market for vegetables is expanding, with rising consumer awareness of healthy eating habits and increasing demand for plant-based foods. Vegetables like spinach, kale, and broccoli are abundant in vitamins A, C, and K, as well as essential minerals like iron, calcium, and magnesium, which are vital for bone health, immune function, and cellular processes (Stephen et al. 2023). Carrots, known for their high beta-carotene content, are a significant source of vitamin A, supporting vision

[1] Division of Vegetable Improvement, ICAR-Indian Institute of Vegetable Research, Varanasi, Uttar Pradesh, India.
[2] Division of Crop Improvement, ICAR-Central Institute for Cotton Research, Post Bag No. 02; P. O. Shankar Nagar, Nagpur-440010, Maharashtra, India.
* Corresponding author: suhaskarkute@gmail.com

DOI: 10.1201/9781003659907-14

and skin health (Sharma et al. 2021). Leafy greens and cruciferous vegetables also contain a variety of polyphenols, which are powerful antioxidants that help reduce inflammation and protect cells from oxidative damage. Fruits, including citrus, berries, and apples, are particularly rich in vitamins C and E, as well as an array of antioxidants such as flavonoids, anthocyanins, and polyphenols. Their bioactive compounds, such as anthocyanins, polyphenols, and essential elements, highlight their importance in functional nutrition and underscore the need for increased consumption of a variety of fruits and vegetables. The continuous increase in global population, climate change, and diminishing arable land present significant challenges to the production of horticultural crops, especially fruits and vegetables. Traditional breeding methods, while historically effective, often lack the precision and speed needed to keep pace with these growing demands due to lengthy breeding cycles and limited genetic diversity (Lamichhane and Thapa 2022). Recently, genome-editing tools such as CRISPR/Cas9, zinc finger nucleases (ZFNs), transcription activator-like effector nucleases (TALENs), and meganucleases have transformed plant biotechnology by offering more targeted and efficient approaches to crop improvement (Bhushan et al. 2016).

These genome-editing technologies allow for precise modifications in the plant genome, enabling the development of crops with enhanced yield, improved nutritional quality, greater disease resistance, and increased tolerance to environmental stresses. Among these tools, CRISPR/Cas9 has shown significant promise in horticultural crops by enabling specific gene edits that improve traits beneficial to crop resilience and productivity. For instance, CRISPR/Cas9 has been applied to increase disease resistance in tomatoes (*Solanum lycopersicum*) by targeting susceptibility genes (Debbarma et al. 2023), as well as to enhance fruit quality in strawberries (*Fragaria × ananassa*) by modifying genes associated with flavor and aroma pathways (Ma et al. 2023). TALENs, ZFNs, and meganucleases, although more complex to design, have also demonstrated success in targeted gene-editing applications, providing additional avenues for improving crop traits in other important horticultural species (Bhagtaney et al. 2023). Globally, countries like China and India lead in the production of fruits and vegetables, making advancements in these crops critical for food security and economic stability, particularly in regions where horticultural production is a major agricultural sector. Incorporating genome-editing tools like CRISPR/Cas9, ZFNs and TALENs into breeding programs for high-demand crops could transform horticultural productivity and sustainability, addressing pressing challenges in agricultural biotechnology (Wan et al. 2021). This chapter aims to summarize recent advancements and applications of genome-editing technologies, particularly CRISPR/Cas, in horticultural crops with a focus on fruits and vegetables. It will explore challenges associated with deploying these technologies and highlight future directions for integrating genome editing into horticultural crop improvement programs.

## 2. CRISPR/Cas Techniques

The CRISPR/Cas9 system was identified in bacteria as a defense mechanism against viral infections. This system uses the Cas9 protein to create double-strand breaks in DNA at a specific location identified by guide RNA or gRNA (Doudna and Charpentier 2014). This cut can be repaired by the cell's repair mechanisms, often resulting in gene disruption or precise gene edits (Figure 1 and Figure 2). CRISPR-Cas technology has revolutionized crop improvement by offering precise, efficient, and targeted gene editing capabilities. Over the past few years, various CRISPR/Cas systems, such as CRISPR/Cas9, CRISPR/Cas12, and newer innovations like CRISPR base editing and prime editing, have been harnessed to enhance crop traits (Gupta and Karkute 2021, Abd-Elsalam et al. 2024) and among them, CRISPR/Cas9 is the most well-known and widely used CRISPR system. These advancements allow for the modification of key genes responsible for traits like disease resistance, drought tolerance, pest resistance, improved yield, and enhanced nutritional content (Figure 3). With the versatility of these systems, CRISPR-based tools are accelerating the development of more resilient, productive, and sustainable crops, while offering new avenues

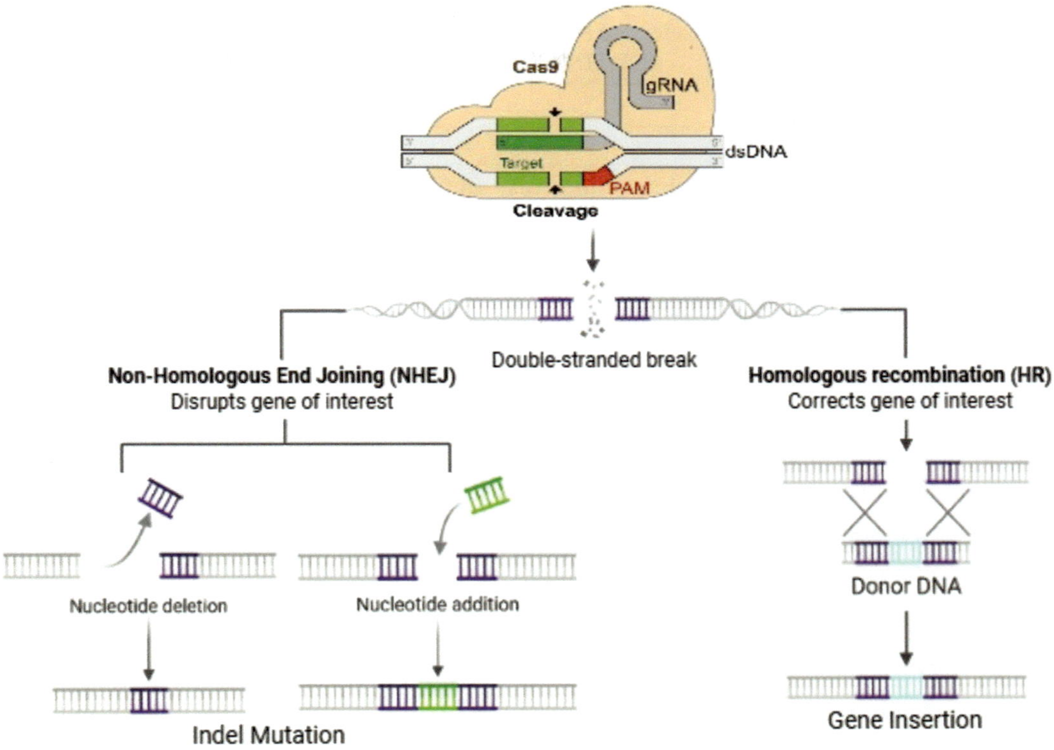


**Figure 1.** An overview of genome editing through CRISPR-Cas9 technique: Cas9 is an RNA-guided endonuclease consisting of two nuclease domains namely HNH and RuvC. The 20nt short guide RNA (sgRNA) sequence determines the target specificity of Cas9. Immediately 5′ of a PAM (Protospacer Adjacent Motif) sequence of the form N20-NGG (or N20-NAG) must be the target sites. The Cas9 nuclease causes double-strand breaks (DSB) at the target region, which can be fixed by the biological system through homologous recombination or non-homologous end joining, which causes gene disruption by indels or gene addition/correction, respectively.

for tackling the challenges posed by climate change, food security, and agricultural sustainability (Ndudzo et al. 2024).

Apart from Cas9, Cas12 is becoming more popular to carry out DNA double-strand breaks due to its different Protospacer Adjacent Motif (PAM) sequence. Cas12 is also capable of producing staggered cuts, which are sometimes more beneficial for certain types of edits. Contrary to Cas9 and Cas12, Cas13 targets RNA rather than DNA. Cas13 proteins bind to RNA and cleave it, providing a powerful tool for RNA silencing or modifying gene expression without altering the DNA sequence itself (O'Connell 2019). Unlike all these three Cas proteins, which target double-stranded molecules, Cas14 is much smaller and can target single-stranded DNA (ssDNA) rather than double-stranded DNA without the requirement of any PAM sequence (Aquino-Jarquin 2019). The array of different Cas proteins with specific properties diversifies their applications, making them a valuable tool for crop improvement.

CRISPR tools are not only limited to introducing InDels, but the advanced versions, like base editing and prime editing, enables precise conversion of one DNA base into another. Base editing uses a modified version of CRISPR-Cas9, where the Cas9 protein is fused with a DNA deaminase enzyme (Kantor et al. 2020). On the other hand, prime editing works by using a reverse transcriptase enzyme combined with a modified Cas9 protein that enables precise edits to be made to a specific location on the genome without creating double-strand breaks (Chen and Liu 2023). CRISPR Epigenome Editing allows researchers to modify the epigenetic marks on DNA without changing the underlying genetic sequence (Goell and Hilton 2021). This is achieved by fusing the

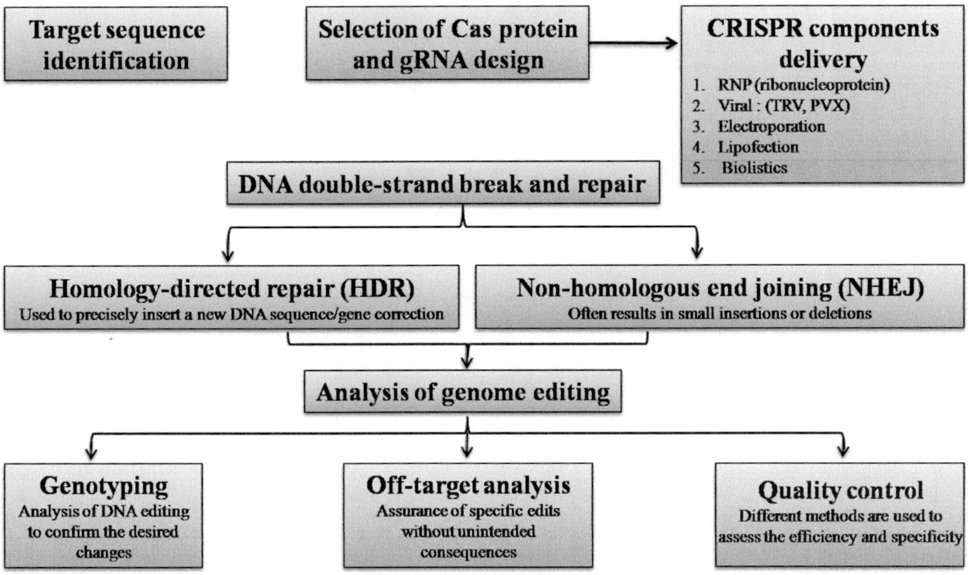


**Figure 2.** Basic steps in genome editing; the genome editing process begins with the identification of a target DNA sequence, followed by the design of a specific guide RNA (gRNA) complementary to the target. A suitable CRISPR-associated (Cas) enzyme is then selected and, if necessary, modified for the intended application. The CRISPR-Cas components are delivered into the plant cells using various transformation methods. The Cas enzyme, guided by the gRNA, introduces a double-strand break at the target site. The resulting DNA break is repaired by the cell through either non-homologous end joining (NHEJ) or homology-directed repair (HDR) mechanisms. Finally, the genome-edited plants are analysed through genotyping, off-target assessment, and quality control before being released for further use.

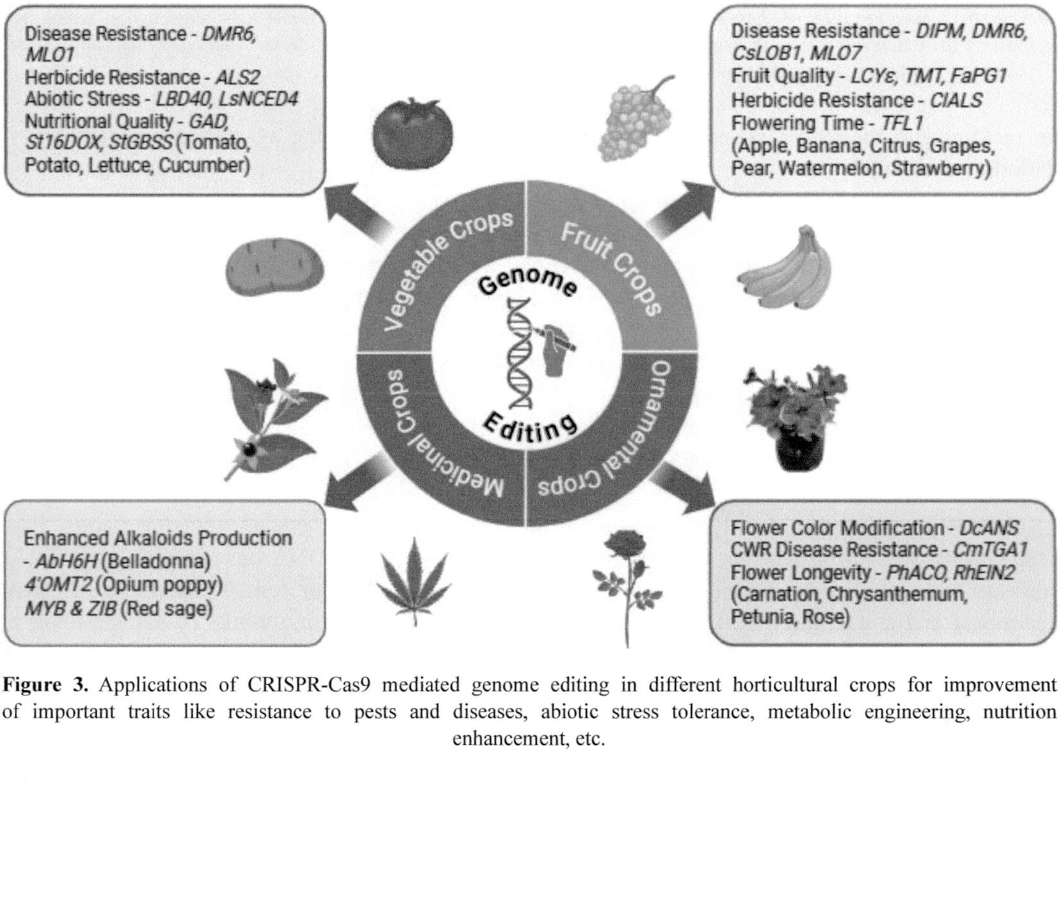


**Figure 3.** Applications of CRISPR-Cas9 mediated genome editing in different horticultural crops for improvement of important traits like resistance to pests and diseases, abiotic stress tolerance, metabolic engineering, nutrition enhancement, etc.

Cas9 protein (or other Cas proteins) to an epigenetic modifier such as a DNA methyltransferase or histone acetyltransferase.

## 3. CRISPR/Cas9 Applications for Improvement of Horticultural Crops

### *3.1 Improvement of Vegetable Crops*

CRISPR/Cas9 technology has enabled precise genetic modifications that were previously unattainable through conventional breeding methods. Genome editing approaches in tomato, potato, carrot, and lettuce have yielded plants with improved pathogen resistance without compromising growth or yield. CRISPR/Cas9 is used not only for improving disease resistance but also for modifying traits like fruit ripening in tomatoes, enhancing nutritional content, and improving environmental adaptability in vegetables, showcasing its ability to tackle various agricultural challenges (Table 1). The precision of CRISPR/Cas9 also makes it invaluable for functional genomics studies in vegetable crops.

The major application of CRISPR/Cas9 technique is in the development of disease resistance in vegetable crops. Virus resistance was among the initial applications of CRISPR/Cas9 where sgRNAs targeting the viral genome were inserted into the plant genome to develop resistance against tomato leaf curl virus and tomato yellow leaf curl virus (Ali et al. 2015, Tashkandi et al. 2018). Further, CRISPR technology was utilized for imparting disease resistance by targeting susceptible genes or negative regulators of disease resistance. Downy mildew resistance (*DMR6*) gene involved in salicylic acid metabolism was recently mutated by CRISPR/Cas9 in tomatoes and the edited plants exhibited broad-spectrum resistance against bacteria, oomycetes, and fungi (Thomazella et al. 2021).

To manage RNA virus diseases, targeting the eukaryotic translation initiation factor 4E (*eIF4E*) has emerged as an effective strategy to confer resistance. This approach was initially used by Chandrasekaran et al. (2016) to develop broad virus resistance against *Cucumber vein yellowing virus, Zucchini yellow mosaic virus,* and *Papaya ring spot mosaic virus-W* in cucumber. Later, a similar strategy has been implemented in several vegetable crops such as potato for potato virus Y (Noureen et al. 2022), tomato for pepper mottle virus (Yoon et al. 2020), melon for moroccan watermelon mosaic virus (Pechar et al. 2022), chinese cabbage for turnip mosaic virus (Liu et al. 2024) and watermelon for zucchini yellow mosaic virus (Li et al. 2024).

CRISPR mediated resistance against bacterial and fungal pathogens has been achieved by targeting susceptibility genes. For instance, as mentioned earlier, the homolog of *DMR6-1* gene in potato was targeted using CRISPR/Cas9 to develop resistance against late blight disease (Kieu et al. 2021). Natural mutants of *Mlo* gene developed resistance against powdery mildew disease and therefore, the *mlo* gene has been targeted in vegetable crops like tomato (Nekrasov et al. 2017) and cucumber (Tek et al. 2022) for inducing powdery mildew resistance. Fusarium wilt in tomato, caused by *Fusarium oxysporum* f. sp. *lycopersici* is a devastating disease in tomato. Resistance against fusarium-wilt was imparted in tomato plants by creating CRISPR/Cas9-mediated double knockout mutants of *TPL1* and *TPL2* genes which directly interacts with SIX8 effector protein of the pathogen (Aalders et al. 2024).

Root parasitic weeds are globally a major threat to tomato production. Strigolactone hormone in plants is responsible for the germination of seeds of weed parasites. To tackle with the root parasitic weed *Phelipanche aegyptiaca*, a genome editing approach is used for knocking out the *CAROTENOID CLEAVAGE DIOXYGENASE 8* (*CCD8*) gene involved in strigolactone synthesis (Bari et al. 2019). Subsequently, *More Axillary Growth 1* (*MAX1*) gene, functioning downstream of *CCD8* was mutated and *max1* mutant tomato lines had reduced Strigolactone levels with resistance to *P. aegyptiaca* (Bari et al. 2021).

In addition to biotic stress, CRISPR/Cas has been used to improve crops for abiotic stress tolerance. In tomatoes, the genome editing approach was successfully used by targeting *NPR1, MAPK3, LBD40* and *ARF*4 genes for drought tolerance, *AGL6* and *CPK28*, for heat tolerance

**Table 1.** List of target genes and traits modification in vegetable crops via CRISPR-Cas9.

| Crop Species | Target Gene | Mutation | Trait Modification | References |
|---|---|---|---|---|
| Tomato | *DMR6-1* | Knockout | Resistance against multiple diseases | Thomazella et al. 2021 |
| Tomato | *SlMlo1* | Knockout | Resistance against powdery mildew | Nekrasov et al. 2017 |
| Tomato | *TPL1 and TPL2* | Knockout | Resistance against Fusarium wilt | Aalders et al. 2024 |
| Tomato | *MAX1* | Knockout | Resistance against *Phelipanche aegytiaca* | Bari et al. 2021 |
| Tomato | *CCD8* | Knockout | Resistance against *Phelipanche aegytiaca* | Bari et al. 2019 |
| Tomato | *MYB12* | Knockout | Pink-coloured tomato | Deng et al. 2018 |
| Tomato | *CrtR-b2 and Psy1* | Knockout | Fruit color | D'Ambrosio et al. 2018 |
| Tomato | *MPK20* | Knockout | Pollen development through controlling sugar and auxin metabolism | Chen et al. 2018 |
| Tomato | *SGR1, LCY-E, Blc, LCY-B1 and LCY-B2* | Knockout | Lycopene content | Li et al. 2018 |
| Tomato | *PL, PG2a, TBG4* | Knockout | Pectin degradation, fruit color and weight | Wang et al. 2019 |
| Tomato | *SlMBP21* | Knockout | Jointless fruit stem | Roldan et al. 2017 |
| Tomato | *SlAGL6, ARF7* | Knockout | Parthenocarpic phenotype | Klap et al. 2016<br>Hu et al. 2023 |
| Tomato | *SlSP5G* | Knockout | Day-length-sensitive flowering | Soyk et al. 2017 |
| Tomato | *SlJAZ2* | Knockout | Disease resistance | Ortigosa et al. 2019 |
| Tomato | *PSY1, MYB12, and SGR1* | Knockout | Fruit color | Yang et al. 2023 |
| Tomato | *SlGAD2 and SlGAD3* | Knockout | ɣ-aminobutyric acid (GABA) content | Nonaka et al. 2017 |
| Tomato | *SlRIN* | Knockout | Fruit ripening | Ito et al. 2015 |
| Potato | *StALS1* | Knockout | Herbicide resistance | Butler et al. 2015 |
| Potato | *St16DOX* | Knockout | Steroidal glycoalkaloid (SGA) biosynthesis | Nakayasu et al. 2018 |
| Potato | *StDMR6* | Knockout | Disease resistance | Kieu et al. 2021 |
| Potato | *StGBSS* | Knockout | Starch composition | Andersson et al. 2018 |
| Potato | *StPPO* | Knockout | Polyphenol oxidation | González et al. 2020 |
| Potato | *eIF4E* | Knockout | Resistance Against Potato Virus Y | Noureen et al. 2022 |
| Sweet potato | *IbGBSSI and IbSBEII* | Knockout | Starch biosynthetic pathway | Wang et al. 2019 |
| Cucumber | *elF4E* | Knockout | Resistance against cucumber vein yellowing virus, zucchini yellow mosaic virus, and papaya ring spot mosaic virus | Chandrasekaran et al. 2016 |
| Cucumber | *CsWIP1* | Knockout | Sex determination | Hu et al. 2017 |
| Cucumber | *CsaMLO* | Knockout | Powdery mildew resistance | Tek et al. 2022 |
| Cabbage | *BoSRK3 and BoMS1* | Knockout | Self-compatibility and male sterility | Ma et al. 2019 |
| Carrot | *flavanone-3-hydroxylase (F3H)* | Knockout | Anthocyanin biosynthesis | Klimek-Chodacka et al. 2018 |
| Carrot | *DcGST1* | Knockout | Orange taproots | Duan et al. 2024 |
| Chinese Kale | *BoaCRTISO* | Gene insertion and replacement | Colour of mutants changed from green to yellow | Sun et al. 2020 |

and *HyPRP1* gene for salt tolerance (Kumar et al. 2023). Further, in-depth understanding of genetic pathways controlling the response to various abiotic stresses will lead to identification of key negative regulators that can be targeted for imparting stress tolerance in vegetable crops. CRISPR/Cas is also crucial in the functional characterization of genes. For example, CRISPR/Cas9-mediated knockout of *BZR1* in tomatoes caused decreased heat tolerance, as evidenced by reduced expression of heat shock proteins and greater sensitivity to high temperatures (Yin et al. 2018). Similarly, the *CBF1* gene, a transcription factor known for its role in cold stress responses, when disrupted by CRISPR/Cas9, leads to decreased chilling stress tolerance in tomato, underscoring its importance in maintaining cold stress resilience (Li et al. 2018).

Herbicide resistance has been a key target for gene editing across various crops, primarily focusing on the acetolactate synthase (ALS) gene family. In tomato, CRISPR/Cas9-mediated editing of the *EPSPS*, *ALS1* and *ALS2* genes has resulted in plants with enhanced tolerance to respective herbicides, enabling more effective weed management without compromising crop health (Yang et al. 2022). Similarly, targeted modification of the *ALS* gene through CRISPR/Cas9 has conferred resistance to herbicides in potato (Butler et al. 2015) and watermelon (Tian et al. 2018). Collectively, these studies highlight the versatility of ALS gene editing as a strategy to confer herbicide resistance across diverse crop species.

Vegetables are a crucial component of the human diet due to their nutritional composition and CRISPR tools have further helped to improve nutritional content and quality of vegetables. Polyphenol oxidase (*PPO*) causes enzymatic browning of vegetables like brinjal and potato, hampering their quality. CRISPR/Cas9-mediated knockout of *PPO* gene was achieved to prevent this enzymatic browning in brinjal (Maioli et al. 2020) and potato (González et al. 2020). γ-amino butyric acid (GABA) helps in lowering the blood pressure and therefore, GABA-rich tomatoes have been developed by editing the *GAD* gene (Nonaka et al. 2017). Sicilian Rouge GABA-rich tomatoes developed by Sanatech Seed are commercially cultivated in Japan (Waltz 2022). Crops like potato and sweet potato with modified starch are developed by knockout of granule bound starch synthase (*GBSS*) and starch branching enzyme (*SBE*) genes (Wang et al. 2019, Zhao et al. 2021). Success of genome editing for quality improvement is most observed in tomato, where several traits like ripening, anthocyanin content, TSS content, etc. have been targeted (Das et al. 2023). To alter the color-related chlorophyll and carotenoid levels in chinese kale, Sun et al. (2020) targeted *BoaCRTISO* gene by CRISPR/Cas9 and developed yellow colored kale. Thus, CRISPR/Cas system has enabled us to improve vegetable crops with higher quality and nutrition.

### *3.2 Improvement of Fruit Crops*

CRISPR/Cas9 system has been utilized in several fruit crops for improvement of various important traits. Biotic stress tolerance is one of the main achievements. Additionally, the technology is used to adjust fruit size, pigment accumulation, and nutritional composition across various species through the manipulation of key biosynthetic genes, demonstrating CRISPR/Cas9's pivotal role in developing more resilient, high-quality fruit crops (Table 2).

To induce resistance to fire blight disease in apple, *MdDIPM4* was edited using CRISPR/Cas9 and the edited plants showed decreased disease symptoms, offering a promising approach to safeguard apple production (Malnoy et al. 2016). They have also targeted *Mlo*-7 gene in grapevine for powdery mildew resistance. On the other hand, for developing resistance against downy mildew in grapevine, mutations in both *VviDMR6-1* and *VviDMR6-2* are carried out by CRISPR/Cas9 (Giacomelli et al. 2023). Botrytis cinerea causes significant yield losses in grapevine and CRISPR-mediated knockout plants of *VvWRKY52* gene showed higher resistance to the disease (Wang et al. 2018). Followed by grapevine, citrus is one of the most extensively used for genome editing to develop resistance to citrus canker and HLB disease by targeting *LOB1* and *DMR6* genes respectively (Wang et al. 2024). Like tomato and potato, DMR6 orthologue in banana was also knocked out through CRISPR/Cas9 and the mutant plants exhibited resistance to Xanthomonas

**Table 2.** List of target genes and traits modification in fruit crops via CRISPR-Cas9.

| Fruit Crop | Target Gene | Mutation Type | Target Trait | Reference |
|---|---|---|---|---|
| Apple | *MdDIPM4* | Knockout | Fire blight resistance | Malnoy et al. 2016 |
| Apple | *MdTFL1* | Knockout | Flowering time | Charrier et al. 2019 |
| Banana | *MaACO1* | Knockout | Ethylene biosynthesis | Hu et al. 2021 |
| Banana | *MaGA20ox2* | Knockout | Semi-dwarf phenotype | Shao et al. 2019 |
| Banana | *DMR6* | Knockout | Bacterial disease resistance | Tripathi et al. 2021 |
| Banana | *LCYε* | Knockout | enhanced β-carotene content | Kaur et al. 2020 |
| Citrus | *CsLOB1* | Knockout | Citrus canker resistance | Su et al. 2023 |
| Citrus | *CsDMR6* | Knockout | HLB, citrus canker disease resistance | Parajuli et al. 2022 |
| Grape | *TMT1 and TMT2* | Knockout | Sugar accumulation | Ren et al. 2021 |
| Grape | *VvDXS1* | Knockout | muscat flavor | Yang et al. 2024 |
| Grape | *VvMLO3* | Knockout | Powdery mildew resistance | Wan et al. 2020 |
| Grape | *VvWRKY52* | Knockout | Resistance to *Botrytis cinerea* | Wang et al. 2018 |
| Kiwifruit | *AcCEN4 and AcCEN* | Knockout | Compact plant with rapid terminal flower and fruit development. | Varkonyi-Gasic et al. 2019 |
| Pear | *PcTFL1* | Knockout | Early Flowering | Charrier et al. 2019 |
| Strawberry | *FaPG1* | Knockout | Fruit firmness | López-Casado et al. 2023 |
| Strawberry | *FvPHO2* | Knockout | Fruit quality with improved anthocyanin, Phosphorous and TSS content | Zhang et al. 2023 |
| Watermelon | *ClALS* | Knockout | Herbicide resistance | Tian et al. 2018 |
| Watermelon | *ClelF4E* | Knockout | Resistance to ZYMV and MWMV | Li et al. 2024, Pechar et al. 2022 |

wilt, a bacterial disease caused by *Xanthomonas campestris* pv. *musacearum* (Xcm) (Tripathi et al. 2021). Genome editing has even been employed in *Theobroma cocoa* where CRISPR/Cas9 was used to knockout the *Non-Expressor of Pathogenesis-Related 3 (TcNPR3)* gene to enhance the plant defense response against *Phytophthora tropicalis* (Fister et al. 2018).

Fruits are primarily consumed for their rich nutrient profiles and therefore, improving quality, nutrition as well as their shelf life for long-distance transportation are key goals of fruit improvement programs. Grapevines with reduced sugar levels have been developed by editing of *TMT1* and *TMT2* genes with the help of a polycistronic tRNA-gRNA system (Ren et al. 2021). Another group has used a prime editing approach to specifically edit *VvDXS1* gene to create muscat flavor in grapevine plants to have higher levels of monoterpenes (Yang et al. 2024). To improve fruit quality in woodland strawberries, *FvPHO2* gene was knocked out using CRISPR/Cas9 technology (Zhang et al. 2023). The mutant lines of strawberry fruits had improved phosphorus, TSS and anthocyanin content. Similarly, the fruit firmness of strawberry fruits was improved by creating CRISPR/Cas9-mediated mutations in the polygalacturonase (*FaPG1)* gene involved in pectin remodeling during fruit softening (López-Casado et al. 2023).

Banana is a highly perishable crop after ripening. Therefore, extending its shelf life is one of the most important objectives in banana breeding. Recently, editing of *ACO1* gene, which encodes 1-aminocyclopropane-1-carboxylate oxidase involved in ethylene biosynthesis in banana, resulted in delayed ripening and extended shelf life (Hu et al. 2021). Efforts were also made to enhance the β-carotene content in bananas by editing the *lycopene epsilon-cyclase* (*LCYε*) gene in "Grand Naine" cultivar (Kaur et al. 2020). The fruit pulp of genome-edited bananas accumulated six-fold higher β-carotene than the non-edited plants. The height of the banana plants is often a constraint in plantation and harvesting. To overcome this issue, it is necessary to breed for dwarf varieties. CRISPR-mediated gene editing of MaGA20ox2 gene successfully created semi-dwarf banana plants (Shao et al. 2019).

Flowering time is a critical factor that influences crop adaptability, breeding cycles, and yield potential. The *TFL1* gene functions as a floral repressor, and its targeted knockout via CRISPR/Cas9 in apple and pear leads to early flowering phenotypes (Charrier et al. 2019). This modification not only shortens the breeding cycle but also holds potential for improving yield stability under variable environmental conditions. Similarly, *CENTRORADIALIS* (*CEN*)-like genes *AcCEN4* and *AcCEN* were mutated by CRISPR/Cas9 in kiwifruit to develop compact plants with rapid terminal flowering and fruit development (Varkonyi-Gasic et al. 2019).

In one of the most recent reports, a novel method for developing agrobacterium-mediated transgene-free genome-edited plants was established in the banana crop by combining embryogenesis and chlorsulfuron selection with the help of base editors (Van den Broeck et al. 2025). This is the first report to produce transgene-free genome-edited lines in sterile crop species like triploid bananas.

### *3.3 Improvement of Medicinal Plants*

CRISPR/Cas9 gene editing has revolutionized the production of medicinal compounds by enabling precise modifications of biosynthetic pathways in plants. This technology allows researchers to enhance the yield of valuable pharmaceuticals, such as artemisinin, morphine, cannabinoids, and tanshinones, by targeting key regulatory and rate-limiting enzymes (Table 3). In addition to optimizing metabolite production, CRISPR/Cas9 offers an effective means for pathway elucidation and the discovery of novel bioactive compounds. Moreover, the use of CRISPR/Cas9 in medicinal plants supports sustainable agriculture practices by reducing reliance on wild harvesting and enabling the domestication of high-value crops. By accelerating breeding cycles and enhancing stress tolerance, CRISPR/Cas9 helps lower production costs, improving healthcare accessibility and creating economic opportunities in regions where these crops are cultivated.

Artemisinin is an important antimalarial compound produced by *Artemisia annua*. In order to increase the artemisinin production, a branch pathway blocking approach was utilized by knockdown of β-farnesene synthase (BFS), caryophyllene synthase (CPS), and squalene synthase (SQS) genes (Lv et al. 2016). The transgenic plants produced 71% more artemisinin in anti-SQS plants, whereas in anti-BFS transgenic plants, artemisinin content was increased by 77%. Multiplex editing by CRISPR/Cas9 enables simultaneous knockout of multiple genes. Thus, complete knockout of the BFS, CPS and SQS genes can be achieved to drastically enhance the production of artemisinin.

CRISPR/Cas technique is a promising tool for metabolic engineering due to its ability to target multiple genes at a time. Opium poppy (*Papaver somniferum* L.) is a widely used medicinal plant for its alkaloids. Alagoz et al. (2016) demonstrated the editing of *4'OMT2* gene involved in the biosynthesis of benzylisoquinoline alkaloids where plants showed reduction in BIA production and the synthesis of novel uncharacterized alkaloid. This study paved the way for engineering the pathways to produce novel bioactive compounds.

Tropane alkaloids are important pharmaceutical compounds produced in *Atropa belladonna*. The *AbH6H* gene codes for hyoscyamine 6β-hydroxylase, an enzyme pivotal for the synthesis of hyoscyamine and its derivatives. Knockout of *AbH6H* via CRISPR/Cas9 results in reduced hyoscyamine conversion to anisodamine nor scopolamine, thereby increasing the hyoscyamine production (Zeng et al. 2021).

*Salvia miltiorrhiza* is used for the treatment of cardiovascular diseases due to its phenolic acids and diterpenoid tanshinones. Hsu et al. (2024) specifically edited transcription factors of MYB and ZIP families regulating metabolic pathways. The knockout plants showed different results based on the nature of the transcription factor. For instance, knockout plants of bZIP1 produced a deeper red phenotype with a higher level of tanshinones. The study demonstrated the development of

**Table 3.** List of target genes and traits modification in medicinal plants via CRISPR-Cas9.

| Crop | Target Gene | Mutation Type | Target Trait | Reference |
|---|---|---|---|---|
| *Atropa belladonna* | *AbH6H* | Knockout | Hyoscyamine synthesis | Zeng et al. 2021 |
| *Papaver somniferum* | *4'OMT2* | Knockout | Biosythesis of benzylisoquinoline alkaloids | Alagoz et al. 2016 |
| *Rehmannia glutinosa* | *RgPDS* | Knockout | Albino plants | Li et al. 2021 |
| *Salvia miltiorrhiza* | *MYB and ZIP* | Knockout | Biosynthetic pathways of phenolic acids and tanshinones | Hsu et al. 2024 |

transgene-free plants by gene editing using protoplast to plant regeneration for enhancing active compounds.

### *3.4 Improvement of Ornamental Plants*

Ornamental plants have extensive scope for their improvement in the form of color, shape and fragrance of flowers, architecture of ornamental trees, color of the leaves, etc. To increase the aesthetic value of the plants and CRISPR/Cas technology has proven worthy to improve these traits in different ornamental plant species (Table 4).

Flower color is the most important and highly sought after trait of ornamental plants. For fine-tuning of the color, genes involved in anthocyanin and carotenoid pathways are the best targets. Genome editing has played an important role in creating variations in color by modulating various pigment synthesis pathways. For example, CRISPR/Cas9-mediated editing of *lavanone 3-hydroxylase* (*F3H*) gene involved in the flavonoid synthesis pathway created flowers with faint blue and pale violet colors in bluewings (*Torenia fournieri*) plants (Nishihara et al. 2018).

White petal Japanese morning glory (*Ipomea nil*) was subjected to CRISPR/Cas9-based editing of the carotenoid cleavage dioxygenase 4 gene, which is involved in the degradation of carotenoids in petals to develop pale yellow petals (Watanabe et al. 2018). The pale-yellow color was the result of a twenty-fold increase in carotenoid content in edited plants compared to non-edited plants. Similarly, editing of *Dihydroflavonol-4-reductase-B* (*DFR-B*) gene from anthocyanin biosynthetic pathway in *Ipomea nil* produced white flowers (Watanabe et al. 2017).

Petunia is the world's most popular bedding plant and is available in several colors. With the help of CRISPR/Cas9 technique, double mutants of *F3HA* and *F3HB* genes were generated,

**Table 4.** List of target genes and traits modification in ornamental plants via CRISPR-Cas9.

| Crop | Target Gene | Mutation Type | Target Trait | Reference |
|---|---|---|---|---|
| Carnation | *DcANS* | Knockout | Flower color mutant | Mori et al. 2024 |
| Chrysanthemum | *CmPDS* | Knockout | Carotenoid biosynthesis | Chen et al. 2024 |
| Chrysanthemum | *CmTGA1* | Knockout | Chrysanthemum White Rust (CWR) disease | Chen et al. 2024 |
| Ipomoea nil | *InCCD4* | Knockout | Petal colour formation | Watanabe et al. 2018 |
| Ipomoea nil | *DFR* | Knockout | Anthocyanin biosynthesis | Watanabe et al. 2017 |
| Petunia | *F3HA or F3HB* | Knockout | Flower color modification | Yu et al. 2021 |
| Petunia | *PhACO* | Knockout | Flower longevity | Xu et al. 2020 |
| Rose | *RhEIN2* | Knockout | Ethylene signaling pathway | Wang et al. 2023 |

which produced light purplish pink flowers (Yu et al. 2021). Recently, the anthocyanidin synthase (ANS) gene was targeted in carnation by electroporation-based genome editing approach using RNPs to develop flower color mutants (Mori et al. 2024).

Apart from ornamental values, enhancing shelf life is also an important goal of breeding programs in flower crops. CRISPR/Cas9 tool has been used in Japanese morning glory to create mutations in *EPHIMERAL1 (EPH1)* transcription factor to delay the petal senescence (Shibuya et al. 2018). Similarly, flower longevity was increase in Petunia by knockout of *1-aminocyclopropane-1-carboxylate oxidase 1* (*ACO1*) gene (Xu et al. 2020) whereas flower opening was completely blocked in rose (*Rosa hybrida*) by knockout of *ETHYLENE INSENSITIVE2* (*RhEIN2*) gene (Wang et al. 2023).

Developing altered flower anatomy has also been attempted with the help of CRISPR tools. A floral homeotic gene *AGAMOUS* (*AG*) was mutated in gentian plants by CRISPR/Cas9 to produce double-flower-type plants (Nishihara et al. 2023). In case of tobacco, mutations in the miR172 target sequence of *PETALOSA* (*PET*) gene were generated using CRISPR/Cas9 and the resultant mutant plants showed supernumerary petaloid structures like a double flower phenotype (Gattolin et al. 2020).

Another important study through CRISPR/Cas9 in flower crops is the functional characterization of gene involved in disease response to chrysanthemum white rust disease caused by *Puccinia horiana.* Chen et al. (2024) introduced mutations to knockout the *CmTGA1* gene in chrysanthemum plants and it was observed that the mutants had reduced resistance to *P. horiana.*

## 4. Opportunities and Limitations Associated with CRISPR/Cas Technologies in Horticultural Crops

Genome editing techniques specifically based on CRISPR/Cas9 system offer endless opportunities for improvement of horticultural crops due to its high efficiency, simple design and ability to virtually target any gene. Further, several countries, including India, have come up with regulatory guidelines for genome-edited crops where a product-based regulatory approach is adopted and edited crops do not need regulation once it is transgene transgene-free (Daniel et al. 2023). Though generation of transgene-free edited lines is not possible through agrobacterium-mediated transformation in clonally propagated crops, alternative strategies, like the use of RiboNucleoProtein (RNP) complex, have been developed and successfully utilized in such crops to develop transgene-free edited lines. One of the most important limitations of the genome editing technique is the requirement for transformation and regeneration of plants, however, tissue culture-independent methodologies, such as in planta transformation and use of viral vectors for genome editing are being adopted to overcome this barrier (Shen et al. 2024). Off-target effects remain so far, the most significant challenge in CRISPR/Cas9-mediated genome editing. However, careful selection of gRNAs using the multiple gRNA designing tools will help in minimizing the off-target effects. Moreover, in crop improvement programs, off-target effects would not be a matter of concern until and unless the edited plant gives the desirable phenotype without any adverse effects on plant growth and yield. Nevertheless, advancements in the field of genome sequencing, transformation technologies, improved programs for gRNA designing, etc., will overcome the existing limitations and revolutionize the horticulture sector in the near future.

## 5. Conclusion

CRISPR/Cas technology has emerged as a revolutionary tool in crop improvement, offering unprecedented precision and efficiency in genetic modification across diverse plant species. Its importance lies in its versatility to address multiple agricultural challenges simultaneously, such as enhancing disease resistance by targeting susceptibility genes like *DMR6* and *MLO1* in vegetables, extending shelf-life in fruits by modifying ripening-related genes, optimizing pharmaceutical compound production in medicinal plants, and manipulating flower pigmentation in ornamentals. The technology's greatest advantage is its ability to accelerate breeding cycles that traditionally take decades, particularly in perennial crops, while expanding the available genetic diversity for improvement. By enabling targeted modifications that enhance stress tolerance, nutritional quality, and productivity, CRISPR/Cas9 represents a critical innovation for developing climate-resilient crops essential for future food security and sustainable agriculture.

Despite its transformative potential, CRISPR/Cas9 faces significant limitations and challenges that must be addressed for its widespread adoption. The evolving and often inconsistent regulatory frameworks across different countries create uncertainty for commercialization pathways, potentially limiting global impact. Technical challenges persist, including off-target effects that necessitate a thorough evaluation of edited plants and variable editing efficiencies across different plant species and tissues. Public acceptance remains a crucial hurdle, with concerns about safety and ethical implications requiring transparent communication and stakeholder engagement. Future research must focus on developing improved delivery methods, and enhancing editing precision through next-generation systems like base editing and prime editing to utilize the full potential of CRISPR/Cas technology for improvement of horticultural crops.

## References

Aalders, T. R., de Sain, M., Gawehns, F., Oudejans, N., Jak, Y. D., Dekker, H. L. et al. (2024). Specific members of the TOPLESS family are susceptibility genes for Fusarium wilt in tomato and Arabidopsis. Plant Biotechnology Journal 22(1): 248–261.

Abd-Elsalam, K. A., Ahmed, A. and Munir, A. (2024). CRISPRized fruit, vegetable, and ornamental crops: A note from editors. pp. 3–20. In CRISPRized Horticulture Crops. Academic Press.

Alagoz, Y., Gurkok, T., Zhang, B. and Unver, T. (2016). Manipulating the biosynthesis of bioactive compound alkaloids for next-generation metabolic engineering in opium poppy using CRISPR-Cas 9 genome editing technology. Scientific Reports 6(1): 30910.

Ali, Z., Abulfaraj, A., Idris, A., Ali, S., Tashkandi, M. and Mahfouz, M. M. (2015). CRISPR/Cas9-mediated viral interference in plants. Genome Biology 16: 1–11.

Andersson, M., Turesson, H., Olsson, N., Fält, A. S., Ohlsson, P., Gonzalez, M. N. et al. (2018). Genome editing in potato via CRISPR-Cas9 ribonucleoprotein delivery. Physiologia Plantarum 164(4): 378–384.

Aquino-Jarquin, G. (2019). CRISPR-Cas14 is now part of the artillery for gene editing and molecular diagnostic. Nanomedicine: Nanotechnology, Biology and Medicine 18: 428–431.

Bari, V. K., Nassar, J. A. and Aly, R. (2021). CRISPR/Cas9 mediated mutagenesis of MORE AXILLARY GROWTH 1 in tomato confers resistance to root parasitic weed Phelipanche aegyptiaca. Scientific Reports 11(1): 3905.

Bari, V. K., Nassar, J. A., Kheredin, S. M., Gal-On, A., Ron, M., Britt, A. et al. (2019). CRISPR/Cas9-mediated mutagenesis of CAROTENOID CLEAVAGE DIOXYGENASE 8 in tomato provides resistance against the parasitic weed Phelipanche aegyptiaca. Scientific Reports 9(1): 11438.

Barrett, C. B., Reardon, T., Swinnen, J. and Zilberman, D. (2022). Agri-food value chain revolutions in low-and middle-income countries. Journal of Economic Literature 60(4): 1316–1377.

Bhagtaney, L. and Sundarrajan, P. (2023). An overview of tools for genome editing: ZFNs, mega nucleases, and TALENs. CRISPR/Cas-Mediated Genome Editing in Plants 37–64.

Bhushan, K., Pratap, D. and Sharma, P. K. (2016). Transcription activator-like effector nucleases (TALENs): An efficient tool for plant genome editing. Engineering in Life Sciences 16(4): 330–337.

Bisbis, M. B., Gruda, N. and Blanke, M. (2018). Potential impacts of climate change on vegetable production and product quality—A review. Journal of Cleaner Production 170: 1602–1620.

Butler, N. M., Atkins, P. A., Voytas, D. F. and Douches, D. S. (2015). Generation and inheritance of targeted mutations in potato (Solanum tuberosum L.) using the CRISPR/Cas system. PloS One 10(12): e0144591.

Chandrasekaran, J., Brumin, M., Wolf, D., Leibman, D., Klap, C., Pearlsman, M. et al. (2016). Development of broad virus resistance in non-transgenic cucumber using CRISPR/Cas9 technology. Molecular Plant Pathology 17(7): 1140–1153.

Charrier, A., Vergne, E., Dousset, N., Richer, A., Petiteau, A. and Chevreau, E. (2019). Efficient targeted mutagenesis in apple and first time edition of pear using the CRISPR-Cas9 system. Frontiers in Plant Science 10: 40.

Chen, L., Yang, D., Zhang, Y., Wu, L., Zhang, Y., Ye, L. et al. (2018). Evidence for a specific and critical role of mitogen-activated protein kinase 20 in uni-to-binucleate transition of microgametogenesis in tomato. New Phytologist 219(1): 176–194.

Chen, P. J. and Liu, D. R. (2023). Prime editing for precise and highly versatile genome manipulation. Nature Reviews Genetics 24(3): 161–177.

Chen, Q., Zhang, X., Jin, R. and Mao, H. (2024). Establishment of a CRISPR/Cas9 gene-editing system for Chrysanthemum morifolium. Ornamental Plant Research 4(1).

D'Ambrosio, C., Stigliani, A. L. and Giorio, G. (2018). CRISPR/Cas9 editing of carotenoid genes in tomato. Transgenic Research 27(4): 367–378.

Daniel, M. A., Sebastin, R., Yu, J. K., Soosaimanickam, M. P. and Chung, J. W. (2023). Enhancing horticultural crops through genome editing: applications, benefits, and considerations. Horticulturae 9(8): 884.

Das, T., Anand, U., Pal, T., Mandal, S., Kumar, M., Radha et al. (2023). Exploring the potential of CRISPR/Cas genome editing for vegetable crop improvement: an overview of challenges and approaches. Biotechnology and Bioengineering 120(5): 1215–1228.

Debbarma, J., Saikia, B., Singha, D. L., Das, D., Keot, A. K., Maharana, J. et al. (2023). CRISPR/Cas9-mediated mutation in XSP10 and SlSAMT genes impart genetic tolerance to fusarium wilt disease of tomato (Solanum lycopersicum L.). Genes 14(2): 488.

Deng, L., Wang, H., Sun, C., Li, Q., Jiang, H., Du, M. et al. (2018). Efficient generation of pink-fruited tomatoes using CRISPR/Cas9 system. Journal of Genetics and Genomics= Yi chuan xue bao 45(1): 51–54.

Dong, J., Gruda, N., Li, X., Cai, Z., Zhang, L. and Duan, Z. (2022). Global vegetable supply towards sustainable food production and a healthy diet. Journal of Cleaner Production 369: 133212.

Doudna, J. A. and Charpentier, E. (2014). The new frontier of genome engineering with CRISPR-Cas9. Science 346(6213): 1258096.

Duan, A. Q., Deng, Y. J., Tan, S. S., Liu, S. S., Liu, H., Xu, Z. S. et al. (2024). DcGST1, encoding a glutathione S-transferase activated by DcMYB7, is the main contributor to anthocyanin pigmentation in purple carrot. The Plant Journal 117(4): 1069–1083.

Fister, A. S., Landherr, L., Maximova, S. N. and Guiltinan, M. J. (2018). Transient expression of CRISPR/Cas9 machinery targeting TcNPR3 enhances defense response in Theobroma cacao. Frontiers in Plant Science 9: 268.

Gattolin, S., Cirilli, M., Chessa, S., Stella, A., Bassi, D. and Rossini, L. (2020). Mutations in orthologous PETALOSA TOE-type genes cause a dominant double-flower phenotype in phylogenetically distant eudicots. Journal of Experimental Botany 71(9): 2585–2595.

Giacomelli, L., Zeilmaker, T., Giovannini, O., Salvagnin, U., Masuero, D., Franceschi, P. et al. (2023). Simultaneous editing of two DMR6 genes in grapevine results in reduced susceptibility to downy mildew. Frontiers in Plant Science 14: 1242240.

Goell, J. H. and Hilton, I. B. (2021). CRISPR/Cas-based epigenome editing: advances, applications, and clinical utility. Trends in Biotechnology 39(7): 678–691.

González, M. N., Massa, G. A., Andersson, M., Turesson, H., Olsson, N., Fält, A. S. et al. (2020). Reduced enzymatic browning in potato tubers by specific editing of a polyphenol oxidase gene via ribonucleoprotein complexes delivery of the CRISPR/Cas9 system. Frontiers in Plant Science 10: 1649.

Gupta, O. P. and Karkute, S. G. (eds.). (2021). Genome Editing in Plants: Principles and Applications. CRC Press.

Gupta, P., Karkute, S. G., Dash, P. K. and Singh, A. K. (2024). Genome editing for improvement of nutrition and quality in vegetable crops. Applications of Genome Engineering in Plants 222–241.

Hsu, C. T., Chiu, C. C., Hsiao, P. Y., Lin, C. Y., Cheng, S., Lin, Y. C. et al. (2024). Transgene-free CRISPR/Cas9-mediated gene editing through protoplast-to-plant regeneration enhances active compounds in Salvia miltiorrhiza. Plant Biotechnology Journal 22(6): 1549.

Hu, B., Li, D., Liu, X., Qi, J., Gao, D., Zhao, S. et al. (2017). Engineering non-transgenic gynoecious cucumber using an improved transformation protocol and optimized CRISPR/Cas9 system. Molecular Plant 10(12): 1575–1578.

Hu, C., Sheng, O., Deng, G., He, W., Dong, T., Yang, Q. et al. (2021). CRISPR/Cas9-mediated genome editing of *MaACO1* (aminocyclopropane-1-carboxylate oxidase1) promotes the shelf life of banana fruit. Plant Biotechnol. J. 19: 654–656.

Hu, J., Li, X. and Sun, T. P. (2023). Four class A AUXIN RESPONSE FACTORs promote tomato fruit growth despite suppressing fruit set. Nature Plants 9(5): 706–719.

Ito, Y., Nishizawa-Yokoi, A., Endo, M., Mikami, M. and Toki, S. (2015). CRISPR/Cas9-mediated mutagenesis of the RIN locus that regulates tomato fruit ripening. Biochemical and Biophysical Research Communications 467(1): 76–82.

Kantor, A., McClements, M. E. and MacLaren, R. E. (2020). CRISPR-Cas9 DNA base-editing and prime-editing. International Journal of Molecular Sciences 21(17): 6240.

Kaur, N., Alok, A., Shivani, Kumar, P., Kaur, N., Awasthi, P. et al. (2020). CRISPR/Cas9 directed editing of lycopene epsilon-cyclase modulates metabolic flux for β-carotene biosynthesis in banana fruit. Metab. Eng. 59: 76–86.

Kieu, N. P., Lenman, M., Wang, E. S., Petersen, B. L. and Andreasson, E. (2021). Mutations introduced in susceptibility genes through CRISPR/Cas9 genome editing confer increased late blight resistance in potatoes. Scientific Reports 11(1): 4487.

Klap, C., Yeshayahou, E., Bolger, A. M., Arazi, T., Gupta, S. K., Shabtai, S. et al. (2017). Tomato facultative parthenocarpy results from Sl AGAMOUS-LIKE 6 loss of function. Plant Biotechnology Journal 15(5): 634–647.

Klimek-Chodacka, M., Oleszkiewicz, T., Lowder, L. G., Qi, Y. and Baranski, R. (2018). Efficient CRISPR/Cas9-based genome editing in carrot cells. Plant Cell Reports 37(4): 575–586.

Knez, M., Ranić, M. and Gurinović, M. (2024). Underutilized plants increase biodiversity, improve food and nutrition security, reduce malnutrition, and enhance human health and well-being. Let's put them back on the plate!. Nutrition Reviews 82(8): 1111–1124.

Kumar, M., Prusty, M. R., Pandey, M. K., Singh, P. K., Bohra, A., Guo, B. et al. (2023). Application of CRISPR/Cas9-mediated gene editing for abiotic stress management in crop plants. Frontiers in Plant Science 14: 1157678.

Lamichhane, S. and Thapa, S. (2022). Advances from conventional to modern plant breeding methodologies. Plant Breed Biotechnol. 10(1): 1–4.

Li, M., Qiu, Y., Zhu, D., Xu, X., Tian, S., Wang, J. et al. (2024). Editing eIF4E in the watermelon genome using CRISPR/Cas9 technology confers resistance to ZYMV. International Journal of Molecular Sciences 25(21): 11468.

Li, R., Zhang, L., Wang, L., Chen, L., Zhao, R., Sheng, J. et al. (2018). Reduction of tomato-plant chilling tolerance by CRISPR–Cas9-mediated SlCBF1 mutagenesis. Journal of Agricultural and Food Chemistry 66(34): 9042–9051.

Li, X., Wang, Y., Chen, S., Tian, H., Fu, D., Zhu, B. et al. (2018). Lycopene is enriched in tomato fruit by CRISPR/Cas9-mediated multiplex genome editing. Frontiers in Plant Science 9: 559.

Li, X., Zuo, X., Li, M., Yang, X., Zhi, J., Sun, H. et al. (2021). Efficient CRISPR/Cas9-mediated genome editing in Rehmannia glutinosa. Plant Cell Reports 40(9): 1695–1707.

Liu, Y., Xin, X., Li, P., Wang, W., Yu, Y., Zhao, X. et al. (2024). Editing of eIF (iso) 4E. c confers resistance against Turnip mosaic virus in Brassica rapa. Horticultural Plant Journal 10(4): 1020–1034.

López-Casado, G., Sánchez-Raya, C., Ric-Varas, P. D., Paniagua, C., Blanco-Portales, R., Muñoz-Blanco, J. et al. (2023). CRISPR/Cas9 editing of the polygalacturonase FaPG1 gene improves strawberry fruit firmness. Horticulture Research 10(3): uhad011.

Lv, Z., Zhang, F., Pan, Q., Fu, X., Jiang, W., Shen, Q. et al. (2016). Branch pathway blocking in Artemisia annua is a useful method for obtaining high yield artemisinin. Plant and Cell Physiology 57(3): 588–602.

Ma, C., Zhu, C., Zheng, M., Liu, M., Zhang, D., Liu, B. et al. (2019). CRISPR/Cas9-mediated multiple gene editing in Brassica oleracea var. capitata using the endogenous tRNA-processing system. Horticulture Research 6.

Ma, Z., Ma, L. and Zhou, J. (2023). Applications of CRISPR/Cas genome editing in economically important fruit crops: recent advances and future directions. Molecular Horticulture 3(1): 1.

Maioli, A., Gianoglio, S., Moglia, A., Acquadro, A., Valentino, D., Milani, A. M. et al. (2020). Simultaneous CRISPR/Cas9 editing of three PPO genes reduces fruit flesh browning in Solanum melongena L. Frontiers in Plant Science 11: 607161.

Malnoy, M., Viola, R., Jung, M. H., Koo, O. J., Kim, S., Kim, J. S. et al. (2016). DNA-free genetically edited grapevine and apple protoplast using CRISPR/Cas9 ribonucleoproteins. Frontiers in Plant Science 7: 1904.

Mori, K., Tanase, K. and Sasaki, K. (2024). Novel electroporation-based genome editing of carnation plant tissues using RNPs targeting the anthocyanidin synthase gene. Planta 259(4): 84.

Nakayasu, M., Akiyama, R., Lee, H. J., Osakabe, K., Osakabe, Y., Watanabe, B. et al. (2018). Generation of α-solanine-free hairy roots of potato by CRISPR/Cas9 mediated genome editing of the St16DOX gene. Plant Physiology and Biochemistry 131: 70–77.

Ndudzo, A., Makuvise, A. S., Moyo, S. and Bobo, E. D. (2024). CRISPR-Cas9 genome editing in crop breeding for climate change resilience: Implications for smallholder farmers in Africa. Journal of Agriculture and Food Research 101132.

Nekrasov, V., Wang, C., Win, J., Lanz, C., Weigel, D. and Kamoun, S. (2017). Rapid generation of a transgene-free powdery mildew resistant tomato by genome deletion. Scientific Reports 7(1): 482.

Nishihara, M., Higuchi, A., Watanabe, A. and Tasaki, K. (2018). Application of the CRISPR/Cas9 system for modification of flower color in Torenia fournieri. BMC Plant Biology 18: 1–9.

Nishihara, M., Hirabuchi, A., Goto, F., Watanabe, A., Yoshida, C., Washiashi, R. et al. (2023). Efficient double-flowered gentian plant production using the CRISPR/Cas9 system. Plant Biotechnology 40(3): 229–236.

Nonaka, S., Arai, C., Takayama, M., Matsukura, C. and Ezura, H. (2017). Efficient increase of Γ-aminobutyric acid (GABA) content in tomato fruits by targeted mutagenesis. Scientific Reports 7(1): 7057.

Noureen, A., Khan, M. Z., Amin, I., Zainab, T. and Mansoor, S. (2022). CRISPR/Cas9-mediated targeting of susceptibility factor eIF4E-enhanced resistance against potato virus Y. Frontiers in Genetics 13: 922019.

O'Connell, M. R. (2019). Molecular mechanisms of RNA targeting by Cas13-containing type VI CRISPR–Cas systems. Journal of Molecular Biology 431(1): 66–87.

Ortigosa, A., Gimenez-Ibanez, S., Leonhardt, N. and Solano, R. (2019). Design of a bacterial speck resistant tomato by CRISPR/Cas9-mediated editing of Sl JAZ 2. Plant Biotechnology Journal 17(3): 665–673.

Parajuli, S., Huo, H., Gmitter Jr, F. G., Duan, Y., Luo, F. and Deng, Z. (2022). Editing the CsDMR6 gene in citrus results in resistance to the bacterial disease citrus canker. Horticulture Research 9: uhac082.

Pechar, G. S., Donaire, L., Gosalvez, B., García-Almodovar, C., Sánchez-Pina, M. A., Truniger, V. et al. (2022). Editing melon eIF4E associates with virus resistance and male sterility. Plant Biotechnology Journal 20(10): 2006–2022.

Ren, C., Liu, Y., Guo, Y., Duan, W., Fan, P., Li, S. et al. (2021). Optimizing the CRISPR/Cas9 system for genome editing in grape by using grape promoters. Horticulture Research 8.

Roldan, M. V. G., Périlleux, C., Morin, H., Huerga-Fernandez, S., Latrasse, D., Benhamed, M. et al. (2017). Natural and induced loss of function mutations in SlMBP21 MADS-box gene led to jointless-2 phenotype in tomato. Scientific Reports 7(1): 4402.

Shao, X., Wu, S., Dou, T., Zhu, H., Hu, C., Huo, H. et al. (2019). Using CRISPR/Cas9 genome editing system to create MaGA20ox2 gene-modified semi-dwarf banana. Plant Biotechnology Journal 18(1): 17.

Sharma, S., Katoch, V., Kumar, S. and Chatterjee, S. (2021). Functional relationship of vegetable colors and bioactive compounds: Implications in human health. The Journal of Nutritional Biochemistry 92: 108615.

Shen, Y., Ye, T., Li, Z., Kimutai, T. H., Song, H., Dong, X. et al. (2024). Exploiting viral vectors to deliver genome editing reagents in plants. Abiotech 5(2): 247–261.

Shibuya, K., Watanabe, K. and Ono, M. (2018). CRISPR/Cas9-mediated mutagenesis of the EPHEMERAL1 locus that regulates petal senescence in Japanese morning glory. Plant Physiology and Biochemistry 131: 53–57.

Soyk, S., Müller, N. A., Park, S. J., Schmalenbach, I., Jiang, K., Hayama, R. et al. (2017). Variation in the flowering gene SELF PRUNING 5G promotes day-neutrality and early yield in tomato. Nature Genetics 49(1): 162–168.

Stephen, J., Manoharan, D. and Radhakrishnan, M. (2023). Immune boosting functional components of natural foods and its health benefits. Food Production, Processing and Nutrition 5(1): 61.

Su, H., Wang, Y., Xu, J., Omar, A. A., Grosser, J. W., Calovic, M. et al. (2023). Generation of the transgene-free canker-resistant Citrus sinensis using Cas12a/crRNA ribonucleoprotein in the T0 generation. Nature Communications 14(1): 3957.

Sun, B., Jiang, M., Zheng, H., Jian, Y., Huang, W. L., Yuan, Q. et al. (2020). Color-related chlorophyll and carotenoid concentrations of Chinese kale can be altered through CRISPR/Cas9 targeted editing of the carotenoid isomerase gene BoaCRTISO. Horticulture Research 7.

Tashkandi, M., Ali, Z., Aljedaani, F., Shami, A. and Mahfouz, M. M. (2018). Engineering resistance against Tomato yellow leaf curl virus via the CRISPR/Cas9 system in tomato. Plant Signaling & Behavior 13(10): e1525996.

Tek, M. I., Calis, O., Fidan, H., Shah, M. D., Celik, S. and Wani, S. H. (2022). CRISPR/Cas9 based mlo-mediated resistance against Podosphaera xanthii in cucumber (Cucumis sativus L.). Frontiers in Plant Science 13: 1081506.

Thomazella, D. P. D. T., Seong, K., Mackelprang, R., Dahlbeck, D., Geng, Y. U., Gill, U. S. et al. (2021). Loss of function of a DMR6 ortholog in tomato confers broad-spectrum disease resistance. Proceedings of the National Academy of Sciences 118(27): e2026152118.

Tian, S., Jiang, L., Cui, X., Zhang, J., Guo, S., Li, M. et al. (2018). Engineering herbicide-resistant watermelon variety through CRISPR/Cas9-mediated base-editing. Plant Cell Rep. 37: 1353–1356. doi:10.1007/s00299-018-2299-0.

Tripathi, J. N., Ntui, V. O., Shah, T. and Tripathi, L. (2021). CRISPR/Cas9-mediated editing of DMR6 orthologue in banana (Musa spp.) confers enhanced resistance to bacterial disease. Plant Biotechnology Journal 19(7): 1291.

Van den Broeck, S., Ngapout, Y., Panis, B. and Vanderschuren, H. (2025). An Agrobacterium-mediated base editing approach generates transgene-free edited banana. New Phytologist.

Varkonyi-Gasic, E., Wang, T., Voogd, C., Jeon, S., Drummond, R. S., Gleave, A. P. et al. (2019). Mutagenesis of kiwifruit CENTRORADIALIS-like genes transforms a climbing woody perennial with long juvenility and axillary flowering into a compact plant with rapid terminal flowering. Plant Biotechnology Journal 17(5): 869–880.

Waltz, E. (2022). GABA-enriched tomato is first CRISPR-edited food to enter market. Nat. Biotechnol. 40(1): 9–11.

Wan, D. Y., Guo, Y., Cheng, Y., Hu, Y., Xiao, S., Wang, Y. et al. (2020). CRISPR/Cas9-mediated mutagenesis of VvMLO3 results in enhanced resistance to powdery mildew in grapevine (Vitis vinifera). Horticulture Research 7.

Wan, L., Wang, Z., Tang, M., Hong, D., Sun, Y., Ren, J. et al. (2021). CRISPR-Cas9 gene editing for fruit and vegetable crops: strategies and prospects. Horticulturae 7(7): 193.

Wang, C., Li, Y., Wang, N., Yu, Q., Li, Y., Gao, J. et al. (2023). An efficient CRISPR/Cas9 platform for targeted genome editing in rose (Rosa hybrida). Journal of Integrative Plant Biology 65(4): 895–899.

Wang, D., Samsulrizal, N. H., Yan, C., Allcock, N. S., Craigon, J., Blanco-Ulate, B. et al. (2019). Characterization of CRISPR mutants targeting genes modulating pectin degradation in ripening tomato. Plant Physiology 179(2): 544–557.

Wang, H., Wu, Y., Zhang, Y., Yang, J., Fan, W., Zhang, H. et al. (2019). CRISPR/Cas9-based mutagenesis of starch biosynthetic genes in sweet potato (*Ipomoea Batatas*) for the improvement of starch quality. International Journal of Molecular Sciences 20(19): 4702.

Wang, S., Zhao, S., Du, M., Mo, K., Pang, S. and Zou, X. (2024). Development and application of gene editing in citrus: how and what to do. Horticulture Advances 2(1): 1–17.

Wang, X., Tu, M., Wang, D., Liu, J., Li, Y., Li, Z. et al. (2018). CRISPR/Cas9-mediated efficient targeted mutagenesis in grape in the first generation. Plant Biotechnology Journal 16(4): 844–855.

Watanabe, K., Kobayashi, A., Endo, M., Sage-Ono, K., Toki, S. and Ono, M. (2017). CRISPR/Cas9-mediated mutagenesis of the dihydroflavonol-4-reductase-B (DFR-B) locus in the Japanese morning glory Ipomoea (Pharbitis) nil. Scientific Reports 7(1): 10028.

Watanabe, K., Oda-Yamamizo, C., Sage-Ono, K., Ohmiya, A. and Ono, M. (2018). Alteration of flower colour in Ipomoea nil through CRISPR/Cas9-mediated mutagenesis of carotenoid cleavage dioxygenase 4. Transgenic Research 27: 25–38.

Xu, J., Kang, B. C., Naing, A. H., Bae, S. J., Kim, J. S., Kim, H. et al. (2020). CRISPR/Cas9-mediated editing of 1-aminocyclopropane-1-carboxylate oxidase1 enhances Petunia flower longevity. Plant Biotechnology Journal 18(1): 287–297.

Yang, S. H., Kim, E., Park, H. and Koo, Y. (2022). Selection of the high efficient sgRNA for CRISPR-Cas9 to edit herbicide related genes, PDS, ALS, and EPSPS in tomato. Applied Biological Chemistry 65(1): 13.

Yang, T., Ali, M., Lin, L., Li, P., He, H., Zhu, Q. et al. (2023). Recoloring tomato fruit by CRISPR/Cas9-mediated multiplex gene editing. Horticulture Research 10(1): uhac214.

Yang, Y., Wheatley, M., Meakem, V., Galarneau, E., Gutierrez, B. and Zhong, G. Y. (2024). Editing VvDXS1 for the creation of muscat flavour in *Vitis vinifera* cv. Scarlet Royal. Plant Biotechnology Journal 22(6): 1610–1621.

Yin, Y., Qin, K., Song, X., Zhang, Q., Zhou, Y., Xia, X. et al. (2018). BZR1 transcription factor regulates heat stress tolerance through FERONIA receptor-like kinase-mediated reactive oxygen species signaling in tomato. Plant and Cell Physiology 59(11): 2239–2254.

Yoon, Y. J., Venkatesh, J., Lee, J. H., Kim, J., Lee, H. E., Kim, D. S. et al. (2020). Genome editing of eIF4E1 in tomato confers resistance to pepper mottle virus. Frontiers in Plant Science 11: 1098.

Yu, J., Tu, L., Subburaj, S., Bae, S. and Lee, G. J. (2021). Simultaneous targeting of duplicated genes in Petunia protoplasts for flower color modification via CRISPR-Cas9 ribonucleoproteins. Plant Cell Reports 40: 1037–1045.

Zeng, L., Zhang, Q., Jiang, C., Zheng, Y., Zuo, Y., Qin, J. et al. (2021). Development of Atropa belladonna L. plants with high-yield hyoscyamine and without its derivatives using the CRISPR/Cas9 system. International Journal of Molecular Sciences 22(4): 1731.

Zhang, C., Liu, Y., Wang, B., Li, H., Zhang, J., Ma, Y. et al. (2023). CRISPR/Cas9 targeted knockout FvPHO2 can increase phosphorus content and improve fruit quality of woodland strawberry. Scientia Horticulturae 317: 112078.

Zhao, X., Jayarathna, S., Turesson, H., Fält, A. S., Nestor, G., Gonzalez, M. N. et al. (2021). Amylose starch with no detectable branching developed through DNA-free CRISPR-Cas9 mediated mutagenesis of two starch branching enzymes in potato. Scientific Reports 11(1): 4311.

# Chapter 12

# Opportunities and Challenges of CRISPR/Cas Systems in Crop Breeding

*Dipak Sarode,*[1,3,5] *Madhavi Purankar,*[1] *Rachayya Devarumath,*[1,5,*] *Kapil Sushir,*[2] *Prashant Kawar*[3] and *C. Appunu*[4]

## 1. Introduction

The revolutionary Clustered Regulatory Interspaced Short palindromic repeats (CRISPR/Cas) system has emerged as a revolutionary technology and has transformed the landscape of crop breeding. Its unprecedented precision and efficiency in targeted genome editing offers opportunity to resolve critical challenges in agriculture, ranging from enhancing yield, nutritional value to improving resilience against various environmental threats. Genome editing, particularly with tools like CRISPR/Cas9, offers significant advantages by enabling rapid, precise, and predictable modifications to plant genomes. Its efficiency, simplicity, and versatility have made it a widely popular method for crop improvement, especially in resistance breeding (Wolter et al. 2019). The CRISPR/Cas9 system can perform gene knockouts, insertions, and replacements, leading to loss-of-function, knockdown, or activation mutants, which can aid in various aspects of crop improvement, such as improvement in yield, improvement in nutritional quality, enhanced tolerance to abiotic and biotic stresses.

Several challenges in agriculture need to be addressed through genome editing technology, such as maintaining global food security, mitigating climate change, and providing tolerance against severely affecting biotic and abiotic stresses. Moreover, genome editing has some advantages such as precision, which is the main limitation of conventional breeding. The world population is expected to reach 10 billion by 2050, there is a need to increase food crop production by 60% to ensure global food security. Improving abiotic stress tolerance and biotic stress resistance, improving the quality of the crop can help meet this demand (Mohidem et al. 2022). Abiotic stresses like drought, salinity, extreme temperatures, and heavy metals are becoming more severe due to climate change,

[1] Molecular Biology and Genetic Engineering Laboratory, Vasantdada Sugar Institute, Pune, Maharashtra 412307, India.
[2] Plant Breeding Section, Vasantdada Sugar Institute, Pune, Maharashtra 412307, India.
[3] ICAR-Directorate of Floricultural Research, Pune, Maharashtra 411005, India.
[4] ICAR-Sugarcane Breeding Institute, Coimbatore 641007, India.
[5] Department of Biotechnology, Savitribai Phule Pune University, Pune, Maharashtra 411007, India.
* Corresponding author: rdevarumath@gmail.com

DOI: 10.1201/9781003659907-15

threatening crop productivity. Genome editing can offer wide solutions for overexpression/repair/removal of the target gene, thereby helping to withstand against impacts. Biotic and abiotic stresses can cause significant yield losses in major crops. For example, drought alone can reduce yield by 50–70% in different crops (Yadav et al. 2023). Additionally, phytopathogens constitute a persistent and pervasive threat to global agricultural productivity, precipitating substantial economic losses annually, with estimates suggesting that up to 40% of yield in economically important crops are lost due to various fungal, viral, and bacterial diseases (Scheben et al. 2017, Pandit et al. 2022).

Traditional breeding methods take a long time to develop new varieties with desired traits. Genome editing allows for more precise and rapid crop improvement. Traditional breeding methods sometimes fall short when trying to introduce resistance genes, especially if they are linked to unwanted traits or are not present in the target species' genetic pool (Yadav et al. 2023). The introgression of desirable traits into elite varieties is frequently accompanied by undesirable traits due to linkage drag, complicating the process. Moreover, the development of new cultivars using these methods is a long-term endeavor, as backcrossing to segregate unwanted changes in progeny is particularly time-consuming (Li et al. 2022). Genome editing technologies like CRISPR/Cas9 allow for targeted modifications to enhance specific stress tolerance traits without introducing foreign DNA.

*Multiple stress tolerance*: Genome editing can potentially improve tolerance to multiple abiotic stresses simultaneously by targeting key regulatory genes (Scheben et al. 2017). Developing stress-tolerant crops through genome editing can contribute to more sustainable and climate-resilient food production systems.

This chapter highlights the strategic role of genome editing in accelerating research for the development of superior crop varieties with a focus on key applications, including CRISPR/Cas-based applications in mitigating abiotic stresses such as drought and salinity, strategies to provide resistance against biotic stresses, herbicide tolerance, crop domestication and haploid induction. It also explores innovations in plant synthetic biology, newly emerging genome editing tools such as CRISPR/Cpf1, base editors, prime editors, and DNA-free editing systems, while addressing the regulatory landscape surrounding genome-edited crops.

## 2. Genome Editing Tools: A Comprehensive Overview with Emphasis on CRISPR-Cas9

Sequence-specific nucleases (SSNs) enhance genomic editing precision by creating targeted DNA double-strand breaks (DSBs). These engineered endonucleases generate DSBs that facilitate site-specific DNA insertion, deletion, and substitution. The transformative finding by Puchta et al. (1993) demonstrated that I-SceI-induced DSBs enhance homologous recombination (HR) in plants, establishing the foundation for precise plant genome engineering. SSNs target predetermined DNA sequences and activate cellular DSB repair mechanisms, primarily non-homologous end joining (NHEJ) or homology-directed repair (HDR) (Puchta 2005). During NHEJ, random DNA insertions or deletions occur at the cut site before free DNA ends reconnect, inactivating gene function (Puchta 2005). In HDR, homologous Donor DNA is introduced at the break site, serving as a repair template. HDR enables new functional gene introduction or mutation correction through replacement of defective sequences (Symington and Gautier 2011). SSNs are classified based on their target sequence recognition and cleavage capabilities. The four primary SSN categories are Engineered Homing Endonucleases (Meganucleases), zinc finger nucleases (ZFNs), Transcription Activator-Like Effector Nucleases (TALENs), and CRISPR/Cas9. Meganucleases, the earliest genome editing tools, recognize up to 18 base pairs target sites and originate from mitochondria (Smith et al. 2006).

Kim et al. (1996) first described zinc finger nucleases (ZFNs) as chimeric proteins combining a DNA-binding domain with a cleavage domain derived from the FokI type II restriction enzyme (Lone et al. 2018). The DNA binding domain contains 3–6 tandemly arranged zinc finger repeats recognizing 9–18 bp sequences, with each unit identifying 3 bp (Miller et al. 1985). Engineered chimeric ZFNs create pairs that generate DSBs to induce precise genetic modifications. ZFNs require FokI nuclease domain pairing, appropriate spacer sequences (5–6 bp), and proper chimeric nuclease alignment for dsDNA cleavage (Urnov et al. 2010). Custom-designed zinc fingers (ZnFs) are created through mutagenesis by modifying DNA-binding and catalytic domains (Bae et al. 2003). TALENs (Transcription Activator-Like Effector Nucleases) are *Xanthomonas*-derived virulence factor proteins comprising a DNA binding domain and a FokI nuclease domain (Christian et al. 2010). Operating as dimers, TALENs bind opposite DNA strands to create DSBs. Their DNA binding domain features 33–35 amino acid repeats with variability introduced by repeat variable di-residues (RVDs). These RVDs enable each module to recognize and bind specific nucleotides—guanine, adenine, cytosine, or thymine (Deng et al. 2012, Bock 2015).

The type II clustered, regularly interspaced, short palindromic repeat (CRISPR/Cas9) system, represents a revolutionary genome editing approach using RNA-guided engineered nucleases. It directs the streptococcus pyogenes-derived Cas9 endonuclease to specific genomic sequences via guide RNA (sgRNA), causing targeted cleavage (Jinek et al. 2012). Unlike meganucleases, ZFNs, and TALENs which utilize protein/DNA interactions for target recognition, CRISPR/Cas9 employs guide RNA (sgRNA). These sgRNAs are short nucleotide sequences (~ 20 nt) targeting specific genomic regions. Cas9 nuclease cleaves the resulting RNA/DNA complex, creating a DSB at the target site containing a conserved protospacer adjacent motif (PAM). Non-homologous end joining (NHEJ) typically repairs these breaks, causing nucleotide insertions or deletions (indels) in protein-coding regions (Basso et al. 2020). This process may induce frameshift mutations or disrupt targeted gene function. CRISPR/Cas9's design simplicity and precision make it a widely adopted genome editing technique. The CRISPR-Cas system is categorized based on signature genes into two classes with six types and thirty-three subtypes. Class I (types I, III, IV) requires multiple Cas proteins working together across 16 subtypes, while Class II (types II, V, VI) functions with single multipurpose Cas enzymes across 17 subtypes (Makarova and Koonin 2015). Type I systems, predominant in archaea and bacteria, feature Cas3 proteins with helicase/nuclease activity across 7 subtypes. Type II systems contain the widely used Cas9 protein with distinct HNH and RuvC cleavage domains (Jinek et al. 2012). Type III systems encode signature Cas10 genes found mainly in archaea. Recent discoveries include the smaller CRISPR-Casϕ system in bacteriophages (Pausch et al. 2020) and promising Cpf1 proteins from *Acidaminococcus* and *Lachnospiraceae bacterium* (Zetsche et al. 2015). Type IV contains Csf proteins with undetermined functions (Makarova and Koonin 2015), while Type V features Cpl proteins that interact with crRNA without requiring additional RNA (Luo et al. 2016).

## 3. Advancements Beyond CRISPR-Cas9 in Crop Improvement

Genome-editing technology has shown considerable potential in agriculture; yet, its widespread implementation remains impeded by issues like low homologous recombination (HR) efficiency, off-target effects, and limitations imposed by certain protospacer adjacent motif (PAM) sequences. Promisingly, ongoing developments are enhancing and diversifying the genome-editing toolset to address these challenges. In recent years, numerous other CRISPR systems have emerged, providing improved specificity and overcoming the constraints of the conventional CRISPR/Cas9 system. These advancements are driving significant progress in agronomy and plant breeding. The subsequent sections examine the significant potential of these developing genome-editing techniques for molecular crop enhancement initiatives.

### *3.1 CRISPR/Cpf1 Family Nuclease System*

The Type II CRISPR/SpCas9 system is esteemed for its simplicity and efficiency; yet, its dependence on a specific 5'-NGG-3' PAM sequence constrains the range of targetable genomic loci. The CRISPR/Cpf system encompasses a wider variety of nucleases with varying PAM requirements and cleavage processes, so greatly enhancing the genome editing arsenal and offering novel prospects for precise genetic alterations in plants. Cas12a (Cpf1), a Class 2 Type V-A enzyme, identifies T-rich PAM sequences and creates staggered double-strand breaks downstream of the PAM site. It employs a shorter guide RNA (~ 43 bp), improving efficiency in gene knock-in methodologies, and has been effectively used in crops such as rice and soybean (Chen et al. 2019, Kim et al. 2017, Xu et al. 2017). While *in vivo* investigations indicate little off-target effects (Teng et al. 2018), *in vitro* assessments have documented non-specific nicking activity, necessitating additional enhancement (Murugan et al. 2020). Cas12b (C2c1), a Class 2 Type V-B nuclease, is a compact enzyme that necessitates T-rich PAMs and elevated temperatures for optimal functionality. It has demonstrated potential for multiplex genome editing and generating substantial deletions in rice and Arabidopsis, without observable off-target consequences (Ming et al. 2020). Thermal optimization is necessary for the extensive application of crops (Teng et al. 2018). Cas13 (C2c2) belongs to the Class 2 Type VI group, which specifically targets single-stranded RNA instead of DNA. This system, categorized into subtypes Cas13a, Cas13b, and Cas13d, functions without stringent PAM requirements and is utilized for RNA virus resistance and transcriptome engineering in plants (Wolter and Puchta 2018, Schindele et al. 2019). Cas14 (Cas12f) is a highly compact, Class 2 Type V nuclease that can cleave single-stranded DNA (ssDNA) independently of PAM requirements. Owing to its diminutive size and specificity, Cas14 has been investigated for the targeting of ssDNA viruses such as *Geminiviridae* and *Nanoviridae* in agricultural crops (Harrington et al. 2018, Khan et al. 2019b). Collectively, these CRISPR/Cpf systems provide a flexible array of instruments for accurate genome and transcriptome modification in crops. Although Cas14 and Cas13 systems circumvent PAM limitations, current research aims to improve the specificity, efficiency, and versatility of these systems for diverse plant genomes (Ren et al. 2021, Xu et al. 2021).

### *3.2 Base Editing*

Base editing (BE) is a revolutionary genome editing technique that facilitates accurate single-nucleotide alterations without causing double-strand breaks (DSBs) or necessitating donor DNA templates. In contrast to conventional CRISPR/Cas systems that depend on homology-directed repair (HDR), base editors (BEs) directly transform one nucleotide into another, providing an effective and streamlined approach for targeted mutagenesis (Chen et al. 2019). At present, there are two primary categories of base editors: cytosine base editors (CBEs) and adenine base editors (ABEs). CBEs transform C•G base pairs into T•A, whereas ABEs convert A•T base pairs into G•C (Komor et al. 2016, Gaudelli et al. 2017). Cytosine base editors (CBEs) employ cytidine deaminases to convert cytosine into uracil, which is then identified as thymine during DNA replication. Conversely, ABEs employ altered TadA adenine deaminases to transform adenine into inosine, which polymerases perceive as guanine (Gaudelli et al. 2020). Both techniques utilize either a Cas9 nickase (nCas9) or a catalytically inactive dead Cas9 (dCas9) to target the deaminase to specific genomic locations. The effectiveness and specificity of base editing are significantly influenced by the availability of suitable PAM sequences, as CRISPR–Cas proteins remain essential for DNA targeting. Both CBE and ABE facilitate transition mutations, but they are incapable of executing transversions, insertions, or deletions. Base editors have been effectively utilized in agricultural applications across various crops, including rice, tomato, maize, wheat, and Arabidopsis, to produce advantageous features (Hess et al. 2017, Zhang et al. 2017). CBEs have been utilized to develop herbicide-resistant variations in watermelon and wheat (Tian et al. 2018, Zong et al. 2018). Despite the complexity of developing ABEs owing to the absence of naturally occurring adenine deaminases that target DNA, directed evolution has facilitated the generation of functional ABEs for plants (Hua et al. 2018). Nonetheless, apprehensions persist regarding off-target effects. Genome-wide investigations have demonstrated that CBEs may cause unintentional mutations, while ABEs have relatively greater specificity (Jin et al. 2019). In response to this, advanced generations of base editors with enhanced precision and expanded targeting capabilities are being developed (Anzalone et al. 2020, Yan et al. 2021). Since numerous advantageous agronomic features are influenced by single-nucleotide polymorphisms (SNPs) rather than extensive insertions or deletions (Henikoff and Comai 2003), base editing is set to emerge as an essential instrument in precise crop enhancement methodologies.

### *3.3 Prime Editing*

Prime editing (PE) is an advanced genome editing approach that facilitates accurate and adaptable DNA alterations in plants without the necessity of double-strand breaks (DSBs) or donor DNA templates. In contrast to conventional homologous recombination (HR), which exhibits low efficiency in plant systems due to reduced HR frequency, and base editors that are restricted to a limited number of transition mutations, prime editing can facilitate all twelve possible base substitutions, in addition to small insertions and deletions (Anzalone et al. 2020). The prime editing system has two fundamental components: a prime editor (PE) and a specifically engineered prime editing guide RNA (pegRNA). The PE is a chimeric protein that integrates a Cas9 nickase (nCas9), which creates a single-strand incision at the target locus, and a reverse transcriptase (RT) enzyme that generates the requisite DNA sequence. The pegRNA comprises a spacer sequence for target identification, a primer binding site (PBS), and a reverse transcription template that encodes the intended modification. Upon entering the cell, the PE-pegRNA complex is directed to the target DNA. The Cas9 nickase creates a nick adjacent to the PAM site, facilitating the annealing of the PBS region of pegRNA to the exposed strand, which enables the RT domain to synthesize the modified DNA strand utilizing the RT template sequence. The newly produced strand is subsequently incorporated into the DNA via cellular repair mechanisms. Prime editing has been effectively utilized in significant crop species, including rice, wheat, maize, tomato, potato, and *Arabidopsis thaliana* for accurate gene modification, illustrating its extensive applicability in plant biotechnology (Lin et al. 2020, Jin et al. 2021, Lu et al. 2021, Gupta et al. 2023, Jiang et al. 2020, Perroud et al. 2022). Notwithstanding its potential, prime editing is still in nascent phases of research in plants. Although stable transgenic lines with specific modifications have been established (Butt et al. 2020), the efficacy of editing and delivery techniques necessitates further refinement for consistent use across various plant species and characteristics. Nonetheless, premier editing presents a robust alternative to current genome editing techniques by facilitating flexible and precise genomic modifications, potentially expediting the creation of enhanced crop varieties with advantageous characteristics.

## 4. Current Applications of CRISPR/Cas Systems in Crop Breeding

Genome editing has transformed crop breeding by enhancing stress tolerance, nutritional quality, and reproductive efficiency. It helps develop biotic and abiotic stress resistance by modifying key genes, accelerates domestication by introducing desirable traits, and improves nutritional content in crops like rice and maize. The technology also enables male sterility for hybrid seed production and facilitates haploid induction, streamlining the creation of pure lines (Figure 1). These advancements address critical challenges in modern agriculture. In the following paragraphs, we discuss the details of the application of this technology in multiple aspects of crop improvement.

### *4.1 Abiotic Stress Tolerance*

#### *4.1.1 Drought Tolerance*

Several strategies, such as knockdown or modification of transcription factors, modification of proton pumps, and modification of promoters, have shown to increase stomatal efficiencies, retention of chlorophyll content and increase in drought tolerance in several crops. In Arabidopsis, Abscisic acid induced transcription repressor (AITR) which works as a feedback regulator of ABA signaling was knocked out and resulted in increased salt and drought tolerance (Chen et al. 2021). Silensing of trehalase gene in Arabidopsis resulted in higher tolerance to drought stress

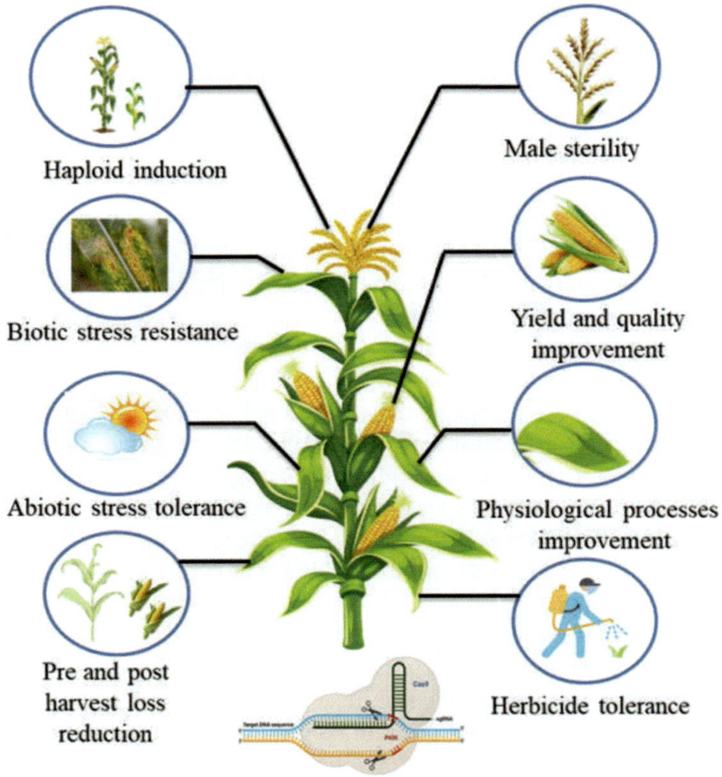


**Figure 1.** Applications of CRISPR/Cas in crop improvement.

(Nuñez-Muñoz et al. 2021). In another report, knock out of a proton pump OST2 also resulted in improved drought tolerance by exhibiting altered stomatal closing (Osakabe et al. 2016). In Arabidopsis, attachment of p65-HSF domain was fused with sgRNA to increase endogenous transcriptional levels in plants. In the same experiment, the system was used for increasing expression of AVP1 (proton pumping pyrophosphatase), which resulted in improved drought tolerance (Park et al. 2017). In rice, the OsNAC6 gene, a transcription factor involved in drought response, has been targeted using CRISPR-Cas9 to enhance drought tolerance in rice. Overexpression of OsNAC6 leads to increased stomatal closure and reduced water loss under drought conditions. Knock out of genes SRL1, SRL2 (Semi rolled leaf 1 and 2) (Liao et al. 2019), OsPYL9 (pyrabactin resistance-like-ABA receptors) (Usman et al. 2020), and ERA1 (Enhanced response to ABA1) (Ogata et al. 2020) genes was found to increase drought tolerance. CRISPR-mediated editing of DST protein (184–305 amino acid deletion) increases drought and salt tolerance in indica rice cultivar (Kumar et al. 2020). In maize, changing of GOS2 promoter region with ARGOS8 promotor was reported to increase drought tolerance (Shi et al. 2017). In tomato, GID 1 gene which acts as gibberellin receptor was altered through CRISPR/Cas 9 and reported to increase xylem proliferation thereby retaining higher leaf water content under drought conditions (Illouz-Eliaz et al. 2020). Similarly, loss of function of 'lateral organ boundaries domain' (LBD40) increased tolerance to drought stress (Liu et al. 2020). Increase in the concentration of GRXS14/15/16/17 (CGFS-type glutaredoxin 14/15/16/17) genes has demonstrated enhanced tolerance to various abiotic stresses, including heat, chilling, drought, heavy metal toxicity, and nutrient deficiency (Kakeshpour et al. 2021). In, chickpea (*Cicer arietinum*) the application of CRISPR-Cas9 technology has been employed to knockout genes like *4CL* and *REV7* for improving drought tolerance, suggesting a potential path for future studies in developing stress-tolerant chickpea varieties (Razzaq et al. 2022).

#### *4.1.2 Salt Tolerance*

In rice, the OsbHLH024 gene, which encodes a transcription factor, has been removed to reduce sodium accumulation in the shoots of rice plants, improving their tolerance to salinity (Alam et al. 2022). Similarly, alteration of the Os RR22 gene, which encodes for a transcription factor, was found to increase salt tolerance (Zhang et al. 2019). Down regulation of ethylene response factor (SIERF) through genome editing in tomato plant has reported to favor root development and root density. It was also found to improve accumulation of sugars and to increase chlorophyll content in these plants. These changes were found mainly due to higher expression of ABA biosynthesis genes and genes related to oxidative stress pathway such as Cu/Zn SOD and MDHR (Bouzroud et al. 2020). In wheat, the *TaHsp101* gene, encoding a heat shock protein, has been targeted to improve heat tolerance. Enhanced expression of *TaHsp101* protects cellular proteins from damage under high-temperature stress.

In rice, cesium tolerance was observed to be increased by inactivation of $K^+$ ion transporter Os HAK1 (Nieves-Cordones et al. 2017). Similarly, editing of gene Osnrsmp also increases rice tolerance to rise.

#### *4.1.3 Herbicide Tolerance*

ALS (Acetolactate synthase) gene editing creating transversion of G-to-T at 1882 position of Os ALS by CRISPR/Cas mediated stem has reported to be successful in creating herbicide resistant rice plants (Wang et al. 2021, Zhang et al. 2021). Homologous recombination of two amino acids (W548L and S627I) in ALS was found to increase tolerance against herbicides (Sun et al. 2016). Another gene, EPSPS (5-enolpyruvylshikimate-3-phosphate synthase), which is inhibited by glyphosate, hampers the formation of aromatic compounds, thereby causing death of weeds. By NHEJ (non-homologus end joining) method intron mediated gene insertion was carried out in EPSPS gene which resulted in glyphosate resistance (Wang et al. 2021b).

### *4.2 Improvement in Biotic Stress Tolerance*

Targeting QTL, that negatively regulate a trait, are easy targets as loss of function of genes can be easily created in the genome editing technology for improving plant biotic stress tolerance (Hua et al. 2019). However, several strategies such as editing of effector-targeted genes, disease susceptibility genes, genes involved in phytohormone signaling, transcription factors, and cell wall modification genes are utilized to improve crops resistance to various diseases owing to need of addressing different pathogenesis routes in different crops. Below are some of the examples which describe the application of different strategies through CRISPR/Cas used against different modes of pathogenesis:

Some pathogens inject effectors into plant cells to suppress immunity. Identifying and modifying the host genes that these effectors target can disrupt the pathogen's strategy and impart resistance towards the pathogen (Andolfo et al. 2016). Genome editing of 'RIN4' in *A. thaliana* has been shown to improve resistance against oomycete Peronospora parasitica and the bacteria *Pseudomonas syringae*. RIN4 acts as effector molecule for both of these organisms (Luo et al. 2009). A similar strategy exists to modify pathogenesis-related (PR) genes. Many genes, such as chitinase, defensin, glucanase, get overexpressed during an attack of pathogen and are proven to provide disease resistance. Targeting upstream regulator region of PR genes is effective strategy to induce overexpression of such PR genes. This can lead to a more coordinated and robust defense response. For example, the modification of Cs LOB1 promoter in citrus through CRISPR/Cas9 increased resistance against citrus canker disease (Peng et al. 2017). Similarly, editing of promoter Os SWEET increases resistance against bacterial blight disease in rice (Oliva et al. 2019). Resistance against tomato yellow leaf curl virus was obtained by targeting gRNA at the sequence encoding viral coat protein or replicase (Tashkandi et al. 2018).

Knocking out of disease susceptibility genes is the most preferred option for improving biotic stress resistance. Mildew Resistance Locus O genes (MLO), Lateral organ boundries (Cs LOB1)

(Jia et al. 2016), Downey mildew resistance (DMR6) (de Toledo et al. 2021) and ethylene responsive factors (ERF922) (Wang et al. 2016) are well-known targets responsible for pathogenesis. Editing of these genes offers precise resistance against the disease. Loss-of-function mutations in MLO genes confer broad-spectrum resistance to powdery mildew in various plant species (Tripathi et al. 2022). Genome editing has been used to create Mlo-B1 targeted deletion in wheat, enhancing their resistance to powdery mildew without imposing penalties in yield attributes (Li et al. 2022). Improvement in disease resistance against *P. syringae, P. capsici,* and *Xanthomonas* spp. was achieved by editing of DMR6-1 (Downey mildew resistance 6-1) gene to generate loss-of-function mutant (Thomazella et al. 2021).

Another strategy of genome editing against pathogenesis is to alter genes involved in pathogenesis signaling: for e.g., altering acid (JA) or ethylene signaling. JA is crucial for defense against necrotrophic pathogens and insects. Ethylene is involved in various stress responses. Modifying JA or ethylene signaling components can enhance resistance to specific pathogens. In tomato, *Pseudomonas syringae* causes bacterial speck disease. It produces coronatine (COR), a mimic of Jasmonic acid (JA), which is required for stomatal opening and colonization during pathogenesis. CRISPR/Cas9-mediated editing of, C-terminus JAZ domain (JAZ2Δjas) enhances stomatal defense without compromising resistance to *P. syringae* (Ortigosa et al. 2019). Transcription factors are key regulators of plant defense responses. Editing specific transcription factor genes can enhance resistance to a range of pathogens. For example, genome editing of WRKY 70 through CRISPR/Cas9 was reported to increase resistance against *Sclerotinia sclerotiorum* (Sun et al. 2018). Pathogens often secrete enzymes that degrade the cell wall. Modifying genes involved in cell wall biosynthesis or reinforcement can enhance resistance. For example, modifying genes involved in lignin biosynthesis can make the cell wall more resistant to degradation (Jhu et al. 2023).

### *4.3 Crop Domestication*

Domestication of crops through selective propagation and inbreeding has resulted in low genetic diversity. Thus, commercial varieties are more susceptible to biotic and abiotic stresses as compared to their wild relatives. Preservation of wild species is important due to their disease resistance and salt-tolerant abilities. Through genome editing technology, precise removal of only undesired characters is possible now, which is more appealing as the genes for resistant characters can be maintained in their original form and also the variety can be directly used for commercial propagation. This technique has been utilized mainly in two crops, rice and tomato (Yu et al. 2021, Li et al. 2018).

In tomato, genes such as Self Pruning (SP) and Self Pruning 5G (SP5G), are crucial for vine-like growth of wild varieties into bushy type growth habit, leading to synchronized fruit ripening and ease of mechanical harvesting. Other genes like CLAVATA3 (SICLV 3) and SIWUS) affecting stem termination and indirectly fruit size and number. In addition, gene SI GGP1 was also edited to increase vit-c content (Li et al. 2018). In wild rice, *Oryza alta* possessing CCDD genome was utilized for domestication. Genes like qSH1 ortholog, An-1, SD1, and GS3were edited, which resulted in reduced seed shattering, decreased seed breaking, and improvement in grain size (Yu et al. 2021).

### *4.4 Improvement in Nutritional Quality*

Rice with higher lysine content has been developed through editing of AK (aspartate kinase) and dihydrodipicolinate synthase (DHDPS) genes. In AK2 mutant, the 449th codon was replaced from Ser (s) to Leu (L ) substitution, whereas in DHDPS1 mutant, 12th codon was substituted from Ser (s) to Asn (N). Editing made it insensitive to the feedback mechanism of lysine, thus accumulation of free lysine content was reported to increase by nearly 58.5-fold higher in AK2 mutant and 39.2-fold higher in DHDPS mutant than wild type rice (Yang et al. 2020). GABA (γ-aminobutyric acid) is a neurotransmiter and has a role in controlling blood pressure and anxiety in humans. In tomato, about 7–15 fold increase in GABA content was carried out by CRISPR/Cas 9-mediated editing of Glutamate Decarboxylase 2/3 GAD2/3 gene (Nonaka et al. 2017). Some people suffer with 'coeliac disease', a type of autoimmune disorder, when ingestion of gluten occurs from food made from wheat, barley and rye. Low-gluten wheat was created by targeting the conserved region near the main immunogenic α-Gliadin 33-mer peptide, thereby reducing gluten content (Jouanin et al. 2020). Similarly, in peanuts three genes Arah1, Ara h2, Arah3 encode for protein that acts as allergic protein in protein. CRISPR/Cas 9 mediated knockout of AraH2 gene, which resulted in low IgEbainding (Alam et al. 2025). In soybean, the FAD2-1A and FAD2B gene code for fatty acid desaturase edited through CRISPR/Casto, prevent the conversion of oleic acid to linoleic acid. Linoleic acid content increases polyunsaturated fatty acid content in soybean oil. Editing of these genes lead to the development of high oleic acid varieties (Fu et al. 2022).

### *4.5 Haploid Induction*

Haploid lines, double haploid lines have become an important agricultural tool, as they can significantly reduce the time required for maintaining pure homozygous lines, which can be used further for hybrid seed production. MTL (Matriniline, also called as Not Like DAD (NLD) are important targets of Haploid induction. It codes for pollen-specific phospholipase, and the loss of function in this gene was reported to trigger haploid induction in maize (Kelliher et al. 2017). In rice, (matrilineal) gene was edited with CRISPR/CAS9 to obtain a haploid inducer line with multiple heading dateMTL. The line was reproduced and maintained till T4 generation by single-seed descent method. About 4–12% haploid induction was reported to be observed in these lines (Wang et al. 2021). A similar strategy has been reported in wheat (Liu et al. 2020b), foxtail millet (Cheng et al. 2021), and sugarcane (Guo et al. 2024).

### *4.6 Male Sterility*

Male sterility is crucial in crop breeding as it prevents self-pollination and ensuring cross-pollination for hybrid seed production, which enhances yield, vigor, and disease resistance. It simplifies hybridization processes, reducing the need for labor-intensive emasculation techniques. CRISPR has demonstrated significant success in generating male sterile lines across various species. For instance, Zhou et al. (2016) achieved transgene-free thermo-sensitive genic male sterile (TGMS) rice lines by knocking out the TMS5 gene, while Li et al. (2017) used a similar strategy to target ZmTMS5 in maize. Similarly, photoperiod-sensitive male sterility in rice was achieved through CRISPR-mediated disruption of the CSA gene (Li et al. 2016). In wheat, deletion of the Ms1 gene resulted in stable male sterility (Okada et al. 2019), and modification of SlSTR1 via CRISPR/Cas9 induced the same outcome in tomato (Du et al. 2020). Functional gene knockouts, such as ZmMs7 (An et al. 2020) and OsROS1 (Xu et al. 2020), further highlight CRISPR's role in controlling male fertility across species.

The ability of multiplex genome editing to target multiple homologous genes simultaneously has enhanced these outcomes. Liu et al. (2022) demonstrated this by creating triple mutants (ZmTGA9-1/-2/-3) in maize, leading to complete male sterility and showcasing the technique's potential for tackling complex traits. Additionally, Singh et al. (2018) and Li et al. (2020) generated male sterile wheat lines by modifying Ms45 and TaNP1, respectively. Beyond inducing sterility, genome editing has also been employed to restore fertility in cytoplasmic male sterile (CMS) systems. By targeting orf79, orf125, and orf312 with mitochondria-targeted TALENs, researchers restored fertility in CMS lines of rice and rapeseed (Kazama et al. 2019, Takatsuka et al. 2022). Furthermore, multiplex genome editing has facilitated simultaneous mutations in TMS5, Pi21, and Xa13, producing thermo-sensitive genic male sterile lines with improved disease resistance (Li et al. 2019). Notably, CRISPR/Cas9 has also been used to edit mitochondrial genes linked to CMS. For example, in tobacco, the CMS-associated mitochondrial gene 'mtatp9' was cleaved using CRISPR/Cas9 with mitochondrial localization signals, resulting in male-sterile mutants characterized by stamen abortion and altered mitochondrial DNA copy numbers (Chang et al. 2023).

## 5. Challenges in CRISPR/Cas Application in Crop

Genome editing has come a long way in the last ten years and has been used successfully in agriculture, but it still has a lot of problems that make it hard to use in breeding programs. It remains hard to design very effective sgRNAs because plant genomes are complicated and changeable. To work consistently across different species, the expression and activity of Cas proteins often need to be improved. Getting CRISPR/Cas systems into plant tissues that don't want to accept them is hard, which makes them less useful. The fear of unintended mutations shows how important it is to be more precise when editing. It usually takes a long time and a lot of work to regenerate edited plants through tissue culture, and it depends on the species, which makes it hard to scale up. Getting rid of transgenes is especially important for getting regulatory approval and public acceptance, especially in polyploid crops with complicated genetics. To get the most out of genome editing for crop improvement, we need to make better sgRNA design tools, better Cas variants, more effective delivery systems, and better regeneration techniques.

### *5.1 Improving CRISPR Efficiency: sgRNA/Cas Expression and Delivery Optimization*

Reliable delivery methods and precise sgRNA and Cas protein expression are the main requirements for increasing the efficacy of CRISPR/Cas-mediated genome modification in plants. Studies show that sgRNAs with about 65% GC content are associated with better editing performance (Ren et al. 2014). Enhanced expression of Streptococcus pyogenes Cas9 (SpCas9) and sgRNA selection utilizing small nuclear RNA promoters, including U3 and U6, particularly endogenous versions, substantially improve modification results in both dicotyledonous and monocotyledonous plants (Lowder et al. 2015, Long et al. 2018). Generally, Cas9 expression operates under RNA polymerase II (Pol II) promoter control, though co-expression with sgRNA beneath a unified Pol II promoter through ribozyme mechanisms has achieved editing rates reaching 30% in rice while supporting multiple target approaches (Tang et al. 2016). Beyond expression enhancement, CRISPR nuclease selection affects modification performance. Type II Cas9, Type V-A Cas12a (Cpf1), and Type V-B Cas12b provide varied editing functionalities. Forms of Cas12a, such as LbCas12a, have outperformed other versions in rice applications and have demonstrated efficacy in altering fruit species, such as citrus and bananas, for traits like coloration and resistance to pathogens (Jia et al. 2019, Wu et al. 2020). Through the use of PTG (polycistronic RNA-gRNA) and Csy4 frameworks, which allow for the simultaneous expression of multiple gRNAs, multiple genome modifications have progressed and produced improved outcomes in a variety of crops, including wheat and tomato (Cermak et al. 2017). Effective multiplexed gene activation can be supported by the modified tRNA–gRNA2.0 system functioning under single Pol II promoter control (Pan et al. 2021). Delivering CRISPR/Cas components effectively remains a critical necessity. Traditional methods, such as particle bombardment and *Agrobacterium*-mediated transformation, are still frequently used (Ozyigit and Yucebilgili Kurtoglu 2020). However, wider application is limited by the limited protoplast regeneration. While nanoparticle-based transformation offers potential solutions for species exhibiting resistance to conventional transformation techniques, viral delivery systems offer transgene-free sgRNA introduction methods in Cas9-expressing varieties (Laforest and Nadakuduti 2022, Demirer et al. 2019). Recent research has developed a novel cut-dip-budding delivery method that circumvents the need for tissue culture and enables CRISPR-mediated editing in both herbaceous and woody species by taking advantage of root-suckering (Cao et al. 2023). In order to advance genome editing across a variety of crop varieties, it is essential to optimize the CRISPR expression system and investigate alternative delivery methods.

### *5.2 Off-Target Effects a Major Challenge in CRISPR/Cas Applications*

Off-target genetic modifications continue to represent one of the most significant constraints in implementing CRISPR/Cas-mediated genome modification technologies. These accidental alterations, occurring at genomic locations with incomplete sequence similarity to the guide RNA (gRNA), can undermine the accuracy and reliability of genetic modification, particularly in crop development and therapeutic applications. Similar issues with non-specific activity arise with base modification, a CRISPR-derived technique that permits single-nucleotide changes without causing double-strand breaks. While the protospacer adjacent motif (PAM) is required to initiate Cas9 attachment, studies using single-molecule visualization and structural analysis have shown that the DNA region adjacent to the PAM, specifically the 3′ terminal of the crRNA target sequence, is essential for preserving nuclease function and R-loop stability (Rees et al. 2017). Unintentional modification of neighboring cytosines in cytosine base modification systems (CBEs) like BE3 can result in unintended single-nucleotide variants (SNVs), which often cause C-to-T changes (Jin et al. 2020). To combat this problem, modified forms such as YEE-BE3 and enhanced precision versions like HF-BE3 have been developed to reduce non-specific modifications in plant applications, such as rice (Rees et al. 2017, Jin et al. 2020).

Improving the accuracy of modifications is a key objective in the development of CRISPR technology. gRNA design refinement is one direct method. According to research, gRNAs with GC contents between 40 and 60 percent exhibit better target-specific activity and fewer off-target interactions (Bae et al. 2014, Ren et al. 2014). Targeting precision is influenced by gRNA structural arrangement in addition to sequence optimization (Naeem et al. 2020). Certain studies report precision enhancements of several magnitude orders. Chemical modifications, such as adding bridged or locked nucleic acids within gRNA sequences, can improve targeting accuracy and strengthen DNA-RNA hybrid stability (Cromwell et al. 2018). A different approach is to temporarily introduce CRISPR/Cas elements as preformed ribonucleoprotein (RNP) assemblies. This has been shown to reduce non-specific activity and stop permanent DNA incorporation in maize and wheat (Svitashev et al. 2016, Liang et al. 2017). To increase target selectivity, other accuracy-focused tools like SpCas9-HF1, Cas12a, and modified deaminases, as well as high-precision SpCas9 (eSpCas9), have been developed (Chen et al. 2017, Zhang et al. 2019, Jin et al. 2020). Despite significant advancements, further development of base modification systems and CRISPR/Cas technologies is still necessary to eradicate unwanted modifications and ensure safe, precise, and reliable genome modifications for a range of species and applications.

### *5.3 Optimizing Genetic Transformation and Regeneration Systems*

For genome modification to work, genetic transformation and plant regeneration must be done well. However, these are still major problems for many plant species. Most of the time, traditional regeneration methods use tissue culture techniques, which can take a long time, require a lot of work, and are prone to poor performance and cellular mosaicism. Another possible approach is to focus CRISPR/Cas system expression on reproductive cells. This makes it more likely that changes will be passed on and less likely that cells will be different from each other. This method, which uses egg cell- or male reproductive tissue-specific promoters to target reproductive cells, has been shown to lower mosaicism in Arabidopsis (Wang et al. 2015). Recent progress has been made in finding developmental control factors that greatly boost the ability of plant regeneration. Transcription regulators like Growth Regulating Factors (GRFs), GRF-Interacting Factors (GIFs), and GRF-GIF fusion proteins have been shown to greatly improve regeneration in a number of crops, such as wheat, rice, and triticale, without having a big impact on plant growth (Debernardi et al. 2020). Transcription regulators like WUSCHEL (WUS) and BABY BOOM (BBM) also help regeneration, but not as much as the other two (Lowe et al. 2016). Since modifying the genome usually changes individual cells, the ability to grow whole plants from single modified cells is essential for making changes that can be passed down. This need also applies to methods of transformation in planta. So, it is very important to improve single-cell regeneration to turn CRISPR changes into plants that work. In many crops, like rice, wheat, and maize, the first attempts at transformation ran into problems because they couldn't regenerate very well. The way CRISPR/Cas components are delivered is another important factor that affects the success and accuracy of modifications, in addition to regeneration. Traditional methods of delivering T-DNA involve adding editing constructs to the genome, which can cause problems with regulation because they create stable, heritable expression. Temporary delivery methods that use RNA or ribonucleoprotein (RNP) complexes can also make genome-modified plants without transgenes. This avoids long biosafety approval processes (Zhang et al. 2019). Temporary delivery also lowers the risk of off-target effects by shortening the time of exposure. In the end, genome modification works only if both the editing components get to the right cells and those cells can grow back. In the future, we should keep working on improving tissue culture protocols, looking into how to use immature seed or meristematic tissue, and finding more ways to use transgene-free editing systems to make sure they are legal and accepted by the public.

## *5.4 DNA-Free Genome Editing Systems*

Conventional genome editing frequently entails the incorporation of DNA constructs containing editing elements into the host genome. Nonetheless, such arbitrary integration may result in undesired genetic modifications, and remnant DNA fragments may still integrate post-degradation, potentially inducing adverse effects (Kim et al. 2014). Furthermore, extended expression of editing nucleases may elevate off-target alterations in plants (Lawrenson et al. 2015, Shan et al. 2015). The incorporation of foreign DNA into the plant genome poses regulatory issues, particularly for genetically modified organisms (Jones 2015). DNA-free genome editing has evolved as a novel technique that mitigates associated hazards, providing a more accurate and regulatory-compliant strategy for enhancing crops. This approach depends on the direct administration of pre-formed CRISPR/Cas9 ribonucleoproteins (RNPs) instead of DNA constructs. Woo et al. (2015) were the pioneers in demonstrating DNA-free genome editing in plants by introducing CRISPR/Cas9 ribonucleoproteins into the protoplasts of Arabidopsis, tobacco, lettuce, and rice. Similarly, Malnoy et al. (2016) successfully altered genes in apple and grape protoplasts. Researchers are looking into alternative delivery methods because many important crops lack efficient protoplast regeneration systems. Using single-stranded DNA oligonucleotides, particle bombardment has proven effective in introducing CRISPR/Cas9 RNPs or RNAs into crops like wheat and maize, resulting in precise gene knock-ins and successful gene knockouts (Zhang et al. 2016, Liang et al. 2017). This technique exhibits high efficiency and very little off-target action. Moreover, combining base editing with DNA-free methods, as demonstrated by Zong et al. (2018) C-to-T conversions in wheat, increases the likelihood of precise and economically viable genome editing in agriculture.

## *5.5 Regulatory Framework and Safety Assessment of Genome-Edited Crops*

The regulation of genome-edited crops, particularly those created with CRISPR-Cas systems, varies considerably among nations. A significant issue is the classification of these crops as genetically modified organisms (GMOs) (Eş et al. 2019). Certain nations, like the European Union (EU), New Zealand, and India, govern genome-edited crops according to the method employed, categorizing them as GMOs regardless of the absence of foreign DNA (Jouanin et al. 2018, Fritsche et al. 2018, Friedrichs et al. 2019). Conversely, nations such as Argentina, Brazil, Canada, Australia, the USA, Japan, and Chile adopt a product-based framework, wherein genome-edited crops devoid of transgenes are not classified as GMOs (Lema 2019, Eriksson et al. 2019, Van et al. 2019). Nations like Nigeria and Kenya are in the process of formulating their genome editing rules (Tripathi et al. 2020), whereas China regulates genome-edited plants under current GMO legislation but has yet to establish specific regulations (Gao et al. 2018). The disparity among nations generates regulatory ambiguity and may impede the marketing and international trade of genome-edited crops (Buchholzer and Frommer 2022). Genome editing is classified into three categories according to the nature of the DNA modification: SDN-1, which effects minor alterations without a DNA template; SDN-2, which employs a short DNA sequence to facilitate a specific modification; and SDN-3, which incorporates larger genetic components (Podevin et al. 2013, Sprink and Wilhelm 2023). These distinctions assist regulatory authorities in assessing the requisite level of risk and regulation. In recent years, numerous governments have revised or implemented new guidelines. India, Japan, China, and the Philippines have enacted laws for genome-edited crops devoid of foreign DNA to be exempt from stringent GMO regulations (Van et al. 2019). In Switzerland, genome-edited crops may receive approval if they demonstrate tangible advantages for the environment, agricultural producers, or consumers. Norway has established a public commission to assess genome editing in agriculture (Buchholzer and Frommer 2022). Despite the precision of genome editing and its resemblance to spontaneous mutations, safety concerns persist. Off-target effects, or unexpected mutations, may arise and require examination during safety evaluations (Wolt et al. 2016). Public knowledge is crucial for fostering consumer trust and endorsement of these technologies. Moreover, concerns over intellectual property rights (IPRs) and access to technology must be resolved to guarantee equitable utilization and benefit-sharing, especially for smallholder farmers and researchers.

## 6. Conclusions

Crop breeding and enhancement have been revolutionized by CRISPR/Cas systems and other genome editing technologies. These tools offer remarkable precision, efficiency, and versatility, enabling accurate modifications that expedite traditional breeding methods. These days, their applications range from enhancing resistance to abiotic and biotic stresses, domestication of crops, and the use of heterosis to haploid induction, the production of male sterile lines, and the improvement in nutritional quality. Genome editing is expanding its ability to alter complex traits while addressing regulatory and biosafety concerns thanks to developing technologies like CRISPR/Cpf1, base editors, prime editors, and DNA-free systems. These techniques allow precise gene alterations without adding foreign DNA, changing the way that genetically modified crops are perceived and controlled globally. Even though there has been a lot of progress, there are still many obstacles to overcome and important future directions. For safe and consistent outcomes, it is essential to improve the precision and reduce off-target effects of genome editing tools. For broader adoption, it will be crucial to develop PAM-independent systems, improve delivery strategies for different crop species, and create traceable modifications. Furthermore, the market introduction of genome-edited crops will be supported by the establishment of thorough, globally coordinated regulatory frameworks, particularly those that distinguish between SDN-1/2 and transgenic SDN-3 events. Recent institutional support and policy changes in countries like India point to encouraging progress. Investment in functional genomics, high-throughput platforms, and collaborative international are required to fully realize the potential of genome editing. Genome editing is ultimately poised to play a crucial role in sophisticated plant breeding, ensuring global food and nutritional security through resilient and sustainable practices.

## References

Alam, M. S., Kong, J., Tao, R., Ahmed, T., Alamin, M., Alotaibi, S. S. et al. (2022). CRISPR/Cas9 mediated knockout of the OsbHLH024 transcription factor improves salt stress resistance in rice (*Oryza sativa* L.). Plants 11: 1184.

An, X., Ma, B., Duan, M., Dong, Z., Liu, R., Yuan, D. et al. (2020). Molecular regulation of ZmMs7 required for maize male fertility and development of a dominant male-sterility system in multiple species. Proc. Natl. Acad. Sci. 117: 23499–23509.

Andolfo, G., Iovieno, P., Frusciante, L. and Ercolano, M. R. (2016). Genome-editing technologies for enhancing plant disease resistance. Frontiers in Plant Science 7: 1813.

Anzalone, A. Z., Koblan, L. W. and Liu, D. R. (2020). Genome editing with CRISPR-Cas nucleases, base editors, transposases and prime editors. Nature Biotechnology 38: 824–844. https://doi.org/10.1038/s41587-020-0561-9.

Bae, K. H., Do Kwon, Y., Shin, H. C., Hwang, M. S., Ryu, E. H., Park, K. S. et al. (2003). Human zinc fingers as building blocks in the construction of artificial transcription factors. Nature Biotechnology 3: 275–80.

Bae, S., Park, J. and Kim, J. S. (2014). Cas-OFFinder: a fast and versatile algorithm that searches for potential off-target sites of Cas9 RNA-guided endonucleases. Bioinformatics 30(10): 1473–5.

Basso, M. F., Arraes, F. B., Grossi-de-Sa, M., Moreira, V. J., Alves-Ferreira, M. and Grossi-de-Sa, M. F. (2020). Insights into genetic and molecular elements for transgenic crop development. Frontiers in Plant Science 11: 509.

Bock, R. (2015). Engineering plastid genomes: methods, tools, and applications in basic research and biotechnology. Annual Review of Plant Biology 66: 211–241.

Bouzroud, S., Gasparini, K., Hu, G., Barbosa, M. A. M., Rosa, B. L., Fahr, M. et al. (2020). Down regulation and loss of auxin response factor 4 function using CRISPR/Cas9 alters plant growth, stomatal function and improves tomato tolerance to salinity and osmotic stress. Genes (Basel) 11(3): E272. doi:10.3390/ genes11030272.

Buchholzer, M. and Frommer, W. B. (2022). An increasing number of countries regulate genome editing in crops. New Phytologist 1; 237.

Butt, H., Rao, G. S., Sedeek, K., Aman, R. and Mahfouz, M. (2020). Engineering herbicide resistance via prime editing in rice. Plant Biotechnol J. 18: 2370–2372. https://doi.org/10.1111/pbi.13399.

Cao, X., Xie, H., Song, M., Lu, J., Ma, P., Huang, B. et al. (2023). Cut–dip–budding delivery system enables genetic modifications in plants without tissue culture. The Innovation 4(1).

Čermák, T., Curtin, S. J., Gil-Humanes, J., Čegan, R., Kono, T. J., Konečná, E. et al. (2017). A multipurpose toolkit to enable advanced genome engineering in plants. The Plant Cell 29(6): 1196–217.

Chang, Y., Liu, B., Jiang, Y., Cao, D., Liu, Y. and Li, Y. (2023). Induce male sterility by CRISPR/Cas9-mediated mitochondrial genome editing in tobacco. Functional & Integrative Genomics 23(3): 205.

Chen, J. S., Dagdas, Y. S., Kleinstiver, B. P., Welch, M. M., Sousa, A. A., Harrington, L. B. et al. (2017). Enhanced proofreading governs CRISPR–Cas9 targeting accuracy. Nature 550(7676): 407–10.

Chen, K., Wang, Y., Zhang, R., Zhang, H. and Gao, C. (2019) CRISPR/Cas genome editing and precision plant breeding in agriculture. Annual Review of Plant Biology 70(1): 667–97. https://doi.org/10.1146/annurev-arplant-050718-100049.

Chen, S., Zhang, N., Zhou, G., Hussain, S., Ahmed, S., Tian, H. et al. (2021). Knockout of the entire family of AITR genes in Arabidopsis leads to enhanced drought and salinity tolerance without fitness costs. BMC Plant Biol. 21: 137. https://doi.org/10.1186/s12870-021-02907-9.

Cheng, Z., Sun, Y., Yang, S., Zhi, H., Yin, T., Ma, X. et al. (2021). Establishing in planta haploid inducer line by edited SiMTL in foxtail millet (*Setariaitalica*). Plant Biotechnol J. 19(6): 1089–1091. https://doi.org/10.1111/pbi.13584.

Christian, M., Cermak, T., Doyle, E. L., Schmidt, C., Zhang, F., Hummel, A. et al. (2010). Targeting DNA double-strand breaks with TAL effector nucleases. Genetics 186(2): 757–761.

Cromwell, C. R., Sung, K., Park, J., Krysler, A. R., Jovel, J., Kim, S. K. et al. (2018). Incorporation of bridged nucleic acids into CRISPR RNAs improves Cas9 endonuclease specificity. Nature Communications 9(1): 1448.

Debernardi, J. M., Tricoli, D. M., Ercoli, M. F., Hayta, S., Ronald, P., Palatnik, J. F. et al. (2020). GRF–GIF chimeric protein improves the regeneration efficiency of transgenic plants. Nature Biotechnology 38(11): 1274–9.

Demirer, G. S., Zhang, H., Goh, N. S., González-Grandío, E. and Landry, M. P. (2019). Carbon nanotube–mediated DNA delivery without transgene integration in intact plants. Nature Protocols (10): 2954–71.

Deng, D., Yan, C., Pan, X., Mahfouz, M., Wang, J., Zhu, J. K. et al. (2012). Structural basis for sequence-specific recognition of DNA by TAL effectors. Science 6069: 720–3.

Du, M., Zhou, K. E., Liu, Y., Deng, L., Zhang, X., Lin, L. et al. (2020). A biotechnology-based male-sterility system for hybrid seed production in tomato. The Plant Journal 102(5): 1090–100.

Eriksson, D., Kershen, D., Nepomuceno, A., Pogson, B. J., Prieto, H., Purn hagen, K. et al. (2019). A comparison of the EU regulatory approach to directed mutagenesis with that of other jurisdictions, consequences for international trade and potential steps forward. New Phytol. 222: 1673–1684. https:// doi. org/10.1111/nph.15627.

Eş, I., Gavahian, M., Marti-Quijal, F. J., Lorenzo, J. M., Khaneghah, A. M., Tsatsanis, C. et al. (2019). The application of the CRISPR-Cas9 genome editing machinery in food and agri cultural science: Current status, future perspectives, and associ ated challenges. Biotechnol Adv. 37: 410–421. https:// doi. org/ 10. 1016/j.biotechadv.2019.02.006.

Friedrichs, S., Takasu, Y., Kearns, P., Dagalliar, B., Oshima, R., Schofield, J. et al. (2019). Meeting report of the OECD conference on "genome editing: applications in agriculture-implications for health, environment and regulation." Transgenic Res. 28: 419 463. https://doi.org/10.1007/s11248-019-00154-1.

Fritsche, S., Poovaiah, C., MacRae, E. and Thorlby, G. (2018). A New Zealand perspective on the application and regulation of gene editing. Front Plant Sci. 9: 1323. https:// doi. org/10. 3389/fpls.2018.01323.

Fu, M., Chen, L., Cai, Y., Su, Q., Chen, Y. and Hou, W. (2022). CRISPR/Cas9-mediated mutagenesis of GmFAD2-1A and/or GmFAD2-1B to create high-oleic-acid soybean. Agronomy 12(12): 3218.

Gao, W., Xu, W. T., Huang, K. L., Guo, M. Z. and Luo, Y. B. (2018). Risk analysis for genome editing-derived food safety in China. Food Control 84: 128–37.

Gaudelli, N. M., Komor, A. C., Rees, H. A., Packer, M. S., Badran, A. H., Bryson, D. I. et al. (2017). Programmable base editing of A•T to G•C in genomic DNA without DNA cleavage. Nature 551: 464–471. https://doi.org/10.1038/nature24644.

Gaudelli, N. M., Lam, D. K., Rees, H. A., Solá-Esteves, N. M., Barrera, L. A., Born, D. A. et al. (2020). Directed evolution of adenine base editors with increased activity and therapeutic application. Nature Biotechnol. 38: 892–900. https://doi.org/10.1038/s41587-020-0491-6.

Guo, Y., Sun, S., Chen, S., Zhang, Y., Wang, W., Deng, Z. et al. (2024). *In vivo* induction of sugarcane (Saccharum spp.) haploids by genome editing. Plant Physiology 196: 731–734.

Gupta, A. B.. Liu, Q. J., Chen and Yang, B. (2023). High-efficiency prime editing enables new strategies for broad-spectrum resistance to bacterial blight of rice. Plant Biotechnology Journal 21(7): 1454–1464.

Harrington, L. B., Burstein, D., Chen, J. S., Paez-Espino, D., Ma, E., White, I. P. et al. (2018). Programmed DNA destruction by miniature CRISPR-Cas14 enzymes. Science 362: 839–842. https:// doi. org/10.1126/scien ce.aav4294.

Henikoff, S. and Comai, L. (2003). Single-nucleotide mutations for plant functional genomics. Annu. Rev. Plant Biol. 54: 375–401.

Hess, G. T., Tycko, J., Yao, D. and Bassik, M. C. (2017). Methods and applications of CRISPR mediated base editing in eukaryotic genomes. Mol Cell 68: 26–43.

Hua, K., Tao, X., Yuan, F., Wang, D. and Zhu, J. K. (2018). Precise A· T to G· C base editing in the rice genome. Molecular Plant 11(4): 627–30.

Hua, K., Zhang, J., Botella, J. R., Ma, C., Kong, F., Liu, B. et al. (2019). Perspectives on the application of genome- editing technologies in crop breeding. Molecular Plant 12: 1047–1059.

Illouz-Eliaz, N., Nissan, I., Nir, I., Ramon, U., Shohat, H. and Weiss, D. (2020). Mutations in the tomato gibberellin receptors suppress xylem proliferation and reduce water loss under water-deficit conditions. Journal of Experimental Botany 71(12): 3603–3612. https://doi.org/10.1093/jxb/eraa137.

Jhu, M. Y., Ellison, E. E. and Sinha, N. R. (2023). CRISPR gene editing to improve crop resistance to parasitic plants. Front. Genome Ed. 5: 1289416. doi: 10.3389/fgeed.2023.1289416.

Jia, H., Orbovic, V., Jones, J. F. and Wang, N. (2016). Modification of the PthA4 effector binding elements in Type I CsLOB1 promoter using Cas9/sgRNA to produce transgenic duncan grapefruit alleviating XccDpthA4: dCsLOB1.3 infection. Plant Biotechnol. J. 14: 1291–1301. doi: 10.1111/pbi. 12495.

Jia, H., Orbović, V. and Wang, N. (2019). CRISPR-LbCas12a-mediated modification of citrus. Plant Biotechnology Journal 17(10): 1928–37.

Jiang, Y. Y., Chai, Y. P., Lu, M. H., Han, X. L., Lin, Q., Zhang, Y. et al. (2020). Prime editing efficiently generates W542L and S621I double mutations in two ALS genes in maize. Genome Biology 21(1): 257.

Jin, S., Fei, H., Zhu, Z., Luo, Y., Liu, J., Gao, S. et al. (2020). Rationally designed APOBEC3B cytosine base editors with improved specificity. Molecular Cell 79(5): 728–40.

Jin, S., Lin, Q., Luo, Y., Zhu, Z., Liu, G., Li, Y. et al. (2021). Genome-wide specificity of prime editors in plants. Nature Biotechnology 39(10): 1292–9.

Jin, S., Zong, Y., Gao, Q., Zhu, Z., Wang, Y., Qin, P. et al. (2019). Cytosine, but not adenine, base editors induce genome-wide off-target mutations in rice. Science 364: 292–295. https://doi.org/10.1126/science.aaw7166.

Jinek, M., Chylinski, K., Fonfara, I., Hauer, M., Doudna, J. A. and Charpentier, E. (2012). A programmable dual-RNA–guided DNA endonuclease in adaptive bacterial immunity. Science 6096: 816–21.

Jones, H. D. (2015). Regulatory uncertainty over genome editing. Nat Plants 14011.

Jouanin, A., Boyd, L. A., Visser, R. G. F. and Smulders, M. J. (2018). Development of wheat with hypoimmunogenic gluten obstructed by the gene editing policy in Europe. Front Plant Sci. 9: 1523. https:// doi. org/10.3389/fpls.2018.01523.

Jouanin, A., Gilissen, L. J. W. J., Schaart, J. G., Leigh, F. J., Cockram, J., Wallington, E. J. et al. (2020). CRISPR/Cas 9 Gene editing of Gluten in wheat to reduce Gluten content and exposure–Reviewing methods to screen for Coeliac Safety. Front Nutr 7: 51. Doi:10.3389/fnut.2020.00051.

Kakeshpour, T., Tamang, T. M., Motolai, G., Fleming, Z. W., Park, J. E., Wu, Q. et al. (2021). CGFS-type glutaredoxin mutations reduce tolerance to multiple abiotic stresses in tomato. PhysiologiaPlantarum 173(3): 1263–79.

Kazama, T., Okuno, M., Watari, Y., Yanase, S., Koizuka, C., Tsuruta, Y. et al. (2019). Curing cytoplasmic male sterility via TALEN-mediated mitochondrial genome editing. Nature Plants 5: 722–730.

Kelliher, T., Starr, D., Richbourg, L., Chintamanani, S., Delzer, B., Nuccio, M. L. et al. (2017). MATRILINEAL, a sperm-specific phospholipase, triggers maize haploid induction. Nature 542: 105–109. https://doi.org/10.1038/nature20827.

Khan, M. Z., Haider, S., Mansoor, S. and Amin, I. (2019b). Targeting plant ssDNA viruses with engineered miniature CRISPR-Cas14a. Trends Biotech 37: 800–804. https:// doi. org/10. 1016/j. tibte ch.2019.03.015.

Kim, H., Kim, S. T., Ryu, J., Kang, B. C., Kim, J. S. and Kim, S. G. (2017). CRISPR/Cpf1-mediated DNA-free plant genome editing. Nat Commun 8: 14406. https://doi.org/10.1038/ncomms14406.

Kim, S., Kim, D., Cho, S. W., Kim, J. and Kim, J. S. (2014). Highly efficient RNA-guided genome editing in human cells via delivery of purified Cas9 ribonucleoproteins. Genome Res. 1012–9.

Kim, Y. G., Cha, J. and Chandrasegaran, S. (1996). Hybrid restriction enzymes: zinc finger fusions to Fok I cleavage domain. Proceedings of the National Academy of Sciences 3: 1156–60.

Komor, A. C., Kim, Y. B., Packer, M. S., Zuris, J. A. and Liu, D. R. (2016). Program mable editing of a target base in genomic DNA without double stranded DNA cleavage. Nature 533: 420–424. https:// doi. org/ 10. 1038/nature17946.

Kumar, S. V. V., Verma, R. K., Yadav, S. K., Yadav, P., Watts, A., Rao, M. V. et al. (2020). CRISPR-Cas9 mediated genome editing of drought and salt tolerance (OsDST) Gene in Indica Mega Rice Cultivar MTU1010. Physiol. Mol. Biol. Plants 26: 1099–1110.

Laforest, L. C. and Nadakuduti, S. S. (2022). Advances in delivery mechanisms of CRISPR gene-editing reagents in plants. Frontiers in Genome Editing 4: 830178.

Lawrenson, T., Shorinola, O., Stacey, N., Li, C., Østergaard, L., Patron, N. et al. (2015). Induction of targeted, heritable mutations in barley and Brassica oleracea using RNA-guided Cas9 nuclease. Genome Biology 1–3. https://doi.org/10.1186/s13059-015-0826-7.

Lema, M. A. (2019). Regulatory aspects of gene editing in Argentina. Transgenic Res 28: 147–150. https:// doi. org/ 10. 1007/s11248-019-00145-2.

Li, J., Zhang, H., Si, X., Tian, Y., Chen, K., Liu, J. et al. (2017). Generation of thermosensitive male-sterile maize by targeted knockout of the ZmTMS5 gene. J. Genet. Genom. 44: 465–468.

Li, Q., Zhang, D., Chen, M., Liang, W., Wei, J., Qi, Y. et al. (2016). Devel opment of japonica photo-sensitive genic male sterile rice lines by editing carbon starved anther using CRISPR/Cas9. J. Genet Genom. 43: 415–419.

Li, S., Shen, L., Hu, P., Liu, Q., Zhu, X., Qian, Q. et al. (2019). Developing disease-resistant thermosensitive male sterile rice by multiplex gene editing. Journal of Integrative Plant Biology 61(12): 1201–5.

Li, S., Lin, D., Zhang, Y., Deng, M., Chen, Y., Lv, B. et al. (2022). Genome-edited powdery mildew resistance in wheat without growth penalties. Nature 602: 455–460. doi:10.1038/s41586-022-04395-9.

Li, T., Yang, X., Yu, Y., Si, X., Zhai, X., Zhang, H. et al. (2018). Domestication of wild tomato is accelerated by genome editing. Nat. Biotechnol. 36: 1160–1163. doi:10.1038/nbt.4273.

Li, Y., Wu, X., Zhang, Y. and Zhang, Q. (2022). CRISPR/Cas genome editing improves abiotic and biotic stress tolerance of crops. Front. Genome Ed. 4: 987817. doi: 10.3389/fgeed.2022.987817.

Liang, Z., Chen, K., Li, T., Zhang, Y., Wang, Y., Zhao, Q. et al. (2017). Efficient DNA-free genome editing of bread wheat using CRISPR/Cas9 ribonucleoprotein complexes. Nature Communications 8(1): 14261. https://doi.org/10.1038/ncomms14261.

Liao, S., Qin, X., Luo, L., Han, Y., Wang, X., Usman, B. et al. (2019). CRISPR/Cas9-induced mutagenesis of *Semi-Rolled Leaf1,2* confers curled leaf phenotype and drought tolerance by influencing protein expression patterns and ROS scavenging in rice (*Oryza sativa* L.). Agronomy 9(11): 728. https://doi.org/10.3390/agronomy9110728.

Lin, Q., Zong, Y., Xue, C., Wang, S., Jin, S., Zhu, Z. et al. (2020). Prime genome editing in rice and wheat. Nat Biotechnol 38: 582–585. https:// doi.org/10.1038/s41587-020-0455-x.

Liu, H., Wang, K., Jia, Z., Gong, Q., Lin, Z., Du, L. et al. (2020b). Efficient induction of haploid plants in wheat by editing of TaMTL using an optimized Agrobacterium-mediated CRISPR system. J. Exp. Bot. 71: 1337–1349. https://doi.org/10.1093/jxb/erz529.

Liu, X., Wu, D., Shan, T., Xu, S., Qin, R., Li, H. et al. (2020). The Trihelix Transcription Factor OsGTc-2 is involved adaption to salt stress in rice. Plant Mol. Biol. 103: 545–560.

Liu, X., Zhang, S., Jiang, Y., Yan, T., Fang, C., Hou, Q. et al. (2022). Use of CRISPR/Cas9-based gene editing to simultaneously mutate multiple homologous genes required for pollen development and male fertility in maize. Cells 11: 43.

Lone, B. A., Karna, S. K., Ahmad, F., Shahi, N. and Pokharel, Y. R. (2018). CRISPR/Cas9 system: a bacterial tailor for genomic engineering. Genetics Research International 1: 3797214.

Long, L., Guo, D. D., Gao, W., Yang, W. W., Hou, L. P., Ma, X. N. et al. (2018). Optimization of CRISPR/Cas9 genome editing in cotton by improved sgRNA expression. Plant Methods 14: 1–9.

Lowder, L. G., Zhang, D., Baltes, N. J., Paul III, J. W., Tang, X., Zheng, X. et al. (2015). A CRISPR/Cas9 toolbox for multiplexed plant genome editing and transcriptional regulation. Plant Physiology 169(2): 971–85.

Lowe, K., Wu, E., Wang, N., Hoerster, G., Hastings, C., Cho, M. J. et al. (2016). Morphogenic regulators Baby boom and Wuschel improve monocot transformation. The Plant Cell 28(9): 1998–2015.

Lu, Y., Tian, Y., Shen, R., Yao, Q., Zhong, D., Zhang, X. et al. (2020). Precise genome modification in tomato using an improved prime editing system. Plant Biotechnology Journal 19(3): 415.

Luo, M. L., Leenay, R. T. and Beisel, C. L. (2016). Current and future prospects for CRISPR-based tools in bacteria. Biotechnology and Bioengineering 5: 930–43.

Luo, Y., Caldwell, K. S., Wroblewski, T., Wright, M. E. and Michelmore, R. W. (2009). Proteolysis of a negative regulator of innate immunity is dependent on resistance genes in tomato and Nicotianabenthamiana and induced by multiple bacterial effectors. Plant Cell 21: 2458–2472. DOI:10.1105/tpc.107.056044.

Makarova, K. S. and Koonin, E. V. (2015). Annotation and classification of CRISPR-Cas systems. CRISPR: Methods and Protocols 47–75.

Malnoy, M., Viola, R., Jung, M. H., Koo, O. J., Kim, S., Kim, J. S. et al. (2016). DNA-free genetically edited grapevine and apple protoplast using CRISPR/Cas9 ribonucleoproteins. Frontiers in Plant Science 7: 1904. https://doi.org/10.3389/fpls.2016.01904.

Miller, J., McLachlan, A. and Klug, A. (1985). Repetitive zinc-binding domains in the protein transcription factor IIIA from Xenopus oocytes. The EMBO Journal 6: 1609–1614.

Ming, M., Ren, Q., Pan, C., He, Y., Zhang, Y., Liu, S. et al. (2020). Crispr Cas12b enables efficient plant genome engineering. Nat Plants 6: 202–208. https://doi.org/10.1038/s41477-020-0614-6.

Mohidem, N. A. J., Hashim, N., Shamsudin, R. and Che Man, H. (2022). Rice for food security: revisiting its production, diversity, rice milling process and nutrient content. Agriculture 12: 741.

Murugan, K., Seetharam, A. S., Severin, A. J. and Sashital, D. G. (2020). CRISPR Cas12a has widespread off-target and dsDNA-nicking effects. J. Biol. Chem. 295: 5538–5553. https:// doi.org/10.1074/jbc.RA120.012933.

Naeem, M., Majeed, S., Hoque, M. Z. and Ahmad, I. (2020). Latest developed strategies to minimize the off-target effects in CRISPR-Cas-mediated genome editing. Cells 9(7): 1608.

Nieves-Cordones, M., Mohamed, S., Tanoi, K., Kobayashi, N. I., Takagi, K., Vernet, A. et al. (2017). Production of low-$Cs^+$ rice plants by inactivation of the $K^+$ transporter OsHAK1 with the CRISPR-Cas system. Plant J. 92: 43–56.

Nonaka, S., Arai, C., Takayama, M., Matsukura, C. and Ezura, H. (2017). Efficient increase of γ-aminobutyric acid (GABA) content in tomato fruits by targeted mutagenesis. Scientific Reports 7(1): 7057.

Nuñez-Muñoz, L., Vargas-Hernández, B., Hinojosa-Moya, J., Ruiz-Medrano and Xoconostle-Cázares, B. (2021). Plant drought tolerance provided through genome editing of the trehalase gene. Plant Signaling & Behavior 16(4). https://doi.org/10.1080/15592324.2021.1877005.

Ogata, T., Ishizaki, T., Fujita, M. and Fujita, Y. (2020). CRISPR/Cas9-targeted mutagenesis of OsERA1 confers enhanced responses to abscisic acid and drought stress and increased primary root growth under nonstressed conditions in rice. PLoS ONE 15(12): e0243376. https://doi.org/10.1371/journal.pone.0243376.

Oliva, R., Ji, C., Atienza-Grande, G., Huguet-Tapia, J. C., Perez-Quintero, A., Li, T. et al. (2019). Broad-spectrum resistance to bacterial blight in rice using genome editing. Nature Biotechnology 37(11): 1344–50.

Okada, A., Arndell, T., Borisjuk, N., Sharma, N., Watson-Haigh, N. S., Tucker, E. J. et al. (2019). CRISPR/Cas9 mediated knockout of Ms1 enables the rapid generation of male sterile hexaploid wheat lines for use in hybrid seed production. Plant Biotechnol J. 17: 1905–1913.

Ortigosa, A., Gimenez-Ibanez, S., Leonhardt, N. and Solano, R. (2019). Design of a bacterial speck resistant tomato by CRISPR/Cas9-mediated editing of SlJAZ2. Plant Biotechnol J. 17(3): 665–673. doi: 10.1111/pbi.13006.

Osakabe, Y., Watanabe, T., Sugano, S. S., Ueta, R., Ishihara, R., Shinozaki, K. et al. (2016). Optimization of CRISPR/Cas9 genome editing to modify abiotic stress responses in plants. Sci Rep. 6: 26685. https://doi.org/10.1038/srep26685.

Ozyigit, I. I. and Yucebilgili Kurtoglu, K. (2020). Particle bombardment technology and its applications in plants. Molecular Biology Reports 47(12): 9831–47.

Pan, C., Wu, X., Markel, K., Malzahn, A. A., Kundagrami, N., Sretenovic, S. et al. (2021). CRISPR–Act3. 0 for highly efficient multiplexed gene activation in plants. Nature Plants 7(7): 942–53.

Pandit, M. A., Kumar, J., Gulati, S., Bhandari, N., Mehta, P., Katyal, R. et al. (2022). Major biological control strategies for plant pathogens. Pathogens 11(2): 273. https://doi.org/10.3390/pathogens11020273.

Park, J. J., Dempewolf, E., Zhang, W. and Wang, Z.-Y. (2017). RNA-guided transcriptional activation via CRISPR/dCas9 mimics overexpression phenotypes in Arabidopsis. PLoS ONE 12(6): e0179410. https://doi.org/10.1371/journal.pone.0179410.

Pausch, P., Al-Shayeb, B., Bisom-Rapp, E., Tsuchida, C. A., Li, Z., Cress, B. F. et al. (2020). CRISPR-CasΦ from huge phages is a hypercompact genome editor. Science 6501: 333–7.

Peng, A., Chen, S., Lei, T., Xu, L., He, Y., Wu, L. et al. (2017). Engineering canker-resistant plants through CRISPR/Cas9-targeted editing of the susceptibility gene CsLOB1 promoter in citrus. Plant Biotechnol J.15(12): 1509–1519. doi: 10.1111/pbi.12733.

Perroud, P. F., Guyon-Debast, A., Veillet, F., Kermarrec, M. P., Chauvin, L., Chauvin, J. E. et al. (2022). Prime Editing in the model plant Physcomitrium patens and its potential in the tetraploid potato. Plant Science 316: 111162.

Podevin, N., Davies, H. V., Hartung, F., Nogué, F. and Casacuberta, J. M. (2013). Site-directed nucleases: a paradigm shift in predictable, knowledge-based plant breeding. Trends in Biotechnology 6: 375–83.

Puchta, H., Dujon, B. and Hohn, B. (1993). Homologous recombination in plant cells is enhanced by *in vivo* induction of double-strand breaks into DNA by a site-specific endonuclease. Nucleic Acids Research 21: 5034–5040.

Puchta, H. (2005). The repair of double-strand breaks in plants: mechanisms and consequences for genome evolution. Journal of Experimental Botany 409: 1–14.

Razzaq, M. K., Hina, A., Gai, J., Raza, G., Akhter, M., Karikari, B. et al. (2022). CRISPR-Cas9 based stress tolerance: New hope for abiotic stress tolerance in chickpea (*Cicerarietinum*). Molecular Biology Reports 49(9): 8977–8985. https://doi.org/10.1007/s11033-022-07391-4.

Rees, H. A., Komor, A. C., Yeh, W. H., Caetano-Lopes, J., Warman, M., Edge, A. S. et al. (2017). Improving the DNA specificity and applicability of base editing through protein engineering and protein delivery. Nature Communications 8(1): 15790.

Ren, Q., Sretenovic, S., Liu, S., Tang, X., Huang, L., He, Y. et al. (2021). PAM-less plant genome editing using a CRISPR–SpRY toolbox. Nat Plants. https://doi. org/10.1038/s41477-020-00827-4.

Ren, X., Yang, Z., Xu, J., Sun, J., Mao, D., Hu, Y. et al. (2014). Enhanced specificity and efficiency of the CRISPR/Cas9 system with optimized sgRNA parameters in Drosophila. Cell Reports 9(3): 1151–1162.

Scheben, A., Wolter, F., Batley, J., Puchta, H. and Edwards, D. (2017). Towards CRISPR/Cas crops–bringing together genomics and genome editing. New Phytologist 216(3): 682–98.

Schindele, A., Dorn, A. and Puchta, H. (2019). CRISPR/Cas brings plant biology and breeding into the fast lane. Curr Opin Biotechnol 61: 7–14. https://doi.org/10.1016/j.copbio.2019.08.006.

Shan, Q., Zhang, Y., Chen, K., Zhang, K. and Gao, C. (2015). Creation of fragrant rice by targeted knockout of the OsBADH2 gene using TALEN technology. Plant Biotechnol J. 791–800. https://doi.org/10.1111/pbi.12312.

Shi, J., Gao, H., Wang, H., Lafitte, H. R., Archibald, R. L., Yang, M. et al. (2017). ARGOS8 variants generated by CRISPR-Cas9 improve maize grain yield under field drought stress conditions. Plant Biotechnol J. 15: 207–216.

Singh, M., Kumar, M., Albertsen, M. C., Young, J. K. and Cigan, A. M. (2018). Concurrent modifications in the three homeologs of Ms45 gene with CRISPR-Cas9 lead to rapid generation of male sterile bread wheat (Triticum aestivum L.). Plant Mol Biol 97: 371–383.

Smith, J., Grizot, S., Arnould, S., Duclert, A., Epinat, J. C., Chames, P. et al. (2006). A combinatorial approach to create artificial homing endonucleases cleaving chosen sequences. Nucleic Acids Research 22: e149.

Sprink, T. and Wilhelm, R. (2023). Genome editing in biotech regulations worldwide. pp. 425–435. In A Roadmap for Plant Genome Editing. Cham Springer Nature Switzerland.

Sun, Q., Lin, L., Liu, D., Wu, D., Fang, Y., Wu, J. et al. (2018). CRISPR/Cas9-mediated multiplex genome editing of the BnWRKY11 and BnWRKY70 Genes in *Brassica napus* L. Int. J. MolSci. 19(9): 2716. doi: 10.3390/ijms19092716.

Svitashev, S., Schwartz, C., Lenderts, B., Young, J. K. and Mark Cigan, A. (2016). Genome editing in maize directed by CRISPR–Cas9 ribonucleoprotein complexes. Nature Communications 7(1): 13274.

Symington, L. S. and Gautier, J. (2011). Double-strand break end resection and repair pathway choice. Annual Review of Genetics 45(1): 247–71.

Takatsuka, A., Kazama, T., Si, A. and Toriyama, K. (2022). TALEN-mediated depletion of the mitochondrial gene orf312 proves that it is a Tadukan-type cytoplasmic male sterility-causative gene in rice. Plant J. 110: 994–1004.

Tang, Xu, Xuelian Zheng, Yiping Qi, Dengwei Zhang, Yan Cheng, Aiting Tang et al. (2016). A single transcript CRISPR-Cas9 system for efficient genome editing in plants. Molecular Plant 6; 9(7): 1088–91.

Tashkandi, M., Ali, Z., Aljedaani, F., Shami, A. and Mahfouz, M. M. (2018). Engineering resistance against Tomato yellow leaf curl virus via the CRISPR/Cas9 system in tomato. Plant Signaling & Behavior 13(10): e1525996.

Teng, F., Cui, T., Feng, G., Guo, L., Xu, K., Gao, Q. et al. (2018). Repurposing CRISPR-Cas12b for mammalian genome engineering. Cell Discov. 4: 63. https:// doi.org/10.1038/s41421-018-0069-3.

Thomazella, D. P., Seong, K., Mackelprang, R., Dahlbeck, D., Geng, Y. U., Gill, U. S. et al. (2021). Loss of function of a DMR6 ortholog in tomato confers broad-spectrum disease resistance. Proceedings of the National Academy of Sciences 118(27): e2026152118.

Tian, S., Jiang, L., Cui, X., Zhang, J., Guo, S., Li, M. et al. (2018). Engineering herbicide-resistant watermelon variety through CRISPR/Cas9-mediated base-editing. Plant Cell Reports 37: 1353–6. https://doi.org/10.1007/s00299-018-2299-0.

Tripathi, L., Dhugga, K. S., Ntui, V. O., Runo, S., Syombua, E. D., Muiruri, S. et al. (2022). Genome editing for sustainable agriculture in Africa. Front. Genome Ed. 4: 876697. doi: 10.3389/fgeed.2022.876697.

Tripathi, L., Ntui, V. O. and Tripathi, J. N. (2020). CRISPR/Cas9-based genome editing of banana for disease resistance. Curr Opin Plant Biol. 56: 118–126. https://doi.org/10.1016/j.pbi.2020.05. 003.

Urnov, F. D., Rebar, E. J., Holmes, M. C., Zhang, H. S. and Gregory, P. D. (2010). Genome editing with engineered zinc finger nucleases. Nature Reviews Genetics 9: 636–46.

Usman, B., Nawaz, G., Zhao, N., Liao, S., Liu, Y. and Li, R. (2020). Precise editing of the *OsPYL9* gene by RNA-guided Cas9 nuclease confers enhanced drought tolerance and grain yield in rice (*Oryza sativa* L.) by regulating circadian rhythm and abiotic stress responsive proteins. Int. J. Mol. Sci. 21: 7854.

Van Vu, T., Sung, Y. W., Kim, J., Doan, D. T. H., Tran, M. T. and Kim, J. Y. (2019). Challenges and perspectives in homology-directed gene tar geting in monocot plants. Rice 12: 95. https:// doi. org/10.1186/s12284-019-0355-1.

Wang, F., Xu, Y., Li, W., Chen, Z., Wang, J., Fan, F. et al. (2021). Creating a novel herbicide-tolerance OsALS allele using CRISPR/Cas9-mediated gene editing. Crop J. 9: 305–312.

Wang, F. J., Wang, C. L., Liu, P. Q., Lei, C. L., Hao, W., Gao, Y. et al. (2016). Enhanced rice blast resistance by CRISPR/Cas9-targeted mutagenesis of the ERF transcription factor gene *OsERF922*. PLoS One 11: e0154027. doi:10. 1371/journal.pone.0154027.

Wang, Z., Wan, L., Xin, Q., Zhang, X., Song, Y., Wang, P. et al. (2021b). Optimizing glyphosate tolerance in rapeseed by CRISPR/Cas9-based geminiviral donor DNA replicon system with Csy4-based single-guide RNA processing. J. Exp. Bot. 72(13): 4796–4808. doi:10.1093/jxb/erab167.

Wang, Z. P., Xing, H. L., Dong, L., Zhang, H. Y., Han, C. Y., Wang, X. C. et al. (2015). Egg cell-specific promoter-controlled CRISPR/Cas9 efficiently generates homozygous mutants for multiple target genes in Arabidopsis in a single generation. Genome Biology 16: 1–2.

Wolt, J. D., Wang, K. and Yang, B. (2016). The regulatory status of genome edited crops. Plant Biotechnol J. 14: 510–518.

Wolter, F. and Puchta, H. (2018). The CRISPR/Cas revolution reaches the RNA world: Cas13, a new Swiss Army Knife for plant biologists. Plant J. 94: 767–775. https://doi.org/10.1111/tpj.13899.

Wolter, F., Schindele, P. and Puchta, H. (2019). Plant breeding at the speed oflight: The power of CRISPR/Cas to generate directed genetic diversity at multiple sites. BMC Plant Biol. 19(1): 176. doi:10.1186/s12870-019-1775-1.

Woo, J. W., Kim, J., Kwon, S. I., Corvalán, C., Cho, S. W., Kim, H. et al. (2015). DNA-free genome editing in plants with preassembled CRISPR-Cas9 ribonucleoproteins. Nature Biotechnology (11): 1162–4. https://doi.org/10.1038/nbt.3389.

Wu, S., Zhu, H., Liu, J., Yang, Q., Shao, X., Bi, F. et al. (2020). Establishment of a PEG-mediated protoplast transformation system based on DNA and CRISPR/Cas9 ribonucleoprotein complexes for banana. BMC Plant Biology 20: 1–0.

Xu, R., Qin, R., Li, H., Li, D., Li, L., Wei, P. et al. (2017). Generation of targeted mutant rice using CRISPR-Cpf1 system. Plant Biotechnol J. 15: 713–717. https://doi.org/10.1111/pbi.12669.

Xu, Y., Wang, F., Chen, Z., Wang, J., Li, W., Fan, F. et al. (2020). CRISPR/Cas9-targeted mutagenesis of the OsROS1 gene induces pollen and embryo sac defects in rice. Plant Biotechnology Journal 18: 1999.

Xu, Z., Kuang, Y., Ren, B., Yan, D., Yan, F., Spetz, C. et al. (2021). SpRY greatly expands the genome editing scope in rice with highly flexible PAM recognition. Genome Biol. 22: 6. https://doi.org/10.1186/s13059-020-02231-9.

Yadav, R. K., Tripathi, M. K., Tiwari, S., Tripathi, N., Asati, R., Chauhan, S. et al. (2023). Genome editing and improvement of abiotic stress tolerance in crop plants. Life 13(7): 1456.

Yan, D., Ren, B., Liu, L., Yan, F. and Zhou, H. (2021). High-efficiency and multiplex adenine base editing in plants using new tada variants. Mol Plant. https://doi.org/10.1016/j.molp.2021.02.007.

Yang, Q. Q., Yu, W. H., Wu, H. Y., Zhang, C. Q., Sun, S. S. and Liu, Q. Q. (2020). Lysine biofortification in rice by modulating feedback inhibition of aspartate kinase and dihydrodipicolinate synthase. Plant Biotechnol J. 19(3): 490–501. doi: 10.1111/pbi.13478.

Yu, H., Lin, T., Meng, X., Du, H., Zhang, J., Liu, G. et al. (2021). A route to *de novo* domestication of wild allotetraploid rice. Cell 184(5): 1156–1170.e14. e1114. doi:10. 1016/j.cell.2021.01.013.

Zetsche, B., Gootenberg, J. S., Abudayyeh, O. O., Slaymaker, I. M., Makarova, K. S., Essletzbichler, P. et al. (2015). Cpf1 is a single RNA-guided endonuclease of a class 2 CRISPR-Cas system. Cell 3: 759–71.

Zhang, A., Liu, Y., Wang, F., Li, T., Chen, Z., Kong, D. et al. (2019). Enhanced rice salinity tolerance via CRISPR/Cas9-targeted mutagenesis of the OsRR22 gene. Mol. Breed. 39: 47.

Zhang, R., Chen, S., Meng, X., Chai, Z., Wang, D., Yuan, Y. et al. (2021). Generating broad-spectrum tolerance to ALS-inhibiting herbicides in rice by base editing. Sci. China Life Sci. 64: 1624–1633.

Zhang, Y. and Gao, C. (2017). Recent advances in DNA-free editing and precise base editing in plants. Emerg. Top Life Sci. 1: 161–8.

Zhang, Y., Liang, Z., Zong, Y., Wang, Y., Liu, J., Chen, K. et al. (2016). Efficient and transgene-free genome editing in wheat through transient expression of CRISPR/Cas9 DNA or RNA. Nature Communications (1): 12617. https://doi.org/10.1038/ncomms12617.

Zhang, Y., Malzahn, A. A., Sretenovic, S. and Qi, Y. (2019). The emerging and uncultivated potential of CRISPR technology in plant science. Nature Plants 5(8): 778–94.

Zhou, H., He, M., Li, J., Chen, L., Huang, Z., Zheng, S. et al. (2016). Development of commercial thermo-sensitive genic male sterile rice accelerates hybrid rice breeding using the CRISPR/Cas9-mediated TMS5 editing system. Sci. Rep. 6: 1–12.

Zong, Y., Song, Q., Li, C., Jin, S., Zhang, D., Wang, Y. et al. (2018). Efficient C-to-T base editing in plants using a fusion of nCas9 and human APOBEC3A. Nature Biotechnology 36(10): 950–3.

# Section IV
# Artificial Intelligence Technology

# Chapter 13

# Introduction to the Application of Artificial Intelligence-based Solutions in Agriculture

*Tavseef Mairaj Shah*

## 1. Introduction

Agriculture is the primary source of food and fiber for the world population and is an increasingly important direct and indirect source of fuel and fertilizers (Balyan et al. 2024). While food and fiber can be self-explanatory to the reader, fuel and fertilizers might need a little explanation. When we say fuels, it does not only include biofuels that compete with food security needs and are hence considered controversial; it includes biogas and biochar oil derived from mainly agricultural waste material. And, when we say agriculture is a source, and not just a consumer of fertilizers, we are talking about biofertilizers and biochar that is derived from agricultural waste or agricultural products or byproducts.

More than 95% of the world's food is sourced from agriculture, in which soil plays a vital role (FAO 2015). Healthy soils are the essential building block of the agrifood value chain that determines both the quality and quantity of agricultural produce. The dependence of agriculture on natural resources like soil and water, and the proper functioning of different ecosystem services, highlights the central nature of agriculture as a life-supporting system on earth. From a socioeconomic perspective as well, agriculture plays a major role, with at least one-third of the world population being directly dependent on agriculture for their livelihoods (FAO 2023).

These salient features and characteristics of agriculture, however, can be attributed to the availability of the different natural resources (in good shape) and the proper functioning of ecosystem services that directly affect the functioning of agriculture. This is, however, no longer the case. Agriculture is facing different challenges that make agriculture as we have known it, impossible and impractical to a large extent. The changing realities have brought new challenges to the fore that restrict the ability of agriculture to perform its normal functioning and to achieve the aims and goals that we have briefly mentioned above.

Environmental Robotics, Institute of Wastewater Management and Water Protection, Hamburg University of Technology, Eissendorfer Strasse 42, 21073 Hamburg, Germany.
Email: tavseef.mairaj.shah@tuhh.de

DOI: 10.1201/9781003659907-17

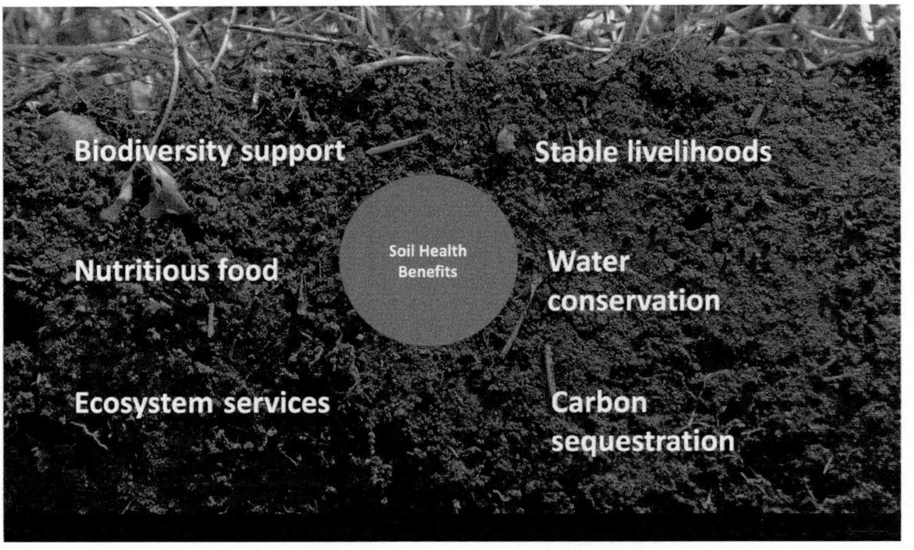


**Figure 1.** Vital functions of healthy soil directly linked to agricultural activity.

Soil forms the core of agricultural activity at the farm level, which is also the part of the agrifood value chain that determines the effects of agricultural activity on the different issues mentioned in this section. The different functions of (healthy) soils associated with agricultural activity and by extension, with the role of artificial intelligence in agriculture, discussed in the following sections, are summarised in Figure 1.

An increasing world population demands a corresponding increase in high quality food supply, which is further complicated by the increase in living standards, leading to further intensification of the associated demands. To a certain extent, this has been seen before, in the 1960s, when hunger had started to threaten the socioeconomic balance of the developing countries. As an answer, agricultural practitioners and scientists came up with a set of practices, technologies, and interventions, which is now collectively known as the 'Green Revolution' (Evenson and Gollin 2003). The Green Revolution, however, had a singular focus, and that was to produce enough food for an increasing population that was lacking nutritious food. The interventions hence included expansion of cultivated areas, introduction of hybrid high-yielding varieties, and expansion of man-made irrigation systems like canals. These interventions ultimately did lead to an increase in the availability of food and hence solved the problem that it had set out to address. But at a cost.

On one hand, agriculture that is currently heavily dependent on agrochemicals for its normal functioning, is the largest consumer of freshwater withdrawals worldwide, and is classified as a major driver of biodiversity loss. On the other hand, activities along the agrifood value chains contribute to one-third of global greenhouse gas emissions; contaminate surface and ground water resources; and have led to one-third of global soil resources being degraded (to a large extent, irreversibly) (Bindraban et al. 2012, Dudley and Alexander 2017, Srivastav et al. 2023).

These negative effects on the environment, climate, and ecosystem services increase the so-called hidden costs associated with the agrifood value chains (Riemer et al. 2023). These include the costs associated with the negative environmental (e.g., land-use change), social (e.g., poverty), and health (e.g., undernourishment) effects of activities along the agrifood value chain. The FAO estimates these costs to be around 12.7 trillion USD, which is almost 10% of the global GDP value (Steven Lord 2023). This puts an easily understandable number to the challenges that agriculture faces today; challenges that need innovative and scalable solutions for sustainable impact. Hereby, it is important to mention that in a deeply traditional sector of agriculture, the solutions proposed need to consider both the traditional knowledge systems that have been functioning for hundreds of years, but also explore the role of the latest technologies in addressing the challenges put forward by the changing realities.

This chapter aims at providing an overview of different AI-based tools that can be (or are already being) leveraged to address pressing challenges faced by the agriculture sector. To illustrate this, different AI-based solutions are enlisted in the following sections. A common thread across these different solutions is that they utilize AI to either enhance the reach of the solutions or to improve the efficiency of the same, while as in some cases AI is indispensable to the very functioning of the solution.

## 2. Agriculture-Climate-Resource Nexus

Agriculture is a source and sink of different environmental and climatic impacts, in that it contributes to them while also being a major casualty of the change in these. To illustrate this, if we take the example of the natural resource of water, we see that agriculture consumes the majority of freshwater withdrawals worldwide, but also contributes to surface water and groundwater contamination which ultimately lead to a decrease in the availability of water for agriculture and for other uses (UNESCO 2023). Similarly, most of the currently conventional agricultural practices lead to a decrease in soil health in the medium or longer run, some even leading to the complete degradation of soil while at the same time soils are the lifeblood of agriculture. On the same lines, climate change induced erratic and extreme weather patterns affect agricultural activity, as well as productivity, while agriculture contributes significantly to greenhouse gas emissions. This situation posits agriculture at a critical place as part of the agriculture-climate-resource nexus. And this nexus forms the core idea of this chapter, where we will see how artificial intelligence can contribute to improving the functioning of this nexus relationship from different perspectives.

The multiple challenges facing the agriculture sector are linked directly or indirectly with changes in the climate dynamics and disruptions in different ecosystem services and natural resource availability (Anwar et al. 2013, Mall et al. 2017, Nelson et al. 2014). The importance of these different nexus relationships and factors that drive these are most highlighted by the irreplaceable contribution agriculture makes to food security, primarily, but also increasingly significant contributions to fiber, fuel, and fertilizer supplies. The fact that current conventional agriculture is no longer sustainably able to fulfill these needs points to the need for innovative solutions that blend tried and tested agricultural systems with state-of-the-art technological approaches like interventions based on artificial intelligence (Aijaz et al. 2025, Méndez-Zambrano et al. 2023, Ryan et al. 2023). Furthermore, this nexus nature of the relationships between different components of climatic, natural resource, and agricultural systems renders linear, unidimensional solutions ineffective, hence necessitating systems solutions that originate from employing a systems perspective of the agriculture-climate-resource nexus.

One of the major challenges facing agricultural systems, and one that is closely linked to the enormous environmental, social, and health costs of agricultural systems, is the negative effects of the heavy dependence of the currently conventional form of agriculture on agrochemicals. In this regard, the above-mentioned solutions could include elements and principles from already existing systems that are being employed at scales smaller than required to drive a transformation of the agricultural systems. These systems, while presenting an alternative as a whole to the currently in convention system of agriculture, could be adapted to different contexts by integrating local traditional knowledge systems and using technological innovations to bridge the knowledge and technological gaps. These different agricultural systems are briefly presented in the table below in comparison to the conventional system. It is pertinent to mention here that there are many overlaps across these different systems, hence the tabulated comparison in Table 1, is only intended to provide the reader with an overview of these different systems (Shah et al. 2023). This will make it easier for the reader to follow the reasoning in the following sections when many of these systems are discussed in the context of different solution packages.

**Table 1.** A brief overview of different proposed transformation pathways in agriculture (CC: Shah et al. 2023).

| **Proposed pathway** | **Transformative element** |
|---|---|
| Biodynamic agriculture | Considering the farm as a whole self-regulating organism where the task is to regenerate soils through bio-stimulants, allowing plants and animals to grow in harmony with cosmic cycles, and producing healthy vital foods. |
| Organic agriculture | Enhancing natural growth processes instead of using external agricultural inputs to deliver healthier nutrition, soils, plants, and animals. |
| Holistic planned grazing | Leveraging the symbiotic relationships between farm animals and grasslands for regenerating land and growing soil and biodiversity. |
| Permaculture | Ecosystem mimicry and spatial landscape strategies to optimize energy and nutrient flows to enhance the diversity, stability and resilience of agricultural systems. |
| Agroecology | An ecological, socio-cultural, technological, economic and political approach to food systems, from production to consumption, with food sovereignty a core principle. |
| Regenerative agriculture | Land and ecosystem regeneration with a focus on soil agroecology. |
| Vertical farming | Indoor farming using controlled inputs of nutrients to increase resilience to weather disruptions and land scarcity. |
| Lab-grown foods | Artificial culturation of animal proteins as an alternative to reduce greenhouse gas emissions associated with meat-based products. |
| Nature-based solutions | Protecting and restoring ecosystems while safeguarding biodiversity. |
| Digital and robotic agriculture | Optimization of food production through digital agriculture to reduce labour intensity and increase yields. |
| Insect farming | Alternative sources of nutrients to take pressure off natural agroecosystems due to intensive footprint of agricultural activities. |

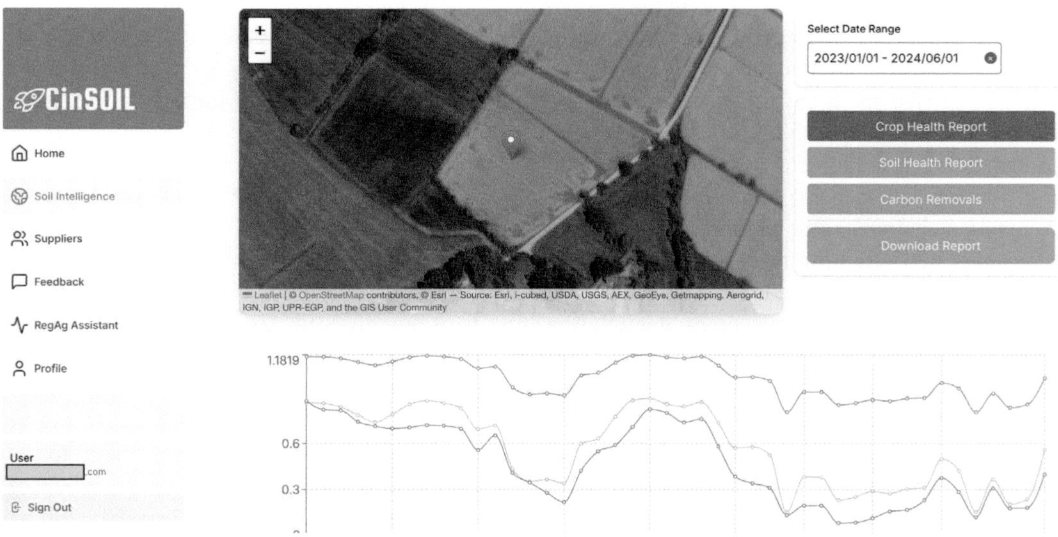


**Figure 2.** The dashboard of an AI-based tool leveraging satellite imagery to different soil and crop health parameters, in particular soil carbon stocks (CC: CinSOIL GmbH, Berlin, Germany).

As can be seen from the comparison of the different alternative agricultural systems, these follow certain core principles aimed at achieving specific goals, for example, ranging from soil health regeneration, production of healthy food, improving agricultural livelihoods, and preserving natural ecosystems. When implemented in isolation, at farm level or even in a group of farms, these systems may face little or no challenges but, in order to base a transformation of entire agrifood value chains on such systems, more synergies between these systems are needed in the implementation pathways. These synergies include, among other things, bridging the existing knowledge and financial gaps by the use of technological innovations, which include solutions based on state-of-the-art technologies based on artificial intelligence (Figure 2).

## 3. Artificial Intelligence: Origins, Definitions, Tools

Artificial intelligence (AI) can broadly be described as the branch of computer science and engineering, which employs advanced algorithms to back machine learning and deep learning models that perform tasks that conventionally require human intelligence. The basic assumption hereby is that the functions of learning, reasoning, and perception that the human intellect processes, can be detailed in a way that can enable a machine to replicate the results or even better emulate the results (Javaid et al. 2023). The flexibility of using different permutations and combinations of data sources as the input makes artificial intelligence suitable to solve a wide range of problems and even the same problem in different ways, potentially leading to optimal output (solutions). Many of the applications of AI have meanwhile become fairly commonplace and unmissable in our daily lives, ranging from face recognition features on our electronic devices to self-driving cars to automatic production lines in different industries.

From the vision of creating intelligence machines spurred by the increasing availability of digital computers based on the assumption of strong analogies between computers and the human brain in the 1950s, AI has come a long way. The term artificial intelligence (AI) was coined by a mathematician called John McCarthy who is credited with the first collaborative effort on studying artificial intelligence at the Dartmouth College in 1956 (Melanie Mitchell 2019). This initial proposal for a collaborative workshop on artificial intelligence contained terminology that is still in vogue in the field, including terms like natural-language processing, neural networks, and machine learning. Although the stated aim of McCarthy's Stanford Artificial Intelligence Project–building a fully intelligent machine in a decade–might not have come to fruition, the examples of the application of AI in agriculture in the following sections gives an idea how far the field has progressed in the past seven decades (Melanie Mitchell 2019).

The role artificial intelligence-based solutions can play in agriculture serve mainly to address the challenges mentioned in the previous section. These challenges can broadly be categorized into the domains of agricultural productivity, food quality, environmental health, resource conservation, and climate protection. The wide spectrum of tools that AI brings to the table makes it possible to design and create solutions to address these myriad challenges facing the agricultural sector. As we will see in the following sections, some of such solutions are already in practice for some time now while some are still under trial at the prototype stage, while a few have not yet moved beyond the ideation stage (Ben Ayed and Hanana 2021, Javaid et al. 2023). The different stages of the development of these various solutions is due to reasons pertaining to necessary data or infrastructure, implementation on the ground, adoption and dis-adoption behavior of the end users.

AI-based solutions cover a range of challenges in the agricultural sector. On the simpler end of the spectrum, this can take the shape of assistance in simple tasks like choosing the more suitable crop types or soil, crop and nutrient management practices based on an analysis of local conditions, expected weather patterns, crop yield predictions, and commodity price forecasts with the help of AI-based decision support systems. On the other end of the spectrum, AI-based tools can make use of machine learning, deep learning, image recognition and processing, artificial neural networks to different ends, for example, real time monitoring of crop performance, soil health, and water stress in the land using remote sensing (Javaid et al. 2023).

In addition to broadly addressing the above-mentioned concrete challenges in the sector, AI can also address issues that have not yet been properly explored and remain abstract. AI-based innovations have the potential to democratize access and initiative in many sectors including agriculture. The farmers of twenty years ago, particularly in the Global South, knew when to do what, based on the regular weather patterns, the stable state of natural resources like soil and water, and the traditional practices that were the norm. Now, on one hand, none of these constants exist anymore and on the other hand, you have a new generation that has little to no interest in agriculture. This is where these solutions could come in and help farmers to adapt to the new realities through monitoring tools, peer to peer learning, and decision support.

## 4. Potential Role of AI in Different Agricultural Sectors

"*The farmers of tomorrow will use their mobile phones as pocket soil laboratories.*", declares a public communication of the European Union-funded multi-university project called AI4SoilHealth (https://ai4soilhealth.eu/). This statement succinctly sums up the role artificial intelligence can play in agriculture, where farmers and landowners represent the main stakeholders and changemakers. The project follows the aims and objectives of the EU's Soil Health Mission for 2030 (European Commission 2020). The following eight objectives of the mission broadly delineate the action areas where the role of AI in agriculture can be the most relevant:

- Reduce desertification
- Conserve soil organic carbon stocks
- Stop soil sealing and increase the re-use of urban soils
- Reduce soil pollution and enhance restoration
- Prevent erosion
- Improve soil structure to enhance soil biodiversity
- Reduce the EU global footprint on soils
- Improve soil literacy in society

The readers can observe that it is mainly these objectives or actions linked to these that form the base of the following sections of this chapter. The following sections are structured around nine different broad topics linked to agriculture and describe the role AI can play in these, exemplified by an ongoing project researching the same or a company working on a solution in the same field.

### *4.1 Ecosystem Monitoring*

The perception of modern agriculture has undergone drastic changes in the past decade. The perspective of food security is no longer the only perspective with which policymakers, businesses, and farmers look at agriculture. The many possible positive and negative contributions of agriculture to many ecosystem services form an integral part of policy about and around agriculture. These different issues or topics that are discussed in close combination with agriculture include soil health, water protection, greenhouse gas emissions, biodiversity conservation (Mijatović et al. 2013, Ortiz et al. 2021). With these additional areas of interest, the monitoring of crop health and yield predictions are not the only applications of AI in agriculture. In fact, these arguably represent the simplest application of AI in agriculture. The more complex and broadly still under development areas of AI applications in agriculture include the monitoring of soil health, carbon and biodiversity parameters, and water dynamics (Garg et al. 2020, Yuzugullu et al. 2024).

The monitoring of these parameters with the help of AI-powered tools not only provides essential insights to the agricultural stakeholders but also opens new potential revenue streams. The monitoring of soil health, for example, includes the monitoring of soil carbon stocks, which could be used to verify and report soil carbon sequestration. These numbers can be used by agribusinesses to offset their carbon emissions or trade carbon units along the agricultural value chain. Similarly, biodiversity monitoring insights can be used to claim incentives linked to biodiversity conservation measures operational in many countries. Such solutions usually use either proximal sensing and measurement tools like sensors, cameras, or drones or remote sensing tools like satellite imagery in combination with machine learning to deliver actionable insights on the issues of carbon and biodiversity.

### *4.2 Nutrient Management*

The use of industrial fertilizers as soil additives is the conventional methodology of maintaining and improving soil nutrient content. Nitrogenous fertilizers, in particular, are widely used to maintain soil nitrogen levels required for optimal plant growth. The invention of the industrial way of production of nitrogenous fertilizers through the Haber-Bosch Process revolutionized how soil nitrogen was approached. The Green Revolution of the 1960s formalized this intending to drastically improve agricultural input to cater to the increasing food needs of the world population (Evenson and Gollin 2003). The use of fertilizers, however, is directly associated with pollution of ground and surface waters, as well as increased greenhouse gas emissions from farm activity. It is pertinent to mention here that roughly half of the quantity of nitrogenous fertilizer applied is not even used by the plants, hence leaving the nutrient cycle and ending up as a pollutant (Wei et al. 2024). In this regard, in certain countries, farmers are required to constantly monitor soil nitrogen levels and nitrogen levels in the groundwater and nearby surface waters, with fines and penalties levied in case the nitrogen levels exceed the regulatory limits.

Modelling of fertilizer application in general and soil nitrogen cycles in particular, adapted to different soil types and adjusted for different inputs provided to the soil and the type of crop grown, can provide the farmers and agricultural companies with insights, resulting in the precise application of (nitrogenous) fertilizers on their farmlands (Li et al. 2025, Takoutsing et al. 2025). This can be complemented by proximal or remote sensing of soil and crop nitrogen content and the nitrous oxide emission rates. These insights can then be used to further adapt the dosage and delivery of fertilizers to the soil, decreasing the overall quantity of fertilizer used, thereby reducing direct costs. Furthermore, this can also lead to a decrease in the quantity of fertilizers lost into the ecosystem, resulting in a reduction in the quantity of nutrients ending up as pollutants.

### *4.3 Food Safety and Security*

As of 2022, nearly two billion people worldwide did not have access to proper nutrition. This includes around one billion people who did not even have access to enough food to eat (United Nations 2022). It is pertinent to mention here the difference between enough food to eat and proper nutrition. One could have enough quantity of food to eat, but that could still be insufficient when it comes to fulfilling daily nutritional needs. While world agriculture currently produces food enough for ten billion people, one-fourth of the world population of eight billion does not have access to proper nutrition (Holt-Giménez et al. 2012, Committee on World Food Security 2022). Apart from wars and conflicts in which starvation is used as a weapon of war, food and nutritional security is also affected by avoidable food losses along the value chain. It is these losses that different initiatives and companies intend to tackle with AI-based tools. Furthermore, in regions where the geographical or weather conditions do not permit agricultural activity in a feasible manner, the use of smart greenhouses or indoor farms can be an approach that can ensure local production of nutritious food (Chen and You 2021, Hati and Singh 2021). Smart indoor or greenhouse farm assistant systems ensure optimal conditions for the growth of plants by controlling parameters like light, temperature, and humidity, in addition to steering the flow of water and nutrients to the plants with the help of sensors and image recognition (Acosta-Coll et al. 2021, Soheli et al. 2022, Ullah et al. 2022).

### *4.4 Plant Protection*

One of the major factors affecting crop performance is disease, which could have a different nature in terms of the causative agents, e.g., bacterial or fungal. Conventionally, certain agrochemicals broadly classified as pesticides have been used to prevent, control, or treat disease incidence in plants and trees. However, before pesticides became a mainstream solution, farmers used to employ traditional methods of disease control, which are now also seen in the different cropping systems being promoted as alternatives to the currently in conventional cropping systems. In this regard, AI can play both an enabling role for these alternative cropping systems and a mitigating role for the conventional cropping systems (Bhardwaj et al. 2025, Fan et al. 2025).

AI-based technologies leveraging proximal and remote sensing to monitor crop health have enabled control of disease at scale. These solutions use technologies like image recognition, neural networks, and machine learning to differentiate between healthy, non-healthy, and vulnerable plants based on different visible physiological features. Furthermore, these solutions also detect, and when necessary, stop the growth of invasive species or weeds that threaten the growth of the main crop (Chauhan et al. 2025, More et al. 2025, Yang et al. 2025).

Deep learning models have been reported to be useful in disease detection in major food crops like maize, with accuracies up to 98.50% reported (Kundu et al. 2022). The use of transfer learning for disease detection in rice crop has also been reported with a higher efficacy compared to conventional machine learning techniques (Sharma et al. 2022). Furthermore, early detection of disease in tomato and grapes with the help of machine learning has also been reported (Fenu and Malloci 2021, Jin et al. 2022). It is expected that the implementation of these detection mechanisms at scale can prevent the disastrous consequences of crop disease on one hand and reduce the use of hazardous agrochemicals used to tackle crop disease on the other.

## *4.5 Weed Detection and Management*

Identification and infestation of weeds is one of the main reasons for yield losses in major food crops like rice, wheat, maize, potato and others (Ganie et al. 2023, Horvath et al. 2023, Masson et al. 2024) (Figure 3). It is hence a challenge that is discussed under a separate section in addition to the discussion on plant protection in the previous section. The conventional method of weed management involves the application of herbicides to tackle the growth of weeds at various growth

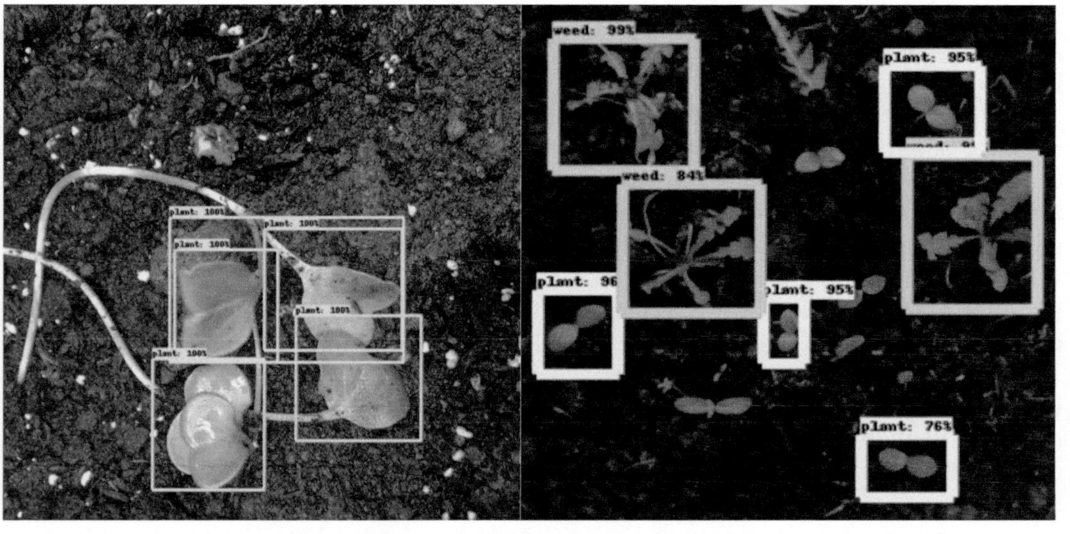


**Figure 3.** Plant and weed identification with using a deep-learning algorithm; left: ground-truth, right: detection (CC: Shah et al. 2021).

stages depending on the type of weeds. The traditional methods of weed management however involves the manual weeding of the crop fields or the use of cover crops to arrest the growth of weeds. These two methodologies of weed management are deemed inadequate for use at a large scale due to different reasons. While as herbicides are deemed unfit due to their harmful environmental and human health impacts, the traditional methods are associated with prohibitive labor requirements. Hence, progressive suggestions include proposals to employ an integrated approach that involves elements of both these methodologies, including the precision application of weedicides (Fernando and Shrestha 2023, Masson et al. 2024). The efficacy of integrated weed management approaches can be significantly improved by the use of AI-based weed detection (Chang et al. 2021, Monteiro and Santos 2022). Machine learning can be employed to use image processing to differentiate between plants (that are intended to stay) and weeds (that are intended to be removed) and subsequently guiding the weeding agent (chemical or mechanical) to undertake the weed removal action (Shah et al. 2021). The use of AI-based tools can on one hand improve the efficiency of the chemical weeding agent, hence leading to a decrease in the use of hazardous herbicides, and on the other hand it can lead to a revival of mechanical weeding practices at scale while requiring minimum labor.

## *4.6 Weather Forecasting*

Agriculture is one of the most affected sectors due to the change in weather patterns and the increasing frequency of extreme weather events attributed to climate change. This includes increased instances of drought, flooding, storms, and non-seasonal/unexpected high precipitation and high or low temperatures that disrupt agricultural activity, leading to complete loss of crop in some cases. These changes do not only affect agricultural activity by themselves but also lead to secondary effects like increased disease incidence and damage to food at different stages of the value chain (Brock and Xepapadeas 2025, Verma et al. 2025). Monitoring systems that leverage artificial intelligence to diagnose disease incidence in time to take countermeasures (possibly automatic in nature) can be a game-changer in this regard (Dolatabadian et al. 2025).

The increasing frequency and severity of extreme weather events due to changes in the climate patterns have made agricultural activity heavily dependent on accurate weather forecasts. Farming used to require experience with different practices and knowledge of weather patterns, based on which different activities would be planned. However, due to the irregular nature of temperature and precipitation changes, the traditional way of following natural weather patterns is no longer enough to conduct agricultural activity optimally. In this regard, insights on weather forecasting are increasingly playing a significant role in agriculture. Advanced weather forecasting services analyze data from different sources and provide farmers and agricultural companies with insights to help them plan the different processes and stages of farming accordingly.

## *4.7 Water Management*

Agriculture is the largest consumer of freshwater worldwide, consuming at least 70% of all freshwater withdrawals worldwide. With increased water scarcity in some regions and pollution of water resources in others, judicious use of water in agriculture has become a necessity. In this regard, certain alternative cropping systems are being promoted that reduce water consumption in agriculture compared to the conventional cropping systems. This includes the System of Rice Intensification, a non-conventional methodology of growing rice that stipulates the alternate wetting

**Figure 4.** FarmBot Genesis gantry and weather station that automatically irrigates the plants based on an analysis of the soil and weather patterns (Creative Commons Attribution 4.0 International, FarmBot).

and drying of the rice fields instead of keeping the rice fields flooded (Ceesay et al. 2006, Shah et al. 2021). Such solutions promote need-based use of water resources rather than the usage patterns dictated by convention.

In this regard, AI-based infrastructure consisting of smart sensors, smart meters, and smart valves can regulate the use of water for irrigation of crops. This can support the practice of need-based use by analyzing the water content of the soil, crop health and water requirements of the crop at a particular growth stage, weather forecasts related to expected precipitation and either provide the user with prompts or automatically release or stop water from going into the fields (Islam et al. 2025, Khalifeh et al. 2025) (Figure 4). This can, on one hand, save water from misuse and on the other hand, it can prevent the negative effects of both water scarcity and water logging on the health and productivity of the crops.

### *4.8 Decision Support Systems*

Given the mounting multiple challenges discussed in the previous sections and the dwindling confidence in traditional methods and solutions, agricultural stakeholders like farmers require data-based insights to support their decision-making process (Akratos et al. 2024). This is to enhance the decision-making process by integrating all possible information and data sources that have a bearing on the different decisions could pan out. These tools can help both conventional farmers as well as farmers wishing to transition to more sustainable cropping systems. For conventional farmers, this could mean support in planning their different activities better through insights on weather, resource availability, and soil health. For transitionary farmers, a decision support system can significantly aid the process of transition by not only providing the previously mentioned insights but also a scenario analysis involving different cropping patterns, practices, projected effects on the different relevant ecosystem services (Kong et al. 2025, Petraki et al. 2025) (Figure 5).

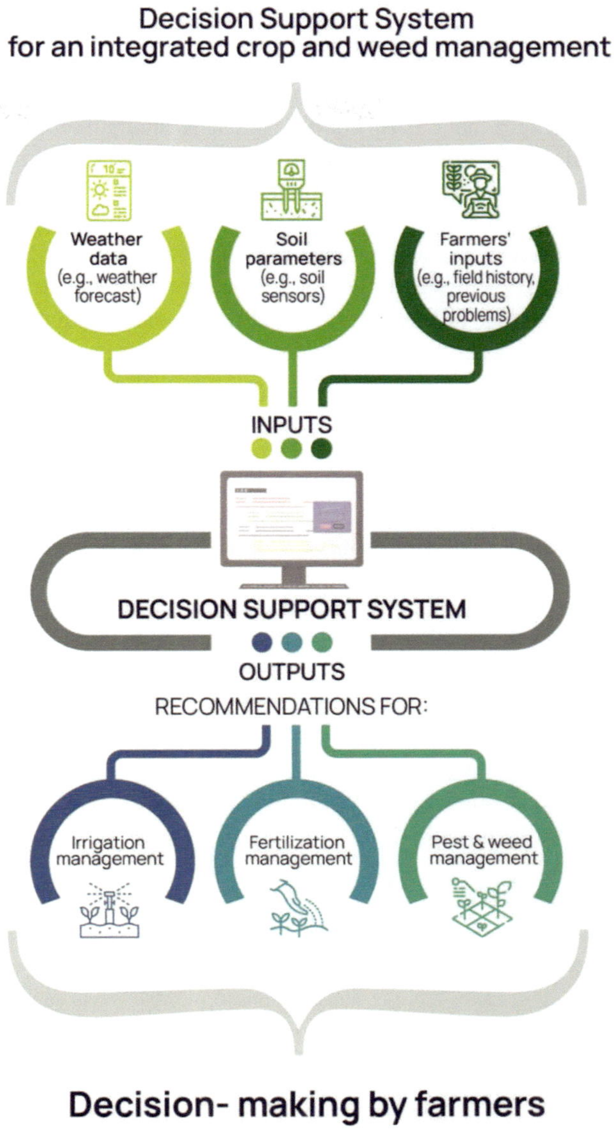


**Figure 5.** Visual example demonstrating the interaction between inputs and outputs to guide farmers in making informed decisions (CC: Petraki et al. 2025).

### *4.9 Trade and Transparency*

Agriculture is a major livelihood provider worldwide, with at least one-third of global population dependent on agriculture for their livelihoods. Agricultural produce is traded across continents and plays a vital role in sustaining food supply chains worldwide. However, it is debatable whether the different stakeholders along the agricultural value chains get their fair share of economic returns. In this regard, initiatives like "Fairtrade" and those that operate farm-to-consumer direct supply lines aim at creating a fairer agricultural business model that provides farmers a better deal for their produce. In such initiatives, traceability is an element that cannot be compromised upon and AI-based tools in combination with technologies like blockchain and internet-of-things play a major role in enhancing transparency along the value chain (Dedeoglu et al. 2023).

## 5. Conclusions

"*Africa doesn't need another .... food app. Let's be clear: tech won't save our food systems. Infrastructure will*". This statement from an agrifood practitioner from a country in Africa points to both the reality of the challenges that our agrifood systems are facing that need addressing and also the fact that technology alone, much less artificial intelligence alone, can address these challenges.

Agriculture today is again at the cusp of revolutionary changes, as it was a century ago at the start of the Third Agricultural Revolution. The role agriculture currently plays and could potentially play in addressing the different socio-economic and environmental challenges, in addition to fulfilling the nutritional needs of the world population are the center of debate at different levels of governance and policymaking. In this regard, the different stakeholders are considering many innovations–technological, social, economic–to address the major issues with agriculture today. The enabling role that artificial intelligence plays is being highlighted with the response ranging from AI being considered as a panacea for all problems agriculture faces today to a cautious criticism of the different technologies.

As discussed in the previous sections, AI-based solutions have the potential to address different challenges and blind spots across the different operational levels of the current agricultural system, ranging from agri-ecosystems monitoring to precision agriculture. Owing to these technological solutions, it appears that AI is necessary to catalyze the required changes in the agricultural system but is not sufficient to bring about the changes on its own. AI can play the role of a so-called super-spreader of innovations, whether social, economic, or technological, by virtue of the possibility of lowering the barriers to access to these innovations across the diversity of stakeholders in agricultural systems. This however should not be mistaken as a matter fact, AI on its own cannot lower the barriers to access, AI on its own cannot solve the problems we have discussed in the previous sections, AI cannot grow food, conserve biodiversity, reduce water pollution, and reduce emissions. However, AI can act as an enabler of these changes. AI can be the enabler of the transformation needed in agriculture today.

## References


Acosta-Coll, M., Anaya, D., Ojeda-Field, L. and Zamora-Musa, R. (2021). Low-cost smart indoor greenhouse for urban farming. Lecture Notes in Computer Science (Including Subseries Lecture Notes in Artificial Intelligence and Lecture Notes in Bioinformatics), 12952 LNCS 120–132. https://doi.org/10.1007/978-3-030-86973-1_9.

Aijaz, N., Lan, H., Raza, T., Yaqub, M., Iqbal, R. and Pathan, M. S. (2025). Artificial intelligence in agriculture: Advancing crop productivity and sustainability. Journal of Agriculture and Food Research 20: 101762. https://doi.org/10.1016/J.JAFR.2025.101762.

Akratos, S., Kagalou, I., Latinopoulos, D., Papaevangelou, V., Mairaj Shah, T., Onyekaozuoro Egwu, C. et al. (2024). Decision support systems based on a multiple-criteria decision analysis to promote a whole-of-resource approach for water management, with a case study of Rural Bengaluru in India. Water 16(12): 1674. https://doi.org/10.3390/W16121674.

Anwar, M. R., Liu, D. L., Macadam, I. and Kelly, G. (2013). Adapting agriculture to climate change: A review. Theoretical and Applied Climatology 113(1–2): 225–245. https://doi.org/10.1007/S00704-012-0780-1/METRICS.

Balyan, S., Jangir, H., Tripathi, S. N., Tripathi, A., Jhang, T. and Pandey, P. (2024). Seeding a sustainable future: navigating the digital horizon of smart agriculture. Sustainability 16(2): 475. https://doi.org/10.3390/SU16020475.

Ben Ayed, R. and Hanana, M. (2021). Artificial intelligence to improve the food and agriculture sector. Journal of Food Quality 2021(1): 5584754. https://doi.org/10.1155/2021/5584754.

Bhardwaj, R., Yadav, A., Sahoo, A., Kumari, P., Singh, L. A., Swapnil, P. et al. (2025). Microalgal-based sustainable bio-fungicides: a promising solution to enhance crop yield. Discover Sustainability 6(1): 1–14. https://doi.org/10.1007/S43621-025-00795-9/FIGURES/3.

Bindraban, P. S., van der Velde, M., Ye, L., van den Berg, M., Materechera, S., Kiba, D. I. et al. (2012). Assessing the impact of soil degradation on food production. Current Opinion in Environmental Sustainability 4(5): 478–488. https://doi.org/10.1016/J.COSUST.2012.09.015.

Brock, W. and Xepapadeas, A. (2025). Land use, climate change and the emergence of infectious diseases: a synthesis. Environmental and Resource Economics 88(3): 795–854. https://doi.org/10.1007/S10640-024-00949-9/FIGURES/15.

Ceesay, M., Reid, W. S., Fernandes, E. C. M. and Uphoff, N. T. (2006). The effects of repeated soil wetting and drying on lowland rice yield with System of Rice Intensification (SRI) methods. International Journal of Agricultural Sustainability 4(1): 5–14. https://doi.org/10.1080/14735903.2006.9686007.

Chang, C. L., Xie, B. X. and Chung, S. C. (2021). Mechanical control with a deep learning method for precise weeding on a farm. Agriculture 11(11): 1049. https://doi.org/10.3390/AGRICULTURE11111049.

Chauhan, A., Alam, A., Khan, M. T., Kumar, A., Bhatt, A. and Maurya, S. K. (2025). A residual deep neural network–based noninvasive system for tomato crop health monitoring, disease type identification, and remedial measures. pp. 177–202. *In*: Singh, A. J., Nikita Gupta, Sanjay Kumar, Sumit Sharma, Subho Upadhyay and Sandeep Kumar (eds.). Artificial Intelligence and Machine Learning Applications for Sustainable Development. CRC Press. https://doi.org/10.1201/9781003581246-7/RESIDUAL-DEEP-NEURAL-NETWORK.

Chen, W. H. and You, F. (2021). Smart greenhouse control under harsh climate conditions based on data-driven robust model predictive control with principal component analysis and kernel density estimation. Journal of Process Control 107: 103–113. https://doi.org/10.1016/J.JPROCONT.2021.10.004.

Committee on World Food Security. (2022). Keynote address by CFS Chair to the Community of Portuguese Speaking Countries. https://www.fao.org/cfs/resources/detail/en/c/1609703/.

Dedeoglu, V., Malik, S., Ramachandran, G., Pal, S. and Jurdak, R. (2023). Blockchain meets edge-AI for food supply chain traceability and provenance. Comprehensive Analytical Chemistry 101: 251–275. https://doi.org/10.1016/BS.COAC.2022.12.001.

Dolatabadian, A., Neik, T. X., Danilevicz, M. F., Upadhyaya, S. R., Batley, J. and Edwards, D. (2025). Image-based crop disease detection using machine learning. Plant Pathology 74(1): 18–38. https://doi.org/10.1111/PPA.14006.

Dudley, N. and Alexander, S. (2017). Agriculture and biodiversity: a review. Biodiversity 18(2–3): 45–49. https://doi.org/10.1080/14888386.2017.1351892.

European Commission. (2020). EU Mission: A Soil Deal for Europe. https://research-and-innovation.ec.europa.eu/funding/funding-opportunities/funding-programmes-and-open-calls/horizon-europe/eu-missions-horizon-europe/soil-deal-europe_en.

Evenson, R. E. and Gollin, D. (2003). Assessing the Impact of the Green Revolution, 1960 to 2000. Science 300(5620): 758–762. https://doi.org/10.1126/SCIENCE.1078710.

Fan, H., Miao, R., Guo, C., Bao, X., He, W., Sun, Y. et al. (2025). Research on the effect of diversified cropping on crop quality: a review. Agriculture 15(5): 456. https://doi.org/10.3390/AGRICULTURE15050456.

FAO. (2015). Healthy soils are the basis for healthy food production | FAO. https://www.fao.org/soils-2015/news/news-detail/en/c/277682/.

FAO. (2023). Almost half the world's population lives in households linked to agrifood systems. https://www.fao.org/newsroom/detail/almost-half-the-world-s-population-lives-in-households-linked-to-agrifood-systems/en.

Fenu, G. and Malloci, F. M. (2021). Forecasting plant and crop disease: an explorative study on current algorithms. Big Data and Cognitive Computing 5(1): 2. https://doi.org/10.3390/BDCC5010002.

Fernando, M. and Shrestha, A. (2023). The potential of cover crops for weed management: a sole tool or component of an integrated weed management system? Plants 12(4): 752. https://doi.org/10.3390/PLANTS12040752.

Ganie, Z. A., Soltani, N., Mckenzie-Gopsill, A. G., Felix, J., Hutchinson, P. J. S., Dille, J. A. J. et al. (2023). Potential potato yield loss from weed interference in the United States and Canada. Weed Technology 37(1): 21–24. https://doi.org/10.1017/WET.2023.5.

Garg, A., Gadi, V. K., Feng, Y. C., Lin, P., Qinhua, W., Ganesan, S. et al. (2020). Dynamics of soil water content using field monitoring and AI: A case study of a vegetated soil in an urban environment in China. Sustainable Computing: Informatics and Systems 28: 100301. https://doi.org/10.1016/J.SUSCOM.2019.01.003.

Hati, A. J. and Singh, R. R. (2021). Smart indoor farms: leveraging technological advancements to power a sustainable agricultural revolution. AgriEngineering 3(4): 728–767. https://doi.org/10.3390/AGRIENGINEERING3040047.

Holt-Giménez, E., Shattuck, A., Altieri, M., Herren, H. and Gliessman, S. (2012). We already grow enough food for 10 billion people... and still can't end hunger. Journal of Sustainable Agriculture 36(6): 595–598. https://doi.org/10.1080/10440046.2012.695331/ASSET//CMS/ASSET/CAA1A048-1F0E-464F-9FCA-A8A421ADCA3F/10440046.2012.695331.FP.PNG.

Horvath, D. P., Clay, S. A., Swanton, C. J., Anderson, J. V. and Chao, W. S. (2023). Weed-induced crop yield loss: a new paradigm and new challenges. Trends in Plant Science 28(5): 567–582. https://doi.org/10.1016/J.TPLANTS.2022.12.014.

Islam, S. Z., Kamarudin, M. S. Bin, Abdullah, M. N. Bin, Mohaffyza, M., Sern, L. C., Othman, M. L. et al. (2025). Sustainable Smart Irrigation System (SIS) using solar PV with rainwater harvesting technique for indoor plants. PLOS ONE 20(3): e0316911. https://doi.org/10.1371/JOURNAL.PONE.0316911.

Javaid, M., Haleem, A., Khan, I. H. and Suman, R. (2023). Understanding the potential applications of Artificial Intelligence in agriculture sector. Advanced Agrochem 2(1): 15–30. https://doi.org/10.1016/J.AAC.2022.10.001.

Jin, X., Yang, H., Li, Z., Huang, C., Yin, D., Domingues, T. et al. (2022). Machine learning for detection and prediction of crop diseases and pests: a comprehensive survey. Agriculture 12(9): 1350. https://doi.org/10.3390/AGRICULTURE12091350.

Khalifeh, A. F., Alqammaz, A., Khasawneh, A. M., Abualigah, L., Darabkh, K. A. and Zinonos, Z. (2025). An environmental remote sensing and prediction model for an IoT smart irrigation system based on an enhanced wind-driven optimization algorithm. Computers and Electrical Engineering 122: 109889. https://doi.org/10.1016/J.COMPELECENG.2024.109889.

Kong, J., Gao, M., Alofaysan, H., Fayziyeva, D. and Liu, Z. (2025). Enhancing urban agriculture networks: A clustering and multicriteria decision-making approach to sustainability indicators and governance. Ecological Indicators 170: 112997. https://doi.org/10.1016/J.ECOLIND.2024.112997.

Kundu, N., Rani, G., Dhaka, V. S., Gupta, K., Nayaka, S. C., Vocaturo, E. et al. (2022). Disease detection, severity prediction, and crop loss estimation in MaizeCrop using deep learning. Artificial Intelligence in Agriculture 6: 276–291. https://doi.org/10.1016/J.AIIA.2022.11.002.

Li, J., Ma, Y., Zhang, J. and Kong, D. (2025). Rapid detection of fertilizer information based on Raman spectroscopy and machine learning. Spectrochimica Acta Part A: Molecular and Biomolecular Spectroscopy 324: 124985. https://doi.org/10.1016/J.SAA.2024.124985.

Mall, R. K., Gupta, A. and Sonkar, G. (2017). Effect of climate change on agricultural crops. Current Developments in Biotechnology and Bioengineering: Crop Modification, Nutrition, and Food Production 23–46. https://doi.org/10.1016/B978-0-444-63661-4.00002-5.

Masson, S., Rueda-Ayala, V., Bragazza, L., Cordeau, S., Munier-Jolain, N. and Wirth, J. (2024). Reducing tillage and herbicide use intensity while limiting weed-related wheat yield loss. European Journal of Agronomy 160: 127284. https://doi.org/10.1016/J.EJA.2024.127284.

Melanie Mitchell. (2019). Artificial intelligence—a guide for thinking humans. In Genetic Programming and Evolvable Machines (Issue 4). Penguin Books Limited. https://doi.org/10.1007/S10710-022-09439-7.

Méndez-Zambrano, P. V., Tierra Pérez, L. P., Ureta Valdez, R. E. and Flores Orozco, Á. P. (2023). Technological innovations for agricultural production from an environmental perspective: a review. Sustainability 15(22): 16100. https://doi.org/10.3390/SU152216100.

Mijatović, D., Van Oudenhoven, F., Eyzaguirre, P. and Hodgkin, T. (2013). The role of agricultural biodiversity in strengthening resilience to climate change: towards an analytical framework. International Journal of Agricultural Sustainability 11(2): 95–107. https://doi.org/10.1080/14735903.2012.691221.

Monteiro, A. and Santos, S. (2022). Sustainable approach to weed management: the role of precision weed management. Agronomy 12(1): 118. https://doi.org/10.3390/AGRONOMY12010118.

More, N., Chadaga, P., Singh, P., Syed, F. A., Mohan, S. and Bulusu, G. (2025). Innovations in crop health monitoring: a deep dive into disease detection. pp. 303–310. *In*: Shailesh, R.. Nikam, Makarand, G.. Kulkarni, Vaibhav, S.. Narwane, Ninad, D. Mehendale and Nilkamal P. More (eds.). Technologies for Energy, Agriculture, and Healthcare. CRC Press. https://doi.org/10.1201/9781003596707-32/INNOVATIONS-CROP-HEALTH-MONITORING-DEEP-DIVE-DISEASE-DETECTION-NILKAMAL-PRATHIK-CHADAGA-PARTHSARTHI-SINGH-FAIZ-ABBAS-SYED-SURAJ-MOHAN-GAUTAM-BULUSU.

Nelson, G. C., Valin, H., Sands, R. D., Havlík, P., Ahammad, H., Deryng, D. et al. (2014). Climate change effects on agriculture: Economic responses to biophysical shocks. Proceedings of the National Academy of Sciences of the United States of America 111(9): 3274–3279. https://doi.org/10.1073/PNAS.1222465110/SUPPL_FILE/SD03.TXT.

Ortiz, A. M. D., Outhwaite, C. L., Dalin, C. and Newbold, T. (2021). A review of the interactions between biodiversity, agriculture, climate change, and international trade: research and policy priorities. One Earth 4(1): 88–101. https://doi.org/10.1016/J.ONEEAR.2020.12.008/ASSET/5F4860DF-5ED1-4162-A4C0-B71B61AB887E/MAIN.ASSETS/GR4.JPG.

Petraki, D., Gazoulis, I., Kokkini, M., Danaskos, M., Kanatas, P., Rekkas, A. et al. (2025). Digital tools and decision support systems in agroecology: benefits, challenges, and practical implementations. Agronomy 15(1): 236. https://doi.org/10.3390/AGRONOMY15010236.

Riemer, O., Shah, T. M. and Müller, A. (2023). The role of true cost accounting in guiding agrifood businesses and investments towards sustainability. In The State of Food and Agriculture 2023. FAO. https://doi.org/10.4060/CC7724EN.

Ryan, M., Isakhanyan, G. and Tekinerdogan, B. (2023). An interdisciplinary approach to artificial intelligence in agriculture. NJAS: Impact in Agricultural and Life Sciences 95(1). https://doi.org/10.1080/27685241.2023.2168568.

Shah, T. M., Nasika, D. P. B. and Otterpohl, R. (2021). Plant and weed identifier robot as an agroecological tool using artificial neural networks for image identification. Agriculture 11(3): 222. https://doi.org/10.3390/AGRICULTURE11030222.

Shah, T. M., Riemer, O., El-Hage Scialabba, N. and Müller, A. (2023). The Agri-Food Systems Transformation Protocol: Mapping Agents and Drivers of Transformation. https://doi.org/10.35435/1.2023.4.

Shah, T. M., Tasawwar, S., Bhat, M. A. and Otterpohl, R. (2021). Intercropping in rice farming under the system of rice intensification—an agroecological strategy for weed control, better yield, increased returns, and social–ecological sustainability. Agronomy 11(5): 1010. https://doi.org/10.3390/AGRONOMY11051010.

Sharma, M., Kumar, C. J. and Deka, A. (2022). Early diagnosis of rice plant disease using machine learning techniques. Archives of Phytopathology and Plant Protection 55(3): 259–283. https://doi.org/10.1080/03235408.2021.2015866.

Soheli, S. J., Jahan, N., Hossain, M. B., Adhikary, A., Khan, A. R. and Wahiduzzaman, M. (2022). Smart greenhouse monitoring system using internet of things and artificial intelligence. Wireless Personal Communications 124(4): 3603–3634. https://doi.org/10.1007/S11277-022-09528-X/METRICS.

Srivastav, A. L., Patel, N., Rani, L., Kumar, P., Dutt, I., Maddodi, B. S. et al. (2023). Sustainable options for fertilizer management in agriculture to prevent water contamination: a review. Environment, Development and Sustainability 26(4): 8303–8327. https://doi.org/10.1007/S10668-023-03117-Z.

Steven Lord. (2023). Hidden costs of agrifood systems and recent trends from 2016 to 2023. https://doi.org/10.4060/cc8581en.

Takoutsing, B., Heuvelink, G. B. M., Aynekulu, E. and Shepherd, K. D. (2025). Modelling and mapping maize yields and making fertilizer recommendations with uncertain soil information. Precision Agriculture 26(1): 1–27. https://doi.org/10.1007/S11119-024-10200-6/METRICS.

Ullah, I., Fayaz, M., Aman, M. and Kim, D. (2022). Toward autonomous farming - a novel scheme based on learning to prediction and optimization for smart greenhouse environment control. IEEE Internet of Things Journal 9(24): 25300–25323. https://doi.org/10.1109/JIOT.2022.3196053.

UNESCO. (2023). Agriculture | UN World Water Development Report 2022. https://www.unesco.org/reports/wwdr/2022/en/agriculture.

United Nations. (2022). Goal 2: Zero Hunger - United Nations Sustainable Development. https://www.un.org/sustainabledevelopment/hunger/.

Verma, K. K., Song, X. P., Kumari, A., Jagadesh, M., Singh, S. K., Bhatt, R. et al. (2025). Climate change adaptation: Challenges for agricultural sustainability. Plant, Cell & Environment 48(4): 2522–2533. https://doi.org/10.1111/PCE.15078.

Wei, Q., Xu, J., Liu, Y., Wang, D., Chen, S., Qian, W. et al. (2024). Nitrogen losses from soil as affected by water and fertilizer management under drip irrigation: Development, hotspots and future perspectives. Agricultural Water Management 296: 108791. https://doi.org/10.1016/J.AGWAT.2024.108791.

Yang, H., Xu, P., Zhang, S., Kim, H. and Shin, I. (2025). Construction of an intelligent analysis system for crop health status based on drone remote sensing data and CNN. IEEE Access. https://doi.org/10.1109/ACCESS.2025.3540279.

Yuzugullu, O., Fajraoui, N., Don, A. and Liebisch, F. (2024). Satellite-based soil organic carbon mapping on European soils using available datasets and support sampling. Science of Remote Sensing 9: 100118. https://doi.org/10.1016/J.SRS.2024.100118.

# Chapter 14

# Deep Learning for Crop Improvement

*Caio Miguel Pantoja Mendonça,*[1] *Vanessa Almeida da Rocha,*[1]
*Linda Grabiely Oliveira de Araújo,*[1] *Josué Manoel Souza Serrão,*[2]
*Vitor Eduardo Lopes Monteiro,*[2] *Daniel Neves Ferreira,*[2]
*Gustavo Guadagnucci Fontanari,*[1,2] *Andrea Komesu,*[3] *Mahendra Rai,*[4]
*Gledson Luiz Salgado de Castro*[5] and *Luiza Helena da Silva Martins*[1,2,*]

## 1. Introduction

An essential concept for understanding deep learning (DL) is Artificial neural networks (ANNs). ANNs are developed based on the architecture of the human brain, aiming to reproduce the processes of neuron connection. These networks allow the learning of new patterns through task training, enabling machine learning from examples and resolving high-level problems (Taye 2023). Therefore, DL allows learning complex features from simple features obtained through data (Muruganandam et al. 2023).

In recent years, DL has gained notable importance in the modernization of agriculture. It is an important component for developing Agriculture 4.0 because it focuses on integrating innovative digital tools in the agricultural sector, interconnecting software, systems, and machinery to increase production efficiency at all stages, from soil preparation to product delivery. Thus, using technologies that unite the physical and digital worlds contributes to improving cultivation methods and increasing production in an environmentally friendly way (de Alcantara et al. 2021).

The agricultural industry has undergone major changes with the arrival of new technologies, which have helped improve functioning and reduce costs. In agriculture, these technologies allow machines to connect to the internet through the Internet of Things, allowing data to be collected in Cloud and near the field. This data is analyzed with the help of artificial intelligence programs, such as deep learning, which makes decisions more accurate (Ferrag et al. 2021). It is believed that

[1] Universidade Federal Rural da Amazônia, UFRA, Programa de Pós-Graduação em Biotecnologia Aplicada a Agropecuária, PPGBAA, Belém, PA, Brazil.
[2] Universidade Federal Rural da Amazônia, UFRA, Instituto de Saúde e Produção Animal – ISPA, Belém, PA, Brazil.
[3] Universidade Federal de São Paulo, UNIFESP, O Instituto do Mar (IMar), Santos, SP, Brazil.
[4] Department of Biotechnology, Sant Gadge Baba Amravati University, Amravati, India.
[5] Universidade Federal Rural da Amazônia, UFRA, Instituto Ciberespacial – ICIBE, Belém, PA, Brazil.
* Corresponding author: luiza.martins@ufra.edu.br

DOI: 10.1201/9781003659907-18

Industry 4.0 will bring a new revolution to agriculture, strengthening Agriculture 4.0 and improving production greatly (Gill et al. 2024).

Currently, deep learning has become a popular approach to solving problems related to image processing, computer vision, and signal processing applications (Gill et al. 2024). In general, models should be trained using datasets that reflect the results based on previous experience. Despite being considered a relatively new data analysis and image processing technique, deep learning has many promising applications and enormous potential in several industries and the agricultural sectors (Hatem et al. 2022).

Recent studies demonstrate several ways in which deep learning can help improve crops and consequently, agriculture. This is mainly because deep neural networks can learn features in an automated and complete way, reducing the need for human intervention in creating attributes. Furthermore, this approach improves the models' performance and generalization capacity. Among the main applications, we can highlight the detection of diseases and pests, the recognition of weeds, the identification of stress signs in plants, phenotypic analysis, genomic prediction, monitoring of plant growth and its use in precision agriculture practices (Coulibaly et al. 2022, Muruganantham et al. 2022, Hati and Singh 2023).

In recent decades, significant advances have been observed in studies aimed at plant phenotyping, especially due to incorporating DL techniques and their respective artificial neural networks (Cai et al. 2024). This approach is superior to the traditional methodologies, and it can provide a more efficient solution for analyzing complex morphological and physiological characteristics (Wang et al. 2023). In this scenario, crop phenotyping (CP) has become essential in Agriculture 4.0 by integrating neural networks, genetic crossing algorithms, and advanced simulation and decision-making support methods (Rui et al. 2024).

An innovative trend is the application of deep learning algorithms in genomic prediction and crop improvement, which can accelerate and improve the accuracy of these processes. Deep learning has been used to improve genomic prediction of crops such as wheat and soybeans, and there are cases where multi-omics data are explored through this tool to promote plant improvement (Gao et al. 2023).

Deep learning (DLL) has been gaining prominence in modern agriculture through the automation and optimization of processes through the analysis of complex data. Given the relevance of this approach to the agricultural sector, this chapter aims to compile applications of existing techniques for crop improvement, such as productivity prediction, phenotyping, and disease detection, using neural networks such as CNN and RNN. In addition, the challenges in the use of these techniques and future perspectives that demonstrate the potential of DL to drive more precise, efficient, and sustainable agriculture will be highlighted.

## 2. Core Concepts of Deep Learning

Machine learning (ML) enables computers to learn through examples and observations. Deep learning (DL), on the other hand, is a more efficient branch of ML that is useful in domains with a large amount of high-dimensional data (Janiesch et al. 2021). The increase in computing capacity and accessibility to massive amounts of data drove the evolution initiated by DL, allowing for more complex and accurate results (Sharifani and Amini 2023).

Artificial neural networks (ANNs) are computational models inspired by the structure and functioning of human neurons. They have the ability to learn from data (Figure 1) when information is fed into them. They can be formed by multiple nonlinear layers, called DL architectures, flexible structures that can be modified according to the context, such as input characteristics and research objectives. Input layers tend to receive data while output layers produce the result, and between these two layers are the so-called hidden layers that are responsible for learning a nonlinear mapping between input and output (Garrido et al. 2021).

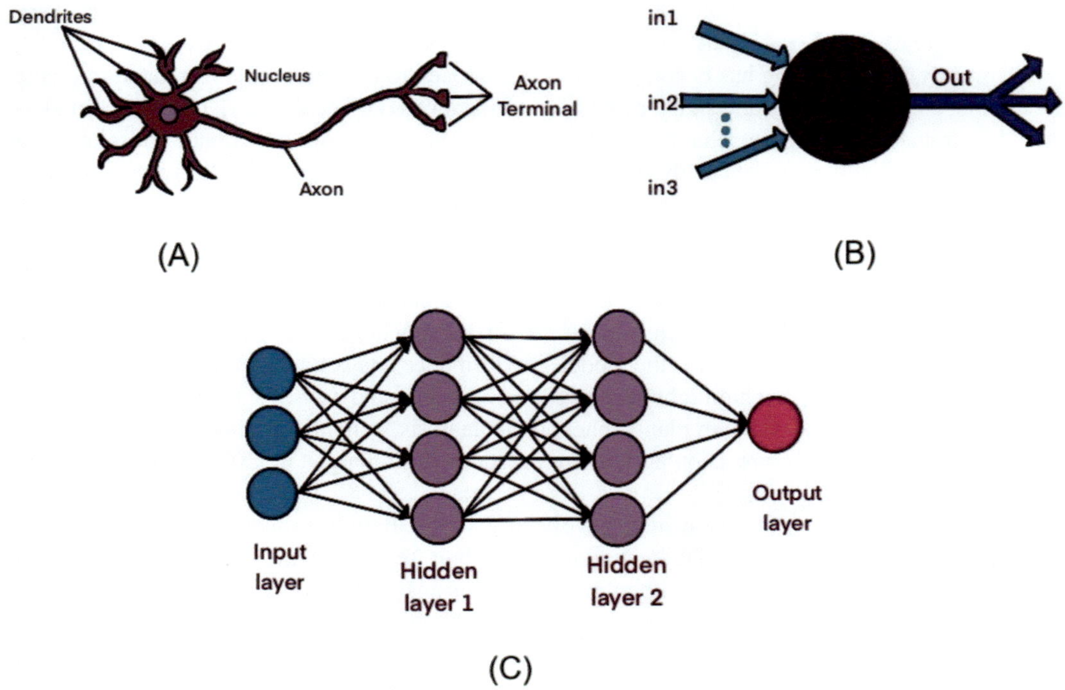


**Figure 1.** Schematic of an artificial neural network based on a human neuron. Source: Own authorship.

Therefore, through the learning and adaptation process, ANN models can handle imprecise data and situations different from those established at the beginning of training. Depending on DL architecture and techniques, they can be classified as unsupervised, supervised, or semi-supervised (Chen et al. 2022).

In supervised learning, input samples and their respective outputs are added, which serve as example patterns. Based on this, neurons are adjusted proportionally to the difference between the obtained and expected outputs. Therefore, this type of learning uses labeled data, which is applied to train and validate models, such as convolutional neural networks and feedforward networks, which are generally taught using labeled data (Hassan et al. 2022).

Unsupervised learning uses unlabeled data to extract patterns from their similarities. This type of learning can detect similarities and patterns based on the intrinsic properties of the data, promoting the recognition and grouping of observations autonomously. An example of an unsupervised task is clustering, which consists of grouping points from a given data set (Raza and Singh 2021).

Autoencoders use unsupervised learning from deep neural networks. Such algorithms are considered representative since the objective here is to learn from representation functions, which are capable of capturing the underlying structures of the data. This allows efficient modeling of the information contained in the input set. In addition, autoencoders play an important role in dimensionality reduction by encoding the data in a reduced representation space, preserving its most relevant characteristics for subsequent analysis or reconstruction tasks (Guo et al. 2021).

As for semi-supervised learning, it establishes an intermediate between supervised and unsupervised learning, using labeled and unlabeled data for the training process, as in the case of FixMatch (Sohn et al. 2020). Some categories that can be highlighted are deep neural networks (DNNs), convolutional neural networks (CNNs), and recurrent neural networks (RNNs).

DNNs are the basis of DL. They have more than one hidden layer, generally structured by advanced neurons compared to simple ANNs, and can perform complex operations and multiple activations in a neuron. These attributes allow DNNs to be fed from raw data input and automatically discover a representation necessary for learning (Janiesch et al. 2021). Therefore, the DNN is a neural network model with several parameters and layers between input and output, making DL appropriate for dealing with larger, more complex data (Sharma et al. 2021). Due to these characteristics, this architecture is often applied in image classification, segmentation, and object detection (Li et al. 2023).

CNNs refer to architectures that can compute large amounts of data and combine the features obtained between the input and output of variables that do not follow a linear pattern. Therefore, the architecture of this network is based on a series of steps that promote the gradual learning of features, following what is established by the modeling task (Dhaka et al. 2021). CNNs are used for feature reduction, selection, and extraction processes, emphasizing the classification of image datasets. For example, their application in precision agriculture to detect foliar diseases in plants has been reported (Pandian et al. 2022).

RNNs are very useful in the processing of sequential data. They are distinguished by their loop schematization, where the result of the first step feeds back into the network to influence the subsequent outcome. Thus, in RNNs, there is feedback from inputs, and the greatest interest in this type of network is due to its temporal processing and its potential to implement adaptive memories; for these reasons, this type of network is widely used in the study of sequential data and lists (Mienye et al. 2024). One type of RNN is the Long Short-Term Memory Network (LSTM), which is used in prediction models due to its ability to model and map temporal information (Ghimire et al. 2021).

As highlighted, digital learning can outperform machine learning in obtaining complex results. It also offers tools and different types of architectures capable of solving various problems. Yet, this form of learning still requires a greater amount of data and is costly in computational terms, but it remains an area of ongoing growth.

## 3. Applications for Deep Learning in Crop Improvement

Deep learning has proven useful in organizing images and finding objects in them. Its strength is that it learns from many photos of different crops, which helps it to recognize everything quickly and accurately (Shamshiri et al. 2024). This new technology can discover and separate crops by analyzing unique details, which is great for differentiating lineages or families of the same plant. In modern agriculture, deep learning is a promising alternative to common visual analysis systems, replacing hand-drawn outlines with a more automated and intelligent approach to identifying objects in the field (Shamshiri et al. 2024).

In the topics below, we will examine each use in detail, seeking to increase our understanding of how deep learning can improve crops while creating alternatives that drive more modern, sustainable, and efficient agriculture.

Figure 2 illustrates a combination of all technologies based on artificial intelligence (neural networks, IoT, deep learning, among others) to help improve and optimize agriculture.

### *3.1 Growth Monitoring*

In the agricultural system, crop performance is important when crops are susceptible to external conditions, and there is competition between weeds and natural predators for resources such as nutrients soil and water. Furthermore, crop performance assets depend on expert observation, an important decision-making tool affecting population and food storage (Jabed and Murad 2024).

In addition, regional variations and historical, climatic, and legal factors have influenced the methods and development of robust approaches that can be sustainable and adaptable (Hadir et al. 2025). Combined with agriculture 4.0, precision agriculture guarantees a revolution in monitoring and automation methods and technologies that ensure crop development and productivity. As one of them, deep learning training methods can be essential (Carraro et al. 2025).

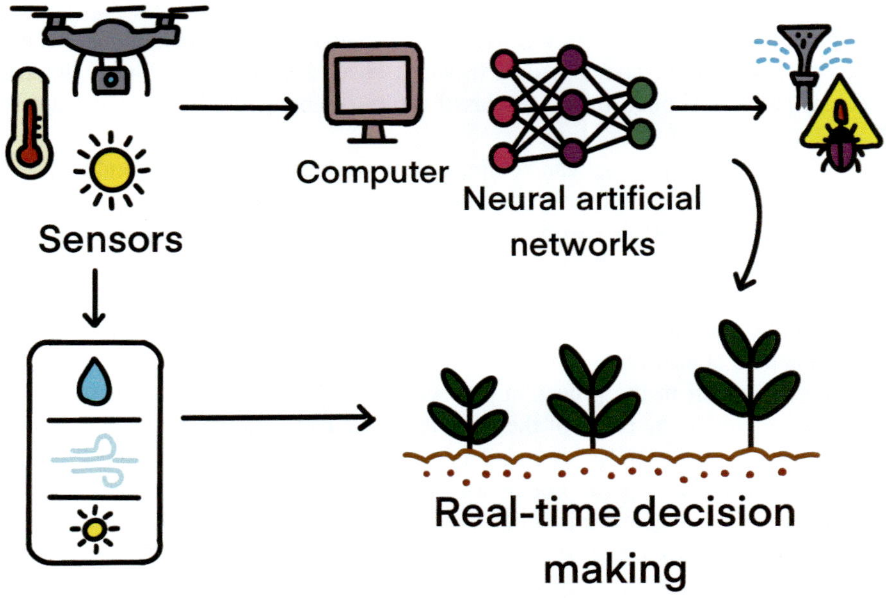


**Figure 2.** This illustration of how deep learning works when applied to agriculture highlights the interaction between sensors, computers, artificial neural networks, and crops for real-time monitoring and decision-making. Source: Own authorship.

The study of changes that occur during seasonal events, such as budding, fruiting, and senescence is called phenology. This field has acquired increasing scientific relevance since phenological events directly influence the fitness of plants and their productive performance and impact fundamental ecological processes, such as species dispersion (Matsoukis 2024). Thus, we can highlight plant phenotyping, which refers to characterizing and quantifying physical and physiological attributes of plants. This tool is very promising for understanding the interactions that occur between genotypes, environment, and agricultural management (Oliveira et al. 2024).

Growth monitoring is one of the tasks that can be optimized using deep learning, being important for controlling and identifying growth stages, health, and species in plants (Wang et al. 2023). In addition, these advances can help regions susceptible to drastic climate changes that affect local agriculture: which suffers from rising temperatures and sudden environmental changes that tend to reduce productivity and harm sustainable development (Mahmoud et al. 2025). The main use of deep learning in monitoring can be summarized as using regression models to estimate and stimulate growth, but this can imply costs, model sizes, and accessibility (Epifani and Caruso 2024).

The methodologies linked to CP enable the selection of appropriate strategies and the detailed characterization of the research environment, especially given the new rural scenario impacted by climate change (Araus et al. 2022). The importance of CP stands out in the monitoring of plant development, enabling the adjustment of characteristics of agronomic interest, which may be associated with the biochemical, functional, or structural aspects of plants, ranging from morphological to molecular levels (Jin et al. 2022).

In this sense, substantial transformations have been observed in image-based phenotyping, driven by the adoption of various convolutional neural network architectures, such as ResNet, SegNet, and AlexNet (Jin et al. 2025). Such models enable a wide range of applications, including leaf counting, panicle segmentation, flower detection, and plant species recognition, in addition to contributing to developing more efficient agricultural management systems (Wang 2004).

The importance of crop phenotyping is related to monitoring plant growth, enabling the adjustment of desirable agricultural traits, considering biochemical, functional or structural aspects, and may encompass not only morphological levels, but also molecular levels, particularly when analyzing genotype-phenotype interactions (Jin et al. 2022). In addition, deep learning can contribute to this by phenotyping weeds and differentiating them from crops (Hasan et al. 2023).

Deep learning can assist crop phenotyping with crop genomic prediction, which uses genetic sequencing to predict offspring, breeding, and the plant itself, associated with the correlation of factors. Its use can be explored in several crops, such as corn and rice, yet it can maintain the same efficiency while being an open source, using software such as the R language (Covarrubias-Pazaran 2016).

Narisetti et al. (2022) developed a tool called DeepShoot. It aims to automate the segmentation and quantification of visible images obtained by cameras positioned on the top and sides of barley, corn, and arabidopsis shoots, originating from phenotyping experiments conducted in a controlled environment. During the tests, the authors observed a progressive improvement in the model's performance throughout the iterations. In this context, the application of DeepShoot has proven more efficient and accurate than approaches based on shallow neural networks, enabling a more agile and precise phenotypic analysis.

Precision agriculture relies heavily on sensors combined with deep learning, artificial intelligence, image processing, and software to reduce energy consumption in rural areas. One such example is AICropCAM, which is found in the studies of Chamara et al. (2023). Wireless sensors integrated into the plantation transmit real-time information with reduced data. However, the search for standardization of images and cost reduction for constructing these systems remains a problem.

The study by Gómez-Zamanillo et al. (2024) aimed to improve and reduce human manipulation in the phenotyping of three pepper species. The results showed that deep learning contributes to improving data matching and the recognition of variations in weight, size, shape, and other characteristics. For this, images obtained from mobile devices in the plantation itself were used.

Thus, deep learning helps improve rural management, producers, and production, verifying the availability of new combined techniques for better variety, breeding, and plantation growth. Improving these techniques guarantees better resource optimization, reduction of expenses, and improvement of cultivation techniques associated with the new technological model, Agriculture 4.0. Thus, deep learning favors new solutions to optimize production and guarantees greater food production.

### *3.2 Stress Detection and Management*

In recent years, deep learning (DL) methods that use images have gained strength in precision agriculture, driving notable improvements in several aspects. Among the most common uses, the following stand out: production forecasting; studying plant characteristics; distinguishing invasive plants; checking soil and environmental quality; improving water use in crops; and identifying signs of plant problems. This last issue is important, as it can greatly affect the quality and quantity of harvested products (Cattivelli et al. 2008, Paul et al. 2025).

The decision about which deep learning model to use depends on the data available, especially the variety of visual cues presented by different types of plant stress. Some stressors cause effects in specific areas, such as localized pests or diseases, while others affect the entire plant or field uniformly, such as a lack of water. Plant stressors can be classified as either biotic or abiotic. Biotic factors include pathogens such as fungi, bacteria, viruses, and pests that attack plants, resulting in various cues such as color changes, wilting, and even plant death (Georgieva and Vassileva 2023).

Identifying plant distress through images requires a set of actions, which include the initial processing of images, the search for relevant attributes, the careful selection of these attributes, and, finally, the categorization of the type of problem. The initial processing of images is vital to improve data quality and simplify pattern recognition by machine learning algorithms. Common practices include removing the background, enhancing contrast, separating elements, changing colors (from RGB to grayscale or HSV), and reducing noise with filters (such as FFT and wavelets). Principal Component Analysis (PCA) and its UPCA variation help simplify the data, highlighting crucial information for identifying stress (Li et al. 2020).

The search for attributes investigates several image dimensions, applying mathematical transformations (Hough, Radon, Haar, among others) to discover specific signs of stress. Subsequently, the classification of images can be done through machine learning to differentiate between the different classes of biotic and abiotic stress. In addition, tools such as the Keras Image Data Generator enable the improvement and standardization of images to optimize the identification of changes in leaves and fruits, contributing to more accurate diagnoses and more effective agricultural decisions (Li et al. 2020).

Convolutional neural networks (CNNs) are essential tools in deep learning for image analysis and computer vision. They excel in visual recognition, tagging, image search, and identifying stress in plants (Azimi et al. 2021). Recent studies have used CNNs and long short-term memory (LSTM) networks to determine plant nutrient deficiencies using large amounts of crop photos (Rojanarungruengporn and Pumrin 2021). The use of advanced techniques such as Mask R-CNN has improved accuracy in separating and collecting information from fruits and leaves (Wang and He 2022).

Chu et al. (2021) created a novel model, Suppression Mask R-CNN, to refine apple identification, achieving an F1 score of 0.905 and an identification time of 0.25 seconds per image. Jia et al. (2020) applied Mask R-CNN to harvesting robots, earning a 97.31% accuracy rate and a 95.70% identification rate. At the same time, Yu et al. (2019) improved strawberry identification with a Mask R-CNN-based method, making the system more reliable in different light conditions.

To predict water shortages, Kaneda et al. (2017)suggested a method based on Support Vector Regression (SW-SVR) and Deep neural networks (DNN), increasing the accuracy of water. Weng et al. (2023) used hyperspectral imaging (HSI) to find drought stress in plant leaves, reaching 95.90% accuracy with SVM and sample matching.

To identify nutrient deficiency, Artificial neural networks (ANN) models were used to classify and predict nutrient deficiency in tomato plants (Tran et al. 2019). Khatoon et al. (2021) developed an AI system to identify diseases in real time, comparing deep neural networks such as DenseNet and ShallowNet, reaching 95.31% accuracy on the test set. Other studies have combined deep learning and meta-learning to analyze hyperspectral images, allowing accurate identification of drought stress and nutrient depletion (Ruan et al. 2023). Transfer learning and pre-trained models (VGG-16, Xception, Inception v3) have been used to identify leaf diseases (Yee-Rendon et al. 2021). The creation of new vegetation indices increased classification accuracy.

Butte et al. (2021) used a Solo UAV drone (3DR) equipped with a Sequoia multispectral camera (Parrot) to capture images of a potato crop that had been partially induced by water stress. The camera captured images in green (550 nm), red (660 nm), red edge (735 nm), and near-infrared (790 nm), as well as high-resolution RGB images (16 MPx). The images were pre-processed and enlarged using data augmentation techniques. The Retina-UNet-Ag model, based on the U-Net and RetinaNet architectures, was developed to detect stressed plants. The dataset was manually labeled, distinguishing healthy (green) and stressed (yellowish) plants. Retina-UNet-Ag demonstrated high accuracy in identifying water stress in potatoes, outperforming other models such as Mask R-CNN, Faster R-CNN, and YOLO v3. However, in monochromatic images, it demonstrated low accuracy due to reduced resolution.

According to studies by Islam et al. (2024), applying computer vision to identify signs of stress in plants faces certain obstacles and restrictions. Despite the potential demonstrated by this technology, some obstacles need to be overcome before it can be effectively applied. Changes in appearance are one of the problems since plants change over time, which complicates the rapid identification of stress. There are also indistinguishable signals, in which several types of stress not caused by living beings can generate similar signals, making it difficult to accurately identify what is affecting the plant. Developing a broad model for different species and cultivation locations is challenging. Added to this is the problematic data since differences in the ways of taking photos and the quality of the images can affect the performance of the programs. It is worth noting that there are limitations in obtaining data that can precisely establish the specific stress conditions to which the plant was subjected, especially because of climatic variations and obstacles in accessing some plantations.

In addition, stress can appear in discrete ways, such as small changes in color, texture, or shape, requiring sophisticated programs to understand these changes. Furthermore, using computer vision systems on a large scale, especially in large crop areas, can be complicated and require significant investment (Islam et al. 2024).

The future of digital agriculture depends on the continued progress of stronger, easier-to-use, and adaptable technologies capable of dealing with the diversity of stress caused by biological and non-biological factors. Thus, deep learning remains a promising partner towards more sustainable, smart, and profitable agriculture.

### *3.3 Yield Prediction*

Recent climate change and population growth may be two main reasons for the increase in food demand and global food uncertainty. This suggests the need for analytical methods for crop yield prediction to sustain and adapt major crops such as wheat, rice, and corn, which are susceptible to these events (Demirhan 2025).

Agricultural yield predictions predict the amount of crops produced in a given area. This process can enable agricultural stakeholders to make data-driven decisions on crop production using their datasets (Subramaniam and Marimuthu 2024).

Environmental factors, such as climate, regional cultivation techniques, weather, growth, and crop quality, influence yield estimates. Researchers categorize these factors into qualitative and quantitative aspects based on the nature of the data provided (Jhajharia et al. 2022).

Classification algorithms can optimize agricultural practices and benefit farmers by reducing losses, improving agrarian productivity performance to increase their resources, and implementing sustainable resource management approaches and methodologies, such as optimizing the allocation of water, fertilizers, and soil resources (Modi et al. 2022).

Recent advances in deep learning have further enhanced these capabilities. For example, studies such as that of Jeong et al. (2024) emphasize how long-term memory (LSTM) can significantly improve rice production forecasting by combining only meteorological and satellite data and integrating a machine learning algorithm to simulate production on the Korean peninsula. The dataset used covered thirteen years of recent data (2010–2022).

LSTM is derived from recurrent neural networks (RNNs), a form of application in the field of deep learning that aims to elucidate analytical responses to activities and benefits from text and image databases (Gaafar et al. 2022).

The thirteen-year dataset used highlights the importance of data diversity and the construction of an efficient dataset, confirming that a comprehensive model is necessary to handle the diversity of agricultural modeling. Integrating systems in conjunction with deep learning can provide a basis for solving problems, such as standardization, monitoring the stability and health of harvest fields, and precision for better integration of sustainable practices that help improve harvests, the environment, and rural producers (Gaafar et al. 2022).

An example of this use would be the study by Suganthi et al. (2025), which consisted of mapping areas in conjunction with Global Positioning System (GPS) and Geographic Information Systems (GIS) remote sensing technology systems to optimize resources; monitor soil quality; growth patterns; crop performance and moisture content using the amount of sunlight, precipitation, and temperature as primary data datasets; connecting them with more independent and interconnected data such as soil properties and climate conditions. In this way, the responsiveness of this integrated model, through its neural networks, reached up to 97% accuracy in certain algorithmic models.

The importance of training and the vast availability of corresponding datasets, their validation, and learning for decision-making and improvement of harvest and yield prediction with more elucidated information that respects data confidentiality, combined with smart agriculture, can be reinforced. The use of deep learning for yield prediction favors the use of satellite images, spatial data, and temporal data of the given area via cell phone applications, being essential in decision-making and resource optimization, in addition to verifying dependent variables in growth and planting (Boukhris et al. 2024)

Another highlight is the use of phytosanitary data, fertilization, irrigation, water quantity, and climate data to improve the growth and rural management of these plantations (Moussaid et al. 2023).

In this way, deep learning's contribution to yield prediction can promote resource optimization and more simplified data visualization, through sensitive variables such as meteorological data and climate, which are intrinsically linked to growth and yield value. This aims to create the best conditions for rural development, estimating and enabling larger harvests.

### *3.4 Genomic Prediction and Breeding*

Genomic prediction, or genomic selection, is an important element of current crop improvement methods. It is a scientific technique whereby one may estimate the performance of individuals with potential for improvement by the genetic constitution (Feng et al. 2024). Through this technique, it is possible to develop a predictive model using a training data set that includes known phenotype and genotype information, making it possible to predict the performance of individuals whose genotype is known but whose phenotype has not yet been observed, which allows early selection based on phenotypic values estimated from genomic data (Ye et al. 2025).

The constant advancement of sequencing technologies has made it possible to obtain genotypes of different plant species with a focus on genetic improvement. This allows genomic prediction to estimate the joint effects of all genes without knowing each one in isolation (Zhang et al. 2022). This strategy provides a major benefit toward improving cultivated species, potentially speeding genetic advancement.

This is achieved via cost minimization during each cycle of improvement, keeping the time between generations to a minimum and optimizing the utilization of human resources against traditional methods (Hickey et al. 2014, Kaler et al. 2022). Furthermore, the need to conduct exhaustive field evaluations is reduced, making the process more efficient and affordable (Sinha et al. 2023).

To keep up with the expected increase in food demand in the coming years, plant breeding has to provide the highest rates of genetic gain to maximize its contribution to increasing agricultural productivity (Montesinos-López et al. 2021). In recent years, different types of (deep) learning methods have been tested regarding how well they perform as genomic prediction tools. The main difference between DL methods and standard statistical learning methods is that DL methods are nonparametric models that provide greater flexibility to trace intricate relationships between data and output (Montesinos-López et al. 2021).

Wang et al. (2023) developed DNNGP (Deep Neural Network Genomic Prediction), a deep learning model that integrates multiomic information in plants to predict phenotypes. With its multilayer architecture, it can identify patterns directly in the raw data, preventing biased results and making the process faster. DNNGP proved more accurate in its predictions than five other methods, including two machine learning and two deep learning approaches, especially when dealing with large volumes of breeding data. In addition, the tool proved efficient in working with large and complex data sets, providing reliable analyses.

In the study by Montesinos-López et al. (2024), a comparative analysis was made between GBLUP models implemented in the BGLR R package and a new multimodal deep learning model developed by the authors. The dataset consisted of 4,464 wheat lines evaluated for five agronomic traits in three or four different environments. As a result, the DL model presented accuracy comparable to that of the GBLUP models for predicting the yield, height, and germination traits, as well as greater prediction accuracy for the maturity and earing traits (Heading).

Gao et al. (2023) developed SoyDNGP, a deep learning-based model that significantly advances soybean trait prediction. Compared to existing methods such as DeepGS and DNNGP, SoyDNGP has a distinct advantage due to its minimal increase in parameter volume and superior predictive accuracy. The tool accurately predicted complex features, demonstrating robust performance across different sample sizes and feature complexities. Furthermore, it can be used across various crops, as the model has also been tested on cotton, corn, rice, and tomatoes.

Gan et al. (2024) integrated deep learning (DL)/machine learning (ML) algorithms, plant phenomics, and whole-genome sequencing data to conduct genomic selection for pest resistance in grapevines. Deep convolutional neural networks (DCNN) were employed to accurately calculate the severity of pest damage on grapevine leaves, achieving a classification accuracy of 95.3% and a correlation coefficient of 0.94 in regression analysis. Through a combination of transcriptome data, they also identified specific pest-resistant genes that play distinct roles in resistance to herbivore attacks.

According to Wang et al. (2025a), although deep learning shows an interesting future in complete genomic prediction, the models tend to operate in an unclear manner, making the analysis and understanding of the relationships between attributes and the results achieved complex. However, there are several possibilities to refine the accuracy of estimates and optimize training procedures. The development of Cropformer, described by Wang et al. (2025b), represents an important technological advance. The group created a deep learning system to evaluate phenotypic aspects of crops and explore tasks performed later, integrating convolutional neural networks with multiple self-attention methods to optimize accuracy. The tool's efficiency was assessed using over twenty characteristics, including tests on five main crops: rice, wheat, tomato, and millet.

Therefore, it is clear that deep learning is a tool with promising applications for genomic prediction and crop improvement, especially considering the global context that demands more efficient and faster agriculture, given the growing demand for food.

### *3.5 Weed and Pest Identification*

Phytosanitary conditions, such as pests and plant diseases, contribute to high losses in agricultural products, impacting the economy, environment, and society. Therefore, these conditions must be detected in the plantation at a high speed using automation (Turkoglu et al. 2022). In addition, with the continuous growth of the world population and the consequent demand for greater agricultural productivity, increasing crop yields has become a global priority. Factors such as light, soil fertility, humidity, pests, diseases, and weed invasion negatively influence productivity (Istiak et al. 2023).

Weeds compete with crops for water, light, space, and nutrients, spreading pests and diseases and compromising crop growth and yield (Istiak et al. 2023). In addition to immediate destruction, insects can propagate or aggravate subsequent contamination caused by viruses, bacteria, and fungi.

The leafhopper *Scaphoideus titanus*, for example, spreads the phytoplasma responsible for the disease "yellowing of grapes." (Constable and Bertaccini 2017). Whiteflies can also be mentioned, representing a serious problem in agriculture, causing losses that can reach billions of dollars in harvests. More than 1,500 different types of whiteflies have been identified, but within this group, the species of the Bemisia tabaci complex are the ones that cause the most damage. This is due to their ability to attack several crops (such as beans, cassava, cotton, potatoes, and tomatoes) and because they transmit more than 400 viruses to plants (Sicat et al. 2022).

Kasinathan et al. (2021) explore machine learning techniques and convolutional neural networks (CNN) to identify and classify pests in crops. They used the Wang (with nine categories) and Xie (with 24 categories) databases, optimizing the images through pre-processing and expansion to refine accuracy. They then evaluated several classification models, such as ANN, SVM, KNN, Naive Bayes, and CNN. CNN stood out with superior performances of 91.5% (Wang) and 90% (Xie). The GrabCut method was used to isolate and highlight the insects' contours, applying it to the Deng and IP102 sets for detection. The algorithm demonstrated high accuracy with efficient execution time. The strategy proves viable for real-time applications in agriculture.

At the São José farm in Campo Grande, dos Santos Ferreira et al. (2017) used convolutional neural networks (CNN) to detect weeds in Brazilian soybean crops. The images obtained by a DJI Phantom 3 Professional drone with a Sony EXMOR RGB camera were processed using the SLIC algorithm for segmentation. After manual analysis and labeling, the images were categorized using the AlexNet model. The result was compared with SVM, AdaBoost, and Random Forest. AlexNet, using a balanced dataset, achieved an overall accuracy of over 90%, correctly classifying 96.3% of the images.

A recent study evaluated a methodology that combines GANs (generative adversarial networks) and transfer learning to identify weeds in agriculture. The proposal included three main stages: pre-training with ImageNet and an agricultural dataset, fine-tuning problem data (including synthetic images generated by GANs), and classification with pre-trained networks such as Xception and Inception. In the classification stage, the Xception network with transfer learning and data augmentation with synthetic images achieved an F1 score of up to 99%, outperforming approaches without GANs. Furthermore, this strategy reduced the number of training epochs required. The results show that combining GANs and transfer learning is effective, especially in scenarios with limited data available (Espejo-Garcia et al. 2021).

Xiao et al. (2022) presented an innovative method for pest identification using hyperspectral images (HSI) combined with deep neural networks. The generated dataset was used to train and test the HPInet model, a deep neural network composed of two main modules: SF, which reduces spectral noise, and ASSF, which adaptively fuses spectral and spatial information.

The architecture uses 3D convolutions at different resolutions without pooling to preserve details. HPInet thus outperformed several traditional methods and deep networks with over 99% overall accuracy.

Conclusively, combining hyperspectral images and deep learning yields an accurate and dependable technique for pest identification, even when the pest is small or visually similar, a breakthrough for phytosanitary agriculture management. Because pests, diseases, and invasive weeds influence production, quick and correct identification of problems caused by them is critical to ensuring food production and agricultural profitability (Xiao et al. 2022).

Advanced technologies, such as convolutional neural networks (CNN), generative adversarial networks (GANs), transfer learning, and hyperspectral images, have effectively identified and distinguished these harmful components. Therefore, incorporating these intelligent tools into agricultural management can significantly reduce losses, favor production growth, and ensure the sustainability of the global food system (Xiao et al. 2022). Table 1 summarizes some studies on Precision Agriculture and deep learning.

Therefore, the various applications of PA and its benefits become evident. However, economic and structural barriers have proven to be challenges for applying new PA technologies, given that, to reach both large and small producers, innovative methods require support and public policies to be implemented (Viana et al. 2024).

**Table 1.** Some studies on precision agriculture, deep learning, and its applications.

| Purpose/Application | Technology/Approach | Main Results | Reference |
|---|---|---|---|
| Definition and benefits of Precision Agriculture (PA) | Techniques based on spatial and temporal variability | Increased economic return and reduced environmental impact | (Oliveira et al. 2020) |
| Integration of PA with Agriculture 4.0 | Big Data, IA, DL | Improvement of the production chain, accurate decision-making | (Liu et al. 2021) |
| Application of UAVs/ drones in PA | Georeferencing, remote sensing, GIS | Mapping, spraying, topography, resource management | (Radoglou-Grammatikis et al. 2020) |
| DL for weed detection in eggplant crops | CNNs (Object Classification and Detection) | 88% accuracy in identifying weeds | (Patel et al. 2023) |
| Communication between sensors via wireless networks in greenhouses | DL for radio wave attenuation | Two-point improvement in accuracy and 1/3 reduction in mean square error | (Cama-Pinto et al. 2023) |
| Soil monitoring system with heat maps | DL for visualization of pH, temperature, and humidity | The error is between 1 and 7%, with efficient maps for agronomic actions | (Rueda-Delgado et al. 2025) |

Source: Own authorship.

## 4. Key Tools and Frameworks for Deep Learning in Crop Science

Deep learning (DL) is a specific branch of machine learning (ML) that emerged in the agriculture and plant science communities due to its versatility in handling large amounts of data and solving complex problems. Deep learning has recently advanced in computer vision and has become a focal point in crop protection, as it can minimize the biases of manual selection of disease point features (Upadhyay et al. 2025). Managing large crops requires careful supervision, including monitoring for diseases that can turn them into waste. In addition, plant disease causes decreased productivity, requiring significant effort and qualified personnel.

Python programming language is an essential tool for the development of Artificial Intelligence (AI) and consequently for the development of Deep Learning (Dhruvitkumar V. Talati 2021). Python language is widely used in research and progress in artificial intelligence, especially due to its simple and understandable syntax, which helps programmers to dedicate themselves to solving challenges without getting lost in the complexities of the language. Another important point is that the language requires fewer lines of code when compared to others, as well as a rich set of integrated libraries and frameworks that make AI development more pragmatic.

The study by Altalak et al. (2022) presents research on recently developed systems in smart agriculture using deep learning techniques. In terms of frameworks, it was found that the main ones used were Keras, TensorFlow, PyTorch, and Deeplearning4j. In addition, it is worth noting that the authors noted the existence of studies in which the models were developed using their frameworks. Furthermore, the authors highlighted that of the articles researched, the most used models were those with a Convolutional Neural Network (CNN) structure. In addition to CNN, other structures mentioned were the LSTM model of the RNN, DNN, and ANN algorithms.

## 5. Challenges and Limitations

Deep learning (DL) emerges as an innovative tool in data analysis and image processing, promoting significant advances in smart agriculture, which covers areas such as efficient water management, soil research, seed analysis, pest and weed control, disease detection, and plant stress (Attri et al. 2023). With deep learning, researchers can overcome the shortcomings of traditional algorithms that rely on artificial resources. Thus, it has been employed in many fields of study, including computer vision, facial recognition, speech recognition, natural language processing, recommendation systems, and speech recognition.

With the world's growing population, the demand for food has also increased a lot. The issue is worsened because the agricultural sector cannot produce due to unpredictable climatic conditions. Therefore, there is a need to apply new innovative and sustainable processes to facilitate the ongoing growth of agriculture with a view to a more efficient and stable future (Waqas et al. 2025). Aside from this, it involves creating algorithms that can operate with limited data sets and consider the variability of climate and soil types within such regions (Liu et al. 2021).

One of the challenges is the limited availability of data. Multimodal datasets that integrate different types of information, such as images, texts, and numerical data, are often lacking. In agriculture, this can include data from satellite images, measurements from sensors in the soil, climate records, and information on management practices (Waqas et al. 2025).

When DL is integrated into technologies such as tractors, drones, and autonomous agricultural robots, we can have precision agriculture, as such systems can perform real-time detection and response to parameters such as pest, disease, or water stress control, and this improves efficiency in crop management, also allowing for faster and more informed decisions (Zhang et al. 2025). As an example of these technologies, we have drones, which are robots capable of monitoring large areas of plantations through DL algorithms that identify problems before they become critical.

The interpretability of models is also a crucial issue, which means that farmers and managers need to understand how an AI model makes its decisions, for example, when recommending the amount of water or fertilizer to be used (Waqas et al. 2025). If a model is not interpretable, this can generate distrust and resistance to adopting new technologies. Studies have shown that the ability to interpret model outputs can improve farmer acceptance and promote the adoption of smart technologies.

Transparency of models helps build trust and foster an environment of collaboration between technology and agricultural practice. This is especially important in a sector such as agriculture, where decisions directly impact production and sustainability (Caruana and Niculescu-Mizil 2006). Finally, the ability to apply solutions in real time is essential. Rapid decisions, such as irrigation or pest control, may be required in agriculture. To achieve this, it is necessary to have algorithms that work efficiently with real-time data and a technological infrastructure that supports this instantaneous analysis (Waqas et al. 2025).

All these challenges are interconnected and reflect the complexity of implementing advanced technologies in agriculture; if we overcome these barriers, it will be easier to make more technological innovations, benefiting industry professionals and promoting more efficient and sustainable agriculture in the future. Several studies indicate that using DL offers transformative potential to address challenges in agriculture, ranging from crop selection to post-harvest management (Waqas et al. 2025).

Precision agriculture technologies can generate positive results for producers, and make adopting more practical agricultural environments possible. Platforms such as Climate Field View and the John Deere Operations Center are examples of how these innovations can be implemented. However, some farmers still resist them due to high implementation costs, lack of infrastructure, and uncertainty about whether this new practice can be effective (Paudel et al. 2023).

## 6. Future Directions

As we talk about DL, the future of smart farming is extremely promising yet highly complex. Climate change and population growth are some of the global issues that can impact crop seasonality, and therefore, there is a necessity to adopt innovative solutions. While the success of DL for agriculture has been fruitful, setbacks must be overcome, such as the lack of availability of multimodal data sets and scalability challenges, data interpretability, and real-time applicability (Waqas et al. 2025). Table 2 summarizes some key items regarding prospects for these technologies in agriculture.

**Table 2.** Future perspectives of deep learning in agriculture.

| Future Perspective | Potential Applications | Technologies Involved | Expected Benefits | Challenges/Observations | Reference |
|---|---|---|---|---|---|
| Autonomous and intelligent agriculture. | Drones and robots for real-time monitoring and intervention. | DL is embedded in UAVs, tractors, and robots. | Faster and more precise actions in crop management. | Cost of implementation and adaptation on small properties. | (Bezas and Foteini Filippidou 2023) |
| Productivity prediction based on genomic and environmental data. | Combination of genotypes, climate, soil, and images | Multimodal DL | More efficient and productive planning. | Need for high-quality data and integration. | (Montesinos-López et al. 2024) |
| Explainability and transparency of models. | Clear explanation of DL system decisions | XAI (Explainable AI) | Greater confidence and adherence from farmers | Complexity of models and difficulty of interpretation | (Bezas and Foteini Filippidou 2023) |
| Democratization of technology. | Simple and accessible interfaces for small and medium producers | DL-based apps and platforms. | Technological inclusion in the field and greater adherence. | Need for digital training and infrastructure. | (Feng et al. 2023) |
| DL as a support for decision-making. | Support tool, not replacement of human labor. | DL algorithms with human interfaces. | Reduced resistance to technology and increased producer appreciation. | Clear communication of the role of DL in the production chain. | (Rehman et al. 2023) |
| Sustainable agriculture. | Optimization of resources (water, fertilizers, energy) and environmental tracking. | DL applied to the rational use of inputs. | Reduced environmental impact and increased efficiency. | Constant monitoring and integrated sensors. | (Liakos et al. 2018) |
| Adoption in different agricultural realities. | Adaptation of DL models to different climates, soils, and production scales. | DL with heterogeneous and local data. | Increased scalability and scope of use. | Technical, economic, and logistical barriers. | (Kumar et al. 2023) |

Source: Own authorship.

The future directions of deep learning in agriculture present a vast potential for transformation and innovation since it can handle complex and large-volume data. DL learning can revolutionize agriculture. Though the future is bright, some issues need to be resolved, such as the need for multimodal data, the scalability of algorithms, and the interpretability of AI models. Adaptability of DL to different climatic conditions and the fact that most farms are small is central to its adoption on a large scale. Additionally, combining it with technologies such as robots and drones allows one to detect and react immediately to some of the problems confronting cultivars, significantly improving the management of crops.

## 7. Case Studies and Success Stories

Deep learning has numerous applications in the agricultural sector. It is one of the tools that stand out in optimizing processes and reaching a new stage of technological development in agricultural production. Due to the large amount of data captured by more modern sensors, it is possible to predict crop yield, identify weeds and pests, and even detect diseases in cultivars. This tool can also contribute to the interpretation of variations in the climate and even in the visual characteristics of plants (Bharman et al. 2022).

Case studies are important to elucidate real experiences and learn how deep learning algorithms work. Technology still has points to be improved, and the agricultural ecosystem presents many variables (Qin et al. 2025). Studying these data together requires complex deep learning models that can discover important data and eliminate interference from many sources. Another problem mentioned is the synchronization and harmonization of data from different sensors. Variations in climate, genetic characteristics, appearance, and regional growth conditions are examples addressed that demonstrate the distinction between existing data.

Three studies, published in 2025, on deep learning applications in agriculture specifically for detecting plant diseases were selected. The studies were based on Convolutional neural networks (CNN). In addition to the studies, two companies that offer services involving deep learning applied to agriculture were cited.

Recent research by Ashurov et al. (2025) shared a new way to identify plant diseases. The idea is to use a convolutional neural network (CNN) with an improved structure. The authors made some adjustments to the network, such as adding Squeeze-and-Excitation and improving the residual connections. This helped the system understand better what matters. To teach the model, they used various information, including several plants and types of diseases, so that the algorithm would work well under any circumstances. The results were very promising, with an accuracy of 98% and an F1 score of 98.2%. Even with all the advantages, the authors believe that it is possible to optimize the model and aim to organize the data better, use more efficient ways to increase the amount of data, and make the model easier to interpret.

Baiju et al. (2025) introduced the MobileH-Transformer, a system that combines the power of convolutional neural networks (CNNs) with transformers. The goal is to efficiently identify diseases that affect plant leaves without requiring much processing power. Tests with known databases show that it performs very well, achieving an F1 score of 97.20% when identifying diseases on corn leaves and 96.80% on parts of the PlantVillage database. In addition to being more accurate than other recent solutions, the model is lightweight, consuming only 0.4 Giga Floating Point Operations (GFLOPs). This allows it to work in real-time on mobile devices, processing 30.5 images per second (FPS).

Baiju et al. (2025) proposed a Slender-CNN model for detecting plant diseases specifically for corn, rice, and wheat crops. The model runs on a carefully designed architecture that includes parallel convolution layers and dimensions that accurately detect damage of different sizes. Furthermore, the authenticated model is identified as a multimode device and distinguishes healthy and diseased classes of single plant species. The model can also detect and search crops for infected ones in agricultural conditions, such as poor environments. The tool has an accuracy of 88.54%,

**Table 3.** Deep learning in agriculture: technologies and real examples.

| Category | Description/Examples | Associated Technologies | Impacts and Benefits | Reference |
|---|---|---|---|---|
| Real-time detection of stresses and pests | Drones, robots, and autonomous tractors monitor and respond to threats on plantations. | DL, computer vision, embedded sensors. | Fast response, greater efficiency, and accurate decisions | (Kumar et al. 2024) |
| Agricultural productivity prediction. | Models with genetic, climate, and soil data. | Multimodal DL, predictive modeling, big data. | Better agricultural management and optimized use of resources. | (Demir et al. 2024) |
| Explainability and Confidence (XAI) | Making DL more understandable and transparent for producers. | XAI, interpretable AI. | Increased trust and technology adoption | (Kobayashi and Alam 2024) |
| Accessibility for small producers | Simple, low-cost interfaces with decision support. | Embedded DL, smart apps. | Technological inclusion in the field. | (Rashid and Kausik 2024) |
| Sustainability and rational use of resources | Optimization of inputs and reduction of losses with DL. | Remote sensing, DL, applied AI. | Sustainable and efficient agriculture. | (Dayioglu and Turker 2021) |
| Climate adaptation and Scalability | Adjusting DL models to different regions and realities. | Adaptive AI, Scalable DL | Increased global adoption and local adaptation | Taranis (2025) |
| Example: Taranis (Israel 2015) | Global agricultural intelligence platform with DL and a bird's eye view. | Computer vision, DL, drones. | Pest identification, carbon, and increased productivity. | Taranis (2025) |
| Taranis Partnerships | Alliances with BASF, ADAMA, DJI, Albo Climate, acquisition of Mavrx | - | Global expansion and diversification of services | Taranis (2025) |
| Example: Bloomfield Robotics (USA, 2019). | Customized plant monitoring with terrestrial imagery | DL, Onboard AI, Continuous Inspection | Accurate plant cycle assessment and decision support. | (Bloomfield 2025) |
| Bloomfield Partnerships | Collaboration with NASA and global tractor manufacturer | | Space monitoring and investment of US$16.2 million | (Bloomfield 2025) |
| General trend | Expansion of DL solutions in Agriculture 4.0. | DL, AI, computer vision, sensing. | Digital transformation in agriculture | (Taranis 2025) (Bloomfield 2025) |

Source: Own authorship.

which surpasses CNN reference models such as VGG19, EfficientNetb6, ResNeXt, DenseNet201, AlexNet, YOLOv5, and MobileNetV3. Table 3 summarizes deep learning in agriculture in real examples.

## Conclusions

Deep learning represents a disruptive innovation in the agricultural scenario, offering intelligent solutions to contemporary challenges in food production. Its applications range from genomic prediction to early disease detection and automating real-time crop monitoring. DL promotes more efficient, precise, and sustainable agriculture by integrating complex and varied data, whether climate, genetic, visual, or environmental. Although there are challenges regarding scalability, cost, and interpretability, continuous advances in technology and accessibility suggest a promising future in which DL will be a fundamental pillar of tomorrow's agriculture.

The integration of deep learning with other tools such as the Internet of Things (IoT) and Big Data is becoming increasingly concrete, demonstrating the potential that these technologies together must revolutionize agriculture. It is important to develop and integrate these new technologies considering the farmers' point of view to ensure that the tools can be practical and applicable to the reality of the field. In addition, it is necessary to consider how the collected data will be protected, ensuring its integrity, and reliability and preventing data leaks or alterations. Although certain concerns and challenges exist, the new revolution in agriculture is a reality and a necessity given the current global scenario, therefore, the positive impacts cannot be ignored and must be welcomed.

## References

A, S. G., Suganthi, S. U., Prinslin, L., Selvi, R. and Prabha, R. (2025). Generative AI in agri: sustainability in smart precision farming yield prediction mapping system based on GIS using deep learning and GPS. Procedia Computer Science 252: 365–380. https://doi.org/10.1016/j.procs.2024.12.038.

Altalak, M., Ammad uddin, M., Alajmi, A. and Rizg, A. (2022). Smart agriculture applications using deep learning technologies: a survey. Applied Sciences 12(12): 5919. https://doi.org/10.3390/app12125919.

Araus, J. L., Kefauver, S. C., Vergara-Díaz, O., Gracia-Romero, A., Rezzouk, F. Z., Segarra, J. et al. (2022). Crop phenotyping in a context of global change: What to measure and how to do it. In Journal of Integrative Plant Biology 64(2): 592–618. John Wiley and Sons Inc. https://doi.org/10.1111/jipb.13191.

Ashurov, A. Y., Al-Gaashani, M. S. A. M., Samee, N. A., Alkanhel, R., Atteia, G., Abdallah, H. A. et al. (2025). Enhancing plant disease detection through deep learning: a Depthwise CNN with squeeze and excitation integration and residual skip connections. Frontiers in Plant Science 15. https://doi.org/10.3389/fpls.2024.1505857.

Attri, I., Awasthi, L. K., Sharma, T. P. and Rathee, P. (2023). A review of deep learning techniques used in agriculture. Ecological Informatics 77: 102217. https://doi.org/10.1016/j.ecoinf.2023.102217.

Azimi, S., Wadhawan, R. and Gandhi, T. K. (2021). Intelligent monitoring of stress induced by water deficiency in plants using deep learning. IEEE Transactions on Instrumentation and Measurement 70: 1–13. https://doi.org/10.1109/TIM.2021.3111994.

Baiju, B. V., Kirupanithi, N., Srinivasan, S., Kapoor, A., Mathivanan, S. K. and Shah, M. A. (2025). Robust CRW crops leaf disease detection and classification in agriculture using hybrid deep learning models. Plant Methods 21(1): 18. https://doi.org/10.1186/s13007-025-01332-5.

Bharman, P., Ahmad Saad, S., Khan, S., Jahan, I., Ray, M. and Biswas, M. (2022). Deep learning in Agriculture: A Review. Asian Journal of Research in Computer Science 28–47. https://doi.org/10.9734/ajrcos/2022/v13i230311.

Bloomfield. (2025, May 1). About Bloomfield Robotics. Https://Bloomfield.Ai/about-Us.

Boukhris, A., Jilali, A. and Sadiq, A. (2024). Satellite imagery, big data, IoT and deep learning techniques for wheat yield prediction in Morocco. Results in Control and Optimization 17: 100489. https://doi.org/10.1016/j.rico.2024.100489.

Butte, S., Vakanski, A., Duellman, K., Wang, H. and Mirkouei, A. (2021). Potato crop stress identification in aerial images using deep learning-based object detection. Agronomy Journal 113(5): 3991–4002. https://doi.org/10.1002/agj2.20841.

Cai, Y., Li, Y., Qi, X., Zhao, J., Jiang, L., Tian, Y. et al. (2024). A deep learning approach for deriving wheat phenology from near-surface RGB image series using spatiotemporal fusion. Plant Methods 20. https://doi.org/10.1186/s13007-024-01278-0.

Cama-Pinto, D., Damas, M., Holgado-Terriza, J. A., Arrabal-Campos, F. M., Martínez-Lao, J. A., Cama-Pinto, A. et al. (2023). A deep learning model of radio wave propagation for precision agriculture and sensor system in greenhouses. Agronomy 13(1): 244. https://doi.org/10.3390/agronomy13010244.

Carraro, A., Saurio, G. and Marinello, F. (2025). Towards rigorous dataset quality standards for deep learning tasks in precision agriculture: A case study exploration. Smart Agricultural Technology 10. https://doi.org/10.1016/j.atech.2024.100721.

Caruana, R. and Niculescu-Mizil, A. (2006). An empirical comparison of supervised learning algorithms. Proceedings of the 23rd International Conference on Machine Learning - ICML '06, 161–168. https://doi.org/10.1145/1143844.1143865.

Cattivelli, L., Rizza, F., Badeck, F.-W., Mazzucotelli, E., Mastrangelo, A. M., Francia, E. et al. (2008). Drought tolerance improvement in crop plants: An integrated view from breeding to genomics. Field Crops Research 105(1–2): 1–14. https://doi.org/10.1016/j.fcr.2007.07.004.

Chamara, N., Bai, G. and Ge, Y. (2023). AICropCAM: Deploying classification, segmentation, detection, and counting deep-learning models for crop monitoring on the edge. Computers and Electronics in Agriculture 215: 108420. https://doi.org/10.1016/j.compag.2023.108420.

Chen, Y., Mancini, M., Zhu, X. and Akata, Z. (2022). Semi-Supervised and Unsupervised Deep Visual learning: A Survey. http://arxiv.org/abs/2208.11296.

Chu, P., Li, Z., Lammers, K., Lu, R. and Liu, X. (2021). Deep learning-based apple detection using a suppression mask R-CNN. Pattern Recognition Letters 147: 206–211. https://doi.org/10.1016/j.patrec.2021.04.022.

Constable, F. and Bertaccini, A. (2017). Worldwide Distribution and Identification of Grapevine Yellows Diseases (pp. 17–46). https://doi.org/10.1007/978-3-319-50648-7_2.

Coulibaly, S., Kamsu-Foguem, B., Kamissoko, D. and Traore, D. (2022). Deep learning for precision agriculture: A bibliometric analysis. In Intelligent Systems with Applications (Vol. 16). Elsevier B.V. https://doi.org/10.1016/j.iswa.2022.200102.

Covarrubias-Pazaran, G. (2016). Genome-assisted prediction of quantitative traits using the R Package sommer. PLOS ONE 11(6): e0156744. https://doi.org/10.1371/journal.pone.0156744.

Dayioglu, M. A. and Turker, U. (2021). Digital transformation for sustainable future - Agriculture 4.0 : A review. Tarım Bilimleri Dergisi. https://doi.org/10.15832/ankutbd.986431.

de Alcantara, I. R., Schimidt, J. G. A., de Freitas Vian, C. E. and Belardo, G. (2021). AGRICULTURE 4.0: Origin, features, and consequences in the world and Brazil. Quaestum 2: 1–14.

Demir, S., Dedeoğlu, M. and Başayiğit, L. (2024). Yield prediction models of organic oil rose farming with agricultural unmanned aerial vehicles (UAVs) images and machine learning algorithms. Remote Sensing Applications: Society and Environment 33: 101131. https://doi.org/10.1016/j.rsase.2023.101131.

Demirhan, H. (2025). A deep learning framework for prediction of crop yield in Australia under the impact of climate change. Information Processing in Agriculture 12(1): 125–138. https://doi.org/10.1016/j.inpa.2024.04.004.

Dhaka, V. S., Meena, S. V., Rani, G., Sinwar, D., Kavita, Ijaz, M. F. and Woźniak, M. (2021). A survey of deep convolutional neural networks applied for prediction of plant leaf diseases. In Sensors (Vol. 21, Issue 14). MDPI AG. https://doi.org/10.3390/s21144749.

Dhruvitkumar V. Talati. (2021). Python: The alchemist behind AI's intelligent evolution. International Journal of Science and Research Archive 3(1): 235–248. https://doi.org/10.30574/ijsra.2021.3.1.0169.

dos Santos Ferreira, A., Matte Freitas, D., Gonçalves da Silva, G., Pistori, H. and Theophilo Folhes, M. (2017). Weed detection in soybean crops using ConvNets. Computers and Electronics in Agriculture 143: 314–324. https://doi.org/10.1016/j.compag.2017.10.027.

Epifani, L. and Caruso, A. (2024). A survey on deep learning in UAV imagery for precision agriculture and wild flora monitoring: Datasets, models and challenges. In Smart Agricultural Technology (Vol. 9). Elsevier B.V. https://doi.org/10.1016/j.atech.2024.100625.

Espejo-Garcia, B., Mylonas, N., Athanasakos, L., Vali, E. and Fountas, S. (2021). Combining generative adversarial networks and agricultural transfer learning for weeds identification. Biosystems Engineering 204: 79–89. https://doi.org/10.1016/j.biosystemseng.2021.01.014.

Feng, W., Gao, P. and Wang, X. (2024). AI breeder: Genomic predictions for crop breeding. New Crops 1: 100010. https://doi.org/10.1016/j.ncrops.2023.12.005.

Ferrag, M. A., Shu, L., Djallel, H. and Choo, K. K. R. (2021). Deep learning-based intrusion detection for distributed denial of service attack in agriculture 4.0. Electronics (Switzerland) 10(11). https://doi.org/10.3390/electronics10111257.

Gaafar, A. S., Dahr, J. M. and Hamoud, A. K. (2022). Comparative analysis of performance of deep learning classification approach based on LSTM-RNN for textual and image datasets. Informatica 46(5). https://doi.org/10.31449/inf.v46i5.3872.

Gan, Y., Liu, Z., Zhang, F., Xu, Q., Wang, X., Xue, H. et al. (2024). Deep learning based genomic breeding of pest-resistant grapevine 1 2. BioRxiv. https://doi.org/10.1101/2024.03.16.585323.

Gao, P., Zhao, H., Luo, Z., Lin, Y., Feng, W., Li, Y. et al. (2023). SoyDNGP: a web-accessible deep learning framework for genomic prediction in soybean breeding. Briefings in Bioinformatics 24(6). https://doi.org/10.1093/bib/bbad349.

Garrido, I., Erazo-Aux, J., Lagüela, S., Sfarra, S., Ibarra-Castanedo, C., Pivarčiová, E. et al. (2021). Introduction of deep learning in thermographic monitoring of cultural heritage and improvement by automatic thermogram pre-processing algorithms. Sensors (Switzerland) 21(3): 1–44. https://doi.org/10.3390/s21030750.

Georgieva, M. and Vassileva, V. (2023). Stress management in plants: examining provisional and unique dose-dependent responses. International Journal of Molecular Sciences 24(6): 5105. https://doi.org/10.3390/ijms24065105.

Ghimire, S., Yaseen, Z. M., Farooque, A. A., Deo, R. C., Zhang, J. and Tao, X. (2021). Streamflow prediction using an integrated methodology based on convolutional neural network and long short-term memory networks. Scientific Reports 11(1). https://doi.org/10.1038/s41598-021-96751-4.

Gill, H. S., Bath, B. S., Singh, R. and Riar, A. S. (2024). Wheat crop classification using deep learning. Multimedia Tools and Applications. https://doi.org/10.1007/s11042-024-18617-x.

Gómez-Zamanillo, L., Galán, P., Bereciartúa-Pérez, A., Picón, A., Moreno, J. M., Berns, M. et al. (2024). Deep learning-based instance segmentation for improved pepper phenotyping. Smart Agricultural Technology 9: 100555. https://doi.org/10.1016/j.atech.2024.100555.

Guo, W., Lin, K. and Ye, W. (2021). Deep embedded k-means clustering. IEEE International Conference on Data Mining Workshops, ICDMW, 2021-December, 686–694. https://doi.org/10.1109/ICDMW53433.2021.00090.

Hadir, A., Adjou, M., Assainova, O., Palka, G. and Elbouz, M. (2025). Comparative study of agricultural parcel delineation deep learning methods using satellite images: Validation through parcels complexity. Smart Agricultural Technology 10. https://doi.org/10.1016/j.atech.2025.100833.

Hasan, A. S. M. M., Diepeveen, D., Laga, H., Jones, M. G. K. and Sohel, F. (2023). Image patch-based deep learning approach for crop and weed recognition. Ecological Informatics 78. https://doi.org/10.1016/j.ecoinf.2023.102361.

Hassan, H., Ren, Z., Zhou, C., Khan, M. A., Pan, Y., Zhao, J. et al. (2022). Supervised and weakly supervised deep learning models for COVID-19 CT diagnosis: A systematic review. In Computer Methods and Programs in Biomedicine (Vol. 218). Elsevier Ireland Ltd. https://doi.org/10.1016/j.cmpb.2022.106731.

Hatem, Y., Hammad, G. and Safwat, G. (2022). Artificial intelligence for plant genomics and crop improvements. Egyptian Journal of Botany 62(2): 291–303. https://doi.org/10.21608/EJBO.2022.83200.1731.

Hati, A. J. and Singh, R. R. (2023). AI-driven pheno-parenting: a deep learning based plant phenotyping trait analysis model on a novel soilless farming dataset. IEEE Access 11: 35298–35314. https://doi.org/10.1109/ACCESS.2023.3265195.

Hickey, J. M., Dreisigacker, S., Crossa, J., Hearne, S., Babu, R., Prasanna, B. M. et al. (2014). Evaluation of genomic selection training population designs and genotyping strategies in plant breeding programs using simulation. Crop Science 54(4): 1476–1488. https://doi.org/10.2135/cropsci2013.03.0195.

Islam, S., Reza, M. N., Samsuzzaman, S., Ahmed, S., Cho, Y. J., Noh, D. H. et al. (2024). Machine vision and artificial intelligence for plant growth stress detection and monitoring: A review. Precision Agriculture Science and Technology 6(1): 33–57. https://doi.org/10.12972/pastj.20240003.

Istiak, Md. A., Syeed, M. M. M., Hossain, M. S., Uddin, M. F., Hasan, M., Khan, R. H. et al. (2023). Adoption of Unmanned Aerial Vehicle (UAV) imagery in agricultural management: A systematic literature review. Ecological Informatics 78: 102305. https://doi.org/10.1016/j.ecoinf.2023.102305.

Jabed, M. A. and Azmi Murad, M. A. (2024). Crop yield prediction in agriculture: A comprehensive review of machine learning and deep learning approaches, with insights for future research and sustainability. In Heliyon (Vol. 10, Issue 24). Elsevier Ltd. https://doi.org/10.1016/j.heliyon.2024.e40836.

Janiesch, C., Zschech, P. and Heinrich, K. (2021). Machine learning and deep learning. Electronic Markets 31(3).

Jeong, S., Ko, J., Ban, J., Shin, T. and Yeom, J. (2024). Deep learning-enhanced remote sensing-integrated crop modeling for rice yield prediction. Ecological Informatics 84: 102886. https://doi.org/10.1016/j.ecoinf.2024.102886.

Jhajharia, K., Mathur, P., Jain, S. and Nijhawan, S. (2022). Crop yield prediction using machine learning and deep learning techniques. Procedia Computer Science 218: 406–417. https://doi.org/10.1016/j.procs.2023.01.023.

Jia, W., Tian, Y., Luo, R., Zhang, Z., Lian, J. and Zheng, Y. (2020). Detection and segmentation of overlapped fruits based on optimized mask R-CNN application in apple harvesting robot. Computers and Electronics in Agriculture 172: 105380. https://doi.org/10.1016/j.compag.2020.105380.

Jin, C., Zhou, L., Pu, Y., Zhang, C., Qi, H. and Zhao, Y. (2025). Application of deep learning for high-throughput phenotyping of seed: a review. Artificial Intelligence Review 58(3). https://doi.org/10.1007/s10462-024-11079-5.

Jin, X., Yang, W., Doonan, J. H. and Atzberger, C. (2022). Crop phenotyping studies with application to crop monitoring. pp. 1221–1223. In Crop Journal (Vol. 10, Issue 5). Institute of Crop Sciences. https://doi.org/10.1016/j.cj.2022.09.001.

Kaler, A. S., Purcell, L. C., Beissinger, T. and Gillman, J. D. (2022). Genomic prediction models for traits differing in heritability for soybean, rice, and maize. BMC Plant Biology 22(1): 87. https://doi.org/10.1186/s12870-022-03479-y.

Kaneda, Y., Shibata, S. and Mineno, H. (2017). Multimodal sliding window-based support vector regression for predicting plant water stress. Knowledge-Based Systems 134: 135–148. https://doi.org/10.1016/j.knosys.2017.07.028.

Kasinathan, T., Singaraju, D. and Uyyala, S. R. (2021). Insect classification and detection in field crops using modern machine learning techniques. Information Processing in Agriculture 8(3): 446–457. https://doi.org/10.1016/j.inpa.2020.09.006.

Khatoon, S., Maruf Hasan, M., Asif, A., Alshmari, M. and Yap, Y.-K. (2021). Image-based Automatic Diagnostic System for Tomato Plants using deep learning. Computers, Materials & Continua 67(1): 595–612. https://doi.org/10.32604/cmc.2021.014580.

Kobayashi, K. and Alam, S. B. (2024). Explainable, interpretable, and trustworthy AI for an intelligent digital twin: A case study on remaining useful life. Engineering Applications of Artificial Intelligence 129: 107620. https://doi.org/10.1016/j.engappai.2023.107620.

Kumar, V., Sharma, K. V., Kedam, N., Patel, A., Kate, T. R. and Rathnayake, U. (2024). A comprehensive review on smart and sustainable agriculture using IoT technologies. Smart Agricultural Technology 8: 100487. https://doi.org/10.1016/j.atech.2024.100487.

Li, Z., Guo, R., Li, M., Chen, Y. and Li, G. (2020). A review of computer vision technologies for plant phenotyping. Computers and Electronics in Agriculture 176: 105672. https://doi.org/10.1016/j.compag.2020.105672.

Li, Z., Li, H. and Meng, L. (2023). Model compression for deep neural networks: a survey. In Computers (Vol. 12, Issue 3). MDPI. https://doi.org/10.3390/computers12030060.

Liu, Y., Ma, X., Shu, L., Hancke, G. P. and Abu-Mahfouz, A. M. (2021). From Industry 4.0 to Agriculture 4.0: Current status, enabling technologies, and research challenges. IEEE Transactions on Industrial Informatics 17(6): 4322–4334. https://doi.org/10.1109/TII.2020.3003910.

Mahmoud, A., Aurangzeb, K., Alhussein, M. and Elbelkasy, M. S. A. (2025). Harnessing deep learning to analyze climate change impacts on crop production. Alexandria Engineering Journal 125: 67–82. https://doi.org/10.1016/j.aej.2025.04.008.

Matsoukis, A. (2024). Phenology: A valuable tool in crop farming. Current Agriculture Research Journal 12(2): 496–497. https://doi.org/10.12944/CARJ.12.2.01.

Mienye, I. D., Swart, T. G. and Obaido, G. (2024). Recurrent neural networks: a comprehensive review of architectures, variants, and applications. Information 15(9): 517. https://doi.org/10.3390/info15090517.

Modi, A., Sharma, P., Saraswat, D. and Mehta, R. (2022). Review of crop yield estimation using machine learning and deep learning techniques. Scalable Computing: Practice and Experience 23(2): 59–80. https://doi.org/10.12694/scpe.v23i2.2025.

Montesinos-López, A., Crespo-Herrera, L., Dreisigacker, S., Gerard, G., Vitale, P., Saint Pierre, C. et al. (2024). Deep learning methods improve genomic prediction of wheat breeding. Frontiers in Plant Science 15. https://doi.org/10.3389/fpls.2024.1324090.

Montesinos-López, O. A., Montesinos-López, A., Pérez-Rodríguez, P., Barrón-López, J. A., Martini, J. W. R., Fajardo-Flores, S. B. et al. (2021). A review of deep learning applications for genomic selection. BMC Genomics 22(1): 19. https://doi.org/10.1186/s12864-020-07319-x.

Moussaid, A., El Fkihi, S., Zennayi, Y., Kassou, I., Bourzeix, F., Lahlou, O. et al. (2023). Citrus yield prediction using deep learning techniques: A combination of field and satellite data. Journal of Open Innovation: Technology, Market, and Complexity 9(2): 100075. https://doi.org/10.1016/j.joitmc.2023.100075.

Muruganandam, S., Joshi, R., Suresh, P., Balakrishna, N., Kishore, K. H. and Manikanthan, S. V. (2023). A deep learning based feed forward artificial neural network to predict the K-barriers for intrusion detection using a wireless sensor network. Measurement: Sensors 25. https://doi.org/10.1016/j.measen.2022.100613.

Muruganantham, P., Wibowo, S., Grandhi, S., Samrat, N. H. and Islam, N. (2022). A systematic literature review on crop yield prediction with deep learning and remote sensing. In Remote Sensing (Vol. 14, Issue 9). MDPI. https://doi.org/10.3390/rs14091990.

Narisetti, N., Henke, M., Neumann, K., Stolzenburg, F., Altmann, T. and Gladilin, E. (2022). Deep learning based greenhouse image segmentation and shoot phenotyping (DeepShoot). Frontiers in Plant Science 13. https://doi.org/10.3389/fpls.2022.906410.

Oliveira, A. J. de, Silva, G. F. da, Silva, G. R. da, Santos, A. A. C. dos, Caldeira, D. S. A., Vilarinho, M. K. C. et al. (2020). POTENCIALIDADES DA UTILIZAÇÃO DE DRONES NA AGRICULTURA DE PRECISÃO/DRONES POTENTIALITY USE IN PRECISION AGRICULTURE. Brazilian Journal of Development 6(9): 64140–64149. https://doi.org/10.34117/bjdv6n9-010.

Oliveira, M. do S. P. de, Mendes, G. G. C., Maciel, A. R. N. A. and Costa, J. B. da. (2024). Fenologia da floração e da frutificação do tucumanzeiro nas condições de Belém, Pará, Brasil. Research, Society and Development 13(7): e13513746428. https://doi.org/10.33448/rsd-v13i7.46428.

Pandian, J. A., Kumar, V. D., Geman, O., Hnatiuc, M., Arif, M. and Kanchanadevi, K. (2022). Plant disease detection using deep convolutional neural network. Applied Sciences (Switzerland) 12(14). https://doi.org/10.3390/app12146982.

Paudel, D., Neupane, R. C., Sigdel, S., Poudel, P. and Khanal, A. R. (2023). COVID-19 pandemic, climate change, and conflicts on agriculture: a trio of challenges to global food security. Sustainability 15(10): 8280. https://doi.org/10.3390/su15108280.

Paul, N., Sunil, G. C., Horvath, D. and Sun, X. (2025). Deep learning for plant stress detection: A comprehensive review of technologies, challenges, and future directions. Computers and Electronics in Agriculture 229: 109734. https://doi.org/10.1016/j.compag.2024.109734.

Qin, Y.-M., Tu, Y.-H., Li, T., Ni, Y., Wang, R.-F. and Wang, H. (2025). Deep learning for sustainable agriculture: a systematic review on applications in lettuce cultivation. Sustainability 17(7): 3190. https://doi.org/10.3390/su17073190.

Radoglou-Grammatikis, P., Sarigiannidis, P., Lagkas, T. and Moscholios, I. (2020). A compilation of UAV applications for precision agriculture. Computer Networks 172: 107148. https://doi.org/10.1016/j.comnet.2020.107148.

Rashid, A. Bin and Kausik, M. A. K. (2024). AI revolutionizing industries worldwide: A comprehensive overview of its diverse applications. Hybrid Advances 7: 100277. https://doi.org/10.1016/j.hybadv.2024.100277.

Raza, R. and Singh, N. K. (2021). A tour of unsupervised deep learning for medical image analysis. Current Medical Imaging 17(9).

Rehman, M. Z. U., Raghuvanshi, D. and Kumar, N. (2023). KisanQRS: A deep learning-based automated query-response system for agricultural decision-making. Computers and Electronics in Agriculture 213: 108180. https://doi.org/10.1016/j.compag.2023.108180.

Rojanarungruengporn, K. and Pumrin, S. (2021). Early stress detection in plant phenotyping using CNN and LSTM architecture. 2021 9th International Electrical Engineering Congress (IEECON) 389–392. https://doi.org/10.1109/iEECON51072.2021.9440342.

Ruan, S., Cang, H., Chen, H., Yan, T., Tan, F., Zhang, Y. et al. (2023). Hyperspectral classification of frost damage stress in tomato plants based on few-shot learning. Agronomy 13(9): 2348. https://doi.org/10.3390/agronomy13092348.

Rueda-Delgado, D., Cuellar-Torres, F., Martinez, D. and Narducci, M. S. (2025). Instrumentation system for monitoring of soil variables in precision agriculture applications. IEEE Access 13: 49777–49787. https://doi.org/10.1109/ACCESS.2025.3550859.

Rui, Z., Zhang, Z., Zhang, M., Azizi, A., Igathinathane, C., Cen, H. et al. (2024). High-throughput proximal ground crop phenotyping systems – A comprehensive review. Computers and Electronics in Agriculture 224.

Shamshiri, R. R., Sturm, B., Weltzien, C., Fulton, J., Khosla, R., Schirrmann, M. et al. (2024). Digitalization of agriculture for sustainable crop production: a use-case review. In Frontiers in Environmental Science (Vol. 12). Frontiers Media SA. https://doi.org/10.3389/fenvs.2024.1375193.

Sharifani, K. and Amini, M. (2023). Machine learning and deep learning: A review of methods and applications. World Information Technology and Engineering Journal 10. https://ssrn.com/abstract=4458723.

Sharma, N., Sharma, R. and Jindal, N. (2021). Machine learning and deep learning Applications—A Vision. Global Transitions Proceedings 2(1): 24–28. https://doi.org/10.1016/j.gltp.2021.01.004.

Sicat, J. P. A., Visendi, P., Sewe, S. O., Bouvaine, S. and Seal, S. E. (2022). Characterization of transposable elements within the Bemisia tabaci species complex. Mobile DNA 13(1): 12. https://doi.org/10.1186/s13100-022-00270-6.

Sinha, D., Maurya, A. K., Abdi, G., Majeed, M., Agarwal, R., Mukherjee, R. et al. (2023). Integrated genomic selection for accelerating breeding programs of climate-smart cereals. Genes 14(7): 1484. https://doi.org/10.3390/genes14071484.

Sohn, K., Berthelot, D., Li, C.-L., Zhang, Z., Carlini, N., Cubuk, E. D. et al. (2020). FixMatch: Simplifying semi-supervised learning with consistency and confidence. 34th Conference on Neural Information Processing Systems (NeurIPS 2020). https://github.com/google-research/fixmatch.

Subramaniam, L. K. and Marimuthu, R. (2024). Crop yield prediction using effective deep learning and dimensionality reduction approaches for Indian regional crops. E-Prime - Advances in Electrical Engineering, Electronics and Energy 8: 100611. https://doi.org/10.1016/j.prime.2024.100611.

Taranis. (2025, April 22). About Us. Https://Www.Taranis.Com/Our-Story/.

Taye, M. M. (2023). Understanding of machine learning with deep learning: architectures, workflow, applications and future directions. In Computers (Vol. 12, Issue 5). MDPI. https://doi.org/10.3390/computers12050091.

Tran, T.-T., Choi, J.-W., Le, T.-T. and Kim, J.-W. (2019). A comparative study of deep CNN in forecasting and classifying the macronutrient deficiencies on development of tomato plant. Applied Sciences 9(8): 1601. https://doi.org/10.3390/app9081601.

Turkoglu, M., Yanikoğlu, B. and Hanbay, D. (2022). PlantDiseaseNet: convolutional neural network ensemble for plant disease and pest detection. Signal, Image and Video Processing 16(2): 301–309. https://doi.org/10.1007/s11760-021-01909-2.

Upadhyay, A., Chandel, N. S., Singh, K. P., Chakraborty, S. K., Nandede, B. M., Kumar, M. et al. (2025). Deep learning and computer vision in plant disease detection: a comprehensive review of techniques, models, and trends in precision agriculture. Artificial Intelligence Review 58(3): 92. https://doi.org/10.1007/s10462-024-11100-x.

Viana, N. C. R. T., Ribeiro, M. M. de L. de O., Tavares, T. D. R. and Tavares, D. H. R. (2024). Inovação na indústria agrícola brasileira: tecnologias de agricultura de precisão. International Journal of Scientific Management and Tourism 10(6): e1197. https://doi.org/10.55905/ijsmtv10n6-013.

Wang, D. and He, D. (2022). Fusion of Mask RCNN and attention mechanism for instance segmentation of apples under complex background. Computers and Electronics in Agriculture 196: 106864. https://doi.org/10.1016/j.compag.2022.106864.

Wang, H., Yan, S., Wang, W., Chen, Y., Hong, J., He, Q. et al. (2025a). Cropformer: An interpretable deep learning framework for crop genomic prediction. Plant Communications. https://doi.org/10.1016/j.xplc.2024.101223.

Wang, H., Yan, S., Wang, W., Chen, Y., Hong, J., He, Q. et al. (2025b). Cropformer: An interpretable deep learning framework for crop genomic prediction. Plant Communications 6(3): 101223. https://doi.org/10.1016/j.xplc.2024.101223.

Wang, K., Abid, M. A., Rasheed, A., Crossa, J., Hearne, S. and Li, H. (2023). DNNGP, a deep neural network-based method for genomic prediction using multi-omics data in plants. Molecular Plant 16(1): 279–293. https://doi.org/10.1016/j.molp.2022.11.004.

Wang, M., Fu, B., Fan, J., Wang, Y., Zhang, L. and Xia, C. (2023). Sweet potato leaf detection in a natural scene based on faster R-CNN with a visual attention mechanism and DIoU-NMS. Ecological Informatics 27.

Wang, X., Zeng, H., Lin, L., Huang, Y., Lin, H. and Que, Y. (2023). Deep learning-empowered crop breeding: intelligent, efficient and promising. Frontiers in Plant Science 14. https://doi.org/10.3389/fpls.2023.1260089.

Wang, Z. (2004). APD: the antimicrobial peptide database. Nucleic Acids Research 32(90001): 590D–592. https://doi.org/10.1093/nar/gkh025.

Waqas Khan, M., Lim, M., Martin, J., Lee, R., Lynn Pokiak, D. and Whalen, D. (2025). Quantifying permafrost degradation processes: real-time analysis using environment-based particle image velocimetry and weather data. Permafrost and Periglacial Processes 36(2): 272–283. https://doi.org/10.1002/ppp.2268.

Weng, S., Ma, J., Tao, W., Tan, Y., Pan, M., Zhang, Z. et al. (2023). Drought stress identification of tomato plant using multi-features of hyperspectral imaging and subsample fusion. Frontiers in Plant Science 14. https://doi.org/10.3389/fpls.2023.1073530.

Xiao, Z., Yin, K., Geng, L., Wu, J., Zhang, F. and Liu, Y. (2022). Pest identification via hyperspectral image and deep learning. Signal, Image and Video Processing 16(4): 873–880. https://doi.org/10.1007/s11760-021-02029-7.

Ye, C., Li, K., Sun, W., Jiang, Y., Zhang, W., Zhang, P. et al. (2025). Biological prior knowledge-embedded deep neural network for plant genomic prediction. Genes 16(4): 411. https://doi.org/10.3390/genes16040411.

Yee-Rendon, A., Torres-Pacheco, I., Trujillo-Lopez, A. S., Romero-Bringas, K. P. and Millan-Almaraz, J. R. (2021). Analysis of New RGB vegetation indices for PHYVV and TMV identification in Jalapeño Pepper (Capsicum annuum) leaves using CNNs-based model. Plants 10(10): 1977. https://doi.org/10.3390/plants10101977.

Yu, Y., Zhang, K., Yang, L. and Zhang, D. (2019). Fruit detection for strawberry harvesting robot in non-structural environment based on Mask-RCNN. Computers and Electronics in Agriculture 163: 104846. https://doi.org/10.1016/j.compag.2019.06.001.

Zhang, Q., Zhang, Q. and Jensen, J. (2022). Association studies and genomic prediction for genetic improvements in agriculture. Frontiers in Plant Science 13. https://doi.org/10.3389/fpls.2022.904230.

Zhang, R., Zhu, H., Chang, Q. and Mao, Q. (2025). A comprehensive review of digital twins technology in agriculture. Agriculture 15(9): 903. https://doi.org/10.3390/agriculture15090903.

# Chapter 15

# Precision Farming and Decision Support Systems, Utilizing Convolutional Networks for Plant Detection

*Yossef Rubalcava-Avila,*[1,*] *Graciela Avila-Quezada,*[2] *Cesar Berzoza-Gaytan*[2] and *Mahendra Rai*[3,4]

## 1. Introduction

Global population growth poses one of the greatest challenges to food production, with projections estimating that the world's population will exceed 9 billion people by 2050. Given this need, precision agriculture emerges as an essential approach to optimize agricultural production by integrating advanced technologies, including sensors, global positioning systems (GPS), geographic information system (GIS), miniaturized computer components, automatic control, in-field and remote sensing, mobile computing, advanced information processing, and telecommunications. These are some of the technologies that have emerged and come together to benefit (Zhang et al. 2002), and improve agricultural yields.

Digital image processing is one of the key technologies within precision agriculture. It consists of the application of computer algorithms to analyze, enhance and extract relevant information from agricultural images (Chen et al. 2023, Jollet et al. 2023). Through these algorithms, it is possible to minimize noise, correct distortions and segment images to highlight specific features of plants and the environment. This type of processing is essential in various agricultural applications, such as detection of disease symptoms in crops, identification of fruits ready for harvest, yield estimation by fruit counting and weight estimation, branch detection for pruning, and pest and disease detection with proximity and remote images (Murad et al. 2023, Wang et al. 2023).

[1] Programming Department, Epoch Games. North Carolina, USA.
[2] Facultad de Ciencias Agrotecnológicas, Universidad Autónoma de Chihuahua, Escorza 900, Chihuahua, 31000 Chihuahua, México.
[3] Department of Biotechnology, SGB Amravati University, Amravati-444 602, Maharashtra, India.
[4] Department of Chemistry, Federal University of Piaui (UFPI), Teresina, Brazil.
* Corresponding author: yossefavila@hotmail.com

DOI: 10.1201/9781003659907-19

Convolutional neural networks (CNNs) have emerged as a useful tool for automated analysis of agricultural images. These networks, a subcategory of deep learning, are particularly effective in visual pattern identification, plant species classification and weed detection. Thanks to their ability to process large volumes of images with high accuracy, CNNs have become a key solution for agricultural management optimization and real-time decision making (Chen et al. 2023). In this chapter, the implementation of CNNs in plant sensing will be discussed in depth, highlighting their practical applications and their impact on agricultural sustainability.

## 2. Enhancing Agricultural Efficiency: The Role of Precision Farming

In modern agriculture, farmers often treat each field as a single unit. While they often recognize variability within fields, they have historically lacked the tools to manage it effectively. As a result, management decisions have generally been based on average conditions, with the hope that inputs would be adequate for most of the field. Precision farming, however, leverages information technology to divide fields into smaller, distinct sections, identifying the unique characteristics of each. This allows farmers to apply inputs precisely where and, in which amounts needed, optimizing for maximum economic return (Stephen 1997). Understanding the applications and benefits of these technologies will enhance one's knowledge and comprehension of the advantages.

Powerful agricultural machines, equipped with key features, provide cutting-edge route planning capabilities that enhance farming activities throughout the year. Through precise crop placement alignment, this technology enhances in-field precision and streamlines guiding line administration (Figure 1). They provide the precise application of seeds, chemicals, and fertilizers, regardless of terrain, with additional support from GPS receiver systems (Koenig 2024).

Automate steering and guidance uses GPS data. Steering guidance technologies keep equipment precisely on track, while planning features use implement data from the first pass to generate guidance lines for all subsequent operations, ensuring optimal path planning and reduced overlaps.

During planting or tillage cycles, these machines ensure in-field accuracy by commanding the tractor to follow the correct guidance line. The equipment also collects extensive data during operations, allowing farmers to analyze trends, make informed decisions, and plan for future seasons. Better outcomes are produced by this data-driven strategy.

Performance and dependability of the next generation are offered by Real-Time Kinematic (RTK) technology. The time required to get complete signal precision is shortened by multi-constellation satellite transmissions. Season-to-season repeatability significantly improves precision by reducing drift in boundaries, coverage, and guidelines. With centimeter-level accuracy, overlaps are minimized, inputs are optimized, and yields are maximized, contributing to more sustainable and profitable farming practices.

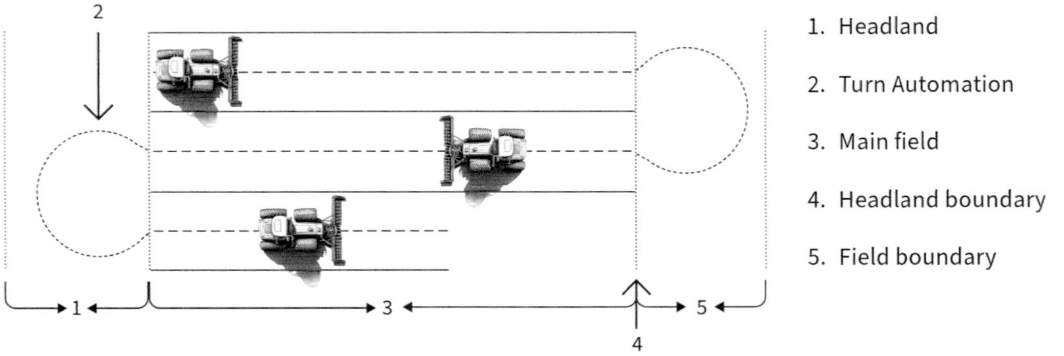


**Figure 1.** Automation guidance planner.

Machine automation further enhances efficiency by enabling autonomous operation in the field, reducing input costs and strain on the operator, which can cause headland inconsistencies, allowing human resources to focus on other essential tasks simultaneously (John Deere 2025).

The rising global demand for food is undoubtedly a key factor driving the development of new policies for modernization in many countries. Additionally, the widespread availability of affordable, accurate global positioning systems (GPS), combined with performance monitors, enables the rapid and efficient recording and mapping of crop yields during harvest. This technology aims to reduce the need for inputs such as fertilizers, seeds, and insecticides while boosting productivity, enhancing profitability, and improving environmental sustainability (Negrete 2017).

## 3. Advancements in Precision Positioning: The Role of RTK Technology in GPS Accuracy

RTK (Real-Time Kinematic) technology processes carrier phase observations for highly accurate positioning. The spatial positions obtained from RTK GPS are precise to the centimeter level, as carrier phase observations yield much greater accuracy than traditional code-based measurements. This method allows positioning to be completed more quickly and with higher precision than other terrestrial techniques (Mekik and Arslanoglu 2009).

The correct time is embedded into the signals that each of these spacecrafts broadcasts, and they are equipped with accurate atomic clocks. This time information is used by a GPS receiver to calculate the separation between each satellite from which a signal is received. As of July 3, 2023, the GPS constellation consisted of thirty-one operational satellites, excluding decommissioned satellites and on-orbit spares. The system ensures at least twenty-four satellites are operational at any time, with additional satellites enhancing reliability (United States Government 2025). By receiving signals from at least four satellites, the receiver can determine its position on Earth through triangulation. However, due to errors in the satellite signals, the calculated position is not always perfectly accurate. Some errors, such as those caused by atmospheric interference, are unavoidable. To improve location accuracy, an additional signal is required to complement the government-based system and correct for these errors (Stephen 1997). When more than four satellites are tracked, additional satellites provide redundancy, allowing for the detection of erroneous measurements, which leads to more reliable position calculations. The position of the satellites at the time of measurement also plays a crucial role in how measurement errors are distributed within the final computed position estimate (Lemmon and Gerdan 1999).

In RTK systems, a mobile "Rover" travels to the locations where spatial positions need to be ascertained, while a stationary "Base Station" is set up for the duration of the observation session. The rover and base station have dual-frequency receivers installed. Through a radio antenna, the base station sends phase observation adjustments to the rover. Correction data is sent to the rover by modern radio systems (Figure 2) (Mekik and Arslanoglu 2009).

A crucial component of the RTK method is data latency, which is the amount of time needed for data to be transmitted from the base station to the distant receiver, followed by software processing and decoding. Depending on several aspects, including the type of data link utilized, the distance between the base and distant stations, and the particular measurement circumstances, latency with current equipment might vary from 0.1 to 2 seconds (El-Mowafy 2000).

There are several benefits to this method. It offers a trustworthy instrument for precisely placing every point. RTK rapidly detects problems such as cycle slippage or loss of signal at the reference station, which hinder kinematic positioning in previous approaches. The survey may proceed without any problems since the system instantly resolves the new integer ambiguity. In real time, known locations may also be staked out in the field with precision down to the centimeter. RTK GPS is anticipated to ultimately supersede the DGPS technology and conventional terrestrial positioning systems because of its considerable benefits in terms of both time and cost-efficiency (Mekik and Arslanoglu 2009).

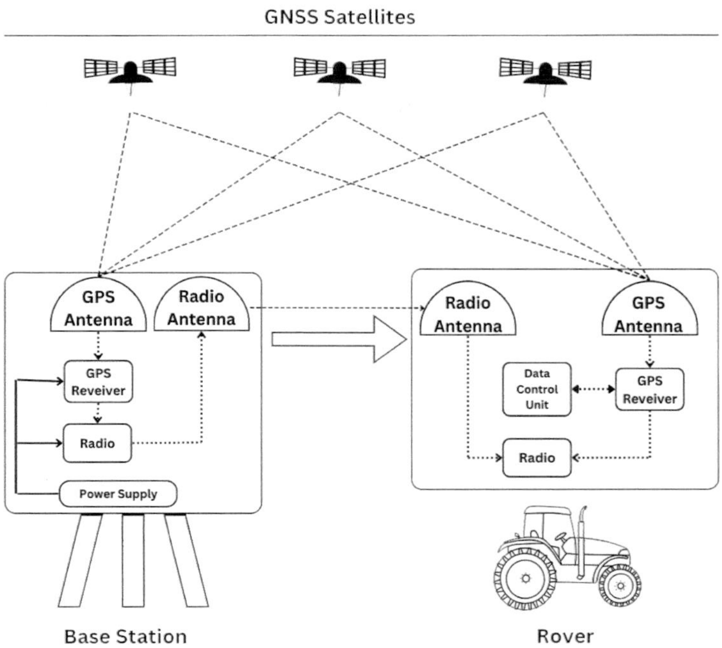


**Figure 2.** Relationship between GNSS satellites, a base station, and a rover.

GNSS satellites transmit signals to both the base station and the rover (tractor). The base station, with a fixed location, calculates signal errors and sends correction data via radio to the rover.

## 4. Precision Farming

Precision agriculture uses advanced technologies to monitor and control agricultural crop variables in order to optimize food production. This discipline integrates tools such as sensors, GPS, data analysis and remote sensing. GPS and Geographic Information Systems (GIS) allow mapping of the agricultural field and measuring spatial variations (Mani et al. 2021), which is essential for integrated crop management, including the application of irrigation, agrochemicals and biocontrol microorganisms with precision, adapting to the specific needs of each area.

Remote sensing through the use of drones or satellites efficiently gathers real-time data on crop conditions, soil health, weather patterns, and plant health (Sishodia et al. 2020). This information enables tailored application of agricultural inputs based on the specific needs of the crops, leading to optimal resource utilization. Furthermore, soil moisture sensors provide immediate data that enhance irrigation management, while yield monitors in harvesting machinery allow for precise assessments of crop productivity (current reference).

These advanced computer systems are essential for effective decision-making in precision agriculture, as they integrate weather data, soil sensor readings, and satellite imagery (Khanal et al. 2017). This integration allows farmers to make informed decisions within complex agricultural environments. Additionally, analyzing both historical and real-time data supports the prediction of crop yields and the forecasting of potential pest and disease outbreaks (Ukhurebor et al. 2022, Mohyuddin et al. 2024).

The application of algorithms and machine learning facilitates the simulation of various agricultural scenarios, allowing the system to predict the outcomes of different farming practices (Sharma et al. 2020). With these simulations, management decisions can be informed, such as adjusting irrigation volumes or fertilizer applications, while estimating the potential results of these practices. Consequently, farmers can make more accurate decisions to optimize resource use and improve overall agricultural efficiency.

Among the limitations of precision agriculture are the high initial costs and the difficulty in handling large amounts of data (Groeneveld et al. 2021). Many agricultural producers do not have the technical knowledge to handle the software. However, with the advancement of AI, machine learning and new sensor technologies, these processes are expected to be optimized and become increasingly accessible to producers.

The future of agriculture will increasingly depend on the development of artificial intelligence and new technologies, enabling more automated and accurate decision making. With these advances, agricultural production will be able to adapt to global challenges and sustainably meet the world's growing demand for food (Ukhurebor et al. 2022).

## 5. Convolutional Neural Networks (CNN)

Crops represent complex environments to analyze because of their agricultural cycles, the variability of their leaves, flowers and fruits, and their phenological stages, foliage or non-uniform backgrounds. In addition, their appearance varies subtly during the day, as the amount and angle of incident solar radiation affects their spectral response. To identify crop diseases, the techniques have been focused on the analysis of reflectance in the visible and near-infrared spectrum, the creation of vegetation-specific indices, and the study of spectral patterns to detect anomalies (Boulent et al. 2019). Therefore, there is a need for techniques that can address these challenges and offer practical solutions for the automated identification of diseases.

CNNs, also known as ConvNet, are an advanced application of artificial intelligence (AI) used for plant detection, gaining significant relevance in precision agriculture (Abdullahi et al. 2017). Convolutional neural networks (CNNs) are very effective in tasks such as object detection, recognition, classification and biometrics. They use convolution layers as filters to create optimized visual representations. Through training, CNNs produce models tailored to specific tasks and can be generalized to process unseen data, offering robustness against varying conditions and variability. However, they require large data sets to train.

CNNs are a class of deep neural networks effective for processing and analyzing images (Traore et al. 2018), making them useful tools for identifying and classifying different plant species, detecting pests, or monitoring crop maturity status, among other applications.

These networks are designed to recognize spatial and hierarchical patterns in images, allowing them to detect objects, edges, textures, and shapes (Baker et al. 2018). Regarding plant detection, CNNs can learn unique visual characteristics of leaves, flowers, or structures of a plant and correctly classify it into a specific category (Gogul and Kumar 2017).

CNNs can be trained to identify different plant species in agricultural fields, aiding in the separation of crops and weeds (Hashemi-Beni et al. 2022). They are also capable of identifying early symptoms of diseases and signs of pests by analyzing images of plants, as they can recognize visual patterns associated with insect damage, necrotic or chlorotic spots, and abnormal growth or leaf development patterns (Sarkar et al. 2023), allowing for timely pest management.

The CNN training and evaluation process enables learning to detect specific patterns in plant images (Jiang and Li 2020). For this purpose, large numbers of plant images are collected, preferably in a single season of the year and under homogeneous light conditions (Dhaka et al. 2021). The images should be of leaves, flowers, roots, or complete plants, and must be labeled for the image can later be associated with the species or condition of the plant (Abade et al. 2021). It is essential to extract, classify and segment the visual characteristics of the images before entering them into the model.

Subsequently, the preprocessed images are fed to the convolutional neural network. Through multiple convolutional layers and resolution reduction, the network learns to identify key plant features, such as textures, edges, and patterns (Maggiori et al. 2016). After training the network with a set of images, its performance is evaluated using an independent dataset. If the accuracy is low, parameters should be adjusted, possibly by increasing the number of images available for training (Kang et al. 2014).

Once trained and validated, the CNN model is applied in the agricultural field on images captured with drone cameras or cell phones. At this point, the network analyzes the images in real time and provides results on the detected plant species (Osco et al. 2021).

Among the advantages of CNNs in agriculture is their ability to perform automated detections of crops and weeds, differentiating them with high precision and reducing the time required for these tasks. In addition, these networks can be applied over large areas of crops continuously, enabling early-stage weed detection (Qu and Su 2024).

Despite the advances and benefits of CNNs in precision agriculture, there are several limitations and challenges in their effective implementation.

*High computational requirements*: Training CNN models requires large amounts of data and considerable computational power, which can make them difficult to implement in resource-limited agricultural settings (Osco et al. 2021).

*Need for large volumes of labeled data*: For CNNs to achieve a high level of accuracy, large sets of accurately labeled images are needed, which can be a costly and laborious process (Dhaka et al. 2021).

*Variability in environmental conditions*: Factors such as illumination, seasonality, and soil quality can affect the accuracy of CNN models, requiring constant adjustments to improve crop and weed detection (Jiang and Li 2020).

*Difficulty in integration with existing agricultural systems*: Adoption of CNNs in agriculture requires advanced technological infrastructure, which can be challenging for farmers without access to these tools (Hashemi-Beni et al. 2022).

*Limited generalizability*: CNN models trained on a specific crop type or condition may not be equally effective in other environments, requiring constant adaptation and retraining (Qu and Su 2024).

*Cost of implementation*: Although CNNs can generate long-term benefits, the initial cost of acquiring specialized hardware and software for implementation remains a barrier to mass adoption (Kang et al. 2014).

Despite these limitations, continued research and development of new methodologies and improved CNN architectures will overcome many of these challenges, making automated plant detection increasingly efficient and accessible to the agricultural sector.

## 6. Applications of CNNs in Precision Agriculture

Convolutional neural networks have proven to be key tools for the optimization of various agricultural practices within the framework of precision agriculture. Their ability to process images and extract visual features has enabled multiple applications that improve productivity and crop management. Some of the main applications include:

*Crop and weed classification and detection*: CNNs enable highly accurate differentiation between cultivated plants and weeds, facilitating localized herbicide application and reducing environmental impact (Hashemi-Beni et al. 2022).

*Crop health monitoring*: Using multispectral imaging and visual pattern analysis, CNNs can detect early symptoms of diseases and nutritional deficiencies in crops, allowing timely intervention and optimization of agrochemical use (Sarkar et al. 2023).

*Agricultural yield estimation*: Through image segmentation and automated fruit and plant counting, CNNs can predict crop yields with high accuracy, facilitating crop planning (Dhaka et al. 2021).

*Harvest automation*: Integrated with agricultural robots and computer vision systems, CNNs can identify harvest-ready fruits and optimize the picking process (Qu and Su 2024).

*Pest and disease detection*: Using drone imagery and proximity sensors, CNNs can identify pest damage, improving decision making for integrated pest management (Osco et al. 2021).

*Optimization of irrigation and fertilization*: Based on the analysis of thermal and multispectral images, these networks can determine the water and nutritional needs of crops, avoiding the excessive use of resources (Jiang and Li 2020).

Advances in computer vision and deep learning technologies have made CNNs increasingly accessible to agriculture, improving the efficiency of agricultural production and reducing long-term operating costs.

According to De Silva and Brown (2023), dataset balancing, filter efficiency, and deep learning models are essential in plant disease detection. In their study, filter performance was shown to be dependent on dataset balancing. The balanced data set revealed that filters capturing visible and near infrared (NIR) segments (like K665, K720) were more effective, highlighting the importance of complete spectral coverage for accurate disease detection. The authors added that among the deep learning models, Vision Transformer (ViT) models, especially ViT-B16, outperformed the others, confirming their superiority in complex tasks such as plant disease identification. The study also noted that very small or large models (like ResNet-50V2 or ViT-L16) tended to fit poorly, while moderately sized models such as ViT-B16 provided the best balance between capability and generalization.

## 7. TOMMicroNet for Disease Detection

Artificial intelligence, assisted by cell phone images, was used in the development of TOMMicroNet, a convolutional neural network (CNN) designed for disease detection by Sentil et al. (2024). To create it, the researchers utilized a 30× microscopic lens attached to a smartphone, generating a dataset of 8,500 images. These images captured thirty-four different tomato plant conditions, including both the upper and underside of leaves, as well as the surface of tomato fruits.

TOMMicroNet consists of fourteen layers specifically trained to classify both biotic and abiotic plant health issues. When compared to six existing pre-trained CNN models, TOMMicroNet demonstrated superior performance, achieving 95% classification accuracy when trained on the complete dataset.

## 8. Conclusions and Future Perspectives

Modern agriculture has been revolutionized by precision farming, integrated with advanced decision support systems (DSS), optimizing resource utilization and improving crop yield. Convolutional neural networks (CNNs) play a crucial role in processing and analyzing agricultural images, making them valuable tools for plant species identification, classification, and rapid disease detection caused by pests and pathogens. Additionally, CNNs facilitate monitoring crop maturity, enhancing decision-making for timely harvesting. The synergy between AI-driven analytics and sensor-based data collection has led to more efficient, cost-effective, and sustainable farming practices. However, challenges such as computational complexity, the need for large annotated datasets, and environmental variability must be addressed for broader adoption.

Future advancements in precision farming and decision support systems (DSS) will be driven by the integration of convolutional neural networks (CNNs) with emerging technologies. The fusion of CNNs with the Internet of Things (IoT) and edge computing will enable real-time, on-field plant detection and decision-making with minimal latency. Moreover, advancements in transfer learning will allow pre-trained deep learning models to be fine-tuned for specific crops and agricultural conditions, reducing reliance on extensive labeled datasets. The incorporation of multispectral and hyperspectral imaging will further enhance plant stress detection, nutrient analysis, and disease prediction with greater precision. AI-powered autonomous farming systems, including drones and robotic machinery, will revolutionize agriculture by performing real-time plant health assessments and targeted interventions. Furthermore, the scalability of CNN-based DSS will be enhanced through cloud-based analytics, providing farmers with remote access to AI-driven insights while integrating multiple data sources. These innovations will contribute to sustainable agriculture by minimizing pesticide use, optimizing water and fertilizer application, and promoting eco-friendly farming practices. However, for widespread adoption, the development of affordable, user-friendly AI tools tailored for smallholder farmers is essential, alongside supportive policies and infrastructure investments. By addressing current challenges and leveraging these technological advancements, CNN-driven precision farming and DSS will play a key role in ensuring food security, sustainability, and efficiency in future agricultural practices.

## References


Abade, A., Ferreira, P. A. and de Barros Vidal, F. (2021). Plant diseases recognition on images using convolutional neural networks: A systematic review. Computers and Electronics in Agriculture 185: 106125.

Abdullahi, H. S., Sheriff, R. and Mahieddine, F. (2017). Convolution neural network in precision agriculture for plant image recognition and classification. pp. 256–272. In 2017 Seventh International Conference on Innovative Computing Technology (INTECH) (Vol. 10). New York: IEEE.

Baker, N., Lu, H., Erlikhman, G. and Kellman, P. J. (2018). Deep convolutional networks do not classify based on global object shape. PLoS Computational Biology 14(12): e1006613.

Boulent, J., Foucher, S., Théau, J. and St-Charles, P.-L. (2019). Convolutional neural networks for the automatic identification of plant diseases. Front. Plant Sci. 10: 941. https://doi.org/10.3389/fpls.2019.00941.

Chen, A., Jacob, M., Shoshani, G. and Charter, M. (2023). Using computer vision, image analysis and UAVs for the automatic recognition and counting of common cranes (Grus grus). Journal of Environmental Management 328: 116948. https://doi.org/10.1016/j.jenvman.2022.116948.

De Silva, M. and Brown, D. (2023). Multispectral plant disease detection with Vision transformer–convolutional neural network hybrid approaches. Sensors 23(20): 8531.

Dhaka, V. S., Meena, S. V., Rani, G., Sinwar, D., Ijaz, M. F. and Woźniak, M. (2021). A survey of deep convolutional neural networks applied for prediction of plant leaf diseases. Sensors 21(14): 4749.

El-Mowafy, A. (2000). Performance analysis of the RTK technique in an urban environment. Australian Surveyor 45(1): 47–54. doi:10.1080/00050353.2000.105.

Gogul, I. and Kumar, V. S. (2017, March). Flower species recognition system using convolution neural networks and transfer learning. pp. 1–6. In 2017 Fourth International Conference on Signal Processing, Communication and Networking (ICSCN). IEEE.

Groeneveld, D., Tekinerdogan, B., Garousi, V. and Catal C. (2021). A domain-specific language framework for farm management information systems in precision agriculture. Precision Agric 22: 1067–1106. https://doi.org/10.1007/s11119-020-09770-y.

Hashemi-Beni, L., Gebrehiwot, A., Karimoddini, A., Shahbazi, A. and Dorbu, F. (2022). Deep convolutional neural networks for weeds and crops discrimination from UAS imagery. Frontiers in Remote Sensing 3: 755939.

Jiang, Y. and Li, C. (2020). Convolutional neural networks for image-based high-throughput plant phenotyping: a review. Plant Phenomics.

John Deere. (2025). Automation 4.0 Activation. https://www.deere.com/en/technology-products/precision-ag-technology/guidance/gen-4-commandcenter-automation-activation/.

Jollet, D., Junker-Frohn, L. V., Steier, A., Meyer-Lüpken, T. and Müller-Linow, M. (2023). A new computer vision workflow to assess yield quality traits in bush bean (*Phaseolus vulgaris* L.). Smart Agricultural Technology 5: 100306. https://doi.org/10.1016/j.atech.2023.100306.

Kang, L., Ye, P., Li, Y. and Doermann, D. (2014). Convolutional neural networks for no-reference image quality assessment. pp. 1733–1740. In Proceedings of the IEEE Conference on Computer Vision and Pattern Recognition.

Khanal, S., Fulton, J. and Shearer, S. (2017). An overview of current and potential applications of thermal remote sensing in precision agriculture. Computers and Electronics in Agriculture 139: 22–32. https://doi.org/10.1016/j.compag.2017.05.001.

Koenig. (2024). Elevate Your Farming Operations With a Complete AutoTrac System. https://blog.koenigequipment.com/elevate-your-farming-operations-with-a-complete-autotrac-system.

Lemmon, T. R. and Gerdan, G. P. (1999). The influence of the number of satellites on the accuracy of RTK GPS positions. Australian Surveyor 44(1): 64–70. https://doi.org/10.1080/00050351.1999.10558774.
Maggiori, E., Tarabalka, Y., Charpiat, G. and Alliez, P. (2016). Convolutional neural networks for large-scale remote-sensing image classification. IEEE Transactions on Geoscience and Remote Sensing 55(2): 645–657.
Mani, P. K., Mandal, A., Biswas, S., Sarkar, B., Mitran, T. and Meena, R. S. (2021). Remote sensing and geographic information system: a tool for precision farming. *In*: Mitran, T., Meena, R. S. and Chakraborty, A. (eds.). Geospatial Technologies for Crops and Soils. Springer, Singapore. https://doi.org/10.1007/978-981-15-6864-0_2.
Mekik Cetin and Arslanoglu Murat. (2009). Investigation on Accuracies of Real Time Kinematic GPS for GIS Applications. 22–35; https://doi.org/10.3390/rs1010022.
Mohyuddin, G., Khan, M. A., Haseeb, A., Mahpara, S., Waseem, M. and Saleh, A. M. (2024). Evaluation of machine learning approaches for precision farming in smart agriculture system-a comprehensive review. IEEE Access.
Murad, N. Y., Mahmood, T., Forkan, A. R. M., Morshed, A., Jayaraman, P. P. and Siddiqui, M. S. (2023). Weed detection using deep learning: A systematic literature review. Sensors 23(7): 3670. https://doi.org/10.3390/s23073670.
Negrete, Jaime. (2017). Precision agriculture in mexico; current status and perspectives. International Journal of Horticulture 7(10): 75–81. doi:10.5376/ijh.2017.07.0010.
Osco, L. P., de Arruda, M. D. S., Gonçalves, D. N., Dias, A., Batistoti, J., de Souza, M. et al. (2021). A CNN approach to simultaneously count plants and detect plantation-rows from UAV imagery. ISPRS Journal of Photogrammetry and Remote Sensing 174: 1–17.
Qu, H. R. and Su, W. H. (2024). Deep learning-based weed–crop recognition for smart agricultural equipment: a review. Agronomy 14(2): 363.
Sarkar, C., Gupta, D., Gupta, U. and Hazarika, B. B. (2023). Leaf disease detection using machine learning and deep learning: Review and challenges. Applied Soft Computing 145: 110534.
Sentil, S., Choudhary, M., Tirsaiwala, M., Rvs, S., Suresh, V. M., Jacob, C. et al. (2024). TOMMicroNet: Convolutional neural networks for smartphone-based microscopic detection of tomato biotic and abiotic plant health issues. Phytopathology® 114(11): 2431–2441.
Sharma, A., Jain, A., Gupta, P. and Chowdary, V. (2020). Machine learning applications for precision agriculture: A comprehensive review. IEEE Access 9: 4843–4873.
Sishodia, R. P., Ray, R. L. and Singh, S. K. (2020). Applications of remote sensing in precision agriculture: A review. Remote Sensing 12(19): 3136. https://doi.org/10.3390/rs12193136.
Stephen W. Searcy and Texas Agricultural Extension Service. (1997). Precision Farming: A New Approach to Crop Management. https://books.google.com.mx/books?id=q2ZFPAAACAAJ.
Traore, B. B., Kamsu-Foguem, B. and Tangara, F. (2018). Deep convolution neural network for image recognition. Ecological Informatics 48: 257–268.
Ukhurebor, K. E., Adetunji, C. O., Olugbemi, O. T., Nwankwo, W., Olayinka, A. S., Umezuruike, C. et al. (2022). Precision agriculture: Weather forecasting for future farming. pp. 101–121. In AI, Edge and Iot-based Smart Agriculture. Academic Press.
United States Government. (2025). URL: https://www.gps.gov/space-segment.
Wang, J., Wang, P., Tian, H., Tansey, K., Liu, J. and Quan, W. (2023). A deep learning framework combining CNN and GRU for improving wheat yield estimates using time series remotely sensed multi-variables. Computers and Electronics in Agriculture 206: 107705. https://doi.org/10.1016/j.compag.2023.107705.
Zhang, N., Wang, M. and Wang, N. (2002). Precision agriculture—a worldwide overview. Computers and Electronics in Agriculture 36(2-3): 113–132. https://doi.org/10.1016/S0168-1699(02)00096-0.

# Chapter 16

# Leveraging Artificial Intelligence for the Detection, Monitoring, and Forecasting of Plant Diseases

*Prasad V. Sherekar*, Divyesha A. Pathak* and *Sanvidhan G. Suke*

## 1. Introduction

The agricultural sector faces significant challenges due to plant diseases, leading to reduced crop yields and food insecurity (Ahmad et al. 2020, Bodiger and Mathew 2023, Kalpana et al. 2024, Alhwaiti et al. 2025). Early detection and accurate identification of plant diseases are crucial for effective management, yet traditional methods—such as manual inspections—are often slow, labor-intensive, and error-prone (Yag and Altan 2022, Ullah et al. 2023, Sentil et al. 2024). Various plant diseases, including those affecting citrus crops, orchids, tobacco, tea, and tomatoes, present threats to global agricultural productivity (Lee et al. 2021, Li et al. 2022, Sentil et al. 2024, Butt et al. 2025, Chen et al. 2025).

Automated disease detection methods, particularly image-based and deep learning (DL) approaches, are increasingly recognized as essential for improving accuracy and efficiency (Ullah et al. 2023, Sentil et al. 2024). However, traditional DL models often struggle with phenotype recognition due to environmental factors and symptom similarity, leading to issues like overfitting. The integration of precision agriculture techniques and intelligent prevention systems offers promising solutions to overcome these limitations (Xiong et al. 2024). In essence, developing reliable, automated, and scalable disease detection systems is vital to safeguarding global food security, optimizing agricultural production, and supporting the economic stability of agriculture-dependent nations.

### *1.1 Important Agriculture Plant Diseases*

The severity of plant diseases varies based on factors such as crop susceptibility, environmental conditions, disease spread, and economic impact. Here is a brief information about various plant diseases.

Department of Biotechnology, Priyadarshini College of Engineering, Hingna Road, Nagpur-440019, Maharashtra, India.
* Corresponding author: sherekar.vprasad@gmail.com

DOI: 10.1201/9781003659907-20

*Flavescence Dorée (FD)*: A grapevine disease caused by *phytoplasmas*, FD is transmitted by *Scaphoideus titanus* leafhoppers. Symptoms include stunted growth, leaf yellowing, delayed ripening, and vine decline. Despite quarantine and pesticide interventions, FD continues to spread across European vineyards, impacting wine production and agricultural economies (Boulent et al. 2020).

*Orchid infections*: Viruses, such as *Odontoglossum ringspot* and *Cymbidium mosaic virus,* cause structural deformities in orchids, negatively affecting their commercial value. The viruses disrupt pigment formation, leading to discolored or malformed flowers. Detection methods such as polymerase chain reactions and serological assays are crucial for managing these infections (Lee et al. 2021).

*Tea Plant Diseases*

*Brown blight*: Caused by *Colletotrichum camelliae*, leading to leaf spots and premature defoliation.

*Target spot*: *Corynespora cassicola* infection results in brown lesions with concentric rings, affecting tea quality.

*Tea coal disease*: Caused by *Exobasidium vexans*, this disease leads to thickened leaves and reduced photosynthetic efficiency.

*Apolygus lucorum*: A major pest causing leaf damage by sucking plant fluids, leading to stunted growth (Li et al. 2022).

*Banana Diseases and Pests*

Various fungal, bacterial, and viral diseases affect banana crops in India, impacting economic stability. Panama disease (*Fusarium oxysporum*) and Black Sigatoka (*Mycosphaerella fijiensis*) severely reduce yield. Proper disease management and pest control strategies are necessary to sustain banana production (Narayanan et al. 2022).

*Wheat Stripe Rust Disease (WRD)*

This fungal disease, caused by *Puccinia striiformis*, severely affects wheat crops by forming yellow stripe-like pustules on leaves, reducing photosynthesis and grain filling. The rapid spread of WRD leads to significant yield losses, exacerbating food insecurity. Climate change and favorable environmental conditions further enhance its epidemic potential (Anwar et al. 2023).

*Tobacco Diseases*

*Angular leaf spot*: Caused by *Pseudomonas syringae*, leading to water-soaked lesions that affect leaf integrity.

*Brown spot*: *Alternaria alternata* infection results in circular necrotic lesions, lowering tobacco leaf quality.

*Wildfire disease*: Caused by *Pseudomonas tabaci*, it produces large necrotic patches, hampering crop yield.

*Frog eye disease*: *Cercospora nicotianae* infection leads to circular brown lesions with lighter centers, reducing tobacco productivity (Chen et al. 2025).

*Rice Diseases*

*Brown spot*: *Bipolaris oryzae* causes leaf discoloration and reduced grain quality.

*Hispa*: An insect pest (*Dicladispa armigera*) that feeds on rice leaves, leading to skeletal damage.

*Leaf smut*: Caused by *Entyloma oryzae*, reducing photosynthesis efficiency.

*Bacterial leaf blight*: *Xanthomonas oryzae* infection leads to water-soaked lesions, severely affecting rice yield.

*Leaf blast*: *Magnaporthe oryzae* produces lesions on rice leaves, leading to major yield losses, especially in humid conditions (Raman and Jayaraman 2025).

These diseases highlight the need for advanced detection, prevention, and control strategies in agriculture to ensure food security and economic stability. Identifying plant diseases requires distinguishing known pathogens from newly emerging threats. Continual learning through AI-based recognition models and image processing techniques aids in early-stage disease detection, improving response strategies and minimizing yield losses.

### *1.2 Need of AI for Plant Diseases Detection*

AI, machine learning (ML), and DL have significantly advanced plant disease detection, aiding farmers in identifying infections caused by viruses, bacteria, and fungi. Remote sensing and mobile technologies further enhance disease monitoring and classification (Abdulridha et al. 2022, Anwar et al. 2023, Sentil et al. 2024, Alhwaiti et al. 2025). The DL techniques, particularly convolutional neural networks (CNNs), have shown promise in plant disease recognition (Sibiya and Sumbwanyambe 2021, Chen et al. 2025). However, challenges such as data scarcity, image noise, classification accuracy, and computational constraints hinder real-world implementation (Albattah et al. 2022, Anwar et al. 2023, Kalpana et al. 2024). The reliance on large-scale labelled datasets also complicates AI adoption in agricultural settings, especially in rural areas (Batool et al. 2025). The emergence of AI-driven solutions enables early disease detection, reducing economic losses and improving food security (Anwar et al. 2023, Faisal et al. 2025). Transfer learning, Internet of Things (IoT) integration, and mobile-based classification methods offer scalable options for precision agriculture (Kalpana et al. 2024, Shafik et al. 2025). Despite these advancements, further research is required to enhance model reliability, address overfitting issues, and ensure adaptability to diverse plant pathogens.

## 2. Fundamentals of AI

AI is a transformative technology that enables computers and other devices to mimic human intelligence, creativity, comprehension, and autonomy (Bini 2018). AI applications span multiple domains, with self-driving cars and virtual assistants like Amazon Alexa and Apple Siri serving as prime examples of how AI-driven systems can autonomously respond to real-time situations without requiring human intervention (Sharma et al. 2021). This autonomy is facilitated by computational algorithms that enable machines to process and analyze data effectively. AI encompasses several goal-oriented subfields, including planning and decision-making, learning, natural language processing, vision, and speech recognition. Among these, ML plays a crucial role in AI by allowing systems to improve their performance through experience and data exposure. ML relies on mathematical models to classify information and predict outcomes. For instance, AI-powered e-commerce platforms use ML algorithms to make personalized product recommendations based on previous customer purchases.

Numerous ML algorithms exist, each suited to different types of data and problem-solving scenarios. These include logistic regression, decision trees (DT), random forests (RF), and support vector machines (SVMs), among others (Table 1). However, neural networks, specifically artificial neural networks (ANNs), are among the most widely utilized ML models. Inspired by the structure and function of the human brain, ANNs consist of interconnected layers of nodes that process and evaluate complex data, recognizing intricate patterns and relationships. One notable application of neural networks is in plant disease detection, where they help distinguish between healthy and

**Table 1.** Concepts and algorithms used in ML.

| **Terms** | **Definition** |
|---|---|
| *Major classes of ML* | |
| Supervised Learning | To enable the algorithm to assess its accuracy, the dataset has been pre-labelled and classed by humans (users). |
| Unsupervised Learning | Utilizing an unlabelled raw dataset, an algorithm finds patterns and relationships in the data without the assistance of users. |
| Reinforcement Learning | The dataset employs a "rewards/punishments" approach, which provides feedback to the algorithm, allowing it to learn from its own trials and errors. |
| *Machine Learning Algorithms* | |
| Linear Regression | Predicts numerical values using a linear relationship between variables. |
| Logistic Regression | Predicts categorical response variables, such as "yes/no" answers to queries. |
| SVM | A technique used for classification and regression tasks. It efficiently handles non-linear classification through the kernel trick, which transforms input data into higher-dimensional feature spaces (Jafar et al. 2024). |
| DT | Predicts outcomes and classifies data by traversing from root to leaf nodes. Each node represents a decision point, with branches showing possible outcomes based on input features. |
| RF | An ensemble of decision trees that classifies new items by aggregating votes from individual trees based on their features. The class receiving the most votes becomes the final classification. |
| K-Nearest Neighbours (KNN) | A non-parametric technique used in pattern recognition and regression analysis. It is a slow learning method that delays computations until classification by using local function approximations. Though simple, KNN efficiently classifies by weighting closer data points more heavily. |
| *Deep Learning Architectures* | |
| CNN | Involves combining an input image (or feature map) with a small matrix called a filter to extract features. This process highlights specific characteristics like edges, shapes, or patterns within the image. |
| Transformer | Consists of an encoder and decoder architecture. The encoder translates the input sequence into a vector, and the decoder converts that vector back into an output sequence. The self-attention mechanism processes the entire data in parallel (Ngugi et al. 2024). |
| Multilayer Perceptron (MLP) | A function of predictors, also called inputs, or independent variables, that minimizes the prediction error of target variables, also called outputs. These models can learn by example (Abdulridha et al. 2022). |
| Inception network | The Inception Module applies multiple filters in parallel, enabling the network to capture different levels of detail within an image. |
| Xception model (Extreme Inception) | Replaces the Inception network's modules with depth-wise separable convolutions, making it more efficient. |

diseased plant parts. By analyzing large datasets of plant images, neural networks can identify subtle variations in texture, color, and structure, aiding in the early detection of plant diseases and improving agricultural efficiency.

## *2.1 Background, Notation, Functioning of Neural Networks*

Each neuron/node of the input layer ($N_1$, $N_2$, $N_3$) takes some value ($x_1$, $x_2$, $x_3$) as input from the dataset. The nodes of the next layer receive a sum of the inputs multiplied by the weights.

At $N_4$: $\sum w_i x_i = w_{11}x_1 + w_{21}x_2 + w_{31}x_3$

At $N_5$: $\sum w_i x_i = w_{12}x_1 + w_{22}x_2 + w_{32}x_3$

An *activation function* (AF) f(x) is used at the nodes of the next layer to introduce non-linearity into the network. This enables the models to learn complex patterns and generate more accurate predictions. In this case, the AF compares the weighted sum of inputs with a threshold value (0), and then decides the output of that node.

For $N_4$ and $N_5$,

Output = f(x) = 1 if $\sum w_i x_i >= 0$

0 if $\sum w_i x_i < 0$

Thus, any node in the next layer can have either 1 or 0 as the output, which will be considered as the final output for an artificial neural network (ANN) similar to that in Figure (1a), or it will be transferred as input to the next layer of neurons as in Figure (1b).

Although there isn't a single universal definition, most experts acknowledge that a neural network with multiple (greater than 2) non-linear layers (those employing Activation Functions) is considered a DL network (Figure 1b), and architectures having greater than 10 non-linear layers is considered Very Deep Network/Learning (Schmidhuber 2015). These numerous layers enable unsupervised learning by automating feature extraction from vast, unlabeled, and unstructured data sets, as well as making their own predictions about what the data represents. CNN are a form of neural network made up of node layers, which include an input layer, one or more hidden layers, and an output layer. Each node connects to another and has a weight and threshold. If the output of any particular node exceeds the defined threshold value, that node is activated and sends data to the next tier of the network. Otherwise, no data is transmitted to the next tier of the network. *Convolution* is a mathematical operation that combines two functions (in this case, the input image and the filter) to produce a third function (the feature map). CNNs are the most widely used DL architectures in object identification using a collection of images, which is what happens while detecting plant diseases from images of leaves or other body parts (Jelali 2024).

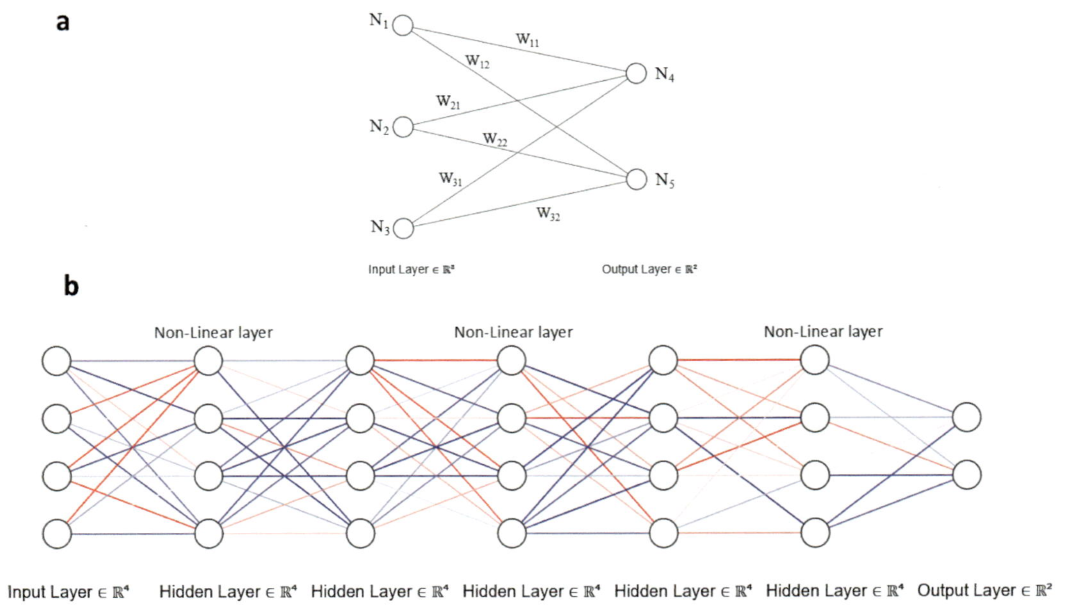


**Figure 1.** (a) Connections between the nodes of a neural network represent the weights between those two nodes. (b) Hidden layers of a deep neural network. Red edges represent positive weights, blue edges represent negative weights, and the intensity of the colors represents the magnitude of the weight.

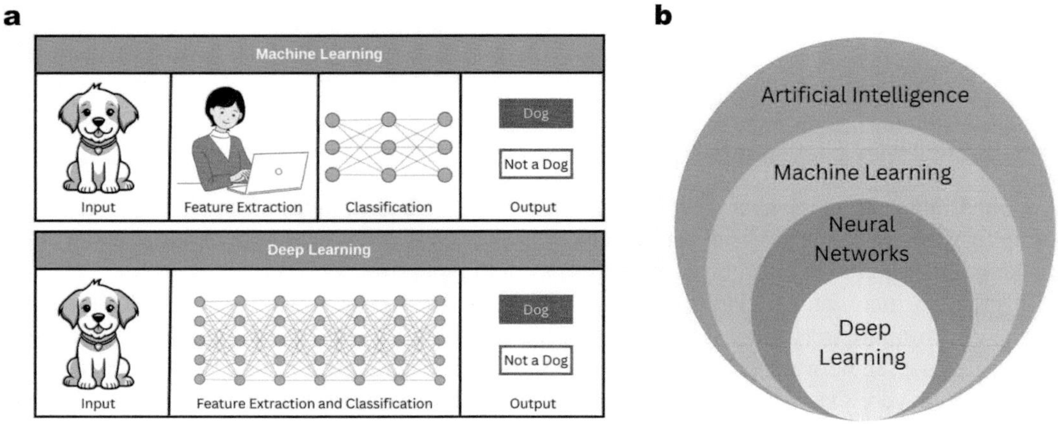


**Figure 2.** (a) Difference between ML and DL based on methodology. (b) Difference between AI, ML, and DL.

*Image processing* is a technique that converts the image in a photo or video frame captured with a camera, scanner, or sensors to a digital format after recording and extracts valuable information from this digital data using a set of algorithms. During these operations, the goal is to obtain descriptive parameters that describe the key data in the image (features). This method is used to define and separate the features to be measured, rectify image faults, improve the visibility of specific features, and threshold them in the background. Image processing can be carried out using ML as well as DL approaches (Figure 2). ML relies heavily on manual feature engineering, domain expertise, and pre-processing, which means you need more human intervention. DL networks learn feature representations automatically, reducing the need for handcrafted features, and have shown higher efficiency and efficacy in this endeavor, managing high-dimensional and complicated data, and understanding more abstract and subtle features that may exist in the image (Shoaib et al. 2023).

*Classification* is a type of supervised learning approach that organizes data points into predefined groupings known as classes. Classifiers (algorithms or models that can carry out the classification task) learn class features from input data and then use those traits to assign possible classes to fresh, unseen data. ML algorithms that enable classification of data include Logistic Regression, DT, RF, SVM, Naive Bayes, and KNN. The most widely used DL algorithm for the classification of diseased plant part images is the CNN.

## 2.2 Types of Image Data

As discussed above, the data that our models learn from is in the form of images. These images can be simple, in the sense, clicked by a DSLR, or any smartphone or webcam that can provide the desired quality of photographs. Such photographs are known as RGB images. RGB represents Red, Green, and Blue. These are the fundamental hues of light. Every color you see on screen is created by combining these three hues. A *pixel* is the smallest unit of any electronic or digital image, similar to what a cell is to an organ. In an RGB image, each pixel contains three values (R, G, and B) that instruct the computer how much of each color to use. A Pure Red pixel is represented as (255, 0, 0), White as (255, 255, 255) – all colors at full intensity, Black is (0, 0, 0) – no color at all, Purple as (128, 0, 128), Yellow as (255, 255, 0) and so on. When we say a filter is being multiplied by the image during the convolution operation, it is basically these pixel values inside the image that are being operated on (Gatou et al. 2024). Advantages of RGB images are:

1. High-spatial resolution.
2. Well-suited for visual analysis.

3. Simple to train models.
4. Requires no expert knowledge.
5. Compatible with existing equipment

*Hyperspectral Imaging* (HIS) combines a conventional imaging system with spectroscopy, enabling the collection of plant physiology data. Such images are three-dimensional (3D), composed of two spatial dimensions (x and y) and one spectral dimension (z). Hyperspectral photographs contain hundreds of wavelength bands close together in a spectral spectrum, creating a dense and colorful cube of information with spatial resolution that can capture several pixels per target (Ferreira et al. 2024). Advantages of HIS images are:

1. This method is non-invasive and non-destructive, unlike molecular analysis, which requires a sample that cannot be reused.
2. Early identification with hyperspectral sensors can aid in disease control and prevent the spread in the field.
3. Capable of identifying and distinguishing diseases even in asymptomatic plants.

Despite these advantages, high prices for equipment and software, and complex procedures have prevented the widespread use of this technology by agricultural specialists.

*Classification tasks* predict which category (class) an input image belongs to. A class is a label that the model learns to predict, helping it identify and diagnose diseases based on image features. For example, in a dataset for detecting diseases in tomato plants (Shoaib et al. 2022a), the classes might be:

- *Healthy*: Images of unaffected, healthy tomato leaves.
- *Early blight*: Leaves exhibiting indications of the fungal disease *Alternaria solani*.
- *Late blight*: Leaves infected with *Phytophthora infestans*, causing rapid deterioration.
- *Leaf mold*: Fungal infection that causes yellow patches on tomato leaves.

## 2.3 Dataset Creation and Organization

The images need to be introduced to the model in a manner that is most effective for its learning. That's why it is organized into a dataset. Any dataset is split, especially in classification problems, into three parts:

*Training set*: The dataset used by our model to identify potential underlying patterns and relationships. It contains the largest percentage of data points, images in this context. It should be unbiased and as representative of the population as feasible.

*Validation set*: The dataset that we utilize to evaluate our model's performance across various model types and hyperparameter settings. It is smaller than the training set.

*Test set*: This set is used to estimate a model's final performance on unseen data following hyperparameter adjustment. It's also useful to see how different models (SVMs, Neural Networks, Random Forests, etc.) compare to each other.

*Data augmentation* is a technique used to virtually increase the diversity and size of a dataset by applying various transformations to the existing data. This is particularly useful in DL, where large amounts of data are required to train robust models. Data augmentation helps prevent overfitting and improves model accuracy by introducing variability in the dataset. Data augmentation can come in handy when it is not possible to create more data, e.g., capturing photographs of a crop after its fruiting season is over, while creating a model to predict diseases that might affect the crop next year/next season. Cropping, rotation, flipping, shearing, changing the brightness level, zooming,

**Table 2.** Open-source plant disease detection datasets.

| **Name of dataset** | **Crops/ Vegetables** | **Size in no. of Images** | **No. of Classes** | **Remarks** | **References** |
|---|---|---|---|---|---|
| Plant Village | Various crops | 54,309 | 38 | Images were taken in lab conditions rather than real fields, which may affect model performance in practice. | Hughes and Salathe 2016 |
| PlantCLEF | Almost the entire plant kingdom | 40,13,695 | 1,44,653 | Training images include those taken from institutes and those from Google and Bing. | Goëau et al. 2023 |
| PlantDoc | Various crops | 2,598 (13 species) | 17 | Due to lack of extensive domain expertise, there are some images in the dataset which can potentially be wrongly classified | Singh et al. 2020 |
| AgriNet dataset | Various crops | 160, 142 | 423 | ---- | Sahili and Awad 2022 |
| DiaMOS Plant | Pear | 3505 | ---- | Images depict 4 leaf stresses and 3 stages of fruit development: fruit set, growth and ripening. | Fenu and Malloci 2021 |
| Citrus fruits and leaves | Citrus fruits | 759 | ---- | Available on Mendeley Data | Rauf et al. 2019 |
| BRACOL | Arabica Coffee | 1747 | ---- | ---- | Krohling et al. 2019 |
| RoCole | Robusta Coffee | 1560 | 4 | Facilitates the evaluation of the performance of machine learning algorithms | Parraga-Alava et al. 2019 |
| Plant Pathology Challenge 2020 | Apple Foliar Diseases | 3644 | ---- | Available on Kaggle | Thapa et al. 2020 |
| NUST Wheat Rust Disease | Wheat | 100 | ---- | Annotated real-world semantic segmentation dataset | Anwar et al. 2023 |
| Plant Diseases and Pests | Rice, wheat, and corn | 1237 | ---- | Images searched online, cleaned manually and included in the database | Wang et al. 2020 |
| Plant Leaf Disease Recognition Dataset | Gourd, Hibiscus, Papaya, Zucchini | 4121 captured + 20956 augmented | ---- | Available on Mendeley Data | Ahmed et al. 2024 |
| Plant Disease Diagnosis Dataset (PDDD) | Fruit, vegetables, crops | 421,133 | 120 | Available on Zenodo | Dong et al. 2023 |

etc., are the transformations usually employed in data augmentation. The previous report provided a comprehensive list of real field datasets, and tips on how to tailor a dataset for recognition of plant diseases (Jelali 2024). There are many open-source plant disease detection datasets provided in numerous studies (Table 2).

## *2.4 Training the Model*

Once the algorithm to be used has been chosen, and the data collected, cleaned and pre-processed, the next step is to train the algorithm. Training simply means exposing your algorithm to your acquired images, so that patterns can be identified among those images mathematically by your algorithm. Accordingly, the algorithm adjusts its weights. Mathematically, this is done by first calculating a Loss Function/Error Function.

Loss/Error = Annotated Input Data (Ground Truth) – Data Predicted by the model

This is the intuitive way of describing the error. Mathematically, it can be in multiple forms like mean squared error, root mean squared error, etc. The model will be able to correctly make classification or perform its desired task properly only when the weights between the nodes (Figure 1) have values that keep the error at its minimum. The weights are first given random values (known as initialization of weights), and the entire data is passed once through the neural network. Then the loss function is calculated. Looking at the value of the loss function, the model tries to change the values of the weight matrix. The updated weights are decided by another algorithm known as the optimizer. Examples of optimizers include Gradient Descent, Stochastic Gradient Descent, Adam, etc. The optimizer provides direction in which the weights update themselves. The data is again passed through the neural network, but this time the weights have a new value. Again, the loss function is calculated, and the optimizer dictates the change of weights. This process is repeated until the error is at its minimum.

Configuration variables known as *Hyperparameters* are established before an ML model's training phase. Instead of learning from the data, they are in charge of the learning process itself. (Model parameters are learned from the data during training and are intrinsic to the model. Neural network weights, linear regression coefficients, and support vectors in SVM are a few examples.) Hyperparameters are essential for fine-tuning a model's performance and have a big impact on its generalization, accuracy, and other metrics. Hyperparameters include:

1. *Learning rate*: Regulates how much the estimated mistake affects the model.
2. *Batch size*: How many training examples are used in a single training cycle?
3. *Number of epochs*: It is the number of times the training dataset will be seen by the model.
4. Train-test split ratio of the input dataset

Tuning hyperparameters entails determining the best values to maximize the model's performance. The ideal tools for this stage are hyperparameter tuning libraries such as Hyperopt, Ray Tune, and Optuna, which use automated algorithms to identify optimal values. The validation dataset is used between training to fine-tune the hyperparameters, i.e., to find their optimal values. This process of using the validation dataset is also known as cross-validation.

## 2.5 Performance Metrics

After the training has been completed, the model is checked against the test dataset that is split from the training dataset beforehand. Commonly used evaluation indices are described below:

$$Accuracy = \frac{Number\ of\ Correct\ Predictions}{Total\ Number\ of\ Predictions} = \frac{TP + TN}{TP + TN + FP + FN}$$

*True Positive* (TP) refers to perfectly predicted positive classes. *False Positive* (FP) refers to negative events that are incorrectly identified as positive. In contrast, *False Negative* (FN) refers to cases in which genuine positive examples were incorrectly categorized as negative (Sujatha et al. 2025). True Negative (TN) refers to those events that are negative in terms of ground truth, and also identified as negative events by the trained algorithm.

$$Precision = \frac{TP}{TP + FP}$$

Precision is the ratio of correctly categorized positive samples (True Positive) to the total number of positively classified samples (correct or incorrect).

$$Recall = \frac{TP}{TP + FN}$$

This is the proportion of correctly classified positive instances to the total number of positive instances. Recall increases with the number of positive samples found.

$$F1\ Score = \frac{2 * Precision * Recall}{Precision + Recall}$$

F1 score is the harmonic mean of precision and recall, and yields a value between 0 and 1. A score of 1 represents the best model, while a score of 0 represents the worst model (Yang et al. 2024a).

*Confusion matrix*: A classification matrix shows the output of a classification algorithm, with each column denoting the actual class of a sample and each row denoting the anticipated class. The number of samples that have been correctly or wrongly classified is shown by the entries in the matrix. It is feasible to obtain insight into the classification algorithm's performance and pinpoint areas for development by looking at the entries in the confusion matrix (Shoaib et al. 2023).

*Area under curve (AUC)*: This statistic measures how well a classification model predicts binary outcomes. It is widely utilized in ML and data analytics. The AUC is determined by plotting the True Positive Rate (TPR) against False Positive Rate (FPR) at various thresholds and calculating the area under the resulting curve.

### *2.6 Model Deployment*

Once the model has been tested and fine-tuned, it is ready to be used in real-life scenarios by users. Incorporating the learned model into the software framework required to operate it is known as Model Deployment. Making the predictions from a trained ML model accessible to users, management, or other systems is the goal of model deployment. In some cases, models are deployed on local devices (e.g., smartphones, IoT devices) rather than centralized servers, when latency, bandwidth, or privacy is a concern. It is frequently vital to have the ability to update the model after its first deployment is finished to ensure that it consistently represents the most recent changes in the environment and data (Paleyes et al. 2022).

### *2.7 Computational Hardware Requirements*

*Processor*

When we say a DL model is computationally expensive, we are referring to the training step and the step where we run the model on any machine to predict something. A normal CPU is not able to carry out the huge number of calculations a neural network usually performs, as they are general-purpose processors, and are designed to execute instructions one after another. Hence, a CPU will not be able to do the computing within the required amount of time. Thus, we need to use co-processors to get this task done. Graphical Processing Units (GPUs) have hundreds or thousands of specialized cores that can simultaneously solve multiple matrix multiplications and large-scale data processing, making them suitable for training and deploying DL models. GPUs are able to handle floating-point operations efficiently, which are crucial for neural networks.

Another kind of processor used specifically for DL is the Neural Processing Unit (NPUs). NPUs are specialized hardware designed specifically for AI and DL workloads. They are optimized for low-power consumption, making them ideal for edge devices (smartphones, cameras, etc.). They have dedicated AI accelerators, improving inference speed and efficiency, and they are designed for on-device AI processing, reducing dependency on cloud-based computation (Tan and Cao 2021). NPUs can offload AI-related tasks from CPUs and GPUs, allowing them to focus on other

**Table 3.** Processors and their use.

| Processor | Use |
|---|---|
| Central Processing Unit (CPU) | General-purpose processors for all-around computing |
| Graphic Processing Unit (GPU) | Excellent for parallel processing, particularly in graphics and AI |
| Neural Processing Unit (NPU) | AI-optimized processors for energy-efficient and fast on-device AI inference |
| Field-Programmable Gate Array (FPGA) | Developers can optimize FPGA hardware for specific deep learning models, improving performance |

operations and improving overall system efficiency. While GPUs are still the dominant choice for AI training, NPUs are becoming essential for efficient AI inference in mobile and embedded systems. Field-Programmable Gate Arrays (FPGAs) are programmable hardware devices that can be modified after manufacturing (Table 3). This versatility enables them to adapt to various calculations, including DL applications. FPGAs are excellent at parallel processing. They can handle numerous operations simultaneously, making them ideal for large-scale matrix operations and neural network computations (Marković et al. 2024).

*Memory and Other System Requirements*

Random Access Memory (RAM) is vital for handling the in-memory computations and temporary storage of data during the training process. For simple ML tasks, such as linear regression, logistic regression, or small-scale clustering and classification, 4 to 8 GB of RAM is recommended. DL models, particularly those involving neural networks, require substantial RAM, especially when training on large datasets or with high-resolution images: Small-scale projects (e.g., simple CNNs for image classification): 16 GB to 32 GB, and Large-scale projects (e.g., training complex models like Inception): 32 GB to 64 GB or more of RAM is recommended.

The DL operations, particularly those requiring high-performance GPUs, produce tremendous heat and necessitate a reliable power source. To avoid overheating and maintain system stability, make sure the system has suitable cooling solutions (e.g., fans, liquid cooling) and a steady power supply.

## *2.8 Tools to Create AI Models*

Although ML is considered as the science that gives computers the ability to learn without being explicitly programmed (Sharma et al. 2021), any AI/ML architecture has to be created via programming. The popular programming languages used for this include Python, MATLAB, C++, Julia, etc. A Python library contains bundles of pre-written code that can be used repeatedly in different programs. TensorFlow is a flexible and scalable python software library that enables efficient computation across various hardware devices, including GPUs and NPUs. It uses dataflow graphs to track computations. This library and related tools enable users to efficiently program and train neural networks and other ML models and deploy them to production (Pang et al. 2020).

Unlike other libraries, such as TensorFlow, which uses a static computational graph, PyTorch allows for the modification of the graph on-the-fly, making it more suitable for research and experimentation (Shoaib et al. 2023). PyTorch also includes a number of pre-built modules, such as convolutional and recurrent layers, which can be simply integrated into a model. This enables rapid prototyping and experimentation with various model designs. Other AI and DL framework generators include Amazon Bedrock and Amazon SageMaker, Caffe (Convolutional Architecture for Fast Feature Embedding), Keras, MXNet, Theano, etc.

## *2.9 Current Trends in DL*

*Image Pre-Processing*

Although a CNN or other deep learning architectures are able to carry out image processing on their own, pre-processing is carried out to enhance image quality, normalize input characteristics, and reduce irrelevant variability, thereby improving the robustness and overall performance of the model. Pre-processing optimizes computing resources and ensures consistent image resolutions compared to a benchmark. Methods include standardization, image size regularization, etc. The image is also adjusted to fit the fixed color scale for optimal analysis and interpretation. The acquired photographs may have noise, blur, low or high illumination, undesirable background, and more. It is vital to efficiently handle raw data and classify diseases using automated methods. The raw data are converted into a specified format and cleaned up to remove noise or distortion. Some of the basic pre-processing operations include:

*Gray-scaling*: Converts color images to grayscale to simplify data and reduce computation.

*Binarization*: Converts grayscale images into black-and-white using thresholding.

*Normalization*: Scales pixel intensity values for improved model performance.

*Noise reduction*: Removes unwanted noise via smoothing and blurring.

The process of dividing a picture into its basic pieces or objects is known as image *Segmentation*. Image segmentation divides an image into many segments to simplify its representation, making it more comprehensible and easier to deal with. Effective segmentation allows for the exact identification and placement of objects inside an image, enabling tasks such as feature extraction, pattern detection, and scene understanding. The similarity approach in image segmentation groups pixels or regions based on their similar properties. Thresholding is a highly successful strategy for segmenting images based on pixel intensity levels. It involves simple computation, enhances contrast, and separates objects. But thresholding loses intensity details and struggles with noise. Clustering-based segmentation partitions the image into groups (clusters) of similar pixels. It identifies patterns, detects outliers, and automates segmentation. But it is influenced by initial cluster centroids, and costly for large datasets. Edge-based segmentation aims to detect the boundaries between several sections in a picture. This approach identifies substantial changes in intensity or color, which usually indicate the presence of edges. Edge detection enhances object contrast, but is sensitive to noise, and ineffective for low contrast (Jafar et al. 2024).

*Transfer Learning*

Transfer learning occurs when knowledge learned from one task or data set is applied to improve model performance on another related task or data set.

*Ensemble Learning*

Combining multiple models to improve overall performance and reduce dependence on a single model.

*Edge Devices*

A DL on Edge involves deploying ML models directly on devices like smartphones, IoT devices, and embedded systems. Edge computing is transforming plant disease detection by enabling real-time analysis directly on local devices. Instead of relying on cloud-based solutions, edge computing processes data closer to the source, reducing latency and bandwidth usage. Edge computing offers several advantages for plant disease detection, making it a valuable tool in precision agriculture.

*Real-time processing*: Edge devices can analyze images of plant leaves instantly, allowing farmers to detect diseases early and take timely action.

*Reduced latency*: Since data is processed locally on edge devices rather than being sent to a distant cloud server, response times are much faster.

*Lower bandwidth usage*: Edge computing minimizes the need for constant internet connectivity, reducing data transmission costs and making it feasible for remote agricultural areas.

*Energy efficiency*: Optimized DL models can run on low-power devices like smartphones and Raspberry Pi, making disease detection more accessible.

*Enhanced accuracy*: Advanced AI models deployed on edge devices can achieve high accuracy in identifying plant diseases, sometimes exceeding 97% precision.

*Scalability*: Farmers can use mobile devices or other edge computing tools to monitor large fields without requiring expensive infrastructure.

Edge devices have limits on memory, computing power, and battery life. Models must be optimized in terms of size and efficiency (Marković et al. 2024).

*Steps in Creation of AI Model*

- Choosing the task to be achieved (e.g., Regression or Classification)
- Choosing the algorithm (e.g., DT or Neural Network)
- Choosing the Activation Function (e.g., Softmax or ReLU)
- Creation of a Dataset
- Image pre-processing to enhance quality of data
- Segmentation to make feature extraction less cumbersome
- Training the model with the dataset

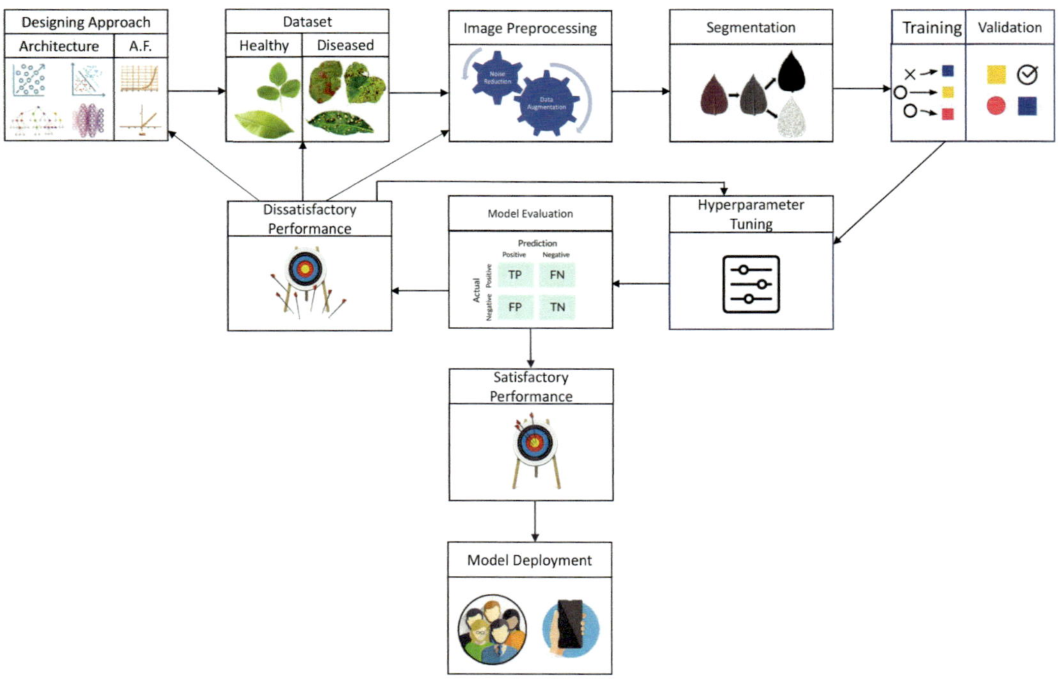


**Figure 3.** General workflow for the creation of an AI model for plant disease classification.

- Hyperparameter Tuning using the training data
- Testing of the model with unknown data
- Deployment in case of satisfactory performance
  OR
- Readjustment of the hyperparameters, architecture of the model, changes in pre-processing, etc., in case of dissatisfactory performance

Collectively, by leveraging neural networks, image processing, and hyperspectral imaging, AI enhances the ability to identify and classify plant diseases with high accuracy. The systematic organization of datasets, data augmentation techniques, and model training contributes to robust AI performance (Figure 3). Evaluating models through precision metrics, including accuracy, recall, F1 score, confusion matrices, and AUC analysis, ensures reliability in real-world applications. Finally, successful model deployment integrates AI seamlessly into agricultural systems, supporting early disease detection, improving crop yield, and promoting sustainable farming practices.

## 3. Recent Advancements in AI-assisted Plant Disease Detection, Monitoring and Forecasting

AI-powered image processing and ML have significantly improved plant disease detection in agriculture. By capturing and analyzing environmental images, researchers can accurately identify diseased regions in crops. These methods have led to the development of efficient diagnostic systems for farmers, enhancing disease management across various crops, such as plum fruit plants (Ahmad et al. 2020), rice (Patil and Kumar 2021, Raman and Jayaraman 2025), watermelon (Abdulridha et al. 2022), sunflower (Dawod and Dobre 2022), strawberry (Jiang et al. 2022, Marco-Detchart et al. 2023), tea (Li et al. 2022), banana (Narayanan et al. 2022), tomato (Shoaib et al. 2022a, Yag and Altan 2022, Marco-Detchart et al. 2023, Yang et al. 2024b), coffee (Topal et al. 2024), etc. Several AI-based methods have been introduced for plant disease detection, disease monitoring and forecasting, with some of the most effective techniques outlined in Table 4 and discussed below.

### *3.1 AI for Plant Disease Detection*

AI and DL models—like CNNs, ResNet, SVM, Mask RCNN, and MD-Unet—were proposed to identify plant diseases. These models achieved high accuracy (up to 99.5%) using real-field images and sensor data, supporting precision agriculture and efficient disease management under diverse conditions. Earlier study presented CNN-based framework, Inception-v3 for plum disease detection in real field conditions and optimized it for resource-limited devices using field-collected image data (Ahmad et al. 2020). The authors developed local feature matching conditional neural adaptive processes (LFM-CNAPS), a meta-learning-based method for detecting previously unseen plant diseases using minimal annotated data, supported by the Miniplantdisease-dataset featuring twenty-six plant species and sixty diseases (Chen et al. 2021). In another study, the severity stages of downy mildew disease in watermelon were effectively classified with an accuracy of 91% by a multilayer perceptron (MLP) classification system using spectral vegetation indices (Abdulridha et al. 2022).

Various studies have used AI to identify foliar plant diseases, but classification has typically relied on analyzing only one side of the leaf. Since some diseases exhibit different symptoms on the lower side, considering both surfaces could improve accuracy and enhance disease detection. In view of this, a recent study introduced CNN-ResNet50 AL-model to detect different leaf diseases of sunflower using upper and lower leaf images and achieved the precision, F1 score, and accuracy above 90% with 100% of recall (Dawod and Dobre 2022). Jiang et al. (2022) developed an image dataset of eight strawberry diseases and analyzed it using a two-head network with Out-of-Distribution (OoD) detection models and Open Set Recognition (OSR) technology. Their modified training approach achieved over 90% accuracy in classifying known diseases and detecting unknown ones (Jiang et al.

**Table 4.** Summary of experimental results of recent studies related to plant disease identification through AI models.

| Plants/ Datasets | Disease class/ Dataset | Classification Technique/Model | Experimental Results (%) | | | | References |
|---|---|---|---|---|---|---|---|
| | | | Precision | Recall | F1 score | Accuracy | |
| Plum fruit plants | Multiple diseases | Inception-v3 | ---- | ---- | ---- | 93.12 | Ahamad et al. 2020 |
| Multiple dataset | Miniplantdisease-Dataset | LFM-CNAPS | ---- | ---- | ---- | 93.90 | Chen et al. 2021 |
| Watermelon | Downy mildew | Multilayer perceptron | ---- | ---- | ---- | 91.00 | Abdulridha et al. 2022 |
| Multiple dataset | Multiple crop leaf diseases | EfficientNetV2-B4 | 99.63 | 99.93 | 99.78 | 99.99 | Albattah et al. 2022 |
| Sunflower | Multiple diseases | CNN ResNet50 model | 91.00 | 100 | 92.00 | 91.00 | Dawod and Dobre 2022 |
| Strawberry | Eight strawberry diseases | Two-head network included in OoD | 91.00 | 94.00 | ---- | 92.00 | Jiang et al. 2022 |
| Tea | Three diseases and *Apolygus lucorum* insect, DSIS | Mask RCNN - Wavelet transform – FRNet | 94.8 | 98.7 | 96.7 | 88.00 | Li et al. 2022 |
| Banana | Xanthomonas wilt, Fusarium wilt, black and yellow Sigatoka | Hybrid CNN-SVM | 97.70 | 97.50 | 97.50 | 98.80 | Narayanan et al. 2022 |
| Pepper, Potato, Tomato | Multiple leaf diseases | CANet - CNN | 95.00 | 91.00 | 93.46 | 92.00 | Shoaib et al. 2022b |
| Tomato | Multi-class leaf diseases | Modified U-Net-InceptionNet | 99.20 | 99.48 | 99.46 | 99.70 | Shoaib et al. 2022a |
| Apple, Grape, Tomato | Multiple leaf diseases | Hybrid FPA-SVM – CNN | 99.52 | 99.60 | 99.56 | 99.55 | Yag and Altan 2022 |
| Multiple dataset | Multiple plan-diseases-severity | LDINET multi-label | 95.28 | 95.29 | 95.28 | 93.51 | Yang et al. 2024a |
| Tomato<br>Grape<br>Strawberry<br>Apple | Blight<br>Esca<br>Scorch<br>Rust | EDGE device-NasNetMobile – CNN | 96.00 | 96.00 | 95.00 | 93.00 | Marco-Detchart et al. 2023 |
| Multiple dataset | Multiple-Eight Class diseases | DeepPlantNet | 91.37 | 91.5 | 91.43 | 97.89 | Ullah et al. 2023 |
| Multiple dataset | Multi-class disease | RST-Nets | 99.94 | 99.96 | 99.95 | 99.95 | Kalpana et al. 2024 |
| Multiple dataset | Multiple leaf disease | DBN-EKELM | 97.00 | 98.10 | 97.80 | 98.20 | Rajasekar et al. 2024 |
| Coffee | Brown leaf spot, Cercospora leaf spot, Leaf miner, Rust | CNN - VGG16 | 96.65 | 93.38 | 94.73 | 95.65 | Topal et al. 2024 |
| Apple<br>Corn<br>Grapes<br>Potatoes<br>Tomato | Multi-class leaf disease | DH-CNN | 99.50 | 99.66 | 99.58 | 99.40 | Yang et al. 2024b |

*Table 4 contd. ...*

*...Table 4 contd.*

| Plants/ Datasets | Disease class/ Dataset | Classification Technique/Model | Experimental Results (%) | | | | References |
|---|---|---|---|---|---|---|---|
| | | | Precision | Recall | F1 score | Accuracy | |
| Peach Strawberry | Bacterial spots, Scorch | YOLOv4 | 98.00 | 99.00 | 99.00 | 98.00 | Alhwaiti et al. 2025 |
| Fourteen plant species | Thirty-eight disease classes | LWDSCSA | 98.30 | 98.90 | 98.58 | 98.58 | Batool et al. 2025 |
| Citrus species (Hybrid Dataset of fruits and leaves) | Canker, Scab, Black spot | DenseNet-201-F-KNN | 99.50 | 99.50 | 99.50 | 99.50 | Butt et al. 2025 |
| | | AlexNet-Q-SVM | 97.15 | 97.30 | 97.20 | 97.36 | |
| Tobacco | Angular leaf spot, Brown spot, Wildfire disease, Frog eye disease | MD-Unet | 96.31 | 95.41 | 95.84 | 94.67 | Chen et al. 2025 |
| Cotton | Mealybug, Anthracnose, Thrips, Blight | ResNet152 | 83.60 | 84.10 | 83.80 | 84.80 | Faisal et al. 2025 |
| Rice | Brown spot Leaf smut Bacterial leaf blight | DB-CGANNet | 98.43 | 98.43 | 98.22 | 98.99 | Raman and Jayaraman 2025 |
| Multiple dataset | Multiple diseases | IX-CNN -RF | 94.20 | 93.90 | 94.10 | 99.73 | Shafik et al. 2025 |
| Rice | Multiple leaf diseases | ANN Softmax AF | 91.40 | 92.20 | 91.60 | 92.15 | Patil and Kumar 2021 |

SVM: support vector machine, LFM-CNAPS: local feature matching conditional neural adaptive processes, OoD: Out of distribution, DSIS: Disease spot and insect spot, FRnet: four channelled residual network, CANet: context-aware convolutional neural network, FPA: flower pollination algorithm, LDI-NET: leaf disease identification network, RST-Nets: Residual Swin transformer networks, DBN-EKELM: deep belief network-based enhanced kernel extreme learning machine, DH-CNN: deep hash convolutional neural networks, PDLM-TK: Plant pest and Disease Lightweight identification Model by fusing Tensor features and Knowledge distillation, YOLOv4: You Only Look Once v4, LWDSCSA: Lightweight Deep Learning model, Depth wise Separable Convolution with Spatial Attention, F-KNN: Fuzzy k-Nearest Neighbors, Q-SVM: Quantum Support Vector Machine, MD-Unet: Multi-scale Residual Dilated Segmentation Model, DB-CGANNet: Dual Branch Convolutional Graph Attention Neural Network, IX: hybrid inception-xception, RF: random forest.

2022). The tea plant (*Camellia sinensis* (L.) O. Kuntze) is a valuable cash crop extensively cultivated in tropical and subtropical regions. To detect tea leaf diseases such as brown blight, target spot, and tea coal and a major pest *Apolygus lucorum* researcher developed an advanced system using Mask RCNN, wavelet transform, and four-channel residual network (FRNet). Mask RCNN segmented affected areas, wavelet transform enhanced image features, and FRNet identified symptoms. Joint disease spot and insect spot (DSIS) on tea leaves was detected with 98.7% of accuracy, ensuring reliable disease and pest detection (Li et al. 2022).

Banana is an important agricultural crop, as about 25.7% of the global banana production is from India. Recently, Narayanan et al. utilized a dataset of 3,500 banana plant images from the field of South India to classify plant health using a two-phase SVM approach. In phase P1, a binary SVM model identifies whether a leaf is healthy or infected based on CNN-extracted features. If infected, phase P2 employs a multiclass SVM to classify the disease into one of four classes, including xanthomonas wilt, fusarium wilt, black and yellow sigatoka. This proposed hybrid model enhances disease detection efficiency in banana plants up to 99% accuracy level (Narayanan et al. 2022). Another study presents a low-cost smart EDGE device using NasNetMobile–CNN for detecting plant diseases such as tomato blight, grape esca, strawberry scorch, and apple rust. By capturing multiple leaf images and integrating data fusion techniques, this device achieves over 90% accuracy in classification. Built with a Raspberry Pi 4 and a four-camera array, it runs a DL model that displays results on an LCD screen and can be mounted on drones, robots, or agricultural machinery for efficient field monitoring (Marco-Detchart et al. 2023). An image pre-processing technique was newly employed for classifying and estimating the severity of coffee leaf diseases, which used an enhanced multivariance product representation (EMPR) to decompose images and construct a new, contrast-enhanced version using a high-dimensional model representation (HDMR). The CNN-VGG16 architecture was applied to these processed images, achieving a 96% accuracy in coffee leaf disease classification and severity estimation (Topal et al. 2024).

Citrus, tobacco, and cotton cultivation are key agricultural industries in South Asia, but they are threatened by severe diseases such as canker, scab, and black spot in citrus, angular leaf spot, brown spot, wildfire, and frog eye disease in tobacco, and mealybug, anthracnose, thrips, and blight in cotton (Butt et al. 2025, Chen et al. 2025, Faisal et al. 2025). Early and precise disease detection is essential to minimizing economic losses and safeguard agricultural productivity in the region. In the present year, 2025, three studies introduced AI-driven models for early disease classification in citrus, tobacco, and cotton crops. Butt et al. (2025) developed an automated DL system using DenseNet-201-F-KNN, achieving 99.5% accuracy in citrus disease detection (Butt et al. 2025). Chen et al. (2025) proposed a Multi-scale Residual Dilated Segmentation Model (MD-Unet) for tobacco leaf disease segmentation, reaching 96.31%, 95.41%, and 95.84% of precision, recall, and F1 score, respectively, with 94.67 of overall accuracy (Chen et al. 2025). Authors collected real image data of cotton disease from the field and pre-processed it for training the various DL models and found that ResNet152 outperformed other models in disease prediction, making it a reliable approach for cotton disease recognition (Faisal et al. 2025).

### *3.2 AI in Disease Monitoring*

Recent studies have advanced AI-based disease monitoring through precise plant disease detection. Techniques like ANN, Spectral Feature Depth Index (SFDI), fluorescence spectroscopy, Self-Organizing Map (SOM), Least Squares Support Vector Machine (LS-SVM), and CNNs achieved high accuracy in detecting diseases in rice, orchids, tomatoes, wheat, and strawberries, enabling rapid, non-destructive monitoring and supporting sustainable, data-driven agriculture.

For classification and prediction of rice diseases across diverse agro-meteorological conditions, researchers utilized ANN with softmax activation functions and achieved the accuracy of 92.15% for rice disease prediction (Patil and Kumar 2021). Bacterial blight causes chlorophyll loss and structural damage in rice leaves, leading to changes in hyperspectral data. The study by Cao et al. (2022) showed that the SFDI accurately predicts Soil and Plant Analysis Development (SPAD) values by capturing these spectral changes. Regular chlorophyll monitoring using SFDI helps in assessing plant health and controlling crop loss from disease (Cao et al. 2022). Likewise, researchers developed a rapid, noncontact AIoT-based platform using fluorescence spectroscopy and an Android smartphone to detect Odontoglossum ringspot virus (ORSV) and Cymbidium mosaic virus (CyMV) infections in orchids. Trained on PCR data, the system identifies infections within 3 seconds with 89% accuracy and 0.96 specificity. It enables efficient large-scale screening in indoor farms, supporting precision agriculture and efficient plant health monitoring and management (Lee et al. 2021).

In noisy field conditions, poor segmentation is one of the major limitations that reduce the performance of traditional image-based plant disease detection models. To overcome this, a non-destructive, rapid, and accurate method for detecting plant defense responses using *in vivo* fluorescence techniques combined with supervised SOM models—supervised Kohonen networks (SKNs), counterpropagation artificial neural networks (CPANNs), and XY-fused networks (XY-Fs) was recently developed. Proposed models effectively classify systemic resistance in tomato plants by exploiting the discriminative patterns in fluorescence kinetics and accuracy of the model was validated through gene expression data. It permits real-time monitoring of resistance induction and offers a cost-efficient tool for plant protection and improved disease management strategies in tomato (Pantazi et al. 2022). Another study presents the Intelligent and Integrated Pest and Disease Management (I-PDM) system, which uses edge computing to automatically detect major greenhouse pests and monitor environmental conditions. Deployed for over 1368 days in tomato and orchid greenhouses, the system enabled long-term data collection and supported real-time, data-driven decisions for integrated pest management (IPM). Results showed up to 50.7% annual reduction in pest populations, demonstrating the system's potential to enhance sustainability and efficiency in IPM beyond traditional approaches (Rustia et al. 2022). Moreover, an AI-driven LS-SVM and Partial Least Squares Discriminant Analysis (PLS-DA) model was developed to classify deoxynivalenol (DON) levels in Fusarium Head Blight (FHB) wheat by fusing Near-Infrared (NIR) and FT-MIR spectral data. Integrated fingerprint information and characteristic wavelength selection methods significantly improved model accuracy and help for agricultural monitoring through enabling precise detection of toxic DON levels (Liang et al. 2023).

### *3.3 AI for Disease Forecasting*

AI is increasingly transforming disease forecasting by enabling accurate and timely predictions of disease outbreaks. Through predictive modelling techniques, AI leverages historical and real-time data to forecast disease trends. Climate and epidemiological modelling integrate environmental variables with disease data to understand how factors like temperature and humidity influence disease spread. Recently, a predictive modelling technique was developed that utilized mathematical models to predict the growth kinetics of *Fusarium graminearum* and the associated accumulation of deoxynivalenol (DON) in stored wheat under varying moisture content and temperature conditions. The model serves as a valuable tool for assessing the risk of fungal contamination and mycotoxin accumulation, providing a scientific basis for developing storage strategies to ensure wheat safety and prevent foodborne hazards (Pei et al. 2022). Furthermore, disease forecasting plays a vital role, as climate profoundly affects the interaction between pathogens and their hosts. Aparecido et al. (2024) explored the impact of climate on coffee diseases, specifically forecasting brown-eye spot caused by *Cercospora coffeicola*. Using agrometeorological data from eight regions in Brazil, researchers calibrated and tested six ML models, including KNN, MLP, SVM, RF, XGBoost, and GradBOOSTING. These models incorporated climatic factors such as temperature, humidity, rainfall, and evapotranspiration to predict disease outbreaks at least one week in advance. XGBoost demonstrated the highest accuracy in high-yielding conditions, while SVM performed better in low-yielding scenarios. These findings highlighted regional disease variations and the effectiveness of AI-driven forecasting, offering valuable tools for managing coffee crop health (Aparecido et al. 2024). Moreover, researchers developed a novel leaf wetness detection system that integrates high-resolution imaging with CNNs. This system effectively classified wetness states based on diurnal and weather-related image variations, employing a divide-and-conquer strategy for disease forecasting. Integrated into the Strawberry Advisory System (SAS), which already uses multi-model data fusion from Campbell Scientific model 237 sensors, the algorithm was found to improve the accuracy of fungicide application decisions. The system achieved a detection accuracy of 95.75%, highlighting its potential to enhance precision disease management (Kondaparthi et al. 2024).

## 4. Limitations and Solutions

AI-based plant disease detection systems, despite their potential, face significant limitations that affect their real-world applicability. Key challenges include the impact of noisy field conditions, such as inconsistent lighting and background clutter, on image quality and segmentation accuracy.

These models often lack generalizability due to reliance on large, annotated datasets and struggle with novel diseases or crop types. Most approaches analyze only the upper side of leaves, potentially missing symptoms on the lower surface. Additionally, high computational demands hinder deployment on edge devices in field settings. Current systems focus on post-symptom detection using RGB imagery, limiting early diagnosis based on physiological changes. Data imbalance and limited annotations for rare diseases further reduce model reliability. Overcoming these limitations is vital for creating efficient, scalable, and real-time AI-driven plant health monitoring tools. Recent advancements emphasize robust, accurate, and field-ready solutions to address these limitations. Techniques like EMPR, HDMR, and wavelet transforms enhance image clarity and reduce noise, while multimodal data fusion combining RGB, spectral, fluorescence, and environmental inputs enriches diagnostic precision. Meta-learning models such as LFM-CNAPS enable effective detection of novel diseases with minimal data. Bidirectional leaf imaging ensures comprehensive symptom recognition, and edge-optimized models like NasNetMobile support real-time analysis on low-cost devices. Integration of physiological signals (e.g., fluorescence spectroscopy, SPAD, hyperspectral imaging) allows for early detection. Predictive modelling using climatic and epidemiological data aids in disease forecasting, while AIoT and cloud systems support scalable, remote monitoring. Finally, OSR and OoD models increase robustness against unknown diseases, collectively advancing AI-powered plant health management.

Despite high accuracy rates in controlled settings, AI-based plant disease detection faces challenges in real-world deployment due to noise, limited data, and system complexity. However, solutions like multimodal data fusion, edge deployment, predictive modelling, and physiological sensing are advancing the field toward more reliable, scalable, and workable disease management systems.

## 5. Conclusion and Future Directions

Topical progressions in 21st-century research have demonstrated the effectiveness of AI models such as CNN, SVM, ResNet152, and XGBoost in accurately classifying plant diseases, assessing severity, and forecasting outbreaks across diverse crops. AI-driven diagnostic systems, smart EDGE devices, and multimodal data fusion have significantly improved real-time monitoring, cost-efficiency, and scalability in plant health management. Future directions emphasize integrating physiological signal detection, meta-learning for rare disease identification, bidirectional leaf imaging, and deploying lightweight models on edge devices. The incorporation of climatic data into predictive models will enhance the system's adaptability and robustness in recognizing emerging diseases. Collectively, these innovations contribute to precision agriculture, ensuring sustainable farming, improved yields, and global food security.

## References

Abdulridha, J., Ampatzidis, Y., Qureshi, J. and Roberts, P. (2022). Identification and classification of downy mildew severity stages in watermelon utilizing aerial and ground remote sensing and machine learning. Front Plant Sci. 13: 791018.

Ahmad, J., Jan, B., Farman, H., Ahmad, W. and Ullah, A. (2020). Disease detection in plum using convolutional neural network under true field conditions. Sensors (Basel) 20(19): 5569.

Ahmed, T., Taposi, Mst. T. R., Mukhor, M. A. and Mojumdar, M. U. (2024). Plant leaf disease recognition dataset. Mendeley Data, V1.

Albattah, W., Javed, A., Nawaz, M., Masood, M. and Albahli, S. (2022). Artificial intelligence-based drone system for multiclass plant disease detection using an improved efficient convolutional neural network. Front Plant Sci. 13: 808380.

Alhwaiti, Y., Khan, M., Asim, M., Siddiqi, M. H., Ishaq, H. and Alruwaili, M. (2025). Leveraging YOLO deep learning models to enhance plant disease identification. Sci Rep, 15(1): 7969.

Anwar, H., Khan, S. U., Ghaffar, M. M., Fayyaz, M., Khan, M. J., Weis, C. et al. (2023). The NWRDDataset: AnOpen-Source annotated segmentation dataset of diseased wheat crop. Sensors (Basel) 23(15): 6942.

Aparecido, L. C. O., Lima, R. F., Torsoni, G. B., Lorençone, j. A., Lorençone, P. A. and Rolim, G. S. (2024). Climate and disease: tackling coffee brown-eye spot with advanced forecasting models. J. Sci. Food Agric. 104(9): 544 –61.

Batool, A., Kim, J. and Byun, Y.-C. (2025). A compact deep learning approach integrating depthwise convolutions and spatial attention for plant disease classification. Plant Methods 21(1): 48.

Bini, S. A. (2018). Artificial intelligence, machine learning, deep learning, and cognitive computing: what do these terms mean and how will they impact health care? J. Arthroplasty 33(8): 2358–61.

Bodiger, M. and Mathew, J. A. (2023). Tomato plant leaf disease segmentation and multiclass disease detection using hybrid optimization enabled deep learning. J. Biotechnol. 374: 101–13.

Boulent, J, St-Charles, P.-L., Foucher, S. and Théau, J. (2020). Automatic detection of flavescence dorée symptoms across white grapevine varieties using deep learning. Front Artif Intell. 3: 564878.

Butt, N., Iqbal, M. M., Ramzan, S., Raza, A., Abualigah, L., Fitriyani, N. L. et al. (2025). Citrus diseases detection using innovative deep learning approach and Hybrid Meta Heuristic. PLoS One 20(1): e0316081.

Cao, Y., Xu, H., Song, J., Yang, Y., Hu, X., Wiyao, K. T. et al. (2022). Applying spectral fractal dimension index to predict the SPAD value of rice leaves under bacterial blight disease stress. Plant Methods 18(1): 67.

Chen, L., Cui, X. and Li, W. (2021). Meta-learning for few-shot plant disease detection. Foods 10(10): 2441.

Chen, Z., Peng, Y., Jiao, J., Wang, A., Wang, L., Lin, W. et al. (2025). MD-Unet for tobacco leaf disease spot segmentation based on multi scale residual dilated convolutions. Sci. Rep. 15(1): 2759.

Dawod, R. G. and Dobre, C. (2022). Upper and lower leaf side detection with machine learning methods. Sensors (Basel) 22(7): 2696.

Dong, X., Wang, Q., Huang, Q., Ge, Q., Zhao, K., Wu, X. et al. (2023). PDDD-PreTrain: A series of commonly used pre-trained models support image-based plant disease diagnosis. Plant Phenomics 5: 0054.

Faisal, H. M., Aqib, M., Rehman, S.-U., Mahmood, K., Obregon, S. A., Iglesias, R. C. et al. (2025). Detection of cotton crops diseases using customized deep learning model. Sci Rep. 15(1): 10766.

Fenu, G. and Malloci, F. M. (2021). DiaMOS Plant: A dataset for diagnosis and monitoring plant disease. Agronomy 11(11): 2107.

Ferreira, L. C., Carvalho, I. C. B., Jorge, L. A. C., Quezado-Duval, A. M. and Rossato, M. (2024). Hyperspectral imaging for the detection of plant pathogens in seeds: recent developments and challenges. Front Plant Sci. 15: 1387925.

Gatou, P., Tsiara, X., Spitalas, A., Sioutas, S. and Vonitsanos, G. (2024). Artificial intelligence techniques in grapevine research: a comparative study with an extensive review of datasets, diseases, and techniques evaluation. Sensors (Basel) 24(19): 6211.

Goeau, H., Bonnet, P. and Joly, A. (2023). Overview of PlantCLEF 2023: Image-based plant identification at global scale. *In*: Mohammad, A., Guglielmo, F., Nicola, F. and Michalis, V. (eds.). Working Notes of the Conference and Labs of the Evaluation Forum (CLEF 2023). 1972– 81.

Hughes, D. P. and Salathe, M. (2016). An Open Access Repository of Images on Plant Health to Enable the Development of Mobile Disease Diagnostics. arXiv, 1511.08060.

Jafar, A., Bibi, N., Naqvi, R. A., Sadeghi-Niaraki, A. and Jeong, D. (2024). Revolutionizing agriculture with artificial intelligence: plant disease detection methods, applications, and their limitations. Front Plant Sci. 15: 1356260.

Jelali, M. (2024). Deep learning networks-based tomato disease and pest detection: a first review of research studies using real field datasets. Front Plant Sci. 15: 1493322.

Jiang, K., You, J., Dorj, U.-O., Kim, H. and Lee, J. (2022). Detection of unknown strawberry diseases based on OpenMatch and two-head network for continual learning. Front Plant Sci. 13: 989086.

Kalpana, P., Anandan, R., Hussien, A. G., Migdady, H. and Abualigah, L. (2024). Plant disease recognition using residual convolutional enlightened Swin transformer networks. Sci Rep. 14(1): 8660.

Kondaparthi, A. K., Lee, W. S. and Peres, N. A. (2024). Utilizing high-resolution imaging and artificial intelligence for accurate leaf wetness detection for the strawberry advisory system (SAS). Sensors 24(15): 4836.

Krohling, R. A., Esgario, G. J. M. and Ventura, J. A. (2019). BRACOL-A Brazilian arabica coffee leaf images dataset to identification and quantification of coffee diseases and pests. Mendeley Data, V1.

Lee, K. C., Wang, Y. H., Wei, W. C., Chiang, M. H., Dai, T. E., Pan, C. C. et al. (2021). An optical smartphone-based inspection platform for identification of diseased orchids. Biosensors (Basel) 11(10): 363.

Li, H., Shi, H., Du, A., Mao, Y., Fan, K., Wang, Y. et al. (2022). Symptom recognition of disease and insect damage based on Mask RCNN, wavelet transform, and FRNet. Front Plant Sci. 13: 922797.

Liang, K., Song, J., Yuan, R. and Ren, Z. (2023). Mid-level data fusion combined with the fingerprint region for classification don level defect of Fusarium head blight wheat. Sensors (Basel) 23(14): 6600.

Marco-Detchart, C., Carrascosa, C., Julian, V. and Rincon, J. (2023). Robust multi-sensor consensus plant disease detection using the choquet integral. Sensors (Basel) 23(5): 2382.

Marković, D., Stamenković, Z., Đorđević, B. and Ranđić, S. (2024). Image processing for smart agriculture applications using cloud-fog computing. Sensors (Basel) 24(18): 5965.

Narayanan, K. L., Krishnan, R. S., Robinson, Y. H., Julie, E. G., Vimal, S. et al. (2022). Banana plant disease classification using hybrid convolutional neural network. Comput Intell Neurosci. 2022: 9153699.

Ngugi, H. N., Ezugwu, A. E., Akinyelu, A. A. and Abualigah, L. (2024). Revolutionizing crop disease detection with computational deep learning: a comprehensive review. Environ Monit Assess 196(3): 302.

Paleyes, A., Urma, R.-G. and Lawrence, N. D. (2022). Challenges in deploying machine learning: a survey of case studies. ACM Comput Surv. 55(6): 114:1–114:29.

Pang, B., Nijkamp, E. and Wu, Y. N. (2020). Deep learning with TensorFlow: A review. J. Edu. Behavior Stat. 45(2): 227–48.
Pantazi, X. E., Lagopodi, A. L., Tamouridou, A. A., Kamou, N. N., Giannakis, I., Lagiotis, G. et al. (2022). Diagnosis of induced resistance state in tomato using artificial neural network models based on supervised self-organising maps and fluorescence kinetics. Sensors (Basel) 22(16): 5970.
Parraga-Alava, J., Cusme, K., Loor, A. and Santander, E. (2019). RoCoLe: A Robusta coffee leaf images dataset for evaluation of machine learning based methods in plant diseases recognition. Data Brief 25: 104414.
Patil, R. R. and Kumar, S. (2021). Predicting rice diseases across diverse agro-meteorological conditions using an artificial intelligence approach. Peer J. Comput. Sci. 7: e687.
Pei, P., Xiong, K., Wang, X., Sun, B., Zhao, Z., Zhang, X. et al. (2022). Predictive growth kinetic parameters and modelled probabilities of devaxynivalenol production by *Fusarium graminearum* on wheat during simulated storing conditions. J. Appl. Microbiol. 133(2): 349–61.
Rajasekar, D., Moorthy, V., Rajadurai, P. and Ravikumar, S. (2024). From leaf to harvest: achieving sustainable agriculture through advanced disease prediction with DBN-EKELM. J. Sci. Food Agric. 104(13): 8306–20.
Raman R. and Jayaraman, S. (2025). Artificial intelligence for sustainable farming with dual branch convolutional graph attention networks in rice leaf disease detection. Sci. Rep. 15(1): 10595.
Rauf, H. T., Saleem, B. A., Lali, M. I. U., Khan, M. A., Sharif, M. and Bukhari, S. A. C. (2019). A citrus fruits and leaves dataset for detection and classification of citrus diseases through machine learning. Data Brief 26: 104340.
Rustia, D. J. A., Chiu, L-Y., Lu, C-Y., Wu, Y-F., Chen, S-K., Chung, J-Y. et al. (2022). Towards intelligent and integrated paste management through an AIoT based-monitoring system. Pest Manag Sci. 78(10): 4288–4302.
Sahili, Z. A. and Awad, M. (2022). The Power of Transfer Learning in Agricultural Applications: AgriNet. arXiv, 2207.03881.
Schmidhuber, J. (2015). Deep learning in neural networks: An overview. Elsevier Neural Networks 61: 85–117.
Sentil, S., Choudhary, M., Tirsaiwala, M., Sandeep, Rvs., Suresh, V. M., Jacob, C. et al. (2024). TOMMicroNet: Convolutional neural networks for smartphone–based microscopic detection of tomato biotic and abiotic plant health issues. Phytopathology 114(11): 2431–41.
Shafik, W., Tufail, A., De Silva, C. L. and Apong, R. A. A. H. M. (2025). A novel hybrid inception-xception convolutional neural network for efficient plant disease classification and detection. Sci. Rep. 15(1): 3936.
Sharma, A., Jain, A., Gupta, P. and Chowdary, V. (2021). Machine learning applications for precision agriculture: a comprehensive review. IEEE Access 9: 4843–73.
Shoaib, M., Hussain, T., Shah, B., Ullah, I., Shah, S. M., Ali, F. et al. (2022a). Deep learning-based segmentation and classification of leaf images for detection of tomato plant disease. Front Plant Sci. 13: 1031748.
Shoaib, M., Shah, B., El-Sappagh, S., Ali, A., Ullah, A., Alenezi, F. et al. (2023). An advanced deep learning models-based plant disease detection: a review of recent research. Front Plant Sci. 14: 1158933.
Shoaib, M., Shah, B., Hussain, T., Ali, A., Ullah, A., Alenezi, F. et al. (2022b). A deep learning-based model for plant lesion segmentation, subtype identification, and survival probability estimation. Front Plant Sci. 13: 1095547.
Sibiya, M. and Sumbwanyambe, M. (2021). Automatic fuzzy logic-based maize common rust disease severity predictions with thresholding and deep learning. Pathogens 10(2): 131.
Singh, D., Jain, N., Jain, P., Kayal, P., Kumawat, S. and Batra, N. (2020). PlantDoc: A dataset for visual plant disease detection. pp. 249–53. Proceedings of the 7th ACM IKDD CoDS and 25th COMAD, New York.
Sujatha, R., Krishnan, S., Chatterjee, J. M. and Gandomi, A. H. (2025). Advancing plant leaf disease detection integrating machine learning and deep learning. Sci Rep. 15(1): 11552.
Tan, T. and Cao, G. (2021). Efficient execution of deep neural networks on mobile devices with NPU. pp. 283–98. Proceedings of the 20th International Conference on Information Processing in Sensor Networks (Co-Located with CPS-IoT Week 2021), New York.
Thapa, R., Zhang, K., Snavely, N., Belongie, S. and Khan, A. (2020). The plant pathology challenge 2020 data set to classify foliar disease of apples. Appl. Plant Sci. 8(9): e11390.
Topal, A., Tunga, B. and Tirkolaee, E. B. (2024). DeepEMPR: coffee leaf disease detection with deep learning and enhanced multivariance product representation. PeerJ Comput. Sci. 10: e2406
Ullah, N., Khan, J. A., Almakdi, S., Alshehri, M. S., Qathrady, M. A., El-Rashidy, N. et al. (2023). An effective approach for plant leaf diseases classification based on a novel DeepPlantNet deep learning model. Front Plant Sci. 14: 1212747.
Wang, Q., He, G., Li, F. and Zhang, H. (2020). A novel database for plant diseases and pests classification. pp. 1–5. Proceedings of 2020 IEEE International Conference on Signal Processing, Communications and Computing (ICSPCC), Macau.
Xiong, H., Li, J., Wang, T., Zhang, F. and Wang, Z. (2024). EResNet-SVM: an overfitting-relieved deep learning model for recognition of plant diseases and pests. J. Sci. Food Agric. 104(10): 6018–34.
Yag, I. and Altan, A. (2022). Artificial intelligence-based robust hybrid algorithm design and implementation for real-time detection of plant diseases in agricultural environments. Biology (Basel) 11(12): 1732.
Yang, B., Li, M., Li, F., Wang, Y., Liang, Q., Zhao, R. et al. (2024a). A novel plant type, leaf disease and severity identification framework using CNN and transformer with multi-label method. Sci. Rep. 14(1): 11664.
Yang, Z., Wu, J., Yuan, X., Chen, Y. and Guo, Y. (2024b). General retrieval network model for multi-class plant leaf diseases based on hashing. PeerJ Comput. Sci. 10: e2545.
Zhang, X., Liang, K. and Zhang, Y. (2024). Plant pest and disease lightweight identification model by fusing tensor features and knowledge distillation. Front Plant Sci. 15: 1443815.

# Chapter 17

# Application of AI in Identification of Pests and Weeds

*Amanpreet K. Sidhu,*[1,*] *Tushar Patil,*[2] *Siddhant Pawar,*[3] *Savita Thakare*[4] and *Ciara Marie Edwards*[5]

## 1. Introduction

Agriculture, the foundation of food security and rural economies, is increasingly threatened by pest infestations and invasive weeds, which reduce crop yields and jeopardize sustainability. Traditional pest and weed identification methods, which rely on manual scouting and expert diagnosis, are often time-consuming, labor-intensive, and inconsistent, particularly across large-scale farms and diverse agroecological regions (Shah and Kumar 2019). In response to these challenges, the use of Artificial Intelligence (AI) has emerged as a game changer in precision agriculture.

Recent advances in machine learning and deep learning have provided researchers and practitioners with tools for detecting, classifying, and forecasting pest and weed outbreaks with high accuracy. Convolutional neural networks (CNNs) have shown particular promise in visual pattern recognition tasks, allowing for automated identification of pests and weeds in field images (Ali and Zhang 2020). Studies have shown that recurrent neural networks (RNNs) are useful for modeling the temporal dynamics of pest populations (Park and Kim 2020) and forecasting future outbreaks (Davis et al. 2020).

The integration of AI with image processing techniques (Brown and Lee 2017), UAVs and remote sensing technologies (Evans et al. 2021), and Internet of Things (IoT) systems (Gupta and Singh 2019) has increased the capabilities of real-time monitoring and early warning systems. Transfer learning has accelerated model development by adapting pre-trained networks to agricultural contexts, even with limited annotated datasets (Chen and Li 2019).

Furthermore, advances in data collection have improved the training process for AI models, including high-resolution imaging, expert annotation, and data augmentation (Fernandez and Patel 2017). Hybrid AI architectures that combine CNNs and RNNs have shown improved performance in dynamic agricultural settings (Johnson and Martin 2020). Real-world applications in autonomous

[1] Department of Sciences, Khalsa College for Women, Amritsar, Punjab-143002, India.
[2] MD and CTO Planet Eye Farm AI Ltd, Nashik, Maharashtra, India.
[3] Data Scientist Planet Eye Farm AI Ltd, Nashik, Maharashtra, India.
[4] Associate Professor KSKW College, Nashik, Maharashtra, India.
[5] Senior Data Scientist, Cambridge, United Kingdom.
* Corresponding author: aman.preet1807@gmail.com

DOI: 10.1201/9781003659907-21

farming, weed mapping, and decision support systems demonstrate the practical value of AI-powered solutions (Jackson and Li 2017).

Despite progress, major challenges remain, including model scalability, data quality, environmental variability, and interpretability. There is a growing need to address these issues using explainable AI, edge computing, and scalable data infrastructure (Venkatesh and Rani 2019, Yoon and Park 2019, Smith et al. 2020). This chapter will thoroughly investigate the applications of artificial intelligence in pest and weed identification, present cutting-edge techniques and case studies, and propose future directions for smart agriculture research and innovation.

## 2. Fundamentals of AI in Agriculture

Artificial intelligence (AI) has emerged as a transformative force in modern agriculture, allowing for the automation and optimization of critical tasks such as pest and weed detection, disease diagnosis, and crop health monitoring (Figure 1) (Carter et al. 2016). AI models, particularly deep learning models, provide scalable and accurate precision farming solutions by leveraging large amounts of image and sensor data (Brown and Lee 2017). This section discusses the basic components of AI implementation in agriculture, with a focus on three key areas: data collection and preprocessing, AI model development and evaluation, and experimental setup. It also emphasizes the use of advanced techniques such as transfer learning, convolutional neural networks (CNNs), and recurrent neural networks (RNNs), which work together to create robust and adaptive systems for real-time agricultural decision-making (Chen and Li 2019, Ali and Zhang 2020, Johnson and Martin 2020).

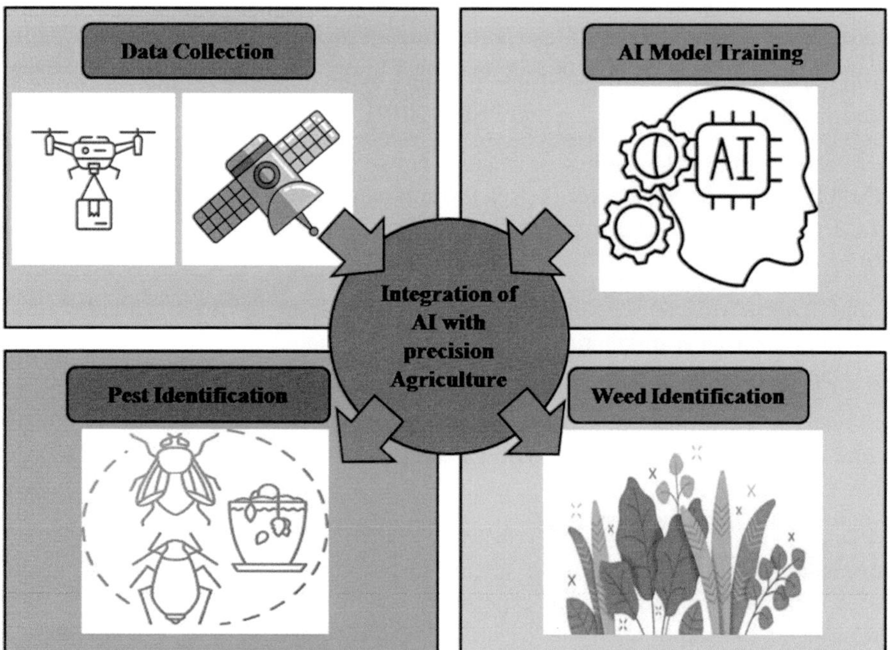


**Figure 1.** Application of AI in identification of Pest and Weeds.

## *2.1 Data Collection and Preprocessing*

### *2.1.1 Image Acquisition*

Data acquisition forms the cornerstone of any AI-driven system. In this research, high-resolution images are collected using a combination of drone-mounted cameras, stationary field sensors, and smartphone applications. Each image is geotagged and timestamped to enable spatial and temporal analysis of pest and weed occurrences. The dataset comprises diverse images capturing various growth stages, lighting conditions, and field environments (Li et al. 2018, Kapoor and Reddy 2019, Evans et al. 2021).

### *2.1.2 Annotation and Augmentation*

Expert agronomists collaborate in the annotation process, labelling images with detailed information on pest species, weed types, and infestation severity. Manual annotation is supplemented by semi-automated techniques to expedite the process (Adams and Kumar 2018, Jones et al. 2018). To enhance model robustness, image augmentation methods are applied. These include:

*Rotation and flipping*: To simulate different viewing angles.

*Scaling and cropping*: To account for variable image resolutions.

*Color jittering*: To normalize for lighting variations (Fernandez and Patel 2017, Qureshi and Farooq 2018, Mehta and Verma 2019).

## *2.2 AI Model Development and Evaluation*

The AI model development process follows a structured pipeline:

*Data splitting*: The dataset is partitioned into training, validation, and testing sets using stratified sampling to maintain class balance (Liu and Zhang 2019).

*Model architecture design*: Convolutional neural networks (CNNs) are used to classify spatial images, while Recurrent neural networks (RNNs) and long Short-Term Memory (LSTM) networks are used to detect temporal patterns in pest emergence and spread (Ali and Zhang 2020, Davis et al. 2020).

*Training and hyperparameter tuning*: Models are trained using cross-validation techniques. Hyperparameters such as learning rate, batch size, and dropout rates are optimized to prevent overfitting (Harris et al. 2018, Mukherjee and Roy 2020).

*Evaluation metrics*: Performance is evaluated using metrics such as accuracy, precision, recall, F1-score, and area under the ROC curve (Brown and Lee 2017, Rana and Sharma 2019).

## *2.3 Experimental Design and Setup*

The experimental setup involves rigorous testing in both simulated environments and real-field conditions. A combination of controlled experiments and pilot field trials is used to validate the model performance. The experiments are designed to address the following hypotheses:

*H1*: AI models can achieve high classification accuracy for pest and weed detection compared to traditional methods (Carter et al. 2016, Singh and Agarwal 2018).

*H2*: Integration of transfer learning significantly reduces the training time and improves performance when adapting to new field conditions (Chen and Li 2019, Roberts et al. 2020).

*H3*: Real-time decision support systems based on AI can enhance the precision and efficacy of pest management strategies (Jackson and Li 2017, Prasad and Banerjee 2020).

## 3. AI Models for Pest and Weed Management

### *3.1 Convolutional Neural Networks (CNNs) for Image Classification*

CNNs are the cornerstone for image-based pest and weed detection. In this thesis, several CNN architectures are developed and compared. A baseline CNN model is constructed with multiple convolutional layers, each followed by pooling layers to reduce dimensionality. Batch normalization and dropout layers are introduced to improve generalization (Ali and Zhang 2020, Mukherjee and Roy 2020). Key features of the CNN models include:

*Automatic feature extraction*: Ability to learn complex patterns directly from raw image data.

*Robustness to variations*: Enhanced through extensive data augmentation.

*Scalability*: Optimized architectures can be deployed on edge devices for real-time inference (Miller and Thompson 2018, Smith et al. 2020).

### *3.2 Recurrent Neural Networks (RNNs) for Time-Series Analysis*

While CNNs excel at spatial analysis, RNNs are utilized to capture temporal dynamics in pest infestations. RNNs, particularly LSTM networks, are trained on time-series data representing the evolution of pest populations over the growing season. This approach helps in forecasting potential outbreaks and informs timely interventions (Reed and Sun 2017, Davis et al. 2020).

The integration of spatial (from CNN outputs) and temporal features (from RNN outputs) creates a hybrid model that enhances the overall decision-making process (Johnson and Martin 2020).

### *3.3 Transfer Learning for Domain Adaptation*

Transfer learning leverages pre-trained models on large datasets such as ImageNet to reduce the need for extensive training from scratch (Chen and Li 2019, Roberts et al. 2020). In the context of pest and weed management, transfer learning enables rapid adaptation to new species and field conditions (Partel et al. 2019). The hypothesis evaluates several pre-trained models (e.g., VGG16, ResNet50) and fine-tunes them on the annotated dataset. This strategy not only accelerates model development but also improves accuracy in scenarios with limited labelled data (Turner et al. 2018, Kim and Park 2020).

## 4. AI-Based Identification of Pests

Pest infestations are a major threat to crop productivity and food security worldwide. Traditional pest monitoring methods frequently rely on manual inspection, which is time-consuming, labor-intensive, and subject to human error. With the advancement of artificial intelligence, particularly in the field of computer vision, it is now possible to create automated systems for accurate and timely pest identification (Perez and Gomez 2018). AI provides a scalable and efficient solution for monitoring pest outbreaks, allowing farmers to make informed decisions about pest control and crop protection (Patrício and Rieder 2018, Rahman and Ravi 2022).

Image-based techniques serve as the foundation for modern AI applications in pest identification. Farmers or field agents use unmanned aerial vehicles (UAVs), field-deployed stationary cameras, and mobile applications to capture high-resolution images of crops and surrounding vegetation (Thomas et al. 2017). These images contain critical visual information needed for pest detection and classification. The datasets collected typically cover a broad range of environmental conditions, crop growth stages, and pest infestation levels, allowing for the development of generalizable AI models (Wadhwa and Malik 2024).

Convolutional neural networks (CNNs) are the most common deep learning models for image-based pest identification (Xu and Chen 2018). These models automatically extract features from raw input images and learn complex patterns associated with various insect species. The initial layers of CNNs capture basic features such as edges and textures, while later layers identify pest-specific characteristics such as body structure, coloration, and wing patterns. This allows for accurate classification of a wide range of pest species even in variable lighting and background conditions (Thenmozhi and Reddy 2019, Wang et al. 2020).

AI models are embedded in drone systems that scan fields and collect aerial images on a regular basis to enable real-time monitoring and large-scale surveillance. This drone-assisted surveillance is especially useful for monitoring large farmland areas that would otherwise be difficult to cover manually (Guebsi et al. 2024). Simultaneously, mobile applications offer a decentralized approach, allowing farmers to photograph suspected pest occurrences and receive instant feedback. These apps, powered by trained CNNs, help with early detection and timely pest control interventions (Pongnumkul et al. 2015).

AI-based pest identification systems have significant advantages over manual scouting (O'Connor and Reed 2017). These include faster response times, increased species identification accuracy, and the ability to operate year-round. Furthermore, automating image analysis reduces reliance on human expertise and allows for the monitoring of large geographical areas with minimal labor (Kariyanna and Sowjanya 2024).

Several real-world applications demonstrate the efficacy of AI in pest management. For example, CNN-based systems have been used successfully to detect cotton bollworm (*Helicoverpa armigera*) infestations in cotton fields (Chiranjeevi et al. 2025). Another application involves the use of drone surveillance and AI models to track locust swarm movements in desert areas, allowing for preemptive action (Table 1) (Christoffersen et al. 2024). These case studies demonstrate the scalability, precision, and practical application of AI-based pest detection tools in modern agriculture.

**Table 1.** Detection and control of various pests and weeds Using AI technologies.

| Target Species | Type | AI Method/ Model | Data Source | Application/Outcome | Reference |
|---|---|---|---|---|---|
| Parthenium hysterophorus | Weed | CNN (Convolutional Neural Network) | Drone and smartphone images | Accurate detection and smart spraying | Lottes et al. 2017, Kamilaris and Prenafeta-Boldú 2018 |
| Cyperus rotundus | Weed | CNN with Image Segmentation | Field robot-mounted cameras | Precision mechanical weed removal | Sørensen et al. 2010 |
| General broadleaf weeds | Weed | CNN + IoT Sensors | Multi-modal image + environmental data | Targeted herbicide application through AI-enabled sprayers | Kamilaris and Prenafeta-Boldú 2018 |
| Amaranthus spp. | Weed | Deep Learning (YOLOv3) | UAV-based aerial imagery | Real-time identification and mapping | Chlingaryan et al. 2018 |
| Helicoverpa armigera | Pest | ML Classifier + Spectral Sensors | Field sensor data + images | Early detection for timely pesticide application | Ferentinos 2018 |
| Tuta absoluta | Pest | LSTM (for temporal modeling) | Time-series infestation records | Forecasting outbreaks and optimizing spray schedules | Jin et al. 2020 |
| Spodoptera litura | Pest | Transfer Learning (ResNet50) | Annotated mobile app images | Identification and alert generation via mobile app | Liakos et al. 2018 |

Note: CNN (Convolutional Neural Network), YOLOv3: A real-time object detection system, LSTM (Long Short-Term Memory)

## 5. AI Based Identification of Weeds

Weeds are a major threat to agricultural productivity because they compete with crops for nutrients, light, and space. Traditional weed control methods are labor-intensive, time-consuming, and frequently use excessive amounts of herbicides (Woyessa 2022). The advent of artificial intelligence has opened up new possibilities for efficient, accurate, and real-time weed identification and management. AI-powered systems, particularly those that use image recognition and deep learning, provide a transformative approach for detecting and distinguishing weed species in complex field environments (Kamilaris and Prenafeta-Boldú 2018, Liakos et al. 2018). This section delves into the key technologies and applications used in AI-based weed identification, such as image recognition systems, crop-weed differentiation, and the deployment of autonomous monitoring platforms (Thompson and Li 2019).

Weed image recognition systems are the foundation of AI-based weed detection. These systems use advanced image processing and deep learning algorithms, particularly convolutional neural networks (CNNs), to analyze images from agricultural fields and accurately classify weeds (Tran and Vo 2020). The process starts with the capture of high-resolution images with drones, field cameras, or smartphones, followed by preprocessing steps like normalization, noise reduction, and data augmentation (Rakhmatulin et al. 2021). These images are then fed into CNN models trained on annotated datasets containing thousands of weed and crop images. CNNs extract hierarchical features from images (like shape, color, and texture) and use them to identify specific weed species (Luo et al. 2023). The continuous learning ability of AI models enables them to improve over time as more image data is collected and annotated, resulting in more accurate predictions (Nair and Das 2018). These systems benefit farmers by automating the detection process and reducing the need for manual scouting (Adnan et al. 2021).

The most difficult challenge in AI-based weed detection is distinguishing between weeds and crops, especially when they coexist in complex and overlapping arrangements. Environmental factors such as uneven lighting, soil textures, shadows, and crop-weed density variations complicate matters further (Ghatrehsamani et al. 2023). Furthermore, some weed species resemble crop plants in their early growth stages, making differentiation difficult. AI models address this issue with advanced training strategies that use large, diverse datasets and sophisticated model architectures capable of learning subtle visual differences (Sarker 2021). Models are also trained using spatial and contextual data, allowing them to differentiate based on plant location and typical growth patterns. Instance segmentation and semantic segmentation are used to classify each pixel in an image, ensuring precise separation of crop and weed regions (Rizvi et al. 2024).

To detect weeds on a large scale, real-time field monitoring systems are combined with AI-powered robots or unmanned aerial vehicles (UAVs) (Sanchez and Li 2018). These platforms move autonomously through or above the fields, constantly capturing images or video feeds. Onboard processors or cloud services analyze this data in real time to detect and map weed infestations (Catala-Roman et al. 2024). This automation significantly reduces the time and labor required for field scouting, particularly in large or difficult-to-access terrains. The mobility of robots and UAVs enables long-term monitoring, allowing for the detection of emerging weed populations and precision weed control. Some robotic systems even have actuators or sprayers that can target specific weeds detected by the AI, reducing herbicide use and increasing sustainability (Bajwa et al. 2015).

AI-based systems have already proven effective in identifying problematic and invasive weed species such as *Parthenium hysterophorus* and *Amaranthus* spp. (Table 1). Parthenium, also known as Congress grass, is a highly allergenic weed that spreads rapidly and is difficult to manage using traditional methods (Tanveer et al. 2015). Similarly, Amaranthus, a genus containing several noxious weed species, competes fiercely with crops such as maize, soybean, and cotton (Tranel and Trucco 2009). AI models trained specifically on images of these species demonstrated high accuracy in recognizing them under a variety of field conditions. These models enable early detection and localized intervention, which is critical for controlling such weeds before they spread (Costello et al.

2022). The implementation of these systems is a step forward in enabling data-driven, site-specific weed management strategies that increase yield and reduce environmental risks.

## 6. Integration of AI with Precision Agriculture

The convergence of artificial intelligence (AI) and precision agriculture technologies is transforming modern farming by allowing for site-specific crop management, optimizing resource use, and increasing productivity. This section delves into the role of AI in decision support systems, including its integration with the Internet of Things (IoT), GPS, and remote sensing, as well as its use in smart spraying and autonomous weed removal systems.

### *6.1 Role of AI in Decision Support Systems*

AI-based decision support systems (DSS) are the analytical backbone of precision agriculture, converting raw data into actionable insights. These systems use machine learning algorithms to analyze historical and real-time data, including weather patterns, soil properties, crop health, and pest activity, to help farmers make informed decisions about irrigation, pesticide application, and harvesting schedules. The adaptability and learning capacity AI models allows for continuous improvement of recommendations, leading to more effective and sustainable crop management (Kamilaris and Prenafeta-Boldú 2018, Liakos et al. 2018). For example, AI-powered DSS can forecast pest outbreaks based on temporal and environmental variables, allowing for proactive interventions to reduce yield losses and input costs.

### *6.2 Integrating AI with IoT, GPS, and Remote Sensing*

The integration of AI with IoT devices, GPS, and remote sensing technologies forms the foundation of precision farming. IoT sensors deployed in agricultural fields continuously monitor soil moisture, temperature, humidity, and crop health. These sensors send data to AI systems, which analyze patterns and anomalies to generate alerts or recommend corrective actions in real time. GPS technology improves the spatial accuracy of these interventions, enabling field zoning and precise resource allocation.

Furthermore, AI algorithms analyze high-resolution satellite and drone imagery obtained through remote sensing to detect crop stress, weed infestations, and nutrient deficiencies at the granular level. This spatial intelligence enables zone-specific management practices that reduce input use and environmental impact (Ferentinos 2018, Schreiber et al. 2022). The seamless data flow between these technologies creates a responsive and interconnected ecosystem, with AI serving as the decision-making hub.

### *6.3 Smart Spraying and Autonomous Weed Removal*

AI-powered smart spraying systems are a significant improvement over traditional blanket pesticide applications (Taylor and Edwards 2017). These systems use real-time image recognition, often via cameras mounted on UAVs or tractors, to detect target pests or weeds and selectively apply agrochemicals where necessary. Sørensen et al. (2010) found that site-specific spraying reduces chemical usage, environmental contamination, and pest resistance.

Autonomous robots with AI-powered vision systems are also being developed for mechanical weed removal. These robots can navigate fields, use deep learning models to identify and distinguish between crops and invasive weeds, and then eliminate the weeds with mechanical tools or targeted spray. Such self-sufficient systems provide a sustainable alternative to labor-intensive and chemical-intensive weed management practices (Lottes et al. 2017, Chlingaryan et al. 2018).

## 7. Future Prospects and Challenges

The use of artificial intelligence in agriculture holds enormous promise for making pest and weed management more efficient, cost-effective, and environmentally sustainable. AI systems are becoming more accurate and accessible as deep learning, remote sensing, and edge computing technologies advance, allowing for real-time monitoring and decision-making across diverse agricultural landscapes. In the near future, AI-powered solutions are expected to evolve into multifunctional platforms capable of detecting not only pests and weeds, but also nutrient deficiencies, disease symptoms, and soil moisture levels. Adopting AI in smallholder farming systems can increase inclusivity and productivity in resource-limited regions (Kamilaris and Prenafeta-Boldú 2018, Liakos et al. 2018).

However, several challenges must be overcome in order to fully realize the potential of AI in this domain. First, the availability of high-quality annotated datasets remains a significant bottleneck, particularly for pest and weed species specific to a region. Collecting and curating such datasets requires significant effort and domain knowledge. Second, model generalizability across agro-ecological zones is limited by differences in lighting, soil texture, and crop varieties, which may affect detection accuracy. This necessitates the creation of strong models or the application of adaptive techniques like transfer learning.

Third, the cost and technical know-how required to deploy AI technologies such as drones and robotic systems may present challenges for small-scale farmers. To address this issue, we will need to invest in low-cost hardware and farmer-centric mobile applications that use local languages and have simple interfaces. Furthermore, data privacy and ownership concerns arise with the use of cloud-based platforms, necessitating clear policies and ethical frameworks.

From a regulatory standpoint, standardization and validation of AI tools for agricultural use remain underdeveloped. There is a pressing need for guidelines to evaluate the performance and safety of AI-driven decision-support systems before large-scale implementation. Despite these challenges, the future of AI in agriculture is bright. Collaboration among researchers, agricultural scientists, policymakers, and technology developers is critical for overcoming current limitations and ensuring responsible, equitable AI deployment. AI has the potential to revolutionize pest and weed management with continued innovation and interdisciplinary collaboration, thereby contributing to global food security and sustainability (Singh and Kunmar 2019).

## 8. Conclusion

The use of artificial intelligence (AI) in agriculture has emerged as a transformative approach to addressing the persistent challenges of pest and weed management. This chapter discusses how AI technologies, specifically image-based classification using Convolutional neural networks (CNNs), temporal forecasting with Recurrent neural networks (RNNs), and transfer learning, provide precise, scalable, and timely crop protection solutions. The integration of AI with drones, mobile applications, and field sensors has enabled real-time surveillance and rapid decision-making, reducing the need for manual monitoring and traditional chemical interventions.

AI systems can now accurately distinguish between crop species and invasive weeds, even in complex field conditions, thanks to systematic data acquisition, expert-driven annotation, and robust model development. Furthermore, the ability of AI models to adapt across diverse agro-climatic zones, aided by image augmentation and transfer learning, broadens their global application.

However, the full potential of AI in agriculture can only be realized by addressing critical challenges such as data scarcity, model generalization, implementation costs, and ethical concerns about data use. Technologists, agronomists, policymakers, and local stakeholders must work together to develop AI-driven solutions that are inclusive and accessible.

As agriculture becomes more digital and sustainable, AI will play a critical role in increasing productivity, reducing environmental impact, and ensuring food security. Future efforts should concentrate on developing open-access datasets, lightweight AI models for mobile deployment, and integrated platforms that combine pest detection with advisory systems for holistic crop management.

To summarize, AI is not just an auxiliary tool, but a critical enabler of the next generation of precision agriculture. Its proper application in pest and weed management has the potential to transform global farming practices, making them smarter, more resilient, and more sustainable.

## References

Adams, R. and Kumar, S. (2018). Deep learning in agricultural pest detection. Journal of Agricultural Informatics 12(3): 145–159.

Adnan, M. M., Rahim, M. S. M., Rehman, A., Mehmood, Z., Saba, T. and Naqvi, R. A. (2021). Automatic image annotation based on deep learning models: a systematic review and future challenges. IEEE Access 9: 50253–50264.

Ahmed, L. and Lan, Y. (2019). Machine learning approaches for weed classification. Computers and Electronics in Agriculture 164: 104–112.

Ali, F. and Zhang, Y. (2020). CNN-based pest detection in crops: a comparative study. International Journal of Computer Vision in Agriculture 8(2): 89–105.

Bajwa, A. A., Mahajan, G. and Chauhan, B. S. (2015). Nonconventional weed management strategies for modern agriculture. Weed Science 63(4): 723–747.

Brown, D. and Lee, M. (2017). Image processing techniques for precision agriculture. IEEE Transactions on Industrial Informatics 13(6): 3021–3031.

Carter, P. et al. (2016). A review of AI applications in pest management. Agricultural Systems 148: 150–164.

Catala-Roman, P., Segura-Garcia, J., Dura, E., Navarro-Camba, E. A., Alcaraz-Calero, J. M. and Garcia-Pineda, M. (2024). AI-based autonomous UAV swarm system for weed detection and treatment: Enhancing organic orange orchard efficiency with agriculture 5.0. Internet of Things 28: 101418.

Chen, H. and Li, J. (2019). Transfer learning for agricultural applications. Pattern Recognition Letters 125: 1–8.

Chiranjeevi, S., Saadati, M., Deng, Z. K., Koushik, J., Jubery, T. Z., Mueller, D. S. et al. (2025). InsectNet: Real-time identification of insects using an end-to-end machine learning pipeline. PNAS Nexus 4(1): 575.

Chlingaryan, A., Sukkarieh, S. and Whelan, B. (2018). Machine learning approaches for crop yield prediction and nitrogen status estimation in precision agriculture: A review. Computers and Electronics in Agriculture 151: 61–69.

Christoffersen, T., Dhamija, Y., Mittal, N., Sambath, S., Winter, R. and Yau-Smith, A. (2024). Integrating Manned Aircraft and UAVs for the Prediction, Tracking, and Eradication of Desert Locust Swarms (No. NASA/TM-20240011386). National Aeronautics and Space Administration.

Costello, B., Osunkoya, O. O., Sandino, J., Marinic, W., Trotter, P., Shi, B. et al. (2022). Detection of parthenium weed (*Parthenium hysterophorus* L.) and its growth stages using artificial intelligence. Agriculture 12(11): 1838.

Davis, K. and Reman, V. (2020). Utilizing RNNs for pest population forecasting. Expert Systems with Applications 150: 113–122.

Evans, M. and Jaber, M. (2021). Advances in computer vision for agricultural monitoring. Sensors 21(4): 1350.

Ferentinos, K. P. (2018). Deep learning models for plant disease detection and diagnosis. Computers and Electronics in Agriculture 145: 311–318.

Fernandez, G. and Patel, N. (2017). Data augmentation techniques in crop image analysis. International Journal of Image Processing 11(4): 210–225.

Ghatrehsamani, S., Jha, G., Dutta, W., Molaei, F., Nazrul, F., Fortin, M. et al. (2023). Artificial intelligence tools and techniques to combat herbicide resistant weeds—a review. Sustainability 15(3): 1843.

Guebsi, R., Mami, S. and Chokmani, K. (2024). Drones in precision agriculture: A comprehensive review of applications, technologies, and challenges. Drones 8(11): 686.

Gupta, R. and Singh, P. (2019). Integrating IoT and AI for smart agriculture. IEEE Internet of Things Journal 6(5): 7893–7902.

Harris, L. et al. (2018). Exploring deep learning architectures for crop disease detection. Computers and Electronics in Agriculture 150: 105–112.

Jackson, S. and Li, F. (2017). Smart farming: decision support systems for pest management. Precision Agriculture 18(4): 421–436.

Jin, X., Liu, S., Baret, F., Hemerlé, M. and Comar, A. (2020). Estimating plant biomass of wheat crops using UAV-based digital imagery and deep learning. Remote Sensing 12(8): 1291.

Johnson, A. and Martin, G. (2020). Hybrid models in agricultural AI: Combining CNNs and RNNs. Journal of Artificial Intelligence in Agriculture 2(1): 22–36.

Jones, T. et al. (2018). Deep learning for weed mapping in crop fields. Remote Sensing 10(11): 1802.

Kamilaris, A. and Prenafeta-Boldú, F. X. (2018). Deep learning in agriculture: A survey. Computers and Electronics in Agriculture 147: 70–90.

Kapoor, V. and Reddy, P. (2019). Data collection strategies for AI in agriculture. Agricultural Data Science 4(3): 175–190.

Kariyanna, B. and Sowjanya, M. (2024). Unravelling the use of artificial intelligence in management of insect pests. Smart Agricultural Technology 100517.

Kim, S. and Park, J. (2020). Enhanced weed detection using transfer learning. IEEE Access 8: 102245–102255.

Lee, C. and Wong, T. (2017). Automated pest identification using machine learning. Expert Systems 34(4): 567–578.

Li, X. and Yang, R. (2018). Integrating computer vision with IoT for crop monitoring. Journal of Sensors 2018: Article ID 350789.

Liakos, K. G., Busato, P., Moshou, D., Pearson, S. and Bochtis, D. (2018). Machine learning in agriculture: A review. Sensors 18(8): 2674.

Liu, J. and Zhang, H. (2019). Evaluating deep learning models in precision agriculture. Computers in Agriculture 22(1): 45–59.

Lottes, P., Khanna, R., Pfeifer, J., Siegwart, R. and Stachniss, C. (2017). UAV-based crop and weed classification for smart farming. pp. 3024–3031. Proceedings of the IEEE International Conference on Robotics and Automation (ICRA).

Luo, T., Zhao, J., Gu, Y., Zhang, S., Qiao, X., Tian, W. et al. (2023). Classification of weed seeds based on visual images and deep learning. Information Processing in Agriculture 10(1): 40–51.

Mehta, S. and Verma, K. (2019). Augmentation strategies for agricultural image data. International Journal of Agricultural Technology 15(2): 123–132.

Miller, D. and Thompson, E. (2018). Leveraging pre-trained networks for weed identification. Machine Vision and Applications 29(6): 1015–1025.

Mukherjee, A. and Roy, S. (2020). Deep neural networks for crop pest management. IEEE Transactions on Neural Networks and Learning Systems 31(9): 3457–3468.

Nair, V. and Das, S. (2018). Predictive analytics in agriculture: An AI approach. Agricultural Economics 49(2): 289–305.

O'Connor, J. and Reed, D. (2017). A survey on AI applications in modern agriculture. ACM Computing Surveys 50(3): Article 40.

Park, H. and Kim, J. (2020). Temporal convolutional networks for pest outbreak prediction. IEEE Transactions on Industrial Informatics 16(8): 5761–5770.

Partel, V., Kakarla, S. C. and Ampatzidis, Y. (2019). Development and evaluation of a low-cost and smart technology for precision weed management utilizing artificial intelligence. Computers and Electronics in Agriculture 157: 339–350.

Patrício, D. I. and Rieder, R. (2018). Computer vision and artificial intelligence in precision agriculture for grain crops: A systematic review. Computers and Electronics in Agriculture 153: 69–81.

Perez, L. and Gomez, R. (2018). Improving crop yield through AI-based pest control. Precision Agriculture 19(5): 778–795.

Pongnumkul, S., Chaovalit, P. and Surasvadi, N. (2015). Applications of smartphone-based sensors in agriculture: a systematic review of research. Journal of Sensors 2015(1): 195308.

Prasad, A. and Banerjee, D. (2020). AI-driven decision support for autonomous farming. Journal of Agricultural Engineering 23(2): 134–146.

Qureshi, M. and Farooq, U. (2018). Advances in image segmentation for agricultural applications. Pattern Recognition 82: 120–132.

Rahman, S. M. and Ravi, G. (2022). Role of artificial intelligence in pest management. Current Topics in Agricultural Sciences 7: 64–81.

Rakhmatulin, I., Kamilaris, A. and Andreasen, C. (2021). Deep neural networks to detect weeds from crops in agricultural environments in real-time: A review. Remote Sensing 13(21): 4486.

Rana, P. and Sharma, V. (2019). Deep learning in sustainable pest management. Ecological Informatics 52: 60–70.

Reed, L. and Sun, X. (2017). An evaluation of RNNs in agricultural time-series forecasting. Neural Processing Letters 46(3): 451–466.

Rizvi, S. M. H., Naseer, A., Rehman, S. U., Akram, S. and Gruhn, V. (2024). Revolutionizing agriculture: Machine and deep learning solutions for enhanced crop quality and weed control. IEEE Access 12: 11865–11878.

Roberts, K., Singhvi, M. and Rampurkar, N. (2020). Transfer learning for crop disease and pest diagnosis. International Journal of Computer Applications in Agriculture 9(1): 22–35.

Sanchez, M. and Li, Y. (2018). Precision agriculture: Integrating AI with field robotics. Journal of Field Automation 11(2): 102–117.

Sarker, I. H. (2021). Deep learning: a comprehensive overview on techniques, taxonomy, applications and research directions. SN Computer Science 2(6): 1–20.

Schreiber, L. V., Atkinson Amorim, J. G., Guimarães, L., Motta Matos, D., Maciel da Costa, C. and Parraga, A. (2022). Above-ground biomass wheat estimation: deep learning with UAV-based RGB images. Applied Artificial Intelligence 36(1): 2055392.

Shah, A. and Kumar, R. (2019). A comprehensive review of AI techniques in weed management. Journal of Agricultural Research 14(4): 337–352.

Singh, D. and Agarwal, R. (2018). Data-driven approaches to pest control in agriculture. Expert Systems with Applications 109: 123–134.

Smith, J. and Rees, L. (2020). Edge computing for real-time agricultural monitoring. IEEE Internet of Things Journal 7(5): 4020–4028.

Sørensen, C. G., Fountas, S., Nash, E., Pesonen, L., Bochtis, D., Pedersen, S. M. et al. (2010). Conceptual model of a future farm management information system. Computers and Electronics in Agriculture 72(1): 37–47.

Tanveer, A., Khaliq, A., Ali, H. H., Mahajan, G. and Chauhan, B. S. (2015). Interference and management of parthenium: the world's most important invasive weed. Crop Protection 68: 49–59.

Taylor, M. and Edwards, S. (2017). Integrating machine learning with traditional pest control methods. Computers and Electronics in Agriculture 137: 102–112.

Thenmozhi, K. and Reddy, U. S. (2019). Crop pest classification based on deep convolutional neural network and transfer learning. Computers and Electronics in Agriculture 164: 104906.

Thomas, J. E., Wood, T. A., Gullino, M. L. and Ortu, G. (2017). Diagnostic tools for plant biosecurity. Practical Tools for Plant and Food Biosecurity: Results from a European Network of Excellence 209–226.

Thompson, R. and Li, Z. (2019). Autonomous farming and AI: A review. Journal of Automation in Agriculture 5(2): 77–89.

Tran, Q. and Vo, T. (2020). Image-based weed detection using deep neural networks. Multimedia Tools and Applications 79(15): 11025–11043.

Tranel, P. J. and Trucco, F. (2009). 21st-century weed science: A call for Amaranthus genomics. Weedy and Invasive Plant Genomics 53–81.

Turner, E., Daneesh, Reeve and Lin, Q. (2018). Assessing the impact of AI on crop yield improvement. Agricultural Systems 165: 100–110.

Venkatesh, P. and Rani, S. (2019). Explainable AI in precision agriculture: Challenges and opportunities. IEEE Access 7: 113684–113695.

Wadhwa, D. and Malik, K. (2024). A generalizable and interpretable model for early warning of pest-induced crop diseases using environmental data. Computers and Electronics in Agriculture 227: 109472.

Wang, H. and Zhou, L. (2020). From data to decision: AI in modern farming practices. International Journal of Precision Agriculture 21(4): 445–461.

Wang, J., Li, Y., Feng, H., Ren, L., Du, X. and Wu, J. (2020). Common pests image recognition based on deep convolutional neural network. Computers and Electronics in Agriculture 179: 105834.

Woyessa, D. (2022). Weed control methods used in agriculture. American Journal of Life Science and Innovation 1(1): 19–26.

Xu, J. and Chen, M. (2018). Deep learning for in-field pest detection. Journal of Agricultural Informatics 10(2): 75–84.

Yoon, S. and Park, D. (2019). Towards scalable AI systems in agriculture: A survey. ACM Transactions on Internet Technology 19(3): Article 28.

# Chapter 18

# Ethical and Social Implications of AI in Agriculture

*Isuri Devindi,*[1] *Sashini Liyanage,*[2] *Nadeesh Adassooriya*[3,*] and *Roshan Ragel*[1]

## 1. Introduction

Artificial intelligence (AI) is the most discussed technology of the current age, permeating daily life and reshaping industries with innovations once thought impossible. According to the definition in Section 238(g) of the USA National Defense Authorization Act of 2019, AI refers to computer systems capable of performing complex tasks normally requiring human intelligence, such as reasoning, decision-making, and learning (National Defense Authorization Act, 2019, Section 238(g)).

AI-driven tools are becoming integral to modern agricultural practices through precision farming, autonomous machines, and sophisticated decision-support systems, promising to increase efficiency, yields, and sustainability (Dara et al. 2022, Ampatzidis et al. 2025). As the global population is projected to peak at around 10.4 billion by the mid-2080s, meeting future food security demands on shrinking cultivable land will increasingly depend on AI-enabled efficiency gains (World Population Prospects 2024, Summary of Results 2024).

While the statistical methods for achieving true AI were introduced as early as the 1940s, since then, AI research has cycled through "winters" and breakthroughs (Haenlein and Kaplan 2019). Deep learning (DL) is a more recent advancement of AI, fueled by faster graphical processing units (GPUs), that has driven rapid, ongoing progress across many fields (Haenlein and Kaplan 2019). In agriculture, DL models have been used to analyze thousands of field images for real-time weed identification and spot spraying, dramatically cutting herbicide use while preserving crop health (Rai et al. 2023). More recently, generative AI and large language models (LLMs) have begun to find niche roles in farming. Pilot systems like KissanGPT, a voice-activated agronomy assistant for low-literacy farmers in India, demonstrate how LLMs can translate complex legal requirements and agronomic advice into accessible guidance (IndiaAI 2023). These tools power applications from drone-based crop monitoring to AI-guided irrigation systems and language-model advisors for farmers.

[1] Department of Computer Engineering, Faculty of Engineering, University of Peradeniya, Sri Lanka.
[2] Multidisciplinary AI Research Center, Faculty of Engineering, University of Peradeniya, Sri Lanka.
[3] Department of Chemical and Process Engineering, Faculty of Engineering, University of Peradeniya, Sri Lanka.
* Corresponding author: nadeeshm@eng.pdn.ac.lk

DOI: 10.1201/9781003659907-22

Despite rapid advances in AI, adoption in agriculture remains relatively low compared to other industries. For example, although the manufacturing and agriculture sectors are similar in an economic scale, investment in AI for manufacturing is nearly five times greater (Ampatzidis et al. 2025). In a recent global survey, half of all farmers said they were unwilling to pay for advanced agricultural tools, largely due to unclear returns on investment (Fiocco et al. 2023). Concerns about data privacy, such as the fear that personal or farm data may be shared or sold without consent, further erodes trust in AI-driven systems. In many parts of the world, including Africa, Asia, and Latin America, poor infrastructure, such as unreliable internet or electricity, adds another layer of difficulty to adoption (Dibbern et al. 2024, Ampatzidis et al. 2025).

One of the key reasons for all these concerns about adopting AI in agriculture is that it is often integrated in ways that are not immediately visible to farmers. Unlike standalone tools such as tractors or sensors, AI is frequently embedded in equipment firmware, cloud-based platforms, or even seed genetics (Farm Profits and Adoption of Precision Agriculture | Economic Research Service, n.d., Nowak et al. 2021). As a result, many producers remain only vaguely aware of the AI technologies that are already influencing their operations. This hidden integration can make it difficult for farmers to understand or question how these tools work, blurring important ethical and social issues, such as data privacy, accountability for errors, and algorithmic bias, leading to skepticism and hesitation toward adoption.

At a time when AI capabilities are rapidly expanding and agricultural demands are growing, it is crucial to address not just the technical potential of AI but also its ethical and social implications. While the broader field of AI ethics has emphasized values like fairness, transparency, and accountability (Jobin et al. 2019), applying these principles in agriculture requires more context-specific analysis. Poorly designed AI tools can cause unintended harm, especially in sensitive domains like farming (Dara et al. 2022).

This chapter explores how AI is currently used in agriculture and highlights key ethical and social considerations. It also reviews relevant governance frameworks and proposes recommendations for ensuring responsible AI deployment in agricultural contexts.

By clarifying how AI technologies operate and highlighting the ethical and social dimensions that often lie in the shadows, this chapter aims to equip stakeholders with the insights needed to build and regulate AI solutions that truly serve the needs of the future of agriculture.

## 2. Current Applications of AI in Agriculture

AI has emerged as a transformative force in agriculture, reshaping traditional farming methods through advanced data processing, pattern recognition, prediction, and automation. AI solutions are powered by various algorithmic paradigms, including discriminative AI (DAI) for classification and detection tasks, generative AI (GAI) for content creation and communication, and hybrid models that integrate real-time sensor data with machine learning (ML) and deep learning (DL) frameworks (Ampatzidis et al. 2025). DAI models such as Support Vector Machines, Random Forests, and neural networks (NNs) focus on learning decision boundaries to classify input data, making them especially effective for tasks such as plant disease detection, pest identification, and crop-type classification (Yadav et al. 2024, Wang et al. 2025, Ampatzidis et al. 2025). GAI models extend these capabilities and can synthesize complex new data by pretraining on massive, diverse datasets (text, images, sensor streams) and then fine-tuning on agriculture-specific tasks (Bi et al. 2023, De Clercq et al. 2024, Ampatzidis et al. 2025). Large language models (LLMs) are a subset of GAI models that produce human-readable text. LLMs enhanced with retrieval-augmented generation (RAG), are increasingly used to support farmer decision-making by converting agricultural data and publications into actionable recommendations.

These AI models are embedded into a variety of agricultural applications. In this chapter, we identify seven distinct application areas of AI in agriculture, as illustrated in Figure 1. Table 1

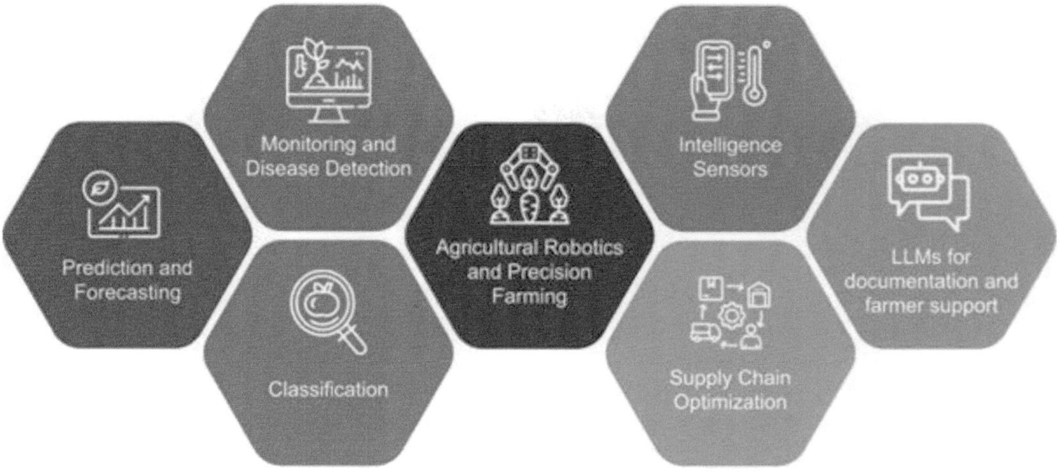


**Figure 1.** Key AI application areas in agriculture.

summarizes current applications of AI in agriculture discussed in the subsequent paragraphs, along with their advantages and limitations.

In prediction and forecasting, AI systems analyze sensor, environmental, and historical data to estimate crop yields, identify droughts, or predict disease outbreaks (Nagar and Nagar 2024). For example, Tamayo-Vera et al. (2025) applied an ensemble of ML models to decades of soil, weather, and NDVI data to forecast potato yields at the postal-code level in eastern Canada. Similarly, Long Short-Term Memory (LSTM)-based neural networks and hybrid convolutional neural network (CNN)-LSTM models have been applied to meteorological data for temperature and wind forecasts (Elshewey et al. 2025). Crop disease and pest forecasting is another growing area: Lee and Yun (2023) trained a deep recurrent model on multi-year environment information of crops (strawberries, peppers, tomatoes, etc.) and predicted risk scores for multiple crop pests and diseases. In practice, platforms like IBM's Watson Decision Platform fuse field sensors and weather forecasts to provide in-season yield predictions and early-warning disease alerts for growers (From Seed to Server: The Evolution of Modern Agriculture, n.d.). These forecasts improve planning (optimizing inputs and labor), reduce waste, and help adapt to climate variability.

Classification-based applications use discriminative models to identify crop diseases, classify plant species, or differentiate between weeds and crops. For example, computer-vision CNNs can grade produce by size, color, or defects. Luo et al. (2024) developed an image-based system to classify the quality of peaches, lemons, apples, and tomatoes into four grades using traditional image processing techniques and a neural network for clustering. Similarly, CNNs trained on leaf images can distinguish dozens of plant diseases with high accuracy, enabling early detection and timely treatment (Mrisho et al. 2020, Gangwar et al. 2024, Shoaib et al. 2023, Adnan and Gölcük 2024, Wang et al. 2025).

Monitoring tasks extend these capabilities by using AI to continuously track field and livestock conditions. Systems integrating satellite images, aerial drone footage, and soil data can generate insights on nutrient deficiencies, moisture stress, or pest infestations in real time, allowing for targeted interventions and reduced chemical usage. Zhu et al. (2024) reviewed the usage of UAV-based deep-learning systems for rapid, accurate, and scalable monitoring of crop health and pest detection. In livestock systems, vision-based AI tools monitor behavior and detect early signs of illness, such as lameness in dairy cows, alerting farmers when an animal's mobility deviates, enabling early intervention (Siachos et al. 2024).

**Table 1.** Current use of AI in agriculture with associated benefits and challenges.

| Application Area | AI Technique Used | Key Technologies | Examples | Impact/Benefit | Challenges | References |
|---|---|---|---|---|---|---|
| Predicting and Forecasting | ML (time-series analysis, regression models) | Satellite imagery, IoT sensors, and weather data | ML-based yield prediction; Sensor-driven irrigation planning; Climate forecast; Pest and crop disease forecast | Optimized planting/ harvest schedules; Proactive risk mitigation | Data quality, model generalization, and model interpretability, multi-source data fusion | Tamayo-Vera et al. 2025 Nagar and Nagar 2024 Elshewey et al. 2025 Lee and Yun 2023 |
| Classification | ML (SVM, Random Forests), DL (CNNs, Vision Transformers) | UAVs, multispectral cameras, hyperspectral imaging | Tomato disease classification; Leaf classification | Accurate diagnosis; Targeted interventions | High computational costs; Need for annotated datasets; Poor generalization to diverse crop types | Mrisho et al. 2020 Adnan and Gölcük 2024 Gangwar et al. 2024 Yadav et al. 2024 |
| Monitoring and Disease Detection | Computer Vision (YOLO, ResNet); DL | Drones, smartphones, IoT probes | Image analysis for leaf health; IoT soil monitoring; Livestock illness monitoring | Timely intervention; Lower chemical use; Improved animal welfare | Sensor calibration: Regional data variation; High cost of imaging systems | Shoaib et al. 2023 Siachos et al. 2024 |
| Intelligent Sensors | IoT, Sensor Fusion; Edge AI | FDR/TDR soil sensors; NPK/pH probes | IBSWS system; EC/NPK/pH sensor-based tools | Optimized irrigation/ fertilization, Reduced environmental impact | Calibration drift, affordability, and sensor durability | Balasooriya et al. 2020, Kim and Park 2021, Yin et al. 2021, Bouhachlaf et al. 2023 |
| Agricultural Robotics & Precision Farming | Reinforcement Learning; IoT; Computer Vision; Edge AI | Autonomous tractors; Harvesting robots | John Deere See & Spray; FRAIL-Bot for fruit harvesting | Labor savings; Reduced resource usage; Precise input application | High initial cost; Adaptability to diverse crops, and complex terrain navigation | Ampatzidis et al. 2025 |
| Supply Chain Optimization | Predictive analytics; NLP | Blockchain; RFID tags; Demand forecasting tools | AI for logistics and demand prediction; Harvest scheduling | Minimized spoilage: Optimized transport; Dynamic pricing/ logistics allocation | Infrastructure gaps; Fragmented data; Tech access for smallholders | Nagar and Nagar 2024, Assimakopoulos et al. 2024 |
| LLMs for Documentation and Support | NLP (GPT, BERT); Chatbots | Mobile apps; Voice assistants | ExtensionBot; KisanGPT; GPT-based legal simplification tools | Reduced admin burden; Better compliance; Aaccess to expert knowledge | Language limitations;Rural internet access; Hallucinations; Contextual accuracy | Ellis 2024, IndiaAI 2023, De Clercq et al. 2024, Ampatzidis et al. 2025 |

Robots equipped with AI algorithms perform tasks such as hoeing, harvesting, and automatic milking with high efficiency. AI-enabled robots can navigate fields, assess fruit ripeness, and perform selective harvesting with minimal supervision. Combined with intelligent sensor networks, these systems support precision agriculture practices, enabling variable rate applications of inputs like water, fertilizer, and pesticides based on site-specific needs (Balasooriya et al. 2020, Kim and Park 2021, Bouhachlaf et al. 2023). For example, research robots like UC Davis's FRAIL-Bot assist human workers in strawberry picking, and industry systems such as John Deere's "See & Spray" use ML-driven vision to spot-spray weeds individually, dramatically reducing herbicide use (Ampatzidis et al. 2025).

In the post-harvest and distribution stages, AI optimizes logistics and supply chains. Forecasting models guide storage and transportation decisions, minimizing spoilage and aligning production with market demand. For instance, AI-powered platforms can dynamically adjust harvesting schedules based on demand projections, helping farmers achieve optimal market timing and pricing (Assimakopoulos et al. 2024, Nagar and Nagar 2024).

Recent advances in GAI and LLMs have further broadened the scope of AI in agriculture by enhancing documentation and advisory services. Notably, recent research demonstrates that GPT-based LLMs can summarize and explain agricultural regulations in plain language, helping farmers understand complex legal requirements (De Clercq et al. 2024, Ampatzidis et al. 2025). One prominent example is ExtensionBot, a domain-specific GAI system developed to support agricultural extension services. Unlike general-purpose LLMs such as ChatGPT, ExtensionBot is trained on a curated dataset of over 360,000 agricultural extension publications and is designed to provide context-specific, science-based answers to practical farming questions (Ellis 2024, Ampatzidis et al. 2025). KissanGPT is another recent digital agronomy assistant that helps Indian farmers with limited literacy skills access personalized farming advice through voice-based interactions (IndiaAI 2023, De Clercq et al. 2024). It supports multiple Indian languages and delivers guidance on crop care, pest control, and market strategies. Although the commercial market for GAI in agriculture is still maturing, the success of initiatives like ExtensionBot and KissanGPT demonstrates the potential of these tools to bridge the gap between knowledge and practice in farming communities.

## 3. Ethical Challenges in Agricultural AI

As AI technologies become increasingly embedded in agricultural systems, from diagnostics and decision support to autonomous machinery and data analytics, they raise a range of ethical questions that extend beyond technical performance. These include issues of fairness, transparency, accountability, sustainability, privacy, misinformation, and cybersecurity. Each of these dimensions carries implications not only for individual farmers but also for food systems, environmental resilience, and rural equity. Figure 2 provides a visual summary of the key ethical challenges addressed in this study, offering a roadmap for the discussion that follows.

### *Bias and Fairness*

Bias in AI refers to systematic errors in decision-making processes that result in unfair or discriminatory outcomes, often arising from unrepresentative training data or flawed algorithm design (Ferrara 2023). Fairness on the other hand, denotes the principle that AI systems should treat all individuals or groups equitably, without preference based on inherent or acquired attributes; fair machine learning aims to avoid such bias in both data and model outputs (Wang et al. 2022).

AI-driven tools in agriculture often reflect the inequalities of their training data and deployment. Models trained on data from large, well-resourced farms can underperform on small-scale farms, effectively excluding subsistence farmers and undercutting local practices (Dara et al. 2022). Similarly, algorithms often carry forward linguistic and cultural biases, making smartphone apps unusable for non-English speakers or areas without the internet.

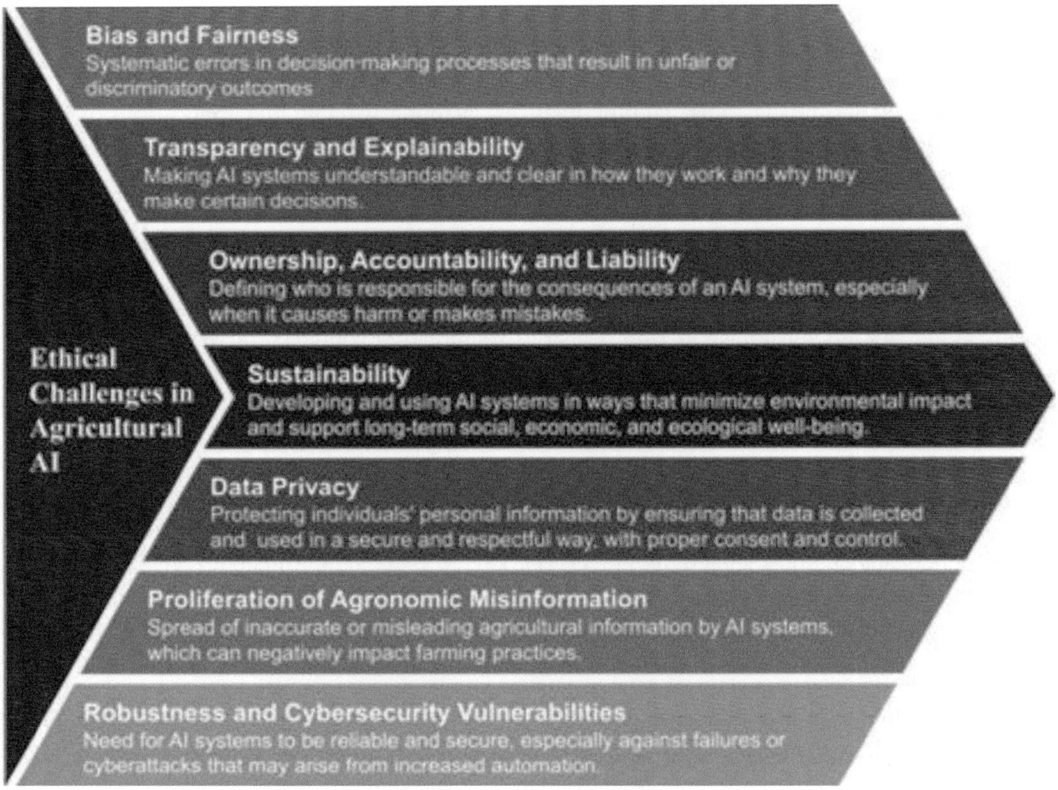


**Figure 2.** Overview of ethical challenges in agricultural AI.

One notable real-world example of algorithm bias is PlantVillage Nuru, a smartphone application that employs artificial intelligence to diagnose cassava diseases by analyzing leaf images. While Nuru achieved moderate accuracy in identifying cassava mosaic disease (CMD) and cassava green mite (CGM) damage (58% and 56% respectively), its performance was notably lower for mild symptoms of cassava brown streak disease (CBSD). This is mainly due to the fact that its training data was heavily skewed toward images with clear, severe symptoms (Mrisho et al. 2020). This highlights how biased datasets can limit diagnostic accuracy and reduce tool reliability for real-world, variable conditions.

### *Transparency and Explainability*

Many AI decision tools operate as opaque "black boxes". Lack of interpretability and understandability of the AI model outcome can undermine transparency, leading farmers to distrust or reject digital tools (Dara et al. 2022, Porfirio 2024). In practice, many agricultural chatbots and sensor systems fail to disclose how they use a farmer's data or come to decisions. In contrast, Digital Green's *Farmer.Chat*, a generative AI-powered chatbot designed to provide smallholder farmers with timely, reliable, and context-specific agricultural advice aimed to solidify trustworthiness by clearly labeling AI-generated advice and using only vetted sources, with guardrails to stop misinformation or dangerous information (such as banned pesticide use) from reaching farmers (Singh et al. 2024).

For responsible deployment, developers should employ explainable AI techniques or interfaces that clarify model logic in farmer-friendly terms (Porfirio 2024). In practice, this involves providing farmers with plain-language model summaries, visual decision trees, or dashboard explanations. Farmers should have the right to ask about the outcome of an AI tool, how the decision was made, what data was used, and which goals are achievable through explainable AI methods (Dara et al.

2022). For example, an expert-driven explainable AI system developed in Europe for detecting multiple agriculture-related hazards uses an explainable XGBoost model together with a probabilistic logistic regression, allowing agronomists to better understand the reasoning behind its outputs (Essenfelder et al. 2025).

### *Ownership, Accountability, and Liability*

It is often unclear who is responsible when AI causes harm. For example, if an automated sprayer dumps excess pesticide and ruins a crop, who will be responsible for the loss of harvest? (Dara et al. 2022). Notably, some companies have even required farmers to waive liability: John Deere made users agree that the company bears no responsibility if there is damage to the farmers' crops, land, profit, machines, or business (Carolan 2019). These practices raise serious ethical concerns about who compensates farmers when digital tools fail. With no clear laws, accountability defaults to power. In many low and middle-income countries, where legal systems are weaker, farmers have little recourse against multinational tech providers. To protect farmers from such risks, robust accountability mechanisms, through enforceable contracts, insurance, or regulation, are essential. International guidelines such as OECD AI principles (AI Principles Overview - OECD.AI n.d.) emphasize that organizations developing, deploying or operating AI should be held accountable for misuse or malfunctions. Without such protections, smallholders remain vulnerable.

### *Sustainability*

While AI can improve resource use efficiency (precision irrigation, targeted inputs), it also poses sustainability risks. Miscalibrated automation could lead to overuse of water or chemicals if models err, potentially polluting soil and waterways (Dara et al. 2022). Vinuesa et al. (2020) caution that rapid AI deployment without oversight could threaten climate goals and biodiversity. Autonomous machines (tractors, drones) add energy and material demands; unchecked robotics might increase soil compaction or emissions. For instance, algorithms that prioritize yield alone might encourage monocultures or overuse of irrigation, undermining soil health (Vinuesa et al. 2020). Ethical deployment requires aligning AI tools with environmental goals (e.g., by optimizing energy use) and monitoring ecological impacts. Importantly, sustainability also means usefulness and affordability for farmers. If an AI tool is too complex or expensive, it fails in rural contexts. It is essential to develop user-centric, reliable, and easy-to-maintain AI systems focusing on smallholder farmers, thus ensuring long-term viability (Ryan 2022).

### *Data Privacy*

AI tools in farming depend on collecting huge amounts of data, from soil moisture and crop images to financial records and personal details, which raises serious privacy concerns. A 2018 American Farm Bureau survey found that 78% of farmers were "concerned" or "extremely concerned" about who could access their agricultural data and how it might be used, highlighting widespread wariness toward data sharing agreements that farmers often neither read nor understand (Janzen 2019). In 2021, an FTC investigator met with Nebraska farmers alleging that John Deere's proprietary software and restrictive contracts prevented them from controlling or repairing their own equipment, illustrating how vendors can leverage data control to lock in users and limit farmer autonomy (*FTC Investigator Meets with Farmers on Deere Data Privacy* 2021). More broadly, the "big data" ecosystem in smart farming exhibits a "privacy paradox"; farmers willingly adopt precision tools yet remain anxious about potential misuse of their data, including unauthorized resale or secondary analytics by tech providers (Amiri-Zarandi et al. 2022).

Technical privacy risks extend beyond contractual issues to the nature of the data itself. Crop-monitoring platforms often collect high-resolution GPS coordinates and socioeconomic metadata; data that, if leaked or sold, could expose land tenure disputes or individual financial hardships (Michler et al. 2022). Furthermore, emerging AI tools like LLM-based agronomy chatbots may coax farmers into revealing trade secrets such as cropping patterns, input usage, and personal demographics with privacy policies (e.g., OpenAI's) permitting the use of user-provided content to train the models that power their services (OpenAI 2023). In parallel, research has demonstrated that earlier LLMs such as GPT-2 can unintentionally reproduce personally identifiable information (phone numbers, email addresses) from their training data, underscoring the risk of model memorization of sensitive personal information (De Clercq et al. 2024).

Implementing privacy by design principles such as data minimization, on-device processing, and explicit opt-in mechanisms can protect farmers' data rights (Gavai et al. 2025).

### *Proliferation of Agronomic Misinformation*

AI's generative capabilities can flood farmers with easily produced but unverified advice or rumors. Simple chatbots or viral social media posts could spread incorrect agronomic information (e.g., false pest treatments, exaggerated climate warnings) under the guise of expertise. These "hallucinations" of AI can have real impacts: a plausible-sounding recommendation to use a banned pesticide could devastate crops or health. Vulnerable rural communities, often lacking strong extension services, are especially at risk if such misinformation circulates unchecked.

In one simulation, individuals with no scientific background used large language models (LLMs) to identify and retrieve information on pathogens that could cause pandemics (Soice et al. 2023). This raises serious questions about the role of LLMs in potentially enabling access to sensitive or dangerous information in domains with high societal impact. In agriculture, this could translate into scenarios where farmers follow seemingly credible but incorrect chatbot-generated recommendations, leading to crop failure or increased reliance on unsustainable practices such as monoculture (De Clercq et al. 2024). In an era where decision-makers increasingly consult AI tools, there is a growing risk that flawed or oversimplified advice from LLMs could influence high-stakes policy, particularly in data-scarce regions.

In addition to accidental errors caused by hallucinations, LLMs may be intentionally exploited to spread agronomic misinformation. Attackers can use prompt injection or "jailbreaking" techniques to bypass safety filters and elicit harmful responses such as instructions for sabotaging farming equipment or misusing agricultural chemicals. This risk is compounded by the threat of "data poisoning," where malicious actors manipulate the training data to embed false information. In the agricultural context, poisoned data could include misleading pesticide recommendations or fake correlations between soil type and crop yield.

To counter these risks, AI systems used in agriculture must integrate rigorous safety mechanisms, including domain-specific guardrails via techniques like retrieval augmented generation (RAG), human-in-the-loop verification, and continuous monitoring for misuse. Without such measures, the widespread adoption of LLMs in agriculture may increase exposure to misinformation with potentially devastating consequences for food security, environmental health, and rural livelihoods.

### *Robustness and Cybersecurity Vulnerabilities*

Smart agriculture platforms combining AI, Internet of Things (IoT) sensors, drones, and edge devices face significant robustness and cybersecurity challenges. Robustness issues arise when AI models misinterpret sensor noise, encounter adversarial inputs, or suffer from model drift under changing environmental conditions. Cybersecurity vulnerabilities in these interconnected platforms often stem from inadequate device authentication schemes, failure to encrypt data-in-transit, and reliance on legacy firmware stacks that lack secure boot or over-the-air (OTA) update mechanisms.

The real-world consequences of these weaknesses are already surfacing. Agricultural drones have been remotely hijacked via unpatched telemetry links, rerouting pesticide application and causing uneven chemical exposures that damage crops (Tychola and Rantos 2025). Ransomware assaults on equipment manufacturers such as the 2022 AGCO breach, which stalled tractor production during the U.S. planting season—further highlight how compromised supply-chain partners can paralyze farming operations (Flowers and Gomes 2022); and attacks on farm-services cooperatives, like the 2021 New Cooperative incident in Iowa, locked out grain elevators and disrupted harvest logistics, threatening regional food distribution networks (Plume and Bing 2021). Beyond operational disruptions, adversaries can manipulate market-forecast inputs, falsifying weather or commodity pricing data to skew AI-driven trading algorithms, exacerbating financial risks for farmer collectives (Gao et al. 2024).

To counter these threats, best practices include rigorous model testing, secure data encryption, and fallback manual controls. Technology providers should simulate attacks to harden systems ("red teaming") and provide farmers with guidelines on secure use.

## 4. Social Implications of AI in Agriculture

From labor dynamics and decision-making autonomy to accessibility, health, and safety, the integration of AI is introducing profound shifts in how agricultural work is organized, who has power, and who stands to benefit or lose. These technologies not only alter the roles and skills required on farms but also affect social equity, linguistic inclusion, and the overall resilience of rural livelihoods. The following sections explore the major social implications of agricultural AI, highlighting the risks that lie ahead.

### *Employment and Labor Dynamics*

The growing capabilities of GAI, particularly LLMs pose notable risks to employment in the agricultural sector. Broadly, advancements in AI are expected to impact around 14% of the global workforce by 2030, with low-skilled jobs like those in agriculture being especially susceptible to automation (Morandini et al. 2023). LLMs could intensify this trend, as they are capable of performing up to 56% of workplace tasks more efficiently, without compromising quality. Language models and robotics could replace roles in advisory services, field scouting, and even some skilled tasks (De Clercq et al. 2024). This raises concerns for rural livelihoods, especially in developing economies where agriculture is a major employer. While new "AI technician" jobs may emerge, there is a risk of exacerbating inequality and youth outmigration if transitions are poorly managed.

Over the past thirty years, the agricultural sector has already seen a loss of nearly 200 million food production jobs. If current developments continue without proactive measures, an additional 120 million jobs may be lost by 2030, mainly in low- and middle-income countries where agricultural employment is a major livelihood source (Brondizio et al. 2023).

At the same time, AI and big data technologies are transforming industries around the world. While these innovations are expected to generate growth and efficiency gains across nearly all sectors, the Agriculture, Forestry, and Fishing industries are seeing relatively lower rates of technological adoption (70%) compared to other fields (World Economic Forum 2025). This indicates a broad, but uneven, integration of advanced tech skills across the global economy. Yet, even in these slower-adopting sectors, there is a growing demand for resilience, flexibility, and agility.

While the impact of generative AI on agriculture is still unfolding, early research shows significant differences across regions. In low-income countries, only about 0.4% of jobs are currently at risk of automation, compared to 5.5% in high-income countries. For occupations in agriculture, forestry, and fishing, up to 8% of tasks are estimated to be at medium to high risk of transformation or replacement due to the growing use of LLMs (Gmyrek et al. 2023).

Moreover, the disruptive potential of these technologies is not limited to manual labor. Other professions, such as those in finance, engineering, and accounting, may also become increasingly vulnerable. Since LLMs can effectively summarize and reproduce publicly available expert knowledge, their use could lower the value placed on specialized skills, even in agricultural domains (Burn-Murdoch 2023, Eloundou et al. 2024).

### *Farmer Autonomy and Decision-Making*

The complex legal language often used by agritech companies and large agribusinesses can leave farmers unaware of the full implications of the agreements they sign when adopting AI technologies (Ryan 2019). As a result, there have been increasing concerns about power imbalances, with some corporations gaining more control over farm operations. For instance, certain companies restrict farmers from repairing or modifying AI-enabled hardware they've purchased, limiting their ability to interact freely with their own equipment, such as tractors or sensors, and effectively reducing their role to that of a dependent user on their own land (Ryan 2019).

Beyond hardware restrictions, farmers are also experiencing a shift in their decision-making power. As AI systems become more embedded in everyday farm management, farmers have begun to internalize the authority of algorithmic recommendations (Gardezi and Stock 2021). These systems, often perceived as offering scientifically validated guidance, can override the farmer's own judgment and experience. In many cases, farmers feel compelled to follow AI-generated advice, either because of the significant financial investment they've made in the technology or due to pressure from the agribusinesses that supply it (Ryan 2020).

The increasing reliance on AI for decision-making can also impact the perceived value of farmers' knowledge. When algorithms act as advisors but lack transparency, farmers may find it difficult to challenge or even understand the rationale behind certain recommendations. This dynamic can reduce their autonomy and deepen inequalities between those who can critically engage with AI systems and those who cannot (Sustainability Directory 2025). To avoid disempowering farmers, AI tools must be designed as supportive aids, not authoritative figures, ensuring that farmers retain the final say in how their farms are managed.

### *Accessibility and Equity*

The integration of AI into agriculture, risks deepening existing socio-economic gaps, particularly between large-scale commercial farms and smallholder farmers. Advanced AI tools are often expensive and require supporting infrastructure, factors that favor wealthier farms and regions. Farmers in low-income areas may lack both the financial resources and digital literacy needed to adopt these tools effectively, preventing them from benefiting equally (De Clercq et al. 2024). To avoid widening this divide, experts recommend making AI systems more accessible through models like "robots-as-a-service," shared data platforms, and targeted training programs for marginalized communities. Ensuring inclusive access is essential to promoting fairness and equity in the global food system.

The emergence of LLMs could further intensify this digital inequality (Sheldrick et al. 2023). In regions with limited connectivity, poor infrastructure, or inadequate digital education, farmers may find themselves excluded from the benefits that LLMs offer. Beyond access issues, there's a risk that these tools could misrepresent or overlook the specific needs of small-scale or non-Western farmers. If trained primarily on data from industrialized agricultural systems, LLMs may struggle to account for diverse soil types, local crop varieties, and weather conditions typical of less-documented regions. These models can produce seemingly logical outputs while actually reflecting biased or incomplete data—a concern echoed across other sectors where algorithmic recommendations have shown hidden racial or geographic biases (Turpin et al. 2023).

The consequences of applying inappropriate or generic AI advice can be serious. Research shows that when farming technologies don't align with local conditions, it not only hinders productivity but can actually worsen global disparities in agricultural output. For smallholder farmers, inaccurate or generalized advice from AI systems may reinforce inefficiencies rather than solve them.

Language is another major barrier. Many LLMs are trained predominantly on English-language data, making them less effective at generating useful outputs in other languages or dialects. This poses challenges for non-English-speaking farmers who may be excluded from meaningful engagement with these systems. Studies have highlighted significant disparities in how LLMs perform across different languages, with dominant global languages typically seeing better results (Nguyen et al. 2023).

Governments and research groups have started to respond to this challenge. For instance, the SEA-LION project led by Singapore is developing a multilingual language model trained on Southeast Asian languages, including Vietnamese and Thai, to better reflect regional linguistic and cultural contexts (SCM 2024). Other researchers are pushing the boundaries of multilingual AI, training models on extremely low-resource languages like Kalamang using only a single grammar book (Tanzer et al. 2023).

Bias in LLMs is not limited to agriculture. In healthcare, for example, models have shown tendencies to associate diseases with specific racial groups, overestimate survival rates in difficult scenarios, and assign longer hospital stays or higher costs based on race (Yang et al. 2024). These cross-sector insights can help agricultural policy-makers anticipate similar risks and design interventions to minimize bias and improve fairness in AI-powered farming systems.

### *Health and Safety*

When deploying AI technologies on farms, ensuring the health and safety of workers is critical. Agriculture is already one of the most dangerous industries, with a worker fatality rate estimated to be 25–78% higher than in many other sectors (Kavanagh et al. 2023). AI has the potential to make farm work safer by minimizing human exposure to hazardous substances like pesticides and other toxic chemicals, which can otherwise have serious health impacts (Rodzalan et al. 2020). Moreover, automation could help reduce the risk of injuries. Since many accidents during interactions between humans and machines are caused by human error, increasing automation could reduce such incidents (Vasconez et al. 2019). Advocates of autonomous machinery argue that these systems could make farming environments significantly safer (Ryan 2019).

However, increased automation introduces new safety risks as well. As farms begin to use driverless tractors, drones, and AI-powered sprayers, it becomes essential to train workers on how to interact safely with this equipment. Malfunctions, such as pesticide sprayers misfiring, could pose new exposure risks. Maintaining safety may require updated regulations, real-time system monitoring, and built-in emergency features to protect workers operating alongside AI systems.

The risks are particularly complex given that agricultural AI operates within dynamic environments involving animals, ecosystems, and unpredictable natural factors. Ensuring that these systems do not cause unintended harm remains a significant challenge (Ryan 2022). Some farmers have voiced concerns that increased use of AI and robotics may lead to riskier farming practices, especially regarding environmental harm, due to reduced human presence and accountability (Stock and Gardezi 2021).

Another growing concern is cybersecurity. As farms become more reliant on interconnected automated systems, they may also become more vulnerable to cyber threats such as hacking, sabotage, or corporate espionage. These threats could compromise not only productivity but also safety (Carolan 2019). Moving forward, integrating robust safety protocols and security measures will be crucial to the responsible adoption of AI in agriculture.

## 5. Regulatory and Policy Considerations

The rapid deployment of AI in agriculture is outpacing existing regulatory frameworks, raising complex legal and ethical questions. Current legislation covering areas such as pesticide use, water rights, data privacy, and labor often fails to anticipate the implications of AI-enabled practices. For example, liability remains unclear in cases where a faulty algorithm causes crop losses or harms soil health. Coordination among agricultural, technological, and legal domains is essential to close these gaps. Without cross-sector collaboration, existing disparities could worsen, especially in smallholder-dominated regions (Ryan et al. 2024).

### *Existing Frameworks Governing AI in Agriculture*

Several high-level frameworks provide ethical guidance on AI deployment in agriculture. The OECD AI Principles (AI Principles Overview - OECD.AI n.d.) and UNESCO's Recommendation on the Ethics of AI (2021) emphasize transparency, accountability, human oversight, and fairness. These foundational documents apply across sectors and are particularly relevant to agriculture, where trust and explainability are essential for adoption (UNESCO 2021).

At the technical level, standards such as IEEE P7010-2020 (impact of AI on well-being) and ISO/IEC TR 5469:2024 (functional safety of AI systems) aim to embed safety and ethical safeguards into AI system design (Ampatzidis et al. 2025). These efforts support value-sensitive design approaches that integrate fairness, privacy, and transparency into the system architecture (Dara et al. 2022).

Governments have also begun to formalize regulations specific to AI in food systems. The EU's proposed AI Act (2023) designates high-risk applications, including autonomous agricultural machinery, and sets stringent standards for risk assessment and documentation. The EU Data Act (2023), applicable from September 2025, establishes data-sharing rules for IoT devices, which include farm sensors and equipment (Ryan et al. 2024). Earlier voluntary measures, such as the EU Code of Conduct on Agricultural Data Sharing (2018), are now being reassessed for alignment with these emerging laws. In the United States, state-level data privacy acts like the California Consumer Privacy Act (2018) and Iowa's Ag Data Privacy Act (2020) recognize farm data as sensitive (Ryan et al. 2024). Many countries, especially in the Global South, lack comprehensive regulations. As a result, agricultural AI remains governed by fragmented laws across environmental, labor, and commercial domains, creating legal uncertainty for developers and users alike.

Industry groups and non-governmental organizations are also contributing to ethical AI governance. The AgGateway consortium maintains the Agricultural Data Application Programming Toolkit (ADAPT) to improve interoperability of farm data formats (AgGateway 2019, Ampatzidis et al. 2025). The Global Open Data for Agriculture and Nutrition (GODAN) initiative champions open data standards, advocating equitable data access for all stakeholders (GODAN n.d.). While many of these frameworks are not legally binding, they significantly contribute to shaping norms for the responsible use of technology, especially in regions where formal regulations are still developing.

### *Challenges in Policy Implementation*

Despite a growing ecosystem of ethical frameworks, translating principles into enforceable regulations remains challenging. Regulatory timelines often lag behind innovation; e.g., the EU Data Act will not take effect until 2025, leaving farmers and developers in a legal gray zone (EU Data Act 2023). Contracts governing AI use and data sharing are often complex, favoring large agritech firms over smallholder farmers (Carbonell 2016).

Moreover, agricultural stakeholders are diverse, with conflicting priorities between small and large-scale producers, public and private sectors, and local and multinational interests. As a result, trust in AI-integrated systems remains low, where farmers may resist adopting new policies or technologies without clear incentives (Dara et al. 2022). Ensuring compliance across jurisdictions, especially where data and equipment cross borders, will require adaptive, multilevel governance that includes farmers, policymakers, researchers, and private developers alike (Ampatzidis et al. 2025).

## 6. Recommendations for Ethical and Socially Sustainable Agricultural AI

### *For Developers and Technology Providers*

*Co-design with farmers and extension agents*: Engage farmers early through pilot projects and demonstration sites to ensure solutions address real needs and local practices.

*Implement explainable AI (XAI)*: Integrate techniques such as SHAP, LIME, or counterfactual explanations so users can see why a nutrient-management or pest-control recommendation was made.

*Optimize for sustainability metrics*: Train models not just to maximize yield but to minimize water use, fertilizer inputs, and carbon footprint, leveraging multi-objective optimization frameworks.

*Mitigate data biases*: Audit datasets to include diverse geographies, farm sizes, organic and conventional practices, and socio-economic contexts to prevent skewed recommendations.

*Robust field-testing under varied conditions*: Collaborate with extension services to validate algorithms across soil types, climates, and cropping systems, and publish third-party reliability audits.

*Local-language and low-bandwidth support*: Deploy lightweight mobile/web apps with offline modes and multilingual interfaces, following successful pilots like Farmer. Chat in four countries (over 15,000 engaged farmers) (Singh et al. 2024).

*Embed feedback and compensation loops*: Build mechanisms for farmers to flag poor recommendations and receive transparent redress, update models rapidly, and compensate documented losses.

### *For Policymakers and Regulators*

*Define clear liability frameworks*: Establish who bears financial responsibility when AI-driven equipment or algorithms cause crop or environmental damage, mandating insurance or compensation funds.

*Mandate farmer data rights*: Require consent-based data collection, ownership, portability, and the right to delete or correct farm data, modeled on GDPR and the EU Data Act (effective 2025).

*Incentivize 'green AI' innovations*: Offer grants, tax credits, or "sustainability labels" for AI tools that demonstrably reduce the environmental footprint (e.g., reduce water use or cut fertilizer and pesticide inputs).

*Invest in rural digital infrastructure*: Expand broadband/5G connectivity in farming regions through public–private partnerships to close the digital divide and enable real-time sensor networks.

*Fund farmer education and extension*: Scale up training programs on AI literacy and critical evaluation of algorithmic advice, integrating them into existing agricultural extension services.

*Require ethical impact assessments*: Before deployment, AI tools should undergo ethics reviews covering fairness, transparency, environmental impact, and farmer welfare, similar to human-AI interaction assessments in other sectors.

*Support open data and interoperability standards*: Endorse initiatives like AgGateway ADAPT and GODAN to foster data-sharing protocols that prevent vendor lock-in and promote innovation.

### *For Farmers and End Users*

*Advocate for transparency*: Demand clear information about what data are collected, how they are used, and the risks of AI systems.

*Participate in co-creation*: Engage with pilot projects, extension networks, and open-source platforms to shape tool development.

*Build digital and AI literacy*: Leverage training programs and peer networks to better understand and manage AI technologies on the farm.

*Collaborate in cooperatives or collectives*: Pool resources and data-sharing agreements to negotiate better service terms with AI providers and share risk across members.

*Adopt iterative evaluation*: Keep records of AI recommendations versus outcomes (yields, input use, costs) to jointly evaluate tools with providers and regulators.

To mitigate risks, a multi-pronged approach is needed.

First, involve farmers and communities in AI deployment – participatory design and feedback can catch issues early.

Second, adopt iterative risk assessments: use ethical impact frameworks to evaluate new AI tools at each stage.

Third, build in technical safeguards: for example, adversarial testing to harden models against malicious inputs, and fallback defaults if AI confidence is low.

Fourth, promote digital literacy: educate stakeholders on AI's limitations and uncertainties.

Finally, encourage diversity in development teams and data sources to counteract bias.

By combining technical rigor (robust algorithms, cybersecurity) with social measures (transparency, training, legal protections), the sector can reduce unintended harms.

## 7. Future Directions

Recent developments in agricultural AI highlight the increasing use of multi-modal sensing, digital twins, Robotics-as-a-Service (RaaS), blockchain traceability, and climate-smart AI alongside traditional machine learning, robotics, and data analytics to improve efficiency, sustainability, and adaptability in farming (Milella et al. 2024, Nagar and Nagar 2024, Gideon et al. 2025, Awais et al. 2025, Hoyle 2025). Multi-modal sensing platforms now fuse high-resolution imagery, soil genomics, weather station data, and even biometric signals from livestock to create richer input streams for predictive models (Toledo et al. 2024). Meanwhile, virtual digital twins of entire farms powered by IoT sensor networks and AI allow growers to simulate weather events, irrigation strategies, and planting schedules before making real-world changes (Awais et al. 2025). The rise of RaaS is lowering barriers to automation by enabling lease-based access to autonomous tractors, drones, and harvesters, spreading costs and making advanced machinery affordable for smaller farms (Milella et al. 2024, Carolan 2019, Vasconez et al. 2019). Integrating blockchain with AI further enhances supply-chain transparency, enabling immutable tracing of produce from field to fork (Nagar and Nagar 2024). On the other hand, climate-smart AI models will gain more importance to recommend crop varieties and management practices resilient to extreme weather patterns, bolstering food security under climate change (Gideon et al. 2025).

Explainable AI (XAI) is becoming a key trend in agricultural technology, focusing on making machine learning decisions transparent and understandable for stakeholders such as farmers, agronomists, and policymakers. It supports trust, accountability, and more informed decision-making in AI-assisted agricultural systems. The recent research like, the AgroXAI systems, exemplify the use of explainable AI (XAI) in agriculture, providing crop recommendations based on soil and weather data. It employs techniques like SHAP and LIME to offer transparent decision-making, aiding farmers in understanding and trusting AI-driven suggestions (Turgut et al. 2024).

With rising population growth and increasing resource limitations, sustainable vertical farming is gaining attention as a promising solution. In these controlled environments, AI plays a key role in automating processes and optimizing resource use. Studies show that AI can boost crop yields and reduce environmental impact, helping tackle issues related to climate change and urban expansion (Chowdhury et al. 2023).

AI is increasingly being used to monitor livestock health, allowing for early detection of diseases and improving animal welfare (Henrietta 2024, Priyadarshini et al. 2024). Automation tools such as robotic feeders and milking systems are streamlining routine tasks in animal husbandry, leading to higher efficiency and lower labor costs (Priyadarshini et al. 2024). Computer vision systems are being applied to observe animal behavior and assess health indicators, including innovative methods like facial recognition for individual livestock. These technologies support better decision-making and more precise interventions (Neethirajan 2017).

One of the most transformative emerging trends in agriculture is AI-driven gene editing, particularly through CRISPR-Cas systems, which enhances crop resilience against climate change-induced stresses like drought, heat, salinity, and pathogens, bolstering food security. AI optimizes CRISPR target selection using machine learning and deep learning to identify precise guide RNAs (gRNAs), minimizing off-target effects for traits such as drought tolerance in rice and wheat (Rajpal et al. 2023). Predictive modeling identifies candidate genes for stress tolerance, as seen in nitrogen uptake in corn (Ahmad et al. 2021), while multiplex editing enables simultaneous modification of multiple genes for complex traits like combined drought and disease resistance in maize (Rai et al. 2025). AI integrates with remote sensing to prioritize climate-smart traits, such as flood-resistant rice (Karavolias et al. 2021), and accelerates breeding programs, exemplified by carbon-sequestering rice varieties (Rajpal et al. 2023). Despite challenges like high costs and regulatory barriers, particularly for smallholder farmers in Africa (Ndudzo et al. 2024), AI-driven CRISPR pipelines are advancing sustainable agriculture by improving precision, efficiency, and crop resilience.

## 8. Conclusion

AI holds great promise for twenty-first-century agriculture, from precision farming to smart advisory tools, but its impact depends on how responsibly it is used. To ensure AI serves everyone fairly, innovation must be guided by strong ethical principles like transparency, accountability, and inclusion. Realizing this vision will require close collaboration among farmers, tech developers, policymakers, and civil society. These stakeholders must act now by updating regulations, supporting farmer education, and investing in ethical research to shape AI into a tool for empowerment, not exclusion. By putting ethics and equity at the center, agriculture can unlock AI's benefits while protecting social and environmental values.

## References

Adnan, A. and Gölcük, A. G. (2024). Classification of tomato diseases using deep learning method. Journal of Intelligent Systems and Internet of Things 14(2): 213–228. https://doi.org/10.54216/jisiot.140217.

AgGateway. (2019). AgGateway case studies: Central Farm Service uses ADAPT to streamline processes in agronomy and operations. https://aggateway.org/Portals/1010/WebSite/About%20Us/CASE%20STUDY%20-%20Central%20Farm%20Services.pdf?ver=2019-02-14-034653-433. [Accessed April 3, 2025].

Ahmad, S., Tang, L., Shahzad, R., Mawia, A. M., Rao, G. S., Jamil, S. et al. (2021). CRISPR-based crop improvements: A way forward to achieve zero hunger. Journal of Agricultural and Food Chemistry 69(30): 8307–8323. https://doi.org/10.1021/acs.jafc.1c02653.

AI Principles Overview - OECD.AI. (n.d.). https://oecd.ai/en/ai-principles.

Amiri-Zarandi, M., Dara, R. A., Duncan, E. and Fraser, E. D. G. (2022). Big data privacy in smart Farming: a review. Sustainability 14(15): 9120. https://doi.org/10.3390/su14159120.

Ampatzidis, Y., Bhandari, M., Ferreyra, A., Gentimis, T., McReynolds, E., Murray, S. C. et al. (2025). AI in Agriculture: Opportunities, Challenges, and Recommendations. https://doi.org/10.62300/iaag042514.

Assimakopoulos, F., Vassilakis, C., Margaris, D., Kotis, K. and Spiliotopoulos, D. (2024). Artificial intelligence tools for the agriculture value chain: status and prospects. Electronics 13(22): 4362. https://doi.org/10.3390/electronics13224362.

Awais, M., Wang, X., Hussain, S., Aziz, F. and Mahmood, M. Q. (2025). Advancing precision agriculture through digital twins and smart farming technologies: a review. AgriEngineering 7(5): 137. https://doi.org/10.3390/agriengineering7050137.

Balasooriya, T. N., Mantri, P. and Suriyampola, P. (2020). IoT-based smart watering system towards improving the efficiency of agricultural irrigation. The Internet of Things 1–7: 1–7. https://doi.org/10.1109/gcaiot51063.2020.9345902.

Bi, L., Wally, O., Hu, G., Tenuta, A. U., Kandel, Y. R. and Mueller, D. S. (2023). A transformer-based approach for early prediction of soybean yield using time-series images. Frontiers in Plant Science 14. https://doi.org/10.3389/fpls.2023.1173036.

Bouhachlaf, L., Benslimane, O. and Hajjaji, S. E. (2023). Monitoring soil elements for irrigation management using Internet of Things (IoT) sensors. World Water Policy 9(4): 756–766. https://doi.org/10.1002/wwp2.12151.

Brondizio, E. S., Giroux, S. A., Valliant, J. C. D., Blekking, J., Dickinson, S. and Henschel, B. (2023). Millions of jobs in food production are disappearing—a change in mindset would help to keep them. Nature 620(7972): 33–36. https://doi.org/10.1038/d41586-023-02447-2.

Burn-Murdoch, J. (2023). Here's what we know about generative AI's impact on white-collar work. Financial Times. https://www.ft.com/content/b2928076-5c52-43e9-8872-08fda2aa2fcf.

Carbonell, I. M. (2016). The ethics of big data in big agriculture. Internet Policy Review 5(1). https://doi.org/10.14763/2016.1.405.

Carolan, M. (2019). Automated agrifood futures: robotics, labor and the distributive politics of digital agriculture. The Journal of Peasant Studies 47(1): 184–207. https://doi.org/10.1080/03066150.2019.1584189.

Chowdhury, H., Argha, D. B. P. and Ahmed, M. A. (2023). Artificial intelligence in sustainable vertical farming. arXiv (Cornell University). https://doi.org/10.48550/arxiv.2312.00030.

Dara, R., Fard, S. M. H. and Kaur, J. (2022). Recommendations for ethical and responsible use of artificial intelligence in digital agriculture. Frontiers in Artificial Intelligence 5. https://doi.org/10.3389/frai.2022.884192.

Data Act. (n.d.). Shaping Europe's Digital Future. https://digital-strategy.ec.europa.eu/en/policies/data-act.

De Clercq, D., Nehring, E., Mayne, H. and Mahdi, A. (2024). Large language models can help boost food production, but be mindful of their risks. Frontiers in Artificial Intelligence 7. https://doi.org/10.3389/frai.2024.1326153.

Dibbern, T., Romani, L. a. S. and Massruhá, S. M. F. S. (2024). Main drivers and barriers to the adoption of Digital Agriculture technologies. Smart Agricultural Technology 8: 100459. https://doi.org/10.1016/j.atech.2024.100459.

Ellis, G. (2024, May 2). OSU developing AI chatbot to answer local Extension questions. Oklahoma State University. https://extension.okstate.edu/articles/2024/extension-bot.html [Accessed 19 May 2025].

Eloundou, T., Manning, S., Mishkin, P. and Rock, D. (2024). GPTs are GPTs: Labor market impact potential of LLMs. Science 384(6702): 1306–1308. https://doi.org/10.1126/science.adj0998.

Elshewey, A. M., Jamjoom, M. M. and Alkhammash, E. H. (2025). An enhanced CNN with ResNet50 and LSTM deep learning forecasting model for climate change decision making. Scientific Reports 15(1). https://doi.org/10.1038/s41598-025-97401-9.

Essenfelder, A. H., Toreti, A. and Seguini, L. (2025). Expert-driven explainable artificial intelligence models can detect multiple climate hazards relevant for agriculture. Communications Earth & Environment 6(1). https://doi.org/10.1038/s43247-024-01987-3.

EU Code of conduct on agricultural data sharing by contractual agreement. (n.d.). EU Code of conduct on agricultural data sharing by contractual agreement. https://fefac.eu/wp-content/uploads/2020/07/eu_code_of_conduct_on_agricultural_data_sharing-1.pdf.

Farm Profits and Adoption of Precision Agriculture | Economic Research Service. (n.d.). https://www.ers.usda.gov/publications/pub-details?pubid=80325.

Ferrara, E. (2023). Fairness and Bias in Artificial intelligence: A brief survey of sources, impacts, and mitigation strategies. Sci. 6(1): 3. https://doi.org/10.3390/sci6010003.

Fiocco, D., Ganesan, V., De La Serrana Lozano, M. G. and Sharifi, H. (2023, February 7). Agtech: Breaking down the farmer adoption dilemma. McKinsey & Company. https://www.mckinsey.com/industries/agriculture/our-insights/agtech-breaking-down-the-farmer-adoption-dilemma [Accessed 02 February 2025].

Flowers, B. and Gomes, N. (2022, May 7). AGCO ransomware attack disrupts tractor sales during U.S. planting season. Reuters. https://www.reuters.com/business/agco-says-some-production-facilities-hit-by-ransomware-attack-2022-05-06/ [Accessed 27 Mar. 2023].

From Seed to Server: The evolution of Modern agriculture. (n.d.). IBM Newsroom. https://newsroom.ibm.com/IBM-watson?item=30660.

FTC investigator meets with farmers on deere data privacy. (2021, September 20). https://communicationsdaily.com/news/2021/09/20/ftc-investigator-meets-with-farmers-on-deere-data-privacy-2109170033 [Accessed 16 May 2025].

Gangwar, A., Rani, G., Dhaka, V. S. and Khandelwal, S. (2024). Selecting a Base Deep Learning Model for Tomato Leaf Disease Classification. 1–6. https://doi.org/10.1109/etncc63262.2024.10767572.

Gao, Y., Camtepe, S. A., Sultan, N. H., Bui, H. T., Mahboubi, A., Aboutorab, H. et al. (2024). Security threats to agricultural artificial intelligence: Position and perspective. Computers and Electronics in Agriculture 227: 109557. https://doi.org/10.1016/j.compag.2024.109557.

Gardezi, M. and Stock, R. (2021). Growing algorithmic governmentality: Interrogating the social construction of trust in precision agriculture. Journal of Rural Studies 84: 1–11. https://doi.org/10.1016/j.jrurstud.2021.03.004.

Gavai, A. K., Bouzembrak, Y., Xhani, D., Sedrakyan, G., Meuwissen, M. P., Souza, R. G. et al. (2025). Agricultural data privacy: emerging platforms & strategies. Food and Humanity 100542. https://doi.org/10.1016/j.foohum.2025.100542.

Gideon Sadikiel Mmbando. (2025). Harnessing artificial intelligence and remote sensing in climate-smart agriculture: the current strategies needed for enhancing global food security. Cogent Food & Agriculture 11(1). doi:https://doi.org/10.1080/23311932.2025.2454354.

Gmyrek, P., Berg, J. and Bescond, D. (2023). Generative AI and jobs: A global analysis of potential effects on job quantity and quality, ILO Working Paper 96 (Geneva, ILO). https://doi.org/10.54394/ FHEM8239.

GODAN. (n.d.). GODAN. [online] Available at: https://www.godan.info/.

Haenlein, M. and Kaplan, A. (2019). A Brief History of artificial intelligence: on the past, present, and future of artificial intelligence. California Management Review 61(4): 5–14. https://doi.org/10.1177/0008125619864925.

Henrietta, H. M. (2024). Artificial intelligence in agriculture: a review of current applications and future trends. Futuristic Trends in Agriculture Engineering & Food Sciences 3 Book 11: 1–6. https://doi.org/10.58532/v3bcag11p1ch1.

Hoyle, R. (2025). Exclusive | Bonsai Robotics Raises $15 Million to Help Combat Farm Labor Shortage. [online] WSJ. Available at: https://www.wsj.com/articles/bonsai-robotics-raises-15-million-to-help-combat-farm-labor-shortage-8556caf5 [Accessed 25 April 2025].

IndiaAI. (2023). How KissanGPT helps Indian farmers earn profit. [online] Available at: https://indiaai.gov.in/article/how-kissangpt-helps-indian-farmers-earn-profit [Accessed 1 May 2025].

Janzen, Todd. (2019). Janzen Schroeder Ag Law. Janzen Schroeder Ag Law, 21 Jan. 2019, www.aglaw.us/janzenaglaw/2019/1/3/farmers-care-about-data.

Jobin, A., Ienca, M. and Vayena, E. (2019). The global landscape of AI ethics guidelines. Nature Machine Intelligence 1(9): 389–399. https://doi.org/10.1038/s42256-019-0088-2.

Karavolias, N. G., Horner, W., Abugu, M. N. and Evanega, S. N. (2021). Application of gene editing for climate change in agriculture. Frontiers in Sustainable Food Systems 5. https://doi.org/10.3389/fsufs.2021.685801.

Kavanagh, M., Lane, M. and Beveridge, J. (2023, September 6). Agriculture safety risks grow as it meets AI. Lexology. https://www.lexology.com/library/detail.aspx?g=b2067f35-c299-4b1b-81af-e01246bdd671.

Kim, H. N. and Park, J. H. (2021). Research trends using soil sensors for precise nutrient and water management in soil for smart farm. Korean Journal of Soil Science and Fertilizer 54(3): 366–382. https://doi.org/10.7745/kjssf.2021.54.3.366.

Lee, S. and Yun, C. M. (2023). A deep learning model for predicting risks of crop pests and diseases from sequential environmental data. Plant Methods 19(1). https://doi.org/10.1186/s13007-023-01122-x.

Luo, D., Luo, R., Cheng, J. and Liu, X. (2024). Quality detection and grading of peach fruit based on image processing method and neural networks in agricultural industry. Frontiers in Plant Science 15. https://doi.org/10.3389/fpls.2024.1415095.

Michler, J. D., Josephson, A., Kilic, T. and Murray, S. (2022). Privacy protection, measurement error, and the integration of remote sensing and socioeconomic survey data. arXiv (Cornell University). https://doi.org/10.48550/arxiv.2202.05220.

Milella, A., Rilling, S., Rana, A., Galati, R., Petitti, A., Hoffmann, M. et al. (2024). Robot-as-a-Service as a new paradigm in precision farming. IEEE Access 12: 47942–47949. https://doi.org/10.1109/access.2024.3381511.

Mmbando, G. S. (2025). Harnessing artificial intelligence and remote sensing in climate-smart agriculture: the current strategies needed for enhancing global food security. Cogent Food & Agriculture 11(1). https://doi.org/10.1080/23311932.2025.2454354.

Morandini, S., Fraboni, F., De Angelis, M., Puzzo, G., Giusino, D. and Pietrantoni, L. (2023). The impact of artificial intelligence on workers' skills: upskilling and reskilling in organisations. Informing Science the International Journal of an Emerging Transdiscipline 26: 039–068. https://doi.org/10.28945/5078.

Mrisho, L. M., Mbilinyi, N. A., Ndalahwa, M., Ramcharan, A. M., Kehs, A. K., McCloskey, P. C. et al. (2020). Accuracy of a smartphone-based object detection model, PlantVillage NURU, in identifying the foliar symptoms of the viral diseases of Cassava–CMD and CBSD. Frontiers in Plant Science 11. https://doi.org/10.3389/fpls.2020.590889.

Nagar, S. and Nagar, D. (2024). Block chain and AI for agriculture. *In*: Futuristic trends in blockchain applications. IIP Series, Volume 3, Book 1, Part 8, Chapter 4. [e-book] e-ISBN: 978-93-6252-632-8. Poddar International College.

National Defense Authorization Act for Fiscal Year 2019, Pub. L. No. 115-232, Section 238(g), 132 Stat. 1636 (2018). Available at: https://www.congress.gov/bill/115th-congress/house-bill/5515/text [Accessed 14 Jan 2025].

Ndudzo, A., Makuvise, A. S., Moyo, S. and Bobo, E. D. (2024). CRISPR-Cas9 genome editing in crop breeding for climate change resilience: Implications for smallholder farmers in Africa. Journal of Agriculture and Food Research 16: 101132. https://doi.org/10.1016/j.jafr.2024.101132.

Neethirajan, S. (2017). Recent advances in wearable sensors for animal health management. Sensing and Bio-Sensing Research 12: 15–29. doi:10.1016/j.sbsr.2016.11.004.

Nguyen, X.-P., Zhang, W., Li, X., Aljunied, M., Tan, Q., Cheng, L. et al. (2023). SeaLLMs-large language models for Southeast Asia. arXiv[preprint] arXiv.2312.00738. doi: 10.48550/arXiv.2312.00738.

Nowak, B. (2021). Precision agriculture: where do we stand? A review of the adoption of precision agriculture technologies on field crops farms in developed countries. Agricultural Research 10(4): 515–522. https://doi.org/10.1007/s40003-021-00539-x.

Plume, K. and Bing, C. (2021). Iowa farm services firm: systems offline due to cybersecurity incident. Reuters. [online] 21 Sep. Available at: https://www.reuters.com/technology/iowa-farm-services-company-reports-cybersecurity-incident-2021-09-20/.

Porfírio, R. P. (2024). Studying the Impact of Explainable AI in Digital Agriculture Solutions. pp. 43–46. doi:https://doi.org/10.1145/3678884.3682046.

Priyadarshini, B., Subhadarshini, S., Pattnayak, A. and Nayak, S. (2024). Artificial Intelligence in Agriculture (pp. 59–71). https://doi.org/10.58532/v3bcag24ch6.

Rai, M. N., Rhodes, B., Jinga, S., Kanchupati, P., Ross, E., Carlson, S. R. et al. (2025). Efficient mutagenesis and genotyping of maize inbreds using biolistics, multiplex CRISPR/Cas9 editing, and Indel-Selective PCR. Plant Methods 21(1). https://doi.org/10.1186/s13007-025-01365-w.

Rai, N., Zhang, Y., Ram, B. G., Schumacher, L., Yellavajjala, R. K., Bajwa, S. et al. (2023). Applications of deep learning in precision weed management: A review. Computers and Electronics in Agriculture 206: 107698. https://doi.org/10.1016/j.compag.2023.107698.

Rajpal, V. R., Sehgal, D., Valluru, R. and Singh, S. (2023). Editorial: Current advances in genomics and gene editing tools for crop improvement in a changing climate scenario. Frontiers in Genetics 14. https://doi.org/10.3389/fgene.2023.1214679.

Rodzalan, S. A., Ong, G. Y. and Mohd Noor, N. N. (2020). A foresight study of artificial intelligence in the agriculture sector in Malaysia. International Journal of Advanced Science and Technology 29(6): 447–462.

Ryan, M. (2019). Ethics of using AI and big data in agriculture: the case of a large agriculture multinational. ORBIT Journal 2(2).

Ryan, M. (2020). Agricultural big data analytics and the ethics of power. Journal of Agricultural and Environmental Ethics 33: 49–69.

Ryan, M. (2022). The social and ethical impacts of artificial intelligence in agriculture: mapping the agricultural AI literature. AI & SOCIETY, [online] 38. doi:https://doi.org/10.1007/s00146-021-01377-9.

Ryan, M., Atik, C., Rijswijk, K., Marc-Jeroen Bogaardt, Maes, E. and Deroo, E. (2024). The future of agricultural data-sharing policy in Europe: stakeholder insights on the EU Code of Conduct. Humanities and Social Sciences Communications [online] 11(1). doi:https://doi.org/10.1057/s41599-024-03710-1.

Sheldrick, S., Chang, S., Kurnia, S. and McKay, D. (2023). Across the Great Digital Divide: Investigating the Impact of AI on Rural SMEs. Atlanta, US: AIS Electronic Library (AISeL).

Shoaib, M., Shah, B., Ei-Sappagh, S., Ali, A., Ullah, A., Alenezi, F. et al. (2023). An advanced deep learning models-based plant disease detection: A review of recent research. Frontiers in Plant Science 14. https://doi.org/10.3389/fpls.2023.1158933.

Siachos, N., Neary, J. M., Smith, R. F. and Oikonomou, G. (2024). Automated dairy cattle lameness detection utilizing the power of artificial intelligence; current status quo and future research opportunities. The Veterinary Journal 304: 106091. https://doi.org/10.1016/j.tvjl.2024.106091.

Singh, N., Wang'ombe, J., Okanga, N., Zelenska, T., Repishti, J., K, J. G. et al. (2024). Farmer.Chat: Scaling AI-Powered Agricultural Services for Smallholder farmers. arXiv (Cornell University). https://doi.org/10.48550/arxiv.2409.08916.

Soice, E. H., Rocha, R., Cordova, K., Specter, M. and Esvelt, K. M. (2023). Can large language models democratize access to dual-use biotechnology? arXiv (Cornell University). https://doi.org/10.48550/arxiv.2306.03809.

Stock, R. and Gardezi, M. (2021). Make bloom and let wither: Biopolitics of precision agriculture at the dawn of surveillance capitalism. Geoforum 122: 193–203. https://doi.org/10.1016/j.geoforum.2021.04.014.

Sustainability Directory. (2025). Ethical Implications of AI-Driven Decision Making in Agriculture → Scenario. [online] Prism → Sustainability Directory. Available at: https://prism.sustainability-directory.com/scenario/ethical-implications-of-ai-driven-decision-making-in-agriculture [Accessed 13 April 2025].

Tamayo-Vera, D., Mesbah, M., Zhang, Y. and Wang, X. (2025). Advanced machine learning for regional potato yield prediction: analysis of essential drivers. Npj Sustainable Agriculture 3(1). https://doi.org/10.1038/s44264-025-00052-6.

Tanzer, G., Suzgun, M., Visser, E., Jurafsky, D. and Melas-Kyriazi, L. (2023). A Benchmark for Learning to Translate a New Language from One Grammar Book. arXiv (Cornell University). https://doi.org/10.48550/arxiv.2309.16575.

Toledo, C. A., Crawford, M. M. and Tuinstra, M. R. (2024). Integrating multi-modal remote sensing, deep learning, and attention mechanisms for yield prediction in plant breeding experiments. Frontiers in Plant Science 15. https://doi.org/10.3389/fpls.2024.1408047.

Turgut, O., Kok, I. and Ozdemir, S. (2024). AgroXAI: Explainable AI-Driven Crop Recommendation System for Agriculture 4.0. arXiv (Cornell University). https://doi.org/10.48550/arxiv.2412.16196.

Turpin, M., Micheal, J., Perez, E. and Bowman, S. R. (2023). Language Models Don't Always Say What They Think: Unfaithful Explanations in Chain-of-Thought Prompting. ArXiv (Cornell University), 1 Jan. 2023, https://doi.org/10.48550/arxiv.2305.04388.

Tychola, K. A. and Rantos, K. (2025). Cyberthreats and security measures in Drone-Assisted agriculture. Electronics 14(1): 149. https://doi.org/10.3390/electronics14010149.

UNESCO. (2021). Recommendation on the Ethics of Artificial Intelligence. [online]. Unesco.org. Available at: https://unesdoc.unesco.org/ark:/48223/pf0000381137.

Vasconez, J. P., Kantor, G. A. and Cheein, F. a. A. (2019). Human–robot interaction in agriculture: A survey and current challenges. Biosystems Engineering 179: 35–48. https://doi.org/10.1016/j.biosystemseng.2018.12.005.

Vinuesa, R., Azizpour, H., Leite, I., Balaam, M., Dignum, V., Domisch, S. et al. (2020). The role of artificial intelligence in achieving the Sustainable Development Goals. Nature Communications 11(1). https://doi.org/10.1038/s41467-019-14108-y.

Wang, S., Xu, D., Liang, H., Bai, Y., Li, X., Zhou, J. et al. (2025). Advances in deep learning applications for plant disease and pest detection: a review. Remote Sensing 17(4): 698. https://doi.org/10.3390/rs17040698,

Wang, X., Zhang, Y. and Zhu, R. (2022). A brief review on algorithmic fairness. Management System Engineering 1(1). https://doi.org/10.1007/s44176-022-00006-z.

World Economic Forum. (2025). The Future of Jobs Report 2025. Geneva, Switzerland: World Economic Forum. Retrieved from https://www.weforum.org/reports/the-future-of-jobs-report-2025/.

World Population Prospects. 2024: Summary of Results. (2024, July 1). DESA Publications. https://desapublications.un.org/publications/world-population-prospects-2024-summary-results [Accessed 12 March, 2025].

Yadav, S., Sharma, S. and Chaudhary, P. (2024). Review on advancement in machine learning and deep learning techniques for crop classification. pp. 593–610. In Soft Computing Research Society eBooks. https://doi.org/10.56155/978-81-955020-7-3-53.

Yang, Y., Liu, X., Jin, Q., Huang, F. and Lu, Z. (2024). Unmasking and quantifying racial bias of large language models in medical report generation. Communications Medicine 4(1). https://doi.org/10.1038/s43856-024-00601-z.

Yin, H., Cao, Y., Marelli, B., Zeng, X., Mason, A. J. and Cao, C. (2021). Soil sensors and plant wearables for smart and precision agriculture. Advanced Materials 33(20). https://doi.org/10.1002/adma.202007764.

Zhu, H., Lin, C., Liu, G., Wang, D., Qin, S., Li, A. et al. (2024). Intelligent agriculture: deep learning in UAV-based remote sensing imagery for crop diseases and pests detection. Frontiers in Plant Science 15. https://doi.org/10.3389/fpls.2024.1435016.

# Index

# Editors' Biography

**Dr. Mahendra Rai** is presently a Senior visiting Professor at the Department of Chemistry, Federal University of Piaui (UFPI), Teresina, Brazil. Formerly, he was Professor and Head of the Department of Biotechnology, SGB Amravati University, Maharashtra State, India. He was a visiting scientist at the University of Geneva, Debrecen University, Hungary; University of Campinas, Brazil; VSB Technical University of Ostrava, Czech Republic; National University of Rosario, Argentina, and the University of São Paulo. He has published more than 450 research papers in national and international journals, with 95 h-index. In addition, he has edited/authored more than seventy-five books owns 6 patents. He has been featured among the top 2% scientists in Nanoscience and Nanotechnology by Stanford University and is ranked 7th nationally in Biology and Biochemistry by Research.com.

**Dr. Graciela Avila Quezada** is a Professor at Faculty of Agrotechnology Science, Universidad Autónoma de Chihuahua, in México. She has been conducting research related to plant pathogens for thirty years, beginning with her undergraduate thesis work focused on the control of *Puccinia graminis* f. sp. *avenae*. She was a visiting scientist at the Volcani Center in Israel; Colegio de Postgraduados, Texcoco, Mexico; SENASICA Laboratories, Ministry of Agriculture in Tecamac, State of Mexico; the California Polytechnic State Univ-San Luis Obispo California (Calpoly); Institut de Recherche pour le Développement IRD Montpellier, France; Universidad Autonoma Chapingo, Mexico State; Universidad de Córdoba, Spain. Dr. Avila was President of the Mexican Society of Phytopathology; Vice-president of the National Phytosanitary Advisory Council (CONACOFI); Executive Secretary of the National System for Research and Technology Transfer (SNITT) of the Minister of Agriculture (Sagarpa); and Director of Academic Cooperation, in the Coordination of Science and Technology, Presidency of Mexico. Besides, she has conducted more than twelve research projects and published more than 70 research papers in national and international journals focused on plant pathogens. She is presently a visiting professor at the Universidad Complutense de Madrid, Spain.